Remote Sensing of Energy Fluxes and Soil Moisture Content

Remote Sensing of Energy Fluxes and Soil Moisture Content

George P. Petropoulos

CRC Press
Taylor & Francis Group
Boca Raton London New York

CRC Press is an imprint of the
Taylor & Francis Group, an **informa** business

CRC Press
Taylor & Francis Group
6000 Broken Sound Parkway NW, Suite 300
Boca Raton, FL 33487-2742

First issued in paperback 2017

ISBN-13: 978-1-4665-0578-0 (hbk)
ISBN-13: 978-1-138-07757-7 (pbk)

Library of Congress Cataloging-in-Publication Data

Remote sensing of energy fluxes and soil moisture content / editor, George P. Petropoulos.
    pages cm
  "A CRC title."
  Includes bibliographical references and index.
  ISBN 978-1-4665-0578-0 (hardcover : alk. paper)
  1. Energy budget (Geophysics)--Remote sensing. 2. Soil moisture--Remote sensing. I. Petropoulos, George P., editor of compilation.

QC809.E6R46 2014
551.5'2530287--dc23

2013034936

Visit the Taylor & Francis Web site at
http://www.taylorandfrancis.com

and the CRC Press Web site at
http://www.crcpress.com

This book is dedicated to my parents Panagiotis & Evgenia, my sister, Konstantina and my partner, Nune, for their love, understanding, and support in the pursuits of my life.

Contents

Section I Controls, Conventional Estimation, and Remote Sensing Methods Overview

Section II Remote Sensing of Surface Energy Fluxes: Algorithms and Case Studies

Section III   Remote Sensing of Soil Surface Moisture: Algorithms and Case Studies

Section IV   Challenges and Future Outlook

Preface

Aims and Scope

This book aims to provide an all-inclusive overview of the state of the art in the methods and modeling techniques employed for deriving spatiotemporal estimates of energy fluxes and soil surface moisture from remote sensing. Overall, the book brings together three types of articles: (1) comprehensive review articles from leading authorities to examine the developments in concepts, methods, and techniques employed in deriving land surface heat fluxes as well as soil surface moisture on field, regional, and large scales, paying particular emphasis to the techniques exploiting Earth Observation (EO) technology; (2) focused articles providing more detailed insights into the principles and operation of some of the most widely applied approaches for the quantification and analysis of surface fluxes and soil moisture with case studies that directly show the great applicability of remote sensing in this field, or articles discussing specific issues in the retrievals of those parameters from space; and (3) focused articles integrating current knowledge and scientific understanding in the remote sensing of energy fluxes and soil moisture, highlighting the main issues, challenges, and future prospects of this technology.

The book integrates decades of research conducted by leading scientists in the field, and it has been designed with different potential users in mind. It can provide an authoritative supplementary textbook for upper division undergraduate and postgraduate students and/or a reference for a burgeoning number of practitioners and professionals who are using EO technology and wish to keep up with new developments in the field. Because of the unique way the book is structured, consisting of four more or less independent units, its use can be adapted to meet the specific needs of different readers leading to its adoption for teaching and research purposes alike.

Book Synopsis

This book is divided into four sections. Section I, *Controls, Conventional Estimation, and Remote Sensing Methods Overview*, provides an overview of the controlling parameters affecting the processes of energy fluxes and soil moisture, reviewing the techniques available in estimating those parameters on field, regional, and large scales using ground instrumentation and/or remote sensing technology. Section II, *Remote Sensing of Surface Energy Fluxes: Algorithms and Case Studies*, exemplifies recent advances in modeling approaches utilizing EO data making available related case studies, including a discussion of the global operational products available and the role of scale in determining those parameters from remote sensing. Section III, *Remote Sensing of Soil Surface Moisture: Algorithms and Case Studies*, presents algorithms and techniques employed in deriving soil surface moisture from remote sensing, also reviewing data assimilation methods, scaling, and filtering approaches and discussing strategies for evaluating global soil moisture products. Section IV, *Challenges and Future Outlook*, provides a discussion on the current status, future trends, and prospects of EO technology in estimating turbulent heat fluxes and soil moisture, underlying the scientific challenges that need to be addressed.

In Section I, Chapters 1 and 2 provide a discussion of the main parameters controlling energy fluxes and soil moisture and on the latest developments in their measurement based on ground instrumentation, outlining the principles of each approach and their relative strengths and limitations. Those two chapters also inform the interested reader on the existing global ground-based observational networks that provide systematically validated *in situ* measurements of energy fluxes, soil moisture, and related parameters. An understanding of these topics is essential in appreciating the challenges and caveats when attempting to establish methods utilizing EO data for deriving spatiotemporal estimates of those parameters. Furthermore, knowledge of these topics is of important practical value in establishing strategies for evaluating the performance of satellite-derived remote sensing estimates by relevant algorithms and of global operational products. Subsequently, Chapters 3, 4, and 5 of this section are focused in providing all-inclusive overviews reporting on the advances of EO technology in deriving turbulent fluxes, soil surface moisture, and radiation fluxes, respectively. In these chapters the methods employed utilizing data acquired from sensors operating at all regions of the electromagnetic spectrum are reviewed providing some information also on relevant operational products available globally. Having an overview of the range of techniques employed and of their relative strengths and limitations is important in being able to appreciate the ability, main knowledge gaps, scientific challenges, and future opportunities towards the use of EO technology in this domain.

In Section II, Chapters 6 to 9 focus on offering more detailed information on the principles and workings of some of the most widely used approaches in estimating latent and sensible heat fluxes from remote sensing. One or two case studies that demonstrate the applicability of the techniques at different places of the world have also been included in each chapter. In particular, in Chapter 6, Teixeira and his colleagues describe SAFER and SEBAL one-layer models and provide as a case study results from their implementation using different types of remote sensing data over different crop types and natural vegetation sites in Brazil. Chapter 7, by Ma and his associates, presents the workings of SEBS and parameterization one-layer methods and illustrates their applicability in deriving regional estimates of energy fluxes for the Tibetan Plateau and northwest China using ASTER multispectral imagery. In Chapter 8, Anderson and her colleagues describe the principles of ALEXI/DisALEXI two-layer modeling systems for retrieving surface energy fluxes, evapotranspiration (ET), and available soil moisture. The authors also present an application of their modeling approach using both high spatial and temporal resolution imagery in monitoring drought, crop condition, and yield. In Chapter 9, Tang and his colleagues present a method for deriving spatiotemporal estimates of evapotranspiration rates based on the satellite-derived surface temperature and vegetation index scatterplot domain. Authors also provide results from a case study comparing the results from this method against *in situ* and some other one-layer modelling approach. In Chapter 10, McCabe and his colleagues provide an overview of recent developments in the estimation of global terrestrial surface heat flux from space. Authors also review community efforts to develop a harmonized and consistent heat flux record for long-term global assessment of land surface evaporation, focusing on intercomparison efforts taking place. They also present the results from the intercomparison of the algorithms they discuss in their chapter discussing the errors and related issues in using these techniques. In Chapter 11, Brunsell and Hu provide a comprehensive discussion of the importance of assessing the spatial scale of the inputs and the impact on the resultant surface energy fluxes and how changes in the spatial scale of controlling variables may alter the spatial scaling properties of the derived fluxes. This is done using as a case study a method based on linking remote sensing observations of surface temperature and vegetation cover with a one-layer model and data they had acquired from the Cloud and Land Surface Interaction Campaign (CLASIC) in central Oklahoma during the summer of 2007. The results from their analysis suggest that it is possible to scale from point measurements of environmental state variables to regional estimates of energy exchanges to obtain an understanding of the spatial relationship between these fluxes and landscape variables.

Section III of the book embodies seven chapters. The first chapter, Chapter 12, prepared by Paloscia and Santi, presents the workings of a method developed by the authors for deriving soil surface moisture from passive microwave measurements potentially at an operational scale. The authors illustrate results from its implementation at case studies in different regions of the world. Subsequently, in Chapter 13, Verhoest and Lievens provide an in-depth discussion on the difficulties encountered in parameterizing surface roughness from field measurements, providing also suggestions for alternative ways in deriving soil moisture from active (i.e., SAR) circumventing roughness parameterization and they demonstrate as a case study the use of one of these methods. Then, in Chapter 14 Temimi and his associates address the potential of the synergistic use of observations in the microwave and infrared domains to monitor soil moisture and inundation from space. Examples of soil moisture–related products that are based on the use of IR observations are cited and discussed by the authors. Authors also present examples of blended products and demonstrate the added value of the use of observations from multiple sensors. In Chapter 15, Sobrino and his associates also focus on synergistic approaches, presenting results from the implementation of a promising technique utilizing optical and thermal infrared observations in deriving estimates of soil surface moisture from remote sensing. Case studies results from the implementation of this method over an agricultural test site in Spain but also further applications to soil moisture and energy fluxes are also provided by the authors. In Chapter 16, Montzka reviews the recent advancements in the assimilation of remotely sensed soil surface moisture into numerical models both for synthetic and real-world studies. He also discusses recent works focusing on the soil moisture profile estimation by assimilation of one remotely sensed soil moisture product and how recent studies deal with different scales in remote sensing soil moisture observations and numerical models. Montzka also commends on the ability of different approaches in estimating model parameters for improved predictions and reviews the impact of soil moisture data assimilation for the improved estimation of further components of the hydrological cycle. In Chapter 17, Brocca and his coauthors provide an excellent overview of the theoretical background related to the different approaches developed for scaling/filtering satellite data, and the problems encountered with using these methods, presenting also results from their evaluation and intercomparison using satellite and *in situ* soil moisture data. To elucidate the salient points of their chapter, they present the significant results from a soil experiment that was designed to study the soil-drying process in relation to surface radiant temperature. Authors demonstrated the use of coarse-scale satellite imagery from ASCAT and AMSR sensors and *in situ* data from validated operational networks in Australia and Africa to test different scaling/filtering approaches. In Chapter 18, Albergel and his colleagues provide an excellent overview of the approaches employed in validating satellite-derived estimates of soil moisture. In this framework, they discuss the theoretical and practical considerations that must be considered in applying each of these approaches, focusing their discussion, in particular, on the description of the statistics most often used to quantify pattern similarity and differences between soil moisture retrieval from space and ground measurements. Subsequently, a relevant case study is presented in their chapter in which *in situ* measurements from the SMOSMANIA and REMEDHUS networks are used to evaluate ASCAT soil moisture product retrievals from space.

In Section IV, Chapter 19, McCabe and his associates describe some of the key issues requiring consideration in the development and use of global surface flux retrievals. The authors also provide some insights on the opportunities and areas on which research needs to focus in the future to advance our current capability in deriving more accurately energy fluxes from remote sensing. Their contribution represents an alternative perspective on some of the issues requiring attention, beyond those often discussed in the literature related to this topic. In Chapter 20, Wagner and his colleagues discuss the main challenges in the retrievals of soil moisture data services from remote sensing using as an example the case of ASCAT products, keeping the discussion to a general level, in order to ensure that it is also of relevance for the other satellite soil moisture monitoring services. Authors conclude their chapter with a discussion of the prospects of operational soil moisture services.

I hope this Preface has successfully provided some insight into the breadth of the topics covered in this book. Users of this book are encouraged to adapt to it and use it the way it best fits their own needs that would help them in understanding the capabilities and potentials of EO technology in the field in which this book is concerned.

Users of this book can inform the editor of any errors, suggestions, or comments at george.petropoulos@aber.ac.uk or petropoulos.george@gmail.com.

George P. Petropoulos

Acknowledgments

I am indebted to a number of people who assisted in realising the publication of this book. I would like to express my deepest thanks to the authors of the different chapters who agreed to contribute to the book despite their already very busy schedules. I would like to express my sincerest gratitude also to the reviewers for their useful and insightful review comments and suggestions that helped improve the book. I am deeply grateful to my father, Panagiotis, my mother, Evgenia, and my sister, Konstantina, for their enduring love and support in all my endeavors. I extend my heartfelt thanks to my partner, Nune, whose faithful support, encouragement, and understanding during all stages of the book preparation was irreplaceable. Finally, I acknowledge the publisher, Taylor & Francis, and in particular, Irma Britton, Kathryn Everett, Robert Sims and Amor Nanas for the fruitful collaboration in accomplishing the preparation of this book.

Closing, I wish to acknowledge below the reviewers for this book in alphabetical order:

Reviewers

Brian Barrett

Luca Brocca

Nathaniel Brunsell

Hywel Griffiths

Weiqiang Ma

Carsten Monzka

Mark Smith

Simonetta Paloscia

Qiuhong Tang

Niko Verhoest

And two anonymous reviewers.

Editor

Dr. George P. Petropoulos is a lecturer in remote sensing and GIS in the Department of Geography and Earth Sciences (DGES) at Aberystwyth University in Wales, United Kingdom. In 1999 he received his BSc degree in Natural Resources Development and Agricultural Engineering from the Agricultural University of Athens, Greece. From 2003 to 2008 he pursued his graduate studies (MSc and PhD degrees) at the University of London, United Kingdom, specializing in remote sensing and Earth observation (EO) modeling. A variant of the methodology he investigated during his PhD has been proposed for the operational retrieval of the surface soil moisture by the National Polar-orbiting Operational Environmental Satellite System (NPOESS) and NASA Space Agencies in a series of satellites due to be launched over the next 12 years. After completing his PhD, Dr. Petropoulos was employed as a Research Associate in the Department of Earth Sciences of the University of Bristol in the United Kingdom for 1.5 years. Since end of 2009, he worked in Greece as a research fellow at different institutions, namely, the Mediterranean Agronomic Institute of Chania (MAICh), the National Observatory of Athens, the Agricultural University of Athens, and the Foundation for Research & Technology.

He is currently an Honorary Research Associate at the Department of Earth Sciences of the University of Bristol, United Kingdom. He is also a guest/adjunct faculty at the Agricultural University of Athens and the MAICh in Greece. He also holds at IGES a Postdoctoral Fellowship from the European Space Agency (2010–2013) and a Marie Curie Fellowship from the European Commission (2013–2017), pursuing research towards the prototyping of energy fluxes and soil surface moisture products from remote sensing at an operational and global scale.

His research work focuses on exploiting EO data alone or synergistically with land surface process models for computing key state variables of the Earth's energy and water budget, including energy fluxes and soil surface moisture. He is also conducting research on the application of remote sensing technology to land cover mapping and its changes that occurred from either anthropogenic activities (e.g., urbanization and mining activity) or natural hazards (mainly floods and fires). In this framework, he is researching and optimizing new image processing methodologies and remote sensing–based operational products, conducting also all-inclusive benchmarking studies on EO products or land surface models, including advanced sensitivity analysis.

He is the author/coauthor of more than 20 peer-reviewed journal articles, more than 55 contributions to international conferences, and more than 10 book chapters. He is also a reviewer in several peer-reviewed journals worldwide. He has also developed fruitful collaborations with key scientists in his area of specialization globally and his work so far has received international recognition via several significant awards he has obtained such as the ESA and Marie Curie fellowships.

Contributors

Clement Albergel
European Centre for Medium-Range
    Weather Forecasts (ECMWF)
Reading, United Kingdom

Martha C. Anderson
US Department of Agriculture
Beltsville, Maryland

Brian W. Barrett
School of Geography and Archaeology
University College Cork (UCC)
Cork, Republic of Ireland

Luís Henrique Bassoi
Embrapa
Petrolina, Pernambuco, Brazil

Amelise Bonhomme
New York City College of Technology
New York, New York

Luca Brocca
Research Institute for Geo-Hydrological
    Protection
National Research Council
Perugia, Italy

Nathaniel A. Brunsell
Department of Geography
University of Kansas
Lawrence, Kansas

Jean-Christophe Calvet
CNRM-GAME
Météo-France
CNRS, UMR 3589
Toulouse, France

Toby N. Carlson
Department of Meteorology
Pennsylvania State University
University Park, Pennsylvania

Jie Cheng
State Key Laboratory of Remote Sensing
    Science
Beijing Normal University and Institute of
    Remote Sensing Applications
Chinese Academy of Sciences
and
College of Global Change and Earth
    System Sciences
Beijing Normal University
Beijing, China

Jaeil Cho
Department of Spatial Information
    Engineering
Pukyong National University
Busan, Korea

Chiara Corbari
Department of Hydraulic
    Environmental and Surveying
    Engineering
Politecnico di Milano
Milan, Italy

Richard de Jeu
VU Amsterdam
Amsterdam, Netherlands

Patricia de Rosnay
European Centre for Medium-Range
    Weather Forecasts (ECMWF)
Reading, United Kingdom

Wouter Dorigo
Department of Geodesy and
  Geoinformation (GEO)
Vienna University of Technology
  (TU Wien)
Vienna, Austria

Ali Ershadi
Water Research Centre
School of Civil and Environmental
  Engineering
University of New South Wales
Sydney, Australia

Julia Figa
EUMETSAT
Darmstadt, Germany

Joshua Fisher
Jet Propulsion Laboratory
California Institute of Technology
Pasadena, California

Belen Franch
Global Change Unit
Image Processing Laboratory
Universitat de València
Burjasot, Spain

Hywel M. Griffiths
Department of Geography and Earth
  Sciences
Aberystwyth University
Aberystwyth, Wales, United Kingdom

Alexander Gruber
Department of Geodesy and
  Geoinformation (GEO)
Vienna University of Technology
  (TU Wien)
Vienna, Austria

Sebastian Hahn
Department of Geodesy and
  Geoinformation (GEO)
Vienna University of Technology
  (TU Wien)
Vienna, Austria

Christopher R. Hain
System Science Interdisciplinary Center
University of Maryland
and
National Oceanic and Atmospheric
  Administration
National Environmental Satellite, Data and
  Information Service
NOAA Center for Weather and Climate
  Prediction
College Park, Maryland

Stefan Hasenauer
Department of Geodesy and
  Geoinformation (GEO)
Vienna University of Technology (TU Wien)
Vienna, Austria

Tao He
Department of Geographical Sciences
University of Maryland
College Park, Maryland

Fernando Braz Tangerino Hernandez
UNESP
Ilha Solteira, São Paulo, Brazil

Leiqiu Hu
Department of Geography
University of Kansas
Lawrence, Kansas

Lei Huang
Institute of Geographic Sciences and
  Natural Resources Research
Chinese Academy of Sciences
Beijing, China

Hirohiko Ishikawa
Disaster Prevention Research Institute
Kyoto University
Uji, Japan

Carlos Jimenez
Laboratoire d'Etude du Rayonnement et de
  la Matière en Astrophysique (LERMA)
Observatoire de Paris
Paris, France

Juan Carlos Jiménez-Muñoz
Global Change Unit
Image Processing Laboratory
Universitat de València
València, Spain

Cezar Kongoli
System Science Interdisciplinary Center
University of Maryland
and
National Oceanic and Atmospheric
    Administration
National Environmental Satellite, Data and
    Information Service
NOAA Center for Weather and Climate
    Prediction
College Park, Maryland

William P. Kustas
US Department of Agriculture
Beltsville, Maryland

Miaoling Liang
Department of Civil and Environmental
    Engineering
Princeton University
Princeton, New Jersey

Shunlin Liang
Department of Geographical Sciences
University of Maryland
College Park, Maryland

and

State Key Laboratory of Remote Sensing
    Science
Beijing Normal University and Institute of
    Remote Sensing Applications
Chinese Academy of Sciences

and

College of Global Change and Earth
    System Sciences
Beijing Normal University
Beijing, China

Hans Lievens
Laboratory of Hydrology and Water
    Management
Ghent University
Ghent, Belgium

Hui Liu
Department of Hydraulic Engineering
Tsinghua University
Beijing, China

H. L. Lopes
UNIVASF
Petrolina, Pernambuco, Brazil

Weiqiang Ma
Key Laboratory of Land Surface Process
    and Climate Change in Cold and Arid
    Regions
Cold and Arid Regions Environmental and
    Engineering Research Institute
Chinese Academy of Sciences
Lanzhou, China

and

Disaster Prevention Research Institute
Kyoto University
Uji, Japan

Yaoming Ma
Laboratory of Tibetan Environment
    Changes and Land Surface Processes
Institute of Tibetan Plateau Research
Chinese Academy of Sciences
Beijing, China

Cristian Mattar
Global Change Unit
Image Processing Laboratory
Universitat de València
València, Spain

and

Laboratory for the Analysis of the
    Biosphere (LAB)
Department of Environmental Sciences
University of Chile
Santiago, Chile

Matthew McCabe
Water Desalination and Reuse Centre
King Abdullah University of Science and
    Technology (KAUST)
Jeddah, Saudi Arabia

Florisa Melone
Research Institute for Geo-Hydrological
    Protection
National Research Council
Perugia, Italy

Diego Miralles
School of Geographical Sciences
University of Bristol
Bristol, United Kingdom

Carsten Montzka
Forschungszentrum Jülich
Institute of Bio- and Geosciences
Jülich, Germany

Tommaso Moramarco
Research Institute for Geo-Hydrological
    Protection
National Research Council
Perugia, Italy

Qiaozhen Mu
Numerical Terradynamics Simulation
    Group
University of Montana
Missoula, Montana

Brigitte Mueller
Institute for Atmospheric and Climate
    Science
ETH Zurich
Zurich, Switzerland

Claudia Notarnicola
EURAC–Institute for Applied Remote
    Sensing
Bolzano, Italy

Simonetta Paloscia
Institute of Applied Physics – National
    Research Council (IFAC – CNR)
Firenze, Italy

George P. Petropoulos
Department of Geography and Earth
    Sciences
Aberystwyth University
Aberystwyth, Wales, United Kingdom

Robert Rabin
National Oceanic and Atmospheric
    Administration (NOAA)
National Severe Storms Laboratory
Norman, Oklahoma

Emanuele Santi
Institute of Applied Physics – National
    Research Council (IFAC – CNR)
Firenze, Italy

Sonia Seneviratne
Institute for Atmospheric and Climate
    Science
ETH Zurich
Zurich, Switzerland

Justin Sheffield
Department of Civil and Environmental
    Engineering
Princeton University
Princeton, New Jersey

José Sobrino
Global Change Unit
Image Processing Laboratory
Universitat de València
València, Spain

Jan Stepinski
NOAA-CREST Institute
City University of New York
New York, New York

Zhongbo Su
Faculty of Geo-Information Science and
    Earth Observation
University of Twente
Enschede, The Netherlands

Qiuhong Tang
Institute of Geographic Sciences and
    Natural Resources Research
Chinese Academy of Sciences
Beijing, China

Antônio Heriberto de Castro Teixeira
Embrapa
Campinas, São Paulo, Brazil

Marouane Temimi
NOAA-CREST Institute
City University of New York
New York, New York

Niko E. C. Verhoest
Laboratory of Hydrology and Water
    Management
Ghent University
Ghent, Belgium

Wolfgang Wagner
Department of Geodesy and
    Geoinformation (GEO)
Vienna University of Technology (TU Wien)
Vienna, Austria

Dongdong Wang
Department of Geographical Sciences
University of Maryland
College Park, Maryland

Morris Scherer-Warren
ANA
Brasília, Distrito Federal, Brazil

Eric Wood
Department of Civil and Environmental
    Engineering
Princeton University
Princeton, New Jersey

Angelika Xaver
Institute of Photogrammetry and Remote
    Sensing
Vienna University of Technology
Vienna, Austria

Pat Jen-Feng Yeh
Department of Civil Engineering
National University of Singapore

Xiwu Zhan
NOAA-NESDIS, Centre for Satellite
    Applications and Research
College Park, Maryland

Xiaotong Zhang
State Key Laboratory of Remote Sensing
    Science
Beijing Normal University and Institute of
    Remote Sensing Applications
Chinese Academy of Sciences
and
College of Global Change and Earth
    System Sciences
Beijing Normal University
Beijing, China

Section I

Controls, Conventional Estimation, and Remote Sensing Methods Overview

1

Turbulent Fluxes of Heat and Moisture at the Earth's Land Surface: Importance, Controlling Parameters, and Conventional Measurement Techniques

George P. Petropoulos, Toby N. Carlson, and Hywel M. Griffiths

CONTENTS

1.1 Introduction

Earth's land surface and atmosphere host a constant exchange of energy, momentum, and water via the turbulent flux of sensible heat (H) and latent heat (LE). LE can be defined as the flux of heat from the Earth's surface to the atmosphere that is associated with evaporation of water at the surface and H as the heat energy transferred between the surface and air along a temperature gradient. LE and evapotranspiration (ET) are used interchangeably, with the only difference being that the former is an energy flux expression of the latter.

LE flux is the single most important mechanism of energy and mass exchange between the hydrosphere, biosphere, and atmosphere, playing a critical role in both the water cycle and energy balance (Sellers et al. 1996; Song et al. 2012). Combined with rainfall and runoff, it controls the availability and distribution of water at the Earth's surface, linking the surface energy balance to the water balance (Fernández-Prieto et al. 2012) and determining the total amount and spatiotemporal distribution of water resources (e.g., Kwast et al. 2009). The importance of the LE flux in the global water cycle is clearly depicted in Figure 1.1. Over land surfaces, approximately 75% of the total precipitation is evapotranspired by the plants and soil. At the regional scale and particularly over arid and semiarid regions, 90% or more of the annual precipitation can be consumed in the process of evapotranspiration. Even in humid regions, one half or more of the water balance can be credited to evapotranspiration (Engman 1991). At a continental scale, roughly two thirds of precipitation is evaporated and slightly more than half of the available net radiation (R_n) is converted into LE flux (Brutsaert 1982).

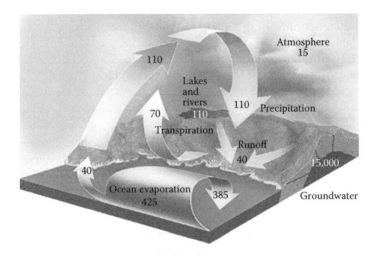

FIGURE 1.1
(See color insert.) Global hydrological cycle in 1000 km³ year⁻¹. Numbers in white are pools (thousands of km³) and numbers in black are fluxes (thousands of km³ year⁻¹). Precipitation over land is 110,000 km³ year⁻¹, two thirds of which (approximately 70,000 km³ year⁻¹) is recycled from plants and the soil by evapotranspiration. The remaining one third (40,000 km³ year⁻¹) is water evaporated from the oceans and transported over land. Total evaporation from the oceans is 10× larger than the flux shown here (425,000 km³ year⁻¹). Approximately 15,000,000 km³ of fresh water is held in groundwater, much of which is not actively exchanged with the Earth's surface. (Adapted from Jackson, R. B. et al., *Ecol. Appl.*, 11, 1027–1045, 2001.)

The estimation of the exchange of both LE and *H* fluxes is of practical value in various studies including but not limited to water resources, agronomy, and meteorology (e.g., Rivas and Caselles 2004; Verstraeten et al. 2008; Li et al. 2011). Measurements of both fluxes are required for monitoring plant water requirements, plant growth, and productivity, as well as for irrigation management and deciding when to carry out cultivation procedures (e.g., Consoli et al. 2006; Glenn et al. 2007; Yang et al. 2010). Furthermore, given the direct relation of LE flux to the water loss from the land surface to the atmosphere, accurate estimation of this parameter is very important for sustainable planning and management of water resources, especially in ecosystems in which water is a limiting resource, such as in many arid and semiarid regions (Mariotto and Gutschick 2010; French et al. 2012). Estimates of both LE and *H* are also required in the numerical modeling and prediction of atmospheric and hydrological cycles and in improving the accuracy of existing weather forecasting models (Jacob et al. 2002), as they act as boundary conditions for the dynamics of the troposphere. For example, the LE flux is of critical importance in meteorology as it is linked directly to changes in the height of the local planetary boundary layer, which can trigger the generation of extreme weather phenomena that may have serious impacts on human life (Billi and Caparrini 2006; Sun et al. 2011).

Quantitative estimates of the turbulent atmospheric fluxes are also essential in many crop yield forecasting models, for example, the widely used Food and Agriculture Organization (FAO) (Allen et al. 1998). Accurate estimation of the regional rates of surface heat fluxes, particularly of LE, is also important for the prediction of desertification and the monitoring of land degradation (McCabe and Wood 2006; Ahmad et al. 2010) as well as to better understand the processes that control ecosystem carbon dioxide (CO_2) exchange and the interactions between parameters in different ecosystem processes (Kustas and Anderson 2009; Petropoulos et al. 2009). In semiarid environments in particular, LE flux over land plays a critical role in both the water circulation within the surface but also in a large number of related physical and ecological variables, such as atmospheric dynamics, subsurface water storage, and vegetation growth (Mutiga et al. 2010).

This chapter aims to, first, provide a background on the principles concerning the description of evaporation and transpiration processes, briefly discussing the parameters affecting their spatiotemporal variation over land surfaces. Second, it aims to provide an overview of the conventional measurement techniques for their estimation, outlining the principles of each method and their relative strengths and limitations. Finally, it aims to inform the interested reader of the existing global ground-based observational networks that at present systematically record and make available validated *in situ* measurements of energy fluxes and related parameters at a global scale. An understanding of all these topics is essential in appreciating the challenges and caveats in the spatiotemporal estimation of energy fluxes from remote sensing–based techniques and to also understand the limitations imposed when comparing such predictions with corresponding *in situ* "reference" measurements.

1.2 Energy Fluxes at the Earth's Surface: Physics and Theory

The term evaporation is used by hydrologists to describe the loss of water from a wet surface through its conversion into its gaseous state, water vapor, and its transport away from the surface into the atmosphere (Ward and Robinson 2000). Evaporation may occur from open water, bare soil, or vegetation surfaces (Novak 2012). In addition to evaporation, there is also direct water use by plants, which is termed transpiration, defined as the process by which water vapor escapes from the living plant, principally the leaves, and enters the atmosphere (Lee 1942). The starting point in the establishment of evaporation and transpiration theory is traced back to 1802 and Dalton's law, which states that "if all the other factors remain constant, evaporation is proportional to the wind speed and the vapor pressure deficit, i.e., the difference between saturation vapor pressure at the temperature of the water surface and the actual vapor pressure of the overlaying air." This theorem was a cornerstone in the advancement of the contemporary studies of evaporation and transpiration.

There is no fundamental difference in the physics of evaporation from water or soil and transpiration by plants. The only difference is in the nature of the controls on the different surfaces (Shuttleworth 1993). Practically, it is usually considered difficult to distinguish between evaporation and transpiration as two separate processes. Wherever vegetation is present, both processes tend to operate together and are generally combined to give the composite term of evapotranspiration (or expressed as its flux—LE). Increasingly, however, it is recognized that being able to distinguish between these two processes is necessary, especially in water-limited, arid, and semiarid regions (Villegas et al. 2010). The following sections briefly discuss the fundamentals of evaporation and transpiration processes, outlining the main parameters affecting their spatiotemporal variation over the Earth's land surface.

1.2.1 The Process of Evaporation

The physics of the evaporation process relate principally to two aspects: (1) the supply of sufficient energy at the evaporation surface in order to increase the separation between the molecules, which is called latent heat of vaporization (λ); and (2) the operation of diffusion processes in the air above the evaporating surface to provide a means for removing the water vapor produced by evaporation. The term LE flux density describes the energy used in the transfer of water vapor molecules from one phase to another. The energy is used to create a phase change, resulting in a mass transfer. The linkages among these parameters can be more easily seen in an expanded mathematical description of the latent heat flux (λE) given as

$$\lambda E = \frac{(\rho \lambda m / P)(e_s - e_a)}{(r_c + r_{av})} \tag{1.1}$$

where λ is the latent heat of vaporization (J kg^{-1}), ρ the density of air (kg m^{-3}), m the ratio of molecular weight of water vapor to that of air (0.622), P the barometric pressure (kPa), e_s the saturation vapor pressure, e_a the actual vapor pressure of the air immediately above the surface, r_c the canopy resistance for water vapor transfer (s m^{-1}), and r_{av} the aerodynamic resistance for water vapor transfer (s m^{-1}).

The energy available for evaporation is given by the simplified version of the surface energy balance equation shown below. For a surface, evapotranspiration must be placed in context of the surface energy balance so that the balance of energy is expressed as

$$\lambda E = R_n - G - H \tag{1.2}$$

where G is the flux of energy to or from the soil (W m^{-2}), and λ is the heat lost by the air when liquid water changes into vapor, called latent heat of vaporization. Consequently, as seen from Equation 1.2, the energy available for evaporation (i.e., for the LE flux) is balanced by energy from several sources: R_n, H, and G fluxes.

In an analogous way, the rate of sensible heat transfer H in air is related to the temperature gradient $\partial T/\partial Z$ via

$$H = -\rho c_p K_H \frac{\partial T}{\partial Z} \tag{1.3}$$

where K_H is the transfer coefficient of heat, ρ is the density of the air, c_p is the specific heat capacity of air. Equation 1.3 for H flux computation requires the measurement of the gradient of temperature at very small distances from the surface. The negative sign in Equation 1.3 is inserted to conform to the convention that energy fluxes away from the evaporating surface are positive.

The partitioning of the available energy at any surface can be described sufficiently using Equations 1.1 and 1.3. In surfaces covered by a vegetation canopy with bare soil patches, the evaporation above the surface will be a mixture of transpiration and direct evaporation, and in areas where the vegetation is dense, transpiration will be expected to be much more important than evaporation direct from the soil.

1.2.2 The Process of Transpiration

Transpiration from a plant occurs as part of its photosynthesis and respiration and is strongly coupled to the rate of photosynthesis by plants because stomata provide the pathway by which carbon dioxide enters and leaves the plant. Hence methods employed to estimate plant transpiration, particularly so from remote sensing observations, are also used to estimate processes that depend on photosynthesis, for example, annual net primary productivity and gross primary productivity (Glenn et al. 2007; Ryu et al. 2011).

The pathway of transpiration is long established and is described in the literature (e.g., Ward 1975; Ward and Robinson 2000; Dingman 2002; Kumagai 2011) and can be summarized as follows. Water enters the plant through the outer layer of epidermal cells of the younger portions of the root system, mainly through the root hairs. Then, from the epidermal cells it moves into the cavities (lumina) of the long cells known as traheides and tracheae. These form the conducting tissue of the vascular system, which transverses roots, stem, and leaves as part of the xylem. Water ascending through these tubes finally reaches the vascular system of the leaves. It then moves through the chlorenchyma at the base of the stomata. From there it leaves the cell walls as vapor and is transferred to the external atmosphere through the stomata.

Thus plants control their water loss by varying the opening of their stomatal apertures. Most of this water is transmitted through the plant and escapes through pores in the leaf system. This is what is known as stomatal transpiration. The rate of transpiration is controlled exactly by the openings or closing of stomata in the leaf. In certain plants, water may also travel upward through a cortex and pith of the root and stem, but the amount is very small except in young plants. Plants also lose water by other mechanisms, but usually this is negligible compared with the loss through the microscopic leaf apertures. For example, although stem and leaves can absorb moisture, the conditions for such absorption seldom occur in nature, and the amount absorbed is so small compared with the loss by transpiration from leaves that it has no significance in the water economy of plants (Penman 1963).

Because transpiration is essentially the same physical process as evaporation, it can be represented by a mass transfer equation, based on Fick's law. However, it should be noted that most efforts for modeling transpiration recast the mass transfer relation for evapotranspiration, using the concept of atmospheric conductance (Meinzer 1942). This is because atmospheric conductance allows for the use of some short circuit analogies for "scaling-up" from a leaf to an entire vegetated surface. Yet, a detailed reference of transpiration modeling approaches is beyond the scope of this discussion.

During the day, a small amount of the water flux, accompanied by sap and nutrients from the ground, is extracted from the stem material, where it is stored during the night. At night, water is again stored in the plant stem for use during the day. The additional transpiration from storage during the day is rather small compared with the total. From the above description of transpiration it is also evident that as far as the plant is concerned, transpiration is a passive process (Briggs 1992). Different species of plants transpire at different rates, but the fundamental controls are the available water in the soil and the plant's ability to transfer water from the soil to its leaves.

1.3 Parameters Controlling Evaporation and Transpiration Over Land

This section attempts to briefly discuss the most important parameters controlling the evaporation and transpiration variations over land surfaces. To facilitate a methodical discussion, the two processes are discussed separately.

1.3.1 Parameters Controlling Evaporation from Land Surfaces

1.3.1.1 Radiation

Incoming solar radiation is the main controlling variable on evaporation as it sets the limits for evaporation and governs the rates at which it can occur by providing the latent heat required to change water from its liquid state to its gaseous state (Matsoukas et al. 2011; Fernández-Prieto et al. 2012). Both solar (shortwave) and terrestrial (long-wave) radiation contribute to this process (Shaw 1994). In addition, radiation is also used by plants for photosynthesis and transpiration. Having said this, the net amount of solar radiation available for ET is itself dependent on latitude, day length, cloud cover, seasonal changes, and air pollution (Briggs and Smithson 1985).

1.3.1.2 Temperature

Incoming solar radiation also controls the surface temperature of water and the air above it, which are closely correlated to the rate of evaporation (see Equation 1.1). This is due to the fact that they control the vapor-holding capacity of the air (Meinzer 1942). A parcel of air at a high temperature will have a higher saturation vapor pressure than a parcel of air at a lower temperature, that is, it will be able to hold more water. Thus evaporation will occur at a faster rate into the warmer air (Shaw 1994; Lui et al. 2011). Water surface temperature is also important because it is the principal control on the rate at which water molecules leave the surface and enter the air above. Changes in water surface temperature may therefore cause an abrupt change in the rate of evaporation in the short term (Ward and Robinson 2000). However, the effect of increasing temperature can be offset by changes in solar radiation, wind speed, and humidity, as has been observed in the recent decrease in pan evaporation rates across the globe (Roderick et al. 2009).

1.3.1.3 Wind

Wind is a key controlling variable on evaporation rate because it enables water molecules to be removed from the ground surface by eddy diffusion (Zheng et al. 2009; McVicar et al. 2012). This allows the vapor pressure gradient to be maintained above the surface as saturated air is moved away from the area in question and air at a different temperature and with a different relative humidity brought in its place. Wind speed is one of the variables that can control the efficiency of this process, but the rate of mixing is also very important and this depends on the turbulence of the air and the rate of change of wind speed with height (Briggs and Smithson 1985). In addition, wind movement (laminar or turbulent) can also be influential. Laminar movement may not have much effect on evaporation, especially over large water bodies where the air temperature (T_a) and relative humidity is uniform with height. However, turbulent wind movement in the canopy layer will remove water vapor from contact with the water surface and increase evaporation (Meinzer 1942).

1.3.1.4 Humidity and Barometric Pressure

Evaporation rate is proportional to the difference between the actual humidity and the saturated humidity at any given temperature (Jones 1997). Humidity therefore directly affects evapotranspiration. However, actual or absolute humidity (the moisture in the air) varies only slightly over the course of a day compared to relative humidity (the moisture present in the air as a percentage of the moisture that would be present in saturated air at the same temperature), which is much more variable, and is a function of temperature (Penman 1968). An increase in the relative humidity of air over an evaporative surface will result in a decrease in evaporation over that surface, as fewer of the water molecules leaving the evaporative surface can be retained in the air. Barometric pressure also affects evaporation rates and can be used as a measure of evaporation differences with altitude (Meinzer 1942; Ji and Zhou 2011).

1.3.1.5 Soil Moisture Content

Soil moisture content (SMC) can affect to what extent actual evaporation reaches potential evaporation (Bougeault 1991; Lockart et al. 2012). As SMC decreases to zero, evaporation decreases rapidly (e.g., Green 1959). It is the SMC of the first top few centimeters of surface soil that is most influential in controlling evaporation from the surface compared to subsoil moisture, because the latter is slow moving and as such may have a negligible effect on evaporation rates from the surface (Ward 1975). Soil texture can also affect evaporation rates from a soil surface (Wesseling et al. 2009), with soils with finer textures and smoother surfaces allowing a higher rate of evaporation (Ward 1975). In addition, soil color can also affect evaporation rates, with darker soils, having lower albedo, absorbing more heat than lighter soils. Different soil color signifies different chemical compositions and thus different aptitudes in water absorption. The resulting increase in surface temperatures may modify evaporation rates at the surface significantly (Ward and Robinson 2000).

1.3.1.6 Vegetation

Except under specific conditions, like in the case of agricultural crops or uniformly distributed forests, vegetated surfaces usually include a mixture of species (e.g., in the case of croplands surrounded by hedgerows). Where the vegetation cover is not continuous, ET is equivalent to the sum of evaporation from the vegetation and evaporation from the soil. In general, vegetation will reduce evaporation from the soil by shading the surface from direct solar radiation and thus decreasing surface temperatures by reducing wind speed, and by raising the relative humidity of the lower layers of the air in the atmosphere (Ward and Robinson 2000; Breshears and Ludwig 2010). Other plant factors, such as vegetation cover density, may also result in significant changes in evaporation rates (Shaw 1994; Villegas et al. 2010).

1.3.2 Parameters Controlling Transpiration by Plants

Evaporation and transpiration are largely controlled by the same factors, but they respond in different ways and to different degrees. Approximately 90% of transpiration loss is determined by the factors that control evaporation (e.g., Meinzer 1942). Thus evaporation rates over a particular surface can be considered to be indicative of the corresponding transpiration rates. Jones (1997) has provided an excellent summary of the key factors affecting transpiration and its contribution to total ET. This section, which outlines the main factors affecting transpiration by plants, follows the early classification of Meinzer (1942), which categorized them as either physiologic or environmental.

1.3.2.1 Physiologic Factors

Physiologic factors that control transpiration include the density and behavior of the stomata, leaf structure, and water content. Stomatal structure varies only slightly according to species. However, the density of stomata on the surface of a leaf is characteristic of each plant species and is influenced by many environmental factors most of which are similar to those which affect evaporation (Ford and Jose 2010; Viliam 2012). The opening and closing of stomata, which controls plant transpiration rates, is affected by many factors, the most important of which are

1. Light intensity (stomata usually open in the light and close in darkness). This is also the reason why nighttime transpiration is negligible, consisting of only 5%–10% of the rate observed during the day.
2. Moisture supply to the leaves (plants' stomata close when guard cells lose turgor, regardless of other influences).
3. T_a, which affects the speed of stomata opening.
4. Humidity, which, when it is high, permits stomata to open wider and remain open longer.

Leaf structure also affects transpiration because it is linked to water concentration in the leaf and the drying out of cell walls surrounding the intercellular air spaces of the leaves. If the leaf has a high water concentration, increased evaporation rates do not produce incipient drying and transpiration will increase (Davie 2002). Furthermore, the stage of the growth of vegetation is another physiological factor affecting plants' transpiration, because plants use more water during periods of active biomass growth and also when root systems have reached maximum spread and efficiency (Meinzer 1942). Fungal diseases of plant leaves have also been found to affect the transpiration rates (Meinzer 1942; Mahlein et al. 2012).

A saturated surface or soil particle has a zero water potential; the potential decreases with decreasing SMC. The decrease from soil to the roots is dependent on a resistance for the water transfer from soil to root, which is sensitively dependent on the soil water content, increasing for decreasing water content and increasing especially rapidly with decreasing SMC when the soil begins to significantly dry out. The soil water potential is also sensitively dependent on soil water content, decreasing with decreasing soil water content. The plant must work increasingly harder with decreasing SMC in order to extract the same amount of water from the soil, but this cannot continue indefinitely without desiccating the plant.

1.3.2.2 Environmental Factors

Environmental factors influencing transpiration include solar radiation, atmospheric and soil conditions, chemicals, and fungal diseases. Solar radiation has a direct accelerating influence on transpiration, as well as secondary effects on stomatal movement. Increased transpiration due to increases in light has been attributed to increasing leaf temperature, and protoplasm and cell wall colloid permeability (Miller 1938). Atmospheric conditions also affect the evaporation rate. T_a is perhaps the single most influential control on transpiration and can cause about 60% of the loss (Dingman 2002). Although wind may have considerable short-term influence, its total effect on transpiration loss during the growing season has been found to be less than 5% (Penman 1963). Furthermore, wind is another parameter that enhances plant photosynthesis, with its main effect to increase the concentration of carbon dioxide at the leaf surface (Lemon 1963), with wind-induced movement in vegetation exposing more leaves to full radiation, so that photoefficiency increases (Hesketh 1961).

Atmospheric humidity can also govern stomatal resistance (that being simply the mathematical inverse of conductance) (Peak and Mott 2011). In a dry atmosphere the difference between the saturation vapor pressure in the leaf and that in the atmosphere can become quite large. In the absence of any external brake, this situation could lead to a rapid desiccation of the plant. In some plants such as legumes, however, an increase in the vapor pressure deficit (the difference between the saturation vapor pressure in the leaf and the atmospheric vapor pressure) causes the stomatal resistance to increase in some plant types. This effect has been referred to as "feed forward" by Farquahar (1978) in that the increase in vapor pressure deficit, which take place during the morning hours as the sun's intensity increases with time, may cause the stomatal resistance to increase and thereby cause a midday stomatal closure.

Soil conditions influence transpiration in two ways—the texture of the soil and the SMC. Texture can influence the speed of soil water movement, and if this is very slow (e.g., in clay soils; Katerji and Mastrorilli 2009), the momentary supply available to the roots at the time of increasing evaporation power may be too small that incipient drying of leaves will occur causing a decrease in transpiration. Abundant water supply will permit transpiration to increase with an increase in evaporating power. Experimental evidence of the effect of SMC on transpiration indicates that a plant will transpire at the maximum rate when the soil has sufficient moisture (e.g., Tromp-van Meerveld and McDonnell 2006) and that increase above this amount does not increase the rate. Last, the use of chemicals applied either to the soil in which the plant is growing or to the leaves in the form of spray may increase or decrease the transpiration rate (Jones 1997).

1.4 Conventional Approaches for Measuring Turbulent Heat Fluxes

Various methods for determining ET (or LE) and/or *H* flux rates over land surfaces based on ground instrumentation have been developed throughout the years. These methods generally either directly estimate actual evapotranspiration (AET) or first calculate potential evapotranspiration (PET) and then attempt to correct this to AET using a range of factors. PET is generally defined as the "water loss, which will occur if at no time there is a deficiency of water in the soil for use of vegetation" (Thornthwaite and Holtzman 1942). PET assumes that the water supply is unlimited and is always enough to supply the requirements of the transpiring vegetation cover. As a result, PET and AET are, in practice, different. However, AET is equal to PET only if there is a constant and adequate supply of water to meet the atmospheric demand that occurs in nature such as open water bodies or in cases where a vegetated surface is constantly moist.

The remainder of this section attempts to provide an overview of the most widely used approaches for measuring directly either AET or PET, discussing the relative strengths and limitations of the different techniques. For convenience, the methods reviewed are grouped into two main categories: (1) direct measurements using ground instrumentation and (2) indirect methods based on the meteorological datasets.

1.4.1 Direct Measurement Methods

1.4.1.1 Evaporation Pans

One of the most widely applied methods for estimating PET includes the use of evaporation pans and tanks (see Figure 1.2a and b), which measure the rate at which water is lost by using a gauge. Pans were originally used to determine free water evaporation, but as PET from surfaces of short vegetation has been found to be similar to free-water evaporation (Linsley et al. 1982; Brutsaert 1982), this method can be applied under such conditions. Among the major advantages of pans is that they can be easily managed, transported to, and installed in most locations. They are also inexpensive and relatively easy to maintain in the field (Jones 1997). However, results in PET estimation from various experimental sites seem to vary according to the size, depth, color, composition, and position of the pan. This means that technically it is not always straightforward to compare results from different sites with this type of instrument (Shaw 1994). One possible source of error in the PET measurement from these devices might be associated with the fact that both the sides of the pan and the water inside will absorb radiation and warm up quicker than in a much larger lake, providing an extra energy source and a greater rate of evaporation (Davie 2002). The pan coefficient has been developed to correct for these effects, and this emphasizes the advantages of using standardized equipment, such as the US Class-A pan.

1.4.1.2 Atmometers

PET can also be measured using atmometers. These consist of a wet, porous ceramic cup, covered with a green fabric placed on a cylinder of distilled water. The green fabric simulates the vegetation canopy through which water evaporates (Andales et al. 2011). The amount of water used by vegetation during a specific time period is then recorded by a graduated scale on the side of the cylinder. Figure 1.2c shows an example of an atmometer.

FIGURE 1.2
(See color insert.) (a) The British Standard Evaporation Tank at Bala, North Wales; (b) a US Class A evaporation pan located in the Maesnant experimental catchment, mid-Wales, surrounded by vegetation representative of the surrounding moorland; (c) an atmometer; (d) diagram of the principles of an evapotranspirometer; and (e) a Popov lysimeter. (a, b, d, and e reprinted from *Global Hydrology: Processes, Resources and Environmental Management*, J. A. A. Jones. Copyright 1997, with permission from Addison Wesley. c is from Andales, A. A. J. et al., Irrigation scheduling: The water balance approach, Colorado State University, Fort Collins, 2011. Available at http://www.etgage.com/04707.html (accessed on October 14, 2012). Photo courtesy of T. Bauder.)

Atmometers have similar advantages to pans and tanks, but one issue to be considered in their use is that care must be taken to ensure that the porous surfaces from which the evapotranspiration takes place are kept clean (Jones 1997). However, compared to tanks and pans, their most significant disadvantage is their inherent sensitivity to wind changes. For this reason, the choice of elevation is very important in the operation of these devices as variable wind speed might produce incorrect measurements (Shaw 1994). To avoid the effect of wind, atmometers are placed in shade, which results in a failure to take account of the effects of insolation on evapotranspiration (Jones 1997).

1.4.1.3 Evapotranspirometers

The use of evapotranspirometers was originally proposed by Thornthwaite (1948) and presents an alternative option in the measurement of PET over land surfaces using ground instrumentation. Figure 1.2d summarizes the principles of evapotranspirometers. Briefly, the evapotranspirometer is a device in which the SMC is maintained at a level that allows water losses to occur at the potential rate. Evapotranspirometers have essentially the same advantages and disadvantages as evaporation pans and atmometers, but their comparative advantage against those is that they can provide more accurate estimates of PET over vegetated surfaces and also that these are not sensitive to changes in wind properties (e.g., Ward 1975).

1.4.1.4 Lysimeters

The ground instruments discussed above all estimate PET rates. However, as AET is more useful for practical applications, various ground instruments are required to provide direct measurements of this parameter. One such instrument is the lysimeter (see Figure 1.2e), which takes the same approach to measurement as the evaporation pan, the fundamental difference between them being that a lysimeter is filled with soil and vegetation as opposed to water and allows percolation through the bottom, that allows the instrument to closely mimic the surrounding soil conditions. This difference is important, as the lysimeter measures AET rather than PET. Many lysimeter types are generally available, and details concerning their construction and operation are given by Brutsaert (1982).

One of the main advantages of lysimeters is that their use allows consumption of water by vegetation to be performed under approximately realistic field conditions. However, a major limitation of those devices is related to their installation, which is very labor intensive and can disturb the area to be measured and its surroundings. Also, they can only measure ET from a very limited area surrounding the plant. As a result, for a realistic and accurate estimate of ET by a lysimeter, it is necessary to ensure that parameters characterizing the vegetation type (e.g., plant density, height, and leaf area of vegetation) measured inside the lysimeter are close to that of the surrounding area (Jones 1997). A further difficulty with lysimeters is when they are applied in areas of poorly watered vegetation (e.g., in natural rainfall areas where precipitation is less than ET). This is because in such cases, variations in soil characteristics (e.g., texture, profile depth, fertility, salinity, and thermal conditions) can make it difficult for ET inside a lysimeter to match the prevailing conditions directly outside the device.

1.4.2 Indirect Measurement Methods Based on Meteorological Datasets

1.4.2.1 Bowen Ratio

The Bowen ratio energy balance (BREB) is a technique used to measure LE and *H* fluxes (as well as CO_2 fluxes). The fluxes are determined by measuring the differences in air temperature (T_a) and vapor pressure (*e*) at two heights in the turbulent boundary layer above the canopy (Rana and Katerji 2000). In a typical BREB tower, the first sensor set is 0.5 m above the canopy height and the second set is 2.0 m above the canopy height (Glenn et al. 2007). The measurement frequency is typically every 2 s and is averaged over a 15- or 20-min time period. In addition to the measurements of T_a and *e*, the R_n and *G* fluxes need to also be computed. Then, by assuming that the turbulent transfer coefficients for heat and moisture are equal, because they are both transported by the same eddies of air over the surface, their ratio (the Bowen ratio β) can be calculated (Bowen 1926) based on the following equation:

$$\beta = \gamma \frac{[T_1 - T_u]}{[e_1 - e_u]} = \frac{H}{\text{LE}} \quad (1.4)$$

where γ is the psychrometric constant, T_l and T_u are the upper and lower temperatures, and e_l and e_u are the lower and upper moisture contents, respectively. Then, the energy balance equation (Equation 1.1) and Equation 1.4 can be combined to solve for LE, as follows:

$$\text{LE} = \frac{(R_n - G)}{(\beta + 1)} \tag{1.5}$$

The BREB method is thus an energy balance method representing the ratio of the H to LE fluxes. This method is considered as an indirect method for LE fluxes estimation, as the latter parameter is measured indirectly from a number of other parameters, and rearranging the energy balance equation. BREB systems are available commercially and used widely for the measurement of LE and H fluxes in different types of ecosystems.

A key advantage of this method is its ability to measure LE fluxes in surfaces where LE is below the PET and also the elimination of the wind or turbulent transfer coefficients. Disadvantages of BREB include the complexity and fragility of the equipment used (i.e., sensors and data loggers), as well as the numerical instability of the equations when β reaches values near –1 (Crago 1996). In addition, it should also be underlined that accurate estimation of both LE and H fluxes by this method is subject to satisfying an important boundary condition, namely, the absence of horizontal energy fluxes (Verstraeten et al. 2008). Implementation of the method is also dependent on the simultaneous acquisition of the R_n and G, and thus additional instrumentation that needs to be also deployed in the field. The method also requires adequate upwind fetch, which can perhaps be considered as another potential limitation restricting its wider use. Furthermore, it should also be noted that the contribution of photosynthesis and advection is neglected from the energy balance equation and thus from the LE and H fluxes estimated by the BREB technique. The fact that as an indirect method of flux estimation it does not allow an internal confirmation of the measurements that it estimates for accuracy can be regarded as another limitation of the method as well (Glenn et al. 2007).

1.4.2.2 Eddy Covariance

One of the most widely used direct methods available for collecting data for the purpose of estimating energy fluxes above a canopy is the eddy covariance method. This technique was originally proposed by Swinbank (1951), who demonstrated that fluxes could be determined from the correlation of temperatures and humidity fluctuations with the vertical component of wind velocity (Figure 1.3). The eddy covariance method explicitly measures the turbulent components of momentum, heat, and moisture, theoretically providing a direct estimate of surface fluxes. A full theoretical background including a discussion of the method is given by Dyer (1961) and Brutsaert (1982). Briefly, the direct eddy flux or eddy transfer method calculates the energy for evaporation from measurements of vertical wind velocity and vapor content of the air at a single point above the evaporation surface. The principles of the method are expressed by

$$\text{LE} = \rho \cdot \overline{w' \cdot q'} = \frac{0.622}{P} \rho \cdot \overline{w' \cdot e'} \tag{1.6}$$

FIGURE 1.3
(See color insert.) Example of instrumentation used in the eddy covariance system installed at one site, named Loobos, located in the Netherlands. (a) The three-dimensional sonic anemometer and fast infrared gas analyzer at the tower; (b) an installed Gill wind master Pro in combination with an open path Li-Cor LI-7500; (c) the data storage device (palmtop PC and PCMCIA cards); and (d) a view of the flux tower showing all the instruments deployed and in operation.

where q' is the instantaneous deviation of specific humidity from mean specific humidity (q), e' is the instantaneous deviation of vapor pressure from mean vapor pressure (e), and w' is the instantaneous deviation of vertical wind velocity from mean vertical wind velocity (w). The overbar indicates means of the products of the instantaneous deviations over 1- to 5-min periods. Residual LE flux can also be computed from the energy balance equation (Equation 1.1), where H flux density is measured by eddy correlation as

$$H = \rho \cdot C_p \cdot \overline{w' \cdot T'} \qquad (1.7)$$

where T' is the instantaneous deviation of T_a from mean temperature (T).

It should be noted that required measurements have to be made at extremely short timescales to account for eddies in vertical motion. This essentially means that very detailed micrometeorological instrumentation is required with all instruments having a very rapid response time (Davie 2002). In practice, evaluation of the method has revealed that systematic errors associated with it include time lags between wind and scalar data owing to travel through sampling tubes and instrument response time, damping of high frequency fluctuations, sensor separation between wind and scalar measurements (Raudkivi 1979), and inability of the sonic anemometer to resolve fine-scale eddies in light winds (Moncrieff et al. 1996). Generally, these types of errors result in underestimation of the turbulent fluxes (e.g., Leuning and King 1992).

A key advantage of this technique is that it provides a direct means of measuring the fluxes without making any kind of assumptions regarding diffusivities or about parameter values, the shape of the vertical velocity profile, atmospheric stability, or the nature of the surface cover (e.g., Dingman 2002). Other advantages are its sound theoretical foundation and the fact that it allows measurement of LE fluxes over potential or non-potential surfaces. Furthermore, as a direct measurement method, the eddy covariance technique allows direct checking of the fluxes estimated by it, which is performed by means of the energy balance closure verification. Disadvantages of the method include the need for complex, fragile, and expensive instrumentation, fetch requirement, and well-trained personnel to obtain accurate results. Also, in terms of the instrumentation, special care has to be taken in their installation. For example, the sonic anemometer used must be plumbed in a way that horizontal wind speed does not add any bias to the flux measurements (Stewart et al. 1996). Furthermore, verification studies of the fluxes estimates acquired by this technique using the energy balance closure approach have shown that the method often underestimates the LE and H fluxes. Closure error has been found to vary seasonally and also interannually, confounding comparisons of LE even at the same site among seasons and years (Glenn et al. 2007). Frequently, measured LE + H are 10% to 30% less than $(R_n - G)$ at eddy covariance towers (Twine et al. 2000; Wilson et al. 2002). It should be noted that there is still a lot of debate on the causes of the lack of closure; yet it is generally admitted that flux measurements by the eddy covariance have an uncertainty or error bound of about 20%–30%. A good discussion of this topic including reference to the methods employed for closing the energy balance can be found for example in Twine et al. (2000), Wilson et al. (2002), Jung et al. (2011), and Aubinet et al. (2012).

At this point it is worthwhile to note that the eddy covariance method has been adopted as the standard method for measuring the turbulent fluxes of heat and moisture as well as of CO_2 and of related micrometeorological parameters over large continental or global ground measurement network sites, with the adaptation including AmeriFlux, AsiaFlux, and OzNet (Australia). This topic is covered in more detail in Section 1.5.

1.4.2.3 Scintillometers

Another method widely employed in the estimation of LE fluxes includes the use of scintillometers (Figure 1.4). This method is based on the physical principle of the propagation of electromagnetic waves in the atmosphere and their disturbance by atmospheric turbulence, known as Monin–Obhukov similarity theory (De Bruin et al. 1995). Essentially, a scintillometer measures the variance of radiation intensity fluctuations. These variations in the refraction index are caused by fluctuations in temperature moisture, pressure, and humidity, as well as their interactions. A large scintillometer consists of a transmitter equipped with disk-shaped arrays of light emitting diodes and a receiver, which measures the perturbed light at a distance up to 5 km. An area-averaged H flux can be derived from this type of measurement and the LE flux can be derived from the energy balance equation.

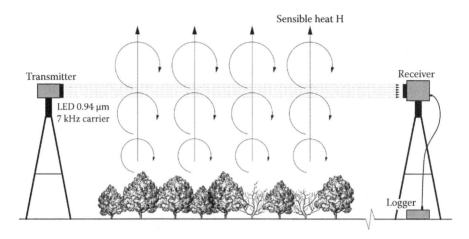

FIGURE 1.4
Operational principle of a scintillometer. (Adapted from Immerzeel, W. W. et al. *FutureWater Report*, 55, 2006.)

The scintillation method is an intermediate method between *in situ* field measurements and large area remote sensing estimates, as it is applicable over distances as long as 5 km, allowing the estimation of surface fluxes at a scale of several kilometers. More specific information on this approach is given by McAneney et al. (1995) and Meijninger and de Bruin (2000).

A key advantage of scintillometers is that they can provide representative estimates of the LE fluxes over large areas with the use of a single instrument, due to the extended spatial averaging of those instruments. Furthermore, as the scintillometer measurement principle is based on evaluations of relative intensity statistics, the system is free of long-term drift and does not require calibration prior to use. Furthermore, use of the optical, contact-free access results in avoiding mechanical interactions that are necessary with conventional instruments and mountings. Lastly, the temporal resolution of scintillometers is an order of magnitude higher than that of point measurements, with typical averaging times of 1–5 min for the fluxes and 10–60 s for the other turbulence statistics, with virtually no statistical noise. However, a major disadvantage is that they are affected by strong turbulence that is referred to as saturation (Savage et al. 2004), which occurs at path lengths of about 250 m (Meijninger 2003).

1.4.2.4 Penman, Penman–Monteith, and Priestley–Taylor Methods

Penman (1948) was the first to show that the mass transfer and energy balance approaches could be combined to obtain an equation for estimating ET without using surface temperature as an input parameter (Dingman 2002). This first model was originally developed for estimating PET over open water. Later, however, Penman and Schofield (1951) refined this original model in order to estimate PET over vegetated surfaces. This was achieved by explicitly incorporating aerodynamics and surface resistance factors to account for variations in ET between vegetation covers that differed in terms of wetness, structure, and physiology.

Monteith (1965) refined the Penman (1948) model so that actual evaporation from a vegetated surface could be estimated. His work involved essentially adding a canopy resistance term (r_c) into the Penman equation so that it takes the following form:

$$\text{AET} = \frac{R_n \Delta + \rho c_p \delta_e / r_a}{\lambda \left(\Delta + \gamma \left(1 + \dfrac{r_c}{r_a} \right) \right)} \qquad (1.8)$$

Δ is the rate of increase in the saturation vapor pressure with temperature, ρ is the density of air, c_p is the specific heat of air at constant pressure, δ_e is the vapor pressure deficit in air, λ is the latent heat of vaporization of water, γ is the psychrometric constant, r_a is the aerodynamic resistance to transport of water vapor, k is the Von Karman constant, d is the displacement height, which replaces the actual surface with a surface that is perceived by the turbulent fluxes, z_0 is the zero plane (i.e., the height within a canopy at which wind speed drops to zero), and u is the wind speed.

Equation 1.8 is generally referred to as the Penman–Monteith equation and is the basic formula used in simple one-dimensional single source descriptions of the evaporation process. Canopy resistance represents the ability of a vegetation canopy to control the rate of transpiration. The Penman–Monteith equation has been termed as a "big-leaf" model because it is one-dimensional and assumes a bulk canopy conductance equal to the parallel sum of the individual leaf stomatal conductances. It has generally been found to work well in a wide range of implementation conditions (Calder 1990; Jensen 1990). However, when vegetation canopies are not homogeneous, the one-dimensional approximation has been found to occasionally lead to significant errors (Ward and Robinson 2000; Verstraeten et al. 2008). A weakness of this model lies also in the difficulty of obtaining adequate measurements of the vegetation factors and especially the r_s (surface resistance), which is a complex function of many climatological and biological factors including radiation, saturation deficit, soil water status, and biomass characteristics (Ward 1975; Monteith 1985). Last but not least, it should be mentioned that the Monteith (1965) approach considers the vegetation canopy as one isothermal leaf and not a set of individual leaves as happens in the real world, which might introduce another source of error (Jones 1997). In practice, however, the Penman–Monteith model has played a very valuable role in the development of conceptual understanding of the complex evapotranspiration processes over heterogeneous surfaces (Ward and Robinson 2000). One of the limitations of this technique is that in the case of sparsely vegetated canopies, the big-leaf assumption of the Penman–Monteith model no longer holds.

A simplification of the Penman–Monteith equation is the Priestley–Taylor equation (Priestley and Taylor 1972), which essentially states that the atmospheric drying power over a wet surface is a constant, multiplied with $\Delta(R_n - G)$:

$$LE = \alpha(R_n - G) \frac{\Delta}{\Delta + \gamma} \qquad (1.9)$$

where α is a constant (-) ranging from 1 to 1.35 for wet surfaces [78], γ is the psychrometric constant (mbar K^{-1}), and Δ is the slope of the vapor pressure curve (mbar K^{-1}).

An operational version of the Penman–Monteith model is currently applied in a large number of countries worldwide as the FAO'56 method. In this method, PET is computed from the Penman–Monteith equation for a hypothetical reference surface defined as "a hypothetical reference crop with an assumed crop height of 0.12 m, a fixed surface resistance of 70s m^{-1} and an albedo of 0.23" (Allen et al. 1998, p. iii). PET is computed based on inputs of meteorological parameters: T_a, humidity, radiation, and wind speed. Subsequently, crop

ET under standard conditions for a specific crop (ET_c) is computed. Perhaps the strongest advantage of this technique is the simplicity in its implementation, which to a large degree justifies its global application. A key limitation of this method is that its implementation can provide results that are not representative over large areas, as some of its input parameters can be highly spatially variable, making the use of this technique inappropriate for estimating spatially distributed ET rates. Last but not least, computation of the actual ET by this method is largely driven by the crop factor input, a parameter that is a function of site-specific environmental factors, crop development stage, and crop type.

1.5 Global Operational Ground Monitoring Networks

The requirement to further our understanding of the processes governing the transfer of heat, mass, and energy between the terrestrial ecosystems and the atmosphere and to improve model predictions of surface–atmosphere exchange has undoubtedly been a priority in recent years (e.g., Ogunjemiyo et al. 2003; Battrick et al. 2006). Data from such "operational" networks not only allow to advance our understanding of the physical processes involved in water and energy exchanges at local scales but also allow us to perform multiscale analysis exploiting either land surface models or remote sensing data, for example, in validating modeling predictions or remote sensing estimates (Dorigo et al. 2011). To this end, the role of Earth observation (EO) is unique, as it allows acquiring spectrally rich information over large regions and at varying spatial and temporal scales that can assist in better understanding Earth system processes. EO provides unique advantages in comparison to conventional methods for deriving spatially distributed estimates of energy fluxes. A discussion on the topic including an overview of the different EO-based methods available in deriving estimates of LE and H fluxes is provided in Chapter 3. In addition, numerous operational products derived from EO data are already distributed globally or their development is underway (see Chapter 10 for a review of these products). The increase in the number of satellite missions and of related operational products that are constantly becoming available requires the implementation of thorough investigation studies examining the accuracy by which the parameters of those EO-based operational products are delivered, including energy fluxes. To this end, reliable *in situ* estimates of energy fluxes and related parameters from observational networks are vital and indispensable. In order to achieve such estimates, various efforts have focused on developing networks of *in situ* validated observations that allow observing, archiving, and distributing systematically and on a global scale key parameters characterizing land surface interaction processes.

FLUXNET (http://fluxnet.ornl.gov/) is at present the largest global network of micrometeorological flux measurement sites that measure the exchanges of carbon dioxide, water vapor, and energy between the biosphere and atmosphere (Baldocchi et al. 1996; Running et al. 1999). Its role is to coordinate regional measurement networks so that ground observations of an array of parameters can be obtained at a global scale, ensuring site-to-site comparability, coordinating simultaneous improvements to existing network plans and the operation of a global archive and distribution center. As a result, FLUXNET provides all the required infrastructure for compiling, archiving, and distributing continuous measurements of carbon dioxide, water vapor, energy fluxes, and many ancillary meteorological, soil, and plant variables submitted by members of regional networks from around the world, such as CarboEuropeIP, AmeriFlux, Fluxnet-Canada, LBA, Asiaflux, Chinaflux,

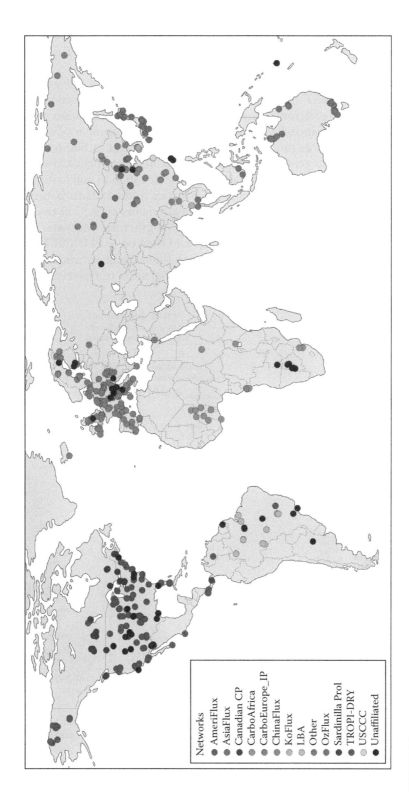

FIGURE 1.5
(See color insert.) Global map of FLUXNET field sites. (Available from http://fluxnet.ornl.gov/introduction. Courtesy of Oak Ridge National Laboratory Distributed Active Archive Center, Oak Ridge, Tennessee.)

Ozflux, CarboAfrica, Koflux, NECC, TCOS-Siberia, and Afriflux. FLUXNET hosts a database of continuous measurements from a total of over 960 site years of data from over 253 eddy covariance measurement sites spread worldwide (Baldocchi 2008). Figure 1.5 illustrates the global map of FLUXNET sites as of October 2011.

A typical tower configuration diagram of a FLUXNET site is depicted in Figure 1.6. In all the sites belonging to FLUXNET, the standard measurement technique used for measuring mass and energy exchange across a horizontal plane between vegetation and the free atmosphere is the eddy covariance. Each network site deploys a number of sensors measuring micrometeorology such as precipitation or wind speed and carbon (i.e., CO_2) flux. The field data are processed to generate 1/2 hourly FLUXNET data files. Other, ancillary data include relatively infrequent measurements of variables such as soil carbon, leaf out dates, vegetation species, canopy height, soil characteristics, and site disturbances such as planting or wildfire. It should also be noted that measurements of SMC at two different soil depths (surface and root zone) are also measured at 30-min time intervals. These data are passed from the sites to their regional networks and then on to FLUXNET. When the data reach FLUXNET, their quality is assessed and gap filled using techniques described by Papale et al. (2006) and Moffat et al. (2007). The spatial scale of observations at each

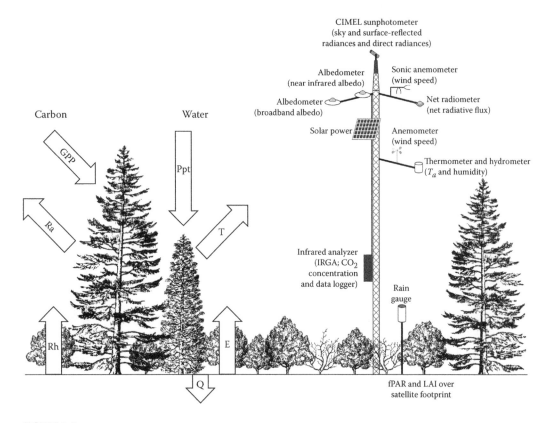

FIGURE 1.6
A generalized FLUXNET tower configuration diagram, showing instrument deployment and key carbon and water fluxes measured. Rh, heterotrophic respiration; Ra, autotrophic respiration; GPP, gross primary production; Q, watershed discharged; E, evaporation; T, transpiration; Ppt, precipitation. (Reprinted from *Remote Sensing of Environment*, 70, S. W. Running et al., A global terrestrial monitoring network, scaling tower fluxes with ecosystem modelling and EOS satellite data, 108–127. Copyright 1999, with permission from Elsevier.)

tower extends through the flux footprint around the tower (ranging between 100 and 1000 m) (Gockede et al. 2004). The FLUXNET community recently began a new synthesis activity the primary goal of which is to add additional years and sites to the database and to process the data with recently developed methods for gap filling, flux partitioning, and uncertainty estimation. Another goal is to share these enhanced data products with other scientific communities under more flexible data policies.

1.6 Conclusions

This chapter has provided an overview of the key principles underlying the processes of evaporation, transpiration, and evapotranspiration, as well as discussing the main parameters influencing their spatiotemporal variation over land surfaces. In addition, an overview covering the range of conventional approaches developed for measuring both LE and H fluxes was provided. As was indicated in the overview of the methods, there are many options to be considered, each having distinct strengths and limitations in their practical use. Generally, the eddy covariance and Bowen ratio flux towers are regarded as the most accurate methods of estimating LE and H fluxes (e.g., Rana and Katerji 2000) with an uncertainty in estimation typically in the order of ~10%–20% (Kustas and Norman 1996). Generally speaking, use of ground instrumentation has certain advantages. Some of those include their ability to provide a relatively direct measurement, as well as their easy installation operation and maintenance of the equipment involved. However, those techniques are often rather complex and labor intensive to operate, whereas their use often requires extensive and sometimes expensive equipment to be deployed in the field, in order to provide only localized estimates of flux measurements.

Various ground-based observational networks have also been developed over time systematically collecting, archiving, and distributing *in situ* data of energy fluxes and related parameters to the interested users' community. FLUXNET is currently the largest of these networks. The key advantages of this network include that it provides both archived and new data of a large number of *in situ* variables characterizing land surface interactions over well-organized ground station networks established in different ecosystems all over the world. The ground data provided are based on uniformly adopted and well-established measurement techniques between the sites, allowing data comparability.

Evidently, one of the main limitations of the conventional methods employed in the estimation of hydrometeorological fluxes as well as of the *in situ* observational networks that are available nowadays includes their inability to provide estimates of energy fluxes over land surfaces regionally and at wider scales of observation. To this end, satellite remote sensing seems to be a promising option. A considerable amount of work has already been done in the last few decades to determine whether remotely sensed imagery can provide accurately spatially explicit maps of these parameters from space, reporting thus far different levels of accuracy. An overview of the EO–based approaches for the estimation of LE and H fluxes is made available in Chapter 3.

Undoubtedly, it is of utmost priority to ensure the continuation of activities supporting the prolongation of long-term measurements of energy fluxes from observational networks such as those of FLUXNET. This will assist us to better understand topics such as how the fluxes of LE and H interplay with other parameters characterizing land surface interaction processes, which is a matter of key scientific priority (Battrick 2006). Also, the rapidly increasing number of remote sensing radiometers being continuously placed in orbit nowadays has necessitated the development of infrastructure for validating remote sensing–derived algorithms and operational products of LE and *H* fluxes. The latter provides a further strong argument toward the support of the continuation and expansion of *in situ* ground observational networks in the future.

Acknowledgments

The preparation of this work was partially supported by the European Space Agency (ESA) Support to Science Element (STSE) under contract STSE-TEBM-EOPG-TN-08-0005. George Petropoulos gratefully acknowledges the financial support provided by the Agency. Dr. Petropoulos and Dr. Griffiths wish to also thank Anthony Smith for his assistance in the preparation of some of the figures included in this chapter.

References

Ahmad, S., A. Karla, and H. Stephen (2010): Estimating soil moisture using remote sensing data: A machine learning approach, *Adv. Water Res.*, 33(1), 69–80.

Allen R., L. Pereira, D. Raes, and M. Smith (1998): Crop evapotranspiration—Guidelines for computing crop water requirements, Irrigation and Drainage Paper 56, Food and Agriculture Organization, Rome.

Andales, A. A., J. L. Chavez, and T. A. Bauder (2011): Irrigation scheduling: The water balance approach, Colorado State University, Fort Collins. Available at http://www.etgage.com/04707.html (accessed on October 14, 2012).

Aubinet, M., T. Vesala, and D. Papale (2012): *Eddy Covariance: A Practical Guide to Measurement and Data Analysis.* Springer, New York.

Baldocchi, D., R. Valentini, W. Oechel, S. Runnig, and R. Dahlman (1996): Strategies for measuring and modelling carbon dioxide and water vapor fluxes over terrestrial ecosystems, *Global Change Biol.*, 2, 159–168.

Baldocchi, D. D. (2008): 'Breathing' of the terrestrial biosphere: Lessons learned from a global network of carbon dioxide flux measurement systems, *Aust. J. Botany*, 56, 1–26.

Battrick, B. (2006): *The Changing Earth. New Scientific Challenges for ESA's Living Planet Programme.* ESA SP-1304. ESA Publications Division, ESTEC, the Netherlands.

Billi, P., and F. Caparrini (2006): Estimating land cover effects on evapotranspiration with remote sensing: A case study in Ethiopian Rift valley, *Hydrol. Sci.*, 51(4), 655–668.

Bougeault, P. (1991): Parameterization schemes of land-surface processes for mesoscale atmospheric models. In *Land Surface Evaporation* (T. J. Schmugge and J. C. Andre, Eds.), Springer-Verlag, New York, pp. 93–120.

Bowen, I. (1926): The ratio of heat losses by conduction and by evaporation from any water surface, *Phys. Rev.*, 27, 779–787.

Breshears, D. D., and J. A. Ludwig (2010): Near-ground solar radiation along the grassland–forest continuum: Tall-tree canopy architecture imposes only muted trends and heterogeneity, *Austral Ecol.*, 35, 31–40.

Briggs, C. L. (1992): A light and electron microscope study of the manutre central cell and egg apparatus of *Solanum nigrum L.* (Solanaceae), *Int. J. Plant Sci.*, 153, 40–48.

Briggs, D., and P. Smithson (1985): *Fundamentals of Physical Geography*, Routledge, London.

Brutsaert, W. (1982): *Evaporation into the Atmosphere: Theory, History and Applications*. D. Reidel, Dordrecht, Holland.

Calder, I. R. (1990): *Evaporation in the Plants*. John Wiley, New York.

Carlson, T. N., W. J. Capehart, and R. R. Gilies (1995): A new look at the simplified method for remote sensing of daily evapotranspiration, *Remote Sens. Environ.*, 54, 161–167.

Cole, L. A., and M. J. Green (1963): Measurements of net radiation over vegetation and of other climatic factors affecting transpiration losses in water catchments, *IASH Publ. Evap.*, 62, 190–202.

Consoli, S., G. Urso, and A. Toscano (2006): Remote sensing to estimate ET-fluxes and the performance of an irrigation district in southern Italy, *Agric. Water Manage.*, 81, 295–314.

Crago, R. D. (1996): Conservation and variability of the evaporative fraction during the daytime, *J. Hydrol.*, 180, 173–194.

Davie, T. (2002): *Fundamentals of Hydrology: Fundamentals of Physical Geography*, Routledge, London.

De Bruin, H. A. R., B. J. J. M. van den Hurk, and W. Kohsiek (1995): The scintillation method tested over a dry vineyard area, *Boundary Layer Meteorol.*, 76, 25–40.

Dingman, S. L. (2002): *Physical Hydrology*, 2nd ed. Prentice-Hall, Englewood Cliffs, New Jersey.

Dorigo, W. A., W. Wagner, R. Hohensinn, S. Hahn, C. Paulik, A. Xaver, A. Gruber et al. (2011): The International Soil Moisture Network: A data hosting facility for global *in situ* soil moisture measurements, *Hydrol. Earth Syst. Sci.*, 15, 1675–1698.

Dyer, A. J. (1961): Measurements of evaporation and heat transfer in the lower atmosphere by an automatic eddy-correlation technique, *Quart. J. R. Meteorol. Soc.*, 87, 401–412.

Engman, E. T. (1991): Application of microwave remote sensing of soil moisture for water resources and agriculture, *Remote Sens. Environ.*, 35, 213–226.

Farquahar, G. D. (1978): Feed forward response of stomata to humidity, *Aust. J. Plant Physiol.*, 5, 787–800.

Fernández-Prieto, D., P. van Oevelen, and W. Wagner (2012): Advances in Earth observation for water cycle science, *Hydrol. Earth Syst. Sci.*, 16, 543–549.

Ford, C. R., and J. M. Vose (2010): Potential changes in transpiration with shifts in species composition following the loss of eastern hemlock in southern Appalachian riparian forests. The 95th ESA Annual Meeting, August 1–6, 2010, Pittsburg, Pennsylvania.

French, A. N., J. G. Alfieri, W. P. Kustas, J. H. Prueger, L. E. Hipps, J. L. Chávez, S. R. Evett et al. (2012): Estimation of surface energy fluxes using surface renewal and flux variance techniques over an advective irrigated agricultural site, *Adv. Water Resour.*, 50, 91–105.

Glenn, E., A. R. Huete, P. L. Nagler, K. K. Hirschboeck, and P. Brown (2007): Integrating remote sensing and ground methods to estimate evapotranspiration, *Crit. Rev. Plant Sci.*, 26, 139–168.

Gockede, M., C. Rebmann, and T. Foken (2004): A combination of quality assessment tools for eddy covariance measurements with footprint modelling for the characterisation of complex sites, *Agric. For. Meteorol.*, 127, 175–188.

Green, F. H. W. (1959): Four years experience in attempting to standardize measurements of potential evapotranspiration in the British Isles and the ecological significance of the results. In *IASH Symposium Hannoversch-Muden*, vol. 1, IEEE, New Brunswick, New Jersey, pp. 92–100.

Hesketh, J. D. (1961): Photosynthesis: Leaf chamber studies with corn, PhD thesis, Cornell, Ithaca, New York.

Immerzeel, W. W., P. Droogers, and A. Gieske (2006): Remote sensing and evapotranspiration mapping: state of the art, *FutureWater Report*, 55.

Jackson, R. B., S. R. Carpenter, C. N. Dahm, D. M. McKnight, R. J. Naiman, S. L. Postel, and S. W. Running (2001): Water in a changing world, *Ecol. Appl.*, 11, 1027–1045.

Jacob, F., M. Weiss, A. Olioso, and A. French (2002): Assessing the narrowband to broadband conversion to estimate visible, near infrared and shortwave apparent albedo from airborne POLDER data, *Agronomie*, 22, 537–546.

Jensen, M. E., R. D. Burman, and R. G. Allen (1990): *Evaporation and Irrigation Water Requirements: Manuals and Reports on Engineering Practice*, no. 70, American Society Civil Engineers, New York.

Ji, Y., and G. Zhou (2011): Important factors governing the incompatible trends of annual pan evapo-ration: Evidence from a small scale region, *Climatic Change*, 106(2), 303–314.

Jones, J. A. A. (1997): *Global Hydrology: Processes, Resources and Environmental Management*, Addison Wesley, Reading, Mass.

Jung, M., M. Reichstein, H. A. Marolis, A. Cescatti, A. D. Richardson, M. A. Arain, A. Arneth et al. (2011): Global patterns of land-atmosphere fluxes of carbon dioxide, latent heat, and sensible heat derived from eddy covariance, satellite, and meteorological observations, *J. Geophys. Res.*, 116(G3), G00J07, doi:10.1029/2010JG001566.

Kalma, J., T. McVicar, and M. McCabe (2008): Estimating land surface evaporation: A review of meth-ods using remotely sensed surface temperature data, *Surv. Geophys.*, 29(4), 421–469.

Katerji, N., and M. Mastrorilli (2009): The effect of soil texture on the water use efficiency of irri-gated crops: Results of a multi-year experiment carried out in the Mediterranean region, *Eur. J. Agron.*, 30(2) 95–100.

Kustas, W., and M. Anderson (2009): Advances in thermal infrared remote sensing for land surface modelling, *Agric. For. Meteorol.*, 149, 2071–2081.

Kustas, W. P., and J. M. Norman (1996): Use of remote sensing for evapotranspiration monitoring over land surfaces, *Hydrol. Sci. J.*, 41(4), 495–516.

Kumagai, T. O. (2011): Transpiration in forest ecosystems. In *Forest Hydrology and Biogeochemistry: Synthesis of Past Research and Future Directions* (D. F. Leyia, D. Carlyle-Moses, and T. Tanaka, Eds.), *Ecol. Stud.*, 216, 389–406.

Lee, C. H. (1942): *Transpiration and Total Evaporation*, McGraw-Hill, New York.

Lemon, E. R. (1963): *Energy and Water Balance of Plant Communities*, Academic Press, New York.

Leuning, R., and K. M. King (1992): Comparison of eddy-covariance measurements of CO2 fluxes by open-path and closed-path CO2 analysers, *Boundary Layer Meteorol.*, 59(3), 297–311.

Li, R., Q. Min, and B. Lin (2009): Estimation of evapotranspiration in a mid-latitude forest using the Microwave Emissivity Difference Vegetation Index (EDVI), *Remote Sens. Environ.*, 113, 2011–2018.

Linsley, R. K., M. A. Kohler, and J. L. H. Paulhus (1982): *Hydrology for Engineers*, 3rd ed., McGraw-Hill, New York.

Liu, X., Y. Luo, D. Zhang, M. Zhang, and C. Liu (2011): Recent changes in pan-evaporation dynamics in China, *Geophys. Res. Lett.*, 38(13), L13404.

Lockart, N., D. Kavetski, and S. W. Franks (2012): On the role of soil moisture in daytime evolution of temperatures, *Hydrol. Process.*, doi: 10.1002/hyp.9525.

Mahlein, A. K., E. C. Oerke, U. Steiner, and H. W. Dehne (2012): Recent advances in sensing plant diseases for precision crop protection, *Eur. J. Plant Pathol.*, 133(1), 197–209.

Mariotto, I., and V. P. Gutschick (2010): Non-Lambertian corrected albedo and vegetation index for estimating land evaporation in a heterogeneous semiarid landscape, *Remote Sens.*, 2, 926–938.

Matsoukas, C., N. Benas, N. Hatzianastassiou, K. G. Pavlakis, K. Kanakidou, and I. Vardavas (2011): Potential evaporation trends over land between 1983–2008: Driven by radiative fluxes or vapour-pressure deficit? *Atmos. Chem. Phys.*, 11(15), 7601–7616.

McAneney, K. G., A. E. Green, and M. Astill (1995): Large-aperture scintillometry: The homogeneous case, *Agric. For. Meteorol.*, 76, 149–162.

McCabe, M. F., and E. F. Wood (2006): Scale influences on the remote estimation of evapotranspira-tion using multiple satellite sensors, *Remote Sens. Environ.*, 105, 271–285.

McVicar, T. R., M. L. Roderick, R. J. Donohue, L. T. Li, T. G. Van Niel, A. Thomas, J. Grieser et al. (2012): Global review and synthesis of trends in observed terrestrial near-surface wind speeds: Implications for evaporation, *J. Hydrol.*, 416, 182–205.

Meijninger, W., and H. A. R. de Bruin (2000): The sensible heat fluxes over irrigated areas in western Turkey determined with a large aperture scintillometer, *J. Hydrol.*, 229, 42–49.

Meijninger, W. M. L. (2003): The sensible heat fluxes over natural landscapes using scintillometry, PhD Dissertation, Wageningen University, the Netherlands.

Meinzer, O. E. (1942): *Hydrology*, McGraw-Hill, New York.

Miller, E. C. (1938): *Plant Physiology*, McGraw-Hill, New York.

Moffat, A., D. Papale, M. Reichstein, D. Y. Hollinger, A. D. Richardson, A. G. Barr, C. Beckstein et al. (2007): Comprehensive comparison of gap-filling techniques for eddy covariance net carbon fluxes, *Agric. For. Meteorol.*, 147, 209–232.

Moncrieff, J. B., Y. Malhi, and R. Leuning (1996): The propagation of errors in long-term measurements of land-atmosphere fluxes of carbon and water, *Global Change Biol.*, 2, 231.

Monteith, J. L. (1965): Evaporation and environment. In *Symposium of the Society for Experimental Biology: The State and Movement of Water in Living Organisms*, vol. 19 (G. E. Fogg, Ed.), Academic Press, San Diego, pp. 205–234.

Monteith, J. L. (1973): *Principles of Environmental Physics*, Edward Arnold, London.

Mutiga, J. K., Z. Su, and T. Woldai (2010): Using satellite remote sensing to assess evapotranspiration: Case study of the upper Ewaso Ng'iro North Basin, Kenya, *Int. J. Appl. Earth Obs. Geoinf.*, 12S, 100–108.

Novák, V. (2012): Evaporation from different surfaces. In *Evapotranspiration in the Soil-Plant-Atmosphere System* (pp. 25–37). Springer, The Netherlands.

Ogunjemiyo, S. O., S. K. Kaharabata, P. H. Schuepp, I. J. MacPherson, R. L. Desjardins, and D. A. Roberts (2003): Methods of estimating CO2, latent heat and sensible heat fluxes from estimates of land cover fractions in the flux footprint, *Agric. For. Meteorol.*, 117, 125–144.

Papale D., M. Reichman, M. Aubinet, E. Canfora, C. Bernhofer, W. Kutsch, B. Longdoz et al. (2006): Towards a standardized processing of Net Ecosystem Exchange measured with eddy covariance technique: Algorithms and uncertainty estimation, *Biogeosciences*, 3, 571–583.

Peak, D., and K. A. Mott (2011): A new, vapour-phase mechanism for stomatal responses to humidity and temperature, *Plant, Cell Environ.*, 34, 162–178.

Penman, H. L. (1948): Natural evaporation from open water, bare soil, and grass, *Proc. R. Soc. London, Ser. A*, 193(1032), 120–145, doi:10.1098/rspa.1948.0037.

Penman, H. L. (1963): *Vegetation and Hydrology*, Tech. Commun. 53, Commonwealth Bureau of Soils, Harpenden, England.

Penman, H. L. (1968): *Vegetation and Hydrology*, translated from Russian, Hydro Meteorological Publishing House, Leningrad.

Penman, H. L., and R. K. Schofield (1951): Some physical aspects of assimilation and transpiration, *Symp. Soc. Exp. Biol.*, 5, 115–129.

Petropoulos, G., T. N. Carlson, M. J. Wooster, and S. Islam (2009): A review of Ts/VI remote sensing based methods for the retrieval of land surface energy fluxes and soil surface moisture, *Prog. Phys. Geogr.*, 33(2), 224–250.

Priestley, C. H. B., and R. J. Taylor (1972): On the assessment of the surface heat flux and evaporation using large-scale parameters, *Mon. Weather Rev.*, AQO, 81–92.

Rana, G., and N. Katerji (2000): Measurement and estimation of actual evapotranspiration in the field under Mediterranean climate: A review, *Euro. J. Agron.*, 13, 125–153.

Raudkivi, A. J. (1979): *Hydrology: An Advanced Introduction to Hydrological Processes and Modelling*, Pergamon Press, Oxford.

Rivas, R., and V. Caselles (2004): A simplified equation to estimate spatial reference evaporation from remote sensing-based surface temperature and local meteorological data, *Remote Sens. Environ.*, 83, 68–76.

Roderick, M. L., M. T. Hobbins, and G. D. Farquhar (2009): Pan evaporation trends and the terrestrial water balance. I. Principles and observations, *Geogr. Compass*, 3(2), 746–760.

Running, S. W., D. D. Baldocchi, D. Turner, S. T. Gower, P. Bakwin, and K. Hibbard (1999): A global terrestrial monitoring network, scaling tower fluxes with ecosystem modelling and EOS satellite data, *Remote Sens. Environ.*, 70, 108–127.

Ryu, Y., D. D. Baldocchi, H. Kobayashi, C. van Ingen, J. Lie, A. T. Black, J. Beringer et al. (2011): Integration of MODIS land and atmosphere products with a coupled-process model to estimate gross primary productivity and evapotranspiration from 1 km to global scales, *Global Biogeochem. Cycles*, 25(4), GB4017.

Savage, M. J., C. S. Everson, G. O. Odhiambo, M. G. Mengistu, and C. Jarmain (2004): Theory and practice of evapotranspiration measurement, with special focus on surface layer scintillometer (SLS) as an operational tool for the estimation of spatially-averaged evaporation, Report 1335/1/04, Water Research Commission, Pretoria.

Sellers P., D. Randall, J. Collatz, J. Berry, C. Field, D. Dazlich, C. Zhang, G. Collelo, and A. Bounous (1996): A revised land surface parameterization (SiB2) for atmospheric GCMs: Model formulation, *J. Clim.*, 9, 676–705.

Shaw, E. (1994): *Hydrology in Practice*, Spon Press, Oxford.

Shuttleworth, W. J. (1993): Evaporation. In *Handbook of Hydrology* (D. R. Maidment, Ed.). McGraw-Hill, New York, pp. 4.1–4.53.

Song, Y., J. Wang, K. Yang, M. Ma, X. Li, Z. Zhang, and X. Wang (2012): A revised surface resistance parameterisation for estimating latent heat flux using remotely sensed data, *Int. J. Appl. Earth Obs. Geoinf.*, 17, 76–84.

Stewart, J. B., E. T. Engman, R. A. Feddes, and Y. Kerr (1996): *Scaling Up in Hydrology Using Remote Sensing*, John Wiley, New York.

Sun, Z., B. Wei, W. Su, W. Shen, C. Wang, D. You, and Z. Liu (2011): Evapotranspiration estimation based on the SEBAL model in the Nansi Lake Wetland of China, *Math. Comput. Model*, 54, 1086–1092.

Swinbank, W. C. (1951): The measurement of vertical transfer of heat and water vapour by eddies in the lower atmosphere, *J. Meteorol.*, 8, 135–145.

Thornthwaite, C. W. (1948): An approach toward a rational classification of climate, *Geograph. Rev.*, 38(1), 55–94.

Thornthwaite, C. W., and B. Holzman (1942): Measurement of evaporation from land and water surfaces, *U.S. Dep. Agric. Tech. Bull.*, 817.

Tromp-van Meerveld, H. J., and J. J. McDonnell (2006): On the interrelations between topography, soil depth, soil moisture, transpiration rates and species distribution at the hillslope scale, *Adv. Water Resour.*, 29(2), 293–310.

Twine, T. E., W. P. Kustas, J. M. Norman, D. R. Cook, P. R. Houser, T. P. Meyers, J. H. Prueger, P. J. Starks, and M. L. Wesely (2000): Correcting eddy-covariance 956 flux underestimates over a grassland, *Agric. For. Meteorol.*, 103, 279–300.

Van den Honert, T. H. (1948): Water transport in plants as a catenary process, *Discuss. Faraday Soc.*, 3, 146–153.

Van der Kwast, J., W. Timmermans, A. Gieskie, Z. Su, A. Olioso, L. Jia, J. Elbers, D. Karssenberg, and S. de Jong (2009): Evaluation of the surface Energy Balance System (SEBS) applied to ASTER imagery with flux-measurements at the SPARC 2004 site (Barrax, Spain), *Hydrol. Earth Sci.*, 13, 1337–1347.

Verstraeten, W. W., F. Veroustraete, and J. Feyen (2008): Assessment of evapotranspiration and soil moisture content across different scales of observation, *Sensors*, 8, 70–117.

Viliam, N. (2012): Evaporation from different surfaces. In *Evapotranspiration in the Soil-Plant-Atmosphere System*, *Prog. Soil Sci.*, Springer, New York, pp. 25–37.

Villegas, J. C., D. D. Breshears, C. B. Zou, and D. J. Law (2010): Ecohydrological controls of soil evaporation in deciduous drylands: How the hierarchical effects of litter, patch and vegetation mosaic cover interact with phenology and season, *J. Arid Environ.*, 74(5), 595–602.

Ward, R. C. (1975): *Principles of Hydrology*, 2nd ed., McGraw-Hill, New York.

Ward, R. C., and M. Robinson (2000): *Principles of Hydrology*, 4th ed., McGraw-Hill, New York.

Wesseling, J. G., C. R. Stoof, C. J. Ritsema, K. Oostindie, and L. W. Dekker (2009): The effect of soil texture and organic amendment on the hydrological behaviour of coarse-textured soils, *Soil Use Manage.*, 25(3), 274–283.

Wilson, M. F., D. D. Baldocchi, M. Aubinet, P. Berbigier, C. Bernhofer, H. Dolman, E. Falge et al. (2002): Surface energy partitioning between latent and sensible heat flux at fluxnet sites, *Water Resour. Res.*, 38, 1294.

Yang, H., Z. Cong, Z. Liu, and Z. Lei (2010): Estimating sub-pixel temperatures using the triangle algorithm, *Int. J. Remote Sens.*, 31(23), 6-47-6060.

Zheng, H., X. Liu, C. Liu, X. Dai, and R. Zhu (2009): Assessing the contribution to pan evaporation trends in Haihe River Basin, China, *J. Geophys. Res.*, 114, D24105.

2

Surface Soil Moisture Estimation: Significance, Controls, and Conventional Measurement Techniques

George P. Petropoulos, Hywel M. Griffiths, Wouter Dorigo,
Angelika Xaver, and Alexander Gruber

CONTENTS

2.1 Soil Moisture: Definition and Significance

Soil moisture is generally defined as the water contained in the unsaturated soil surface of the Earth, derived from rainfall, from snowmelt, or by capillary attraction from groundwater. Soil moisture content (SMC) is a significant component of climatological, hydrological, and ecological systems. Classic estimates of global soil moisture are approximately 70×10^3 km^3 (0.005% of the Earth's total volume of water; Jones 1997), with a renewal time of 280 days (Wetzel 1983). It has long been recognized as a key state variable of the global energy and water cycle due to its control on exchanges of energy and matter and physical processes, in particular, the partitioning of available energy at the Earth's surface into latent (LE) and sensible (H) heat exchange with the atmosphere. SMC also directly impacts the exchanges of trace gases on land, including carbon dioxide (Seneviratne et al. 2010), and strongly influences feedback between the land surface and climate, which, in turn, influences the dynamics of the atmosphere boundary layer and thus weather and global climate (Patel et al. 2009).

Hydrologically, the water stored on land is a key variable controlling numerous key land surface and feedback processes within the climate system. The degree of prior saturation is an important control on river catchment response to rainfall or snowmelt and subsequent flood generation, primarily by partitioning rainfall into infiltration and runoff (Zribi et al. 2005; Penna et al. 2011; Radatz et al. 2012) and also by contributing to runoff itself (Jones 1997). Overland flow will be larger and will occur more quickly on wetter soils and in catchments where areas of saturated soils (e.g., in topographic lows and near watercourses) are more extensive. Knowledge of the spatial distribution of soil moisture can therefore aid us in determining the potential for infiltration, overland flow, floods, and erosion as well as the resultant impacts on streams, reservoirs, infrastructure, and, most importantly, human life (Hebrard et al. 2006). In addition, it can inform sustainable water resources management, the study of ecosystems and ecological processes (Choi et al. 2009), plant water requirements, plant growth and productivity, as well as irrigation management and deciding when to carry out cultivation procedures (e.g., Glenn et al. 2007; Yang et al. 2010). This is particularly true in ecosystems of many arid and semiarid regions in which water is a limiting resource (Mariotto and Gutschick 2010). In a period in which climatic changes are leading to significant changes in the hydrological cycle, which, in turn, impacts on the quantity and quality of food and water available to human society, knowledge of such an important store of freshwater is essential.

SMC is most commonly expressed as either a dimensionless ratio of two masses or two volumes or given as a ratio of a mass per unit volume. These dimensionless ratios can be reported either as decimal fractions or percentages. Soil moisture has traditionally been considered to exist in an unsaturated zone of aeration in which soil pores contain more air than water (Jones 1997). Water enters this zone as a result of the processes listed above through infiltration and capillary action and exits via evapotranspiration or by vertical percolation across the water table into a zone of saturation or groundwater (Jones 1997). Although SMC tends to refer to water in storage, water can drain laterally through the zone of aeration via a process called throughflow.

SMC can also be characterized as a combination of surface SMC (defined as the water contained within the first 5 cm of the soil depth) and root zone SMC (defined as the water content contained below 5 cm of soil depth; e.g., Hillel 1998; Seneviratne et al. 2010; Figure 2.1). In practice, often only a fraction of soil moisture is relevant or measureable, as it exists in a heterogeneous matrix of solid material, both organic and inorganic (Shaw 1994). Thus

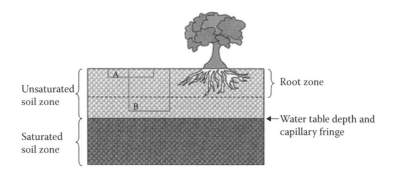

FIGURE 2.1
The saturated and unsaturated soil zones. A and B denote two distinct soil moisture volumes. (Adapted from *Earth-Science Reviews*, 99, S. I. Seneviratne et al., Investigating soil moisture-climate interactions in a changing climate: A review, 125–161. Copyright 2010, with permission from Elsevier.)

soil moisture needs to be considered with regard to a given soil volume. One measure commonly used is volumetric soil moisture θ (m³ H₂O/m⁻³ soil) in a given soil volume V (e.g., volumes A or B in Figure 2.1). Two key parameters often used to inform practical applications of SMC (especially in agricultural systems) are field capacity (defined as the maximum volume of water that a soil can hold) and soil moisture deficit (the amount of water required to raise SMC to field capacity).

This chapter aims to first provide an overview of the main parameters controlling SMC variations, as understanding this is essential in order to be able to appreciate the main limitations and challenges in deriving and validating SMC from remote sensing observations. Second, it aims to provide an all-inclusive overview of the conventional SMC estimation methods and to discuss key operational ground-based observational networks that currently provide such data. This allows an appreciation of the absolute accuracy by which SMC can be measured in the field. This is important from the perspective of remote sensing–related dataset validation by using *in situ* estimates as reference datasets for satellite-derived estimates of SMC. Furthermore, knowledge of available infrastructure providing validated *in situ* observations of soil moisture and related parameters on a consistent basis is of paramount importance in establishing strategies for validating satellite-derived remote sensing estimates and assessing the performance of relevant operational products available globally.

2.2 Environmental Variables Controlling SMC

SMC is generally a highly variable parameter both temporally and spatially, especially at the soil surface. Previous studies have noted that the most important parameters influencing the spatial variability of SMC are topography, soil properties, vegetation type and density, mean moisture content, depth to water table, precipitation depth, solar radiation, and other meteorological factors (e.g., Famiglietti et al. 1998; Herbrard et al. 2006). Soil moisture variability, however, always reflects a combination of the effects of more than one of the above factors and the dominant parameter(s) vary according to the soil wetness state (e.g., Grayson et al. 1997; Gomez-Plaza et al. 2001; Herbrard et al. 2006; Liancourt et al. 2012). The

significance of the exact relationship between SMC and the different factors is variable
and difficult to quantify precisely. This section aims to briefly discuss the effect of each of
the different factors affecting surface soil moisture. Detailed discussions on the topic can
be found, for example, in the works of Famiglietti et al. (1998) and Herbrard et al. (2006).

2.2.1 Climatological and Meteorological Factors

Variability in surface SMC at a catchment scale is strongly influenced by a variety of climatological and meteorological factors, including incoming solar radiation, wind, humidity, and,
most importantly, precipitation. Variations of incoming solar radiation and wind can both
influence the rate of evapotranspiration from soils, either increasing or decreasing SMC. At
its most simple state, the characteristics of surface runoff, subsurface flow, and soil moisture
depend on the characteristics of precipitation (phase, intensity, duration, etc.; Sivapalan et al.
1987; Famiglietti et al. 1998; Salvucci 2001). Reynolds (1970c) was the first to propose that variability in surface SMC might be largest after rainfall because the effects of soil heterogeneity
would be at their maximum, whereas the opposite would occur after a prolonged dry period.

   A significant number of studies have examined the complex interrelationships between
and cumulative effects of multiple climatological and environmental factors on the distribution of surface SMC (Bell et al. 1980; Robinson and Dean 1993; Nyberg 1996; Famiglietti
et al. 1998; Western et al. 1999; Wu et al. 2012). These climatological factors, of course,
strongly influence the dominant vegetation and soil type and the type of land use that can
be utilized in that location and are, in turn, themselves influenced by those factors as well
as by topography. In general, precipitation patterns (in conjunction with other meteorological factors) will dominate patterns of SMC at the watershed scale, but other factors will
become more important at smaller scales (Vinnikov et al. 1996; Crow et al. 2012).

2.2.2 Topography

Topography-related parameters that affect the distribution of SMC in the top soil layer
include slope, aspect, curvature, specific contributing area, and relative elevation. Slope
influences processes such as infiltration, subsurface drainage, and runoff. Aspect and
slope have been shown to have a direct control on the solar irradiance received, which,
in turn, affects the rate of evapotranspiration from the land surface and, as a result, soil
moisture (Hills and Reynolds 1969; Moore et al. 1988; Nyberg 1996; Huang et al. 2011).
Land surface curvature is a measure of the landscape convexity or concavity and influences the convergence of overland flow. Areas characterized by high curvature tend to be
characterized by a larger heterogeneity in SMC than areas in which plan curvature is low
(Moore et al. 1988). The specific contributing area is defined as the upslope surface area
that drains through a unit length of contour on a hill slope. This parameter controls the
potential volume of subsurface moisture that flows from a particular point on the land
surface, affecting the distribution of soil surface moisture (e.g., Nyberg 1996). Generally
speaking, locations with larger contributing areas are expected to be wetter in comparison
to areas of smaller contributing areas (Famiglietti et al. 1998). Last but not least, relative
elevation (so-called slope location) affects soil surface moisture directly by affecting the
degree to which orographic precipitation contributes to SMC as well as indirectly due to
its effect on soil water redistribution (Famiglietti et al. 1998). Many studies have shown
an inversely proportional relationship between soil moisture and relative elevation (e.g.,
Crave and Gauscuel-Odoux 1997). For example, Hawley et al. (1983) examined the influence of variations in vegetation, soil properties, and topography on the distribution of soil

moisture. Their results indicated that relative elevation was the dominant control on local soil moisture variability, whereas the presence of vegetation cover tended to decrease the variations in SMC explained by topography. In another study, Nyberg (1996) evaluated the relationship between surface SMC and a number of parameters including relative elevation and slope. Their study only reported high positive correlations between soil moisture and slope and elevation.

2.2.3 Soil Properties

The composition of soils varies enormously both spatially and temporally but almost always includes material in the solid phase (including both organic and inorganic material), liquid phase (including solutes), and gaseous phase (including oxygen, carbon dioxide, and nitrogen in varying proportions) (Smithson et al. 2008). These solid organic and inorganic components of the soil form the soil structure. The inorganic solid matter of soil is composed of various rock decompositions, clasts, and minerals in different sizes and composition. The diameters of sand particles range between 2 and 0.02 mm, those of silt particles between 0.02 and 0.002 mm, while those of clay particles have diameters smaller than 0.002 mm (Kuruku et al. 2009). A number of studies have documented that surface SMC is closely correlated to the soil properties (e.g., Niemann and Edgell 1993; Crave and Gauscuel-Odoux 1997; Gao et al. 2011; Atchley and Maxwell 2011). Key soil surface properties influencing the concentration and spatiotemporal distribution of moisture in the soil include the soil texture, organic matter content, and soil macroporosity. Texture, in particular, can control the nature of water transmission and retention in the soil. Coarsely textured soils with a high proportion of sand will drain better than finely textured soils such as clays, and as such, will have a lower water-holding capacity and lower SMC. This influence has been shown to occur in large-scale studies in the United States (Panciera 2009) where soil moisture patterns reflect varying soil texture.

In addition, the organic matter content of soils directly influences soil albedo and dielectric properties. Soils with a smaller proportion of decomposed organic substances are characterized by a higher albedo (and thus reflectance) especially in the near-infrared and visible parts of the electromagnetic spectrum, whereas soils with a higher proportion of decomposed organic matter have a lower albedo in all wavelengths ranging between 0.5 and 2.3 mm (Kuruku et al. 2009). Thus, by controlling soil albedo, soil organic matter influences evaporation rates from the soil surface and hence the SMC of the surface layer (Famiglietti et al. 1998). Analysis of soil color can provide information about numerous other soil characteristics (texture, organic matter content, natural drainage condition, aeration, and the phenomena of washing and accretion; Kuruku et al. 2009) and as such is commonly used as a morphological specification for characterizing soils.

2.2.4 Vegetation

The influence of vegetation cover, especially its type, density, and uniformity (Crow et al. 2012), on surface SMC variability is another parameter that has been extensively documented since as early as the 1950s when an experimental study by Lull and Reinhart (1955) underlined the strong effect of vegetation cover in the regional variability of surface SMC, which increased as a function of decreasing vegetation cover. Since then, other authors have investigated the bidirectional relationship between vegetation cover and SMC (e.g., Hawley et al. 1983; Francis et al. 1986; Liancourt et al. 2012). The presence and amount of vegetation influence the concentration of surface SMC by adding organic matter to the soil surface layer and also by extracting water from the soil to be used for vegetation transpiration. Furthermore, the presence of vegetation cover influences soil moisture via the throughfall pattern and shading of the soil layer that is imposed by the vegetation canopy, which, in turn, influences the rate of evaporation from the soil and soil hydraulic conductivity via the impact of root activity (Famiglietti et al. 1998; Atchley and Maxwell 2011).

2.2.5 Land Use

Land use is also very influential in determining the spatial variability of SMC, mainly because of its influence on vegetation cover and associated impacts on infiltration and runoff rates and evapotranspiration processes, with a more pronounced effect exhibited during the growing season (e.g., Fu and Gulinck 1994; Fu and Chen 2000; Qiu et al. 2001; Zhao et al. 2011). Furthermore, many studies have shown that the influence of land use on surface SMC, expressed mainly via transpiration, can even eliminate the effects of topography-related parameters (i.e., aspect; e.g., Ng and Miller 1980; Herbrard et al. 2006). Land use also influences the spatiotemporal variations of key soil surface characteristics, such as topsoil structure, soil crusting, and soil cover by vegetation and other materials, which, as previously mentioned, strongly affect the spatial variation of SMC (Le Bissonnais et al. 2005).

2.3 Conventional Approaches for the Measurement of Soil Moisture

Given the importance of soil moisture in such a wide variety of physical processes and in the management of resources and land use, various methods have been developed for directly measuring this parameter in the field or by analyzing soil samples under laboratory conditions. Overviews of these different techniques have been given by Robock (2000), Robinson et al. (2008), Verstraeten et al. (2008), Dorigo et al. (2011a) and Dobriyal et al. (2012). Some of these works (e.g., Robock 2000; Verstraeten et al. 2008) have classified the existing methods into the following broad categories: gravimetric, nuclear-based, electromagnetic, tensiometer-based, hygrometric, and emerging techniques.

This section aims to provide an overview of the conventional techniques currently available for measuring soil moisture and also to present a few promising emerging measurement techniques. Examples of instruments used, based on the different techniques reviewed herein, are presented in Figure 2.2. For efficiency, and in common with the reviews mentioned above, this overview will follow the categorization adopted previously by others. Each of the techniques is reviewed herein by means of a simple description, outlining its relative advantages and disadvantages. Understanding these not only provides a good basis for understanding estimates of SMC obtained using remote sensing techniques reviewed herein, but also allows appreciating the potential limitations when validating the satellite-derived estimates. Verstraeten et al. (2008) in their recent overview of the available conventional methods employed in estimating soil moisture concluded that even today there is not a single, clearly superior method suitable under all circumstances for measuring SMC.

As Dorigo et al. (2011a) noted, even when one technique is consistently employed, SMC measurements can be strongly influenced by several other factors including, for example,

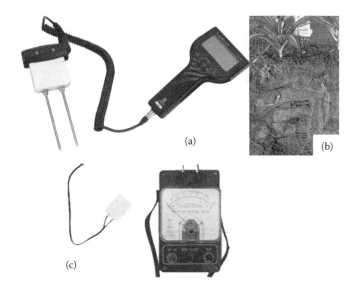

FIGURE 2.2
(See color insert.) Examples of instruments used to measure soil moisture content: (a) a HydroSense II time domain reflectometer (Campbell Scientific; http://www.campbellsci.com/hs2); (b) a sensor by Decagon Devices (http://www.decagon.com/), based on frequency domain technology; and (c) a gypsum block and sensor (http://vfd.ifas.ufl.edu/gainesville/irrigation/gypsum_block_probe.shtml).

instrument calibration, the installation conditions (e.g., installation depth and sensor placement), as well as the spatial resolution, geographical coverage, and representativeness of the measurements obtained. When the aim is to validate regional estimates of SMC derived, for example, from remote sensing observations, given the high spatiotemporal variability of SMC, a large number of SMC measurements from different points are required in order to obtain a value of SMC representative of the study area (Robock 2000). Strategies of upscaling point measurements of SMC to validate large-scale estimates derived from remote sensing have been described by Crow et al. (2012). Alternatively, one or a few sensors can be placed at locations that are representative for the area covered by the satellite footprint (Brocca et al. 2010a,b).

Generally, required accuracy in the measurement of SMC is subject to the application considered each time. Yet an absolute accuracy of about 4% volume by volume (vol vol^{-1}) in SMC is generally satisfactory for a large range of applications (Engman 1992; Walker and Houser 2004; Sequin and Itier 1983; Calvet and Noilhan 2000). Even though most sensor manufacturers claim to achieve this accuracy requirement, these specifications are usually made for a default calibration, which is only valid for "typical conditions" achieved under laboratory conditions. A field-specific calibration and accuracy assessment is indispensable to exploit the potential quality of the sensor and to quantify the actual accuracy. Table 2.1 summarizes the advantages, disadvantages, cost, possible scale of operation, response time, and examples of instruments used to measure soil moisture.

2.3.1 Gravimetric Methods

Gravimetry [sometimes referred to as the (thermo-) gravimetric method] refers to the measurement of soil water content by measuring the difference in weight between a soil

TABLE 2.1

Overview of Conventional Methods for Measuring Soil Moisture in the Field

Method	Strengths	Limitations	Response Time
Gravimetry	High level of accuracy	Destructive Repeat sampling not possible Time and labor intensive High level of possible uncertainty	24 h
Nuclear methods			
Neutron scattering	Straightforward implementation Nondestructive Enables measurement at several depths High level of accuracy	Radiation hazard Time intensive Low sensitivity in upper 20 cm of soil profile	1–2 min
Gamma attenuation	Nondestructive Operation can be automated	Low accuracy Requires high level of expertise Restricted to soil depth of 2.5 cm	Instantaneous
Nuclear magnetic resonance	As for neutron scattering	As for neutron scattering	Instantaneous
Electromagnetic methods			
Resistive sensors	Enables measurement of SMC over extended time periods	Low accuracy Requires individual calibration Limited lifespan of instrument	2–3 h
Capacitive sensors	High level of accuracy Enables SMC measurement at any soil depth	Questionable long-term stability	Instantaneous
TDR	Nondestructive Easy to use Operation can be automated at multiple points High level of accuracy	Environment-sensitive, especially in saline soils	Instantaneous
FDR	Nondestructive	Environment-sensitive	Instantaneous
Tensiometer techniques	Nondestructive High level of accuracy Easy to install, operate and maintain for extended periods	Only allow indirect estimation of SMC Fragile Automated operation impractical	2–3 h
Hydrometric techniques	Low maintenance Operation can be automated	High level of calibration Sensors deteriorate with time as they interact with soil Only provides estimation of SMC	< 3 min
Emerging techniques			
Cosmic rays	Nondestructive	Requires measurement of meteorological variables and calibration	Instantaneous
Distributed temperature sensing	Can be used in remote environments	Difficulty of installing at consistent depths	Instantaneous

sample before and after drying. This is the oldest and most direct method currently used for SMC measurement and remains the standard against which other methods are calibrated and compared (Zazueta and Xin 1994; Verstraeten et al. 2008; Huang et al. 2011). Implementation of this technique is based on extracting a soil sample from the field which is then transferred to a soil analysis laboratory, where it is put into a drying oven at 105°C for a period of 24–48 h. The concentration of water in the soil is determined by subtracting the oven-dry weight from the initial field weight. The difference in mass gives the total soil moisture in the sample, which is converted to volumetric units using the density of the soil. Details of the principles and implementation of this technique are given by Reynolds (1970a,b).

The key advantages of this approach are that it is easily and straightforwardly implemented using low cost technology and equipment. Nevertheless, the method has several disadvantages. It is a destructive method (because soil samples must be removed from the field) that precludes repeat sampling from exactly the same point. Furthermore, its implementation is generally time consuming and labor intensive, requiring the availability of sampling equipment, weighing scale, and an oven, and is difficult to use in rocky environments (Dobriyal et al. 2012). Last but not least, some authors (e.g., Baker and Allmaras 1990) have identified that uncertainty can be introduced in soil moisture estimation by this technique, because of the very precise determination of sample volume required in the conversion from gravimetric to volumetric water content.

2.3.2 Nuclear Techniques

2.3.2.1 Neutron Scattering

In this method, the amount of water in a volume of soil is estimated by measuring the amount of hydrogen it contains, expressed as a percentage. Because most hydrogen atoms in the soil are components of water molecules, the backscatter of the thermalized neutrons from a radioactive source emitted and measured by a detector in the probe directly corresponds to water content in the soil. A neutron probe can measure total soil water content if it is properly calibrated by gravimetric sampling. Depth probes and surface probes are available that measure the SMC at the required depth, or in the uppermost layer, respectively (Dobriyal et al. 2012). More details on this technique implementation can be found, for example, in the works of Goodspeed (1981) and Robock et al. (2000).

The key advantages of this method are that it is relatively easy and straightforward to use. Furthermore, it is a nondestructive technique that enables the measurement of soil moisture distribution profiles at several depths. The method is also accurate—a properly calibrated instrument is capable of an accuracy of better than ±0.02 in volumetric water content (Baker and Allmaras 1990)—and capable of measuring surface SMC in real-time conditions (Dorigo et al. 2011a). Disadvantages of this technique include the high cost of equipment purchase and the radiation hazard involved. Additionally, neutron probes require proper calibration according to each soil type in which they will be used, which, in practice, increases the time it takes to collect data. Furthermore, and perhaps most importantly, they have been shown to be insensitive in measuring soil moisture near the surface (top 20 cm) (Zazueta and Xin 1994) because fast neutrons can escape into the atmosphere (Luebs et al. 1968). This technique is not common when frequent and automated observations are required, and its use has been proven most useful in measuring relative soil moisture differences rather than absolute SMC (Dorigo et al. 2011a).

2.3.2.2 Gamma Attenuation

Another radioactive technique employed in field-based estimation of SMC includes the use of gamma attenuation. The operation of this technique is based on the assumption that scattering and absorption of gamma rays are correlated in their path to the matter density and also that the specific gravity of the soil remains relatively constant as soil moisture changes (Zazueta and Xin 1994). This technique takes measurements in the soil profile and requires two parallel access tubes, one for the radioactive source and one for the detector of primary photons, in order to measure wet density changes in soil from which it is possible to determine SMC. A number of radioisotopes have been used for this purpose, with 137 Caesium being the one most commonly used (Baker and Allmaras 1990). A more detailed discussion on the operation of this method and its principles can be found in the works of Gardner (1986) and Nofziger (1978).

The key advantages of this technique include the fact that it is a nondestructive method that is able to provide the average water content for the profile depth. Furthermore, the operation of the method can be easily automated allowing the user to map the temporal changes in soil water content. Additionally, this technique is much easier to calibrate as it does not have to be site specific. However, its implementation requires the use of relatively expensive instrumentation and a greater level of user expertise. In addition, the use of this method is restricted to estimating SMC of a 2.5-cm-thick sample of soil thickness (but at a very high resolution) and measurements are affected by bulk density changes (Dobriyal et al. 2012; Pires et al. 2005).

2.3.2.3 Nuclear Magnetic Resonance

Nuclear magnetic resonance is another technique used for measuring the volumetric soil water content based on the use of radioactivity. This technique subjects water in the soil to both a static and an oscillating magnetic field perpendicular to each other. A radio frequency detection coil, turning capacitor, and electromagnetic coil are used as sensors to measure the spin echo and free induction decays. This technique can discriminate between bound and free water in the soil. Stafford (1988) provides more details concerning the method's operation principles as well as its relative strengths and weaknesses. Generally, advantages and disadvantages of this technique are similar to those of the neutron scattering technique discussed earlier.

2.3.3 Electromagnetic (or Dielectric Constant) Methods

Some techniques for measuring soil moisture are based on the electromagnetic properties of water. Water's permanent dipole moment (the displacement of positive and negative molecular charge related to the position of the hydrogen atoms relative to the oxygen atoms) is high in comparison with other materials and as such water has a high dielectric constant of ~80 (Robinson et al. 2008). These measurement techniques depend on the effect of water on the bulk dielectric properties of the soil (given that other soil constituents have dielectric constants of less than 5 when measured between 30 MHz and 1 GHz). In a mixture of water and dry soil, the resulting dielectric constant is between these two extremes, thus offering a mechanism for detecting the water content in the soil (Gardner et al. 2001). The dielectric properties of materials such as soil, composed of materials with different dielectric properties, are dependent on frequency, largely due to the fact that more processes occur in such heterogeneous material with different microgeometries, in comparison with homogeneous material.

Values measured in this way are subsequently related through calibration to SMC. A common characteristic of all of these techniques is that they require calibration with gravimetric samples. Broadly speaking, four types of approaches have been developed based on this principle: the resistive sensor (gypsum), the capacitive sensor, the time domain reflectometry (TDR), and the frequency domain reflectometry (FDR). The following sections provide an overview of the operation of those techniques, including a discussion of their advantages and disadvantages. A more detailed discussion on the use of those techniques is given by Verstraeten et al. (2008).

2.3.3.1 Resistive Sensor (Gypsum)

Porous blocks of gypsum are used in one of the most common dielectric constant techniques employed for measuring SMC in the field. The device consists of a porous block made of gypsum or fiberglass containing two electrodes linked to a wire lead. When the device is buried into the soil surface, water will enter or exit the block until the matric potential of the block and the soil are the same. Then, the electrical conductivity of the block to the matric potential for any particular soil is calculated using a calibration curve. The main advantage of this technique is that it is a low-cost solution, allowing the measurement of SMC in the same location in the field over extended periods of time, although this is limited by the dissolution and degradation of the block (Dobriyal et al. 2012). Nonetheless, the key disadvantage of this technique is that it requires individual calibration of the porous blocks for each location and for each measurement interval, which limits the gypsum block life span (Zazueta and Xin 1994). The accuracy of this method is affected by both salt and temperature (Dobriyal et al. 2012).

2.3.3.2 Capacitive Sensor

Another option for measuring SMC through its effect on the dielectric constant is by measuring the capacitance between two electrodes implanted in the soil using a system of probes (e.g., Dean et al. 1987). A frequency excitation is usually given to the probe installed that enables the measurement of the dielectric constant that is directly proportional to the moisture content in the soil layer. An advantage of the use of this technique is that it can measure water content at any soil depth. Also, it can achieve a relatively high precision in soil moisture estimation when ionic concentration of the soil is constant over time. One of the main disadvantages of this method is that the long-term stability of the system calibration is questionable, as well as its high cost.

2.3.3.3 Time Domain Reflectometry

Another technique widely applied for measuring soil moisture based on soil electrical conductivity measurements belonging to the dielectric group of methods is the time domain reflectometer (TDR; Taylor 1955). TDR is a method that uses a device that propagates a high-frequency transverse electromagnetic wave along a cable attached to a parallel conducting probe inserted into the soil. The signal is reflected from one probe to the other before being returned to the meter that measures the time elapsed between sending the pulse and receiving the reflected wave. Assuming that the cable and waveguide length are known, the propagation velocity, which is inversely proportional to the dielectric constant, can be directly related to SMC. This provides a measurement of the average volumetric water content along the length of the waveguide.

Among the most important advantages of this technique are that it is nondestructive to the study site and is not labor intensive (Dobriyal et al. 2012). The ability to automate TDR measurement and to multiplex many waveguides through one instrument (Baker and Allmaras 1990) are further advantages of TDR because they allow unattended measurement at multiple points, either on a scheduled interval or in response to events such as rainfall. When it is properly calibrated and installed, it is a highly accurate method for measuring SMC (Baker and Allmaras 1990; Verstraeten et al. 2008).

2.3.3.4 Frequency Domain Reflectometry

FDR is similar to TDR but estimates SMC through measuring changes in the frequency of a signal as a result of soil dielectric properties (Dobriyal et al. 2012). An electrical circuit using a capacitor and an oscillator measures changes in the resonant frequency and indicates variations in SMC. The main advantage of this method is that it is nondestructive, but in comparison with TDR, it can provide less accurate results due to sensitivity to soil characteristics (e.g., salinity and temperature) and also has a limited scale of use (Dobriyal et al. 2012).

2.3.4 Tensiometer Techniques

Tensiometers are devices that measure the tension or the energy with which water is held by the soil and are comprised of water-filled plastic tubes with hollow ceramic tips attached on one end and a vacuum gauge and airtight seal on the other. These tubes are installed into the soil at the depth at which the soil moisture measurement is required. At this depth, water in the tensiometer eventually comes to pressure equilibrium with the surrounding soil through the ceramic tip. When the soil dries, soil water is pulled out through the tip into the soil, creating a tension or vacuum in the tube. As the soil is rewetted, the tension in the tube is reduced, causing water to reenter the tip, reducing the vacuum. Tensiometers are available commercially in many different types of configurations and are inexpensive, nondestructive, and easy to install and operate satisfactorily in the saturated range. If properly maintained, they can operate in the field for long time periods. Another important advantage of using tensiometers is that they can allow measurement of the water table elevation and/or soil water tension when a positive or negative gauge is installed. However, they are only able to provide direct measurements of the soil water suction, allowing an indirect estimation of SMC. Furthermore, tensiometers are fragile and require care during their installation and maintenance in the field (Dukes et al. 2010). Automated measurements are possible but at a high cost, and they are not electronically stable.

2.3.5 Hygrometric Techniques

Because the thermal inertia of a porous medium depends on moisture content, soil surface temperature can be used as an indication of SMC. For this purpose, electrical resistance hydrometers that utilize chemical salts and acids, aluminum oxide, electrolysis, thermal principles, and white hydrosol are used to measure relative humidity (RH). The resistance of the resistive element measured is a function of RH and allows the SMC to be inferred. More detailed information on the use of these devices can be found in the work of Wiebe et al. (1977). This technique enables SMC measurements to be taken over large areas, and this is perhaps its most important advantage. Important disadvantages include that hydrometers comprise a very large, complex, and expensive system, making their use impractical.

2.3.6 Emerging Techniques

In addition to the established techniques described above, novel techniques for measuring soil moisture are currently being developed (Ochsner et al. in press). Two of the most important are cosmic ray hydrometeorology (Zreda et al. 2012) and distributed temperature sensing (Gao et al. 2011; Striegl and Loheide 2012). The former utilizes the fact that the density of low-energy cosmic ray neutrons in the atmosphere is inversely correlated with soil moisture and measures the neutrons emitted by cosmic rays in air and soil using a stationary cosmic ray soil moisture probe (neutronavka) (Zreda et al. 2012). This system has the distinct advantage of being able to obtain estimates of soil moisture over areas of a few hundred square meters. Limitations include the need to isolate the signal of SMC from other sources of hydrogen source cosmic rays such as some minerals (e.g., clay), vegetation, and organic matter as well as surface and atmospheric water. Measurements of these variables are required in order to calibrate the soil moisture measurements. Currently, a large monitoring network of cosmic ray probes is being set up in the United States and other parts of the world (Zreda et al. 2012).

Distributed temperature sensing uses fiber optic cables that can extend in excess of 50 km in order to measure changes in soil thermal conductivity, which is a function of soil moisture and ambient temperature. The main advantages are the large spatial extent and resolution (1–2 m) that this technique offers and that low power requirements mean that it can be used in remote environments. Disadvantages include the difficulty of placing the fibers at consistent depths and locations and monitoring diurnal changes in soil temperature (Striegl and Loheide 2012).

2.4 Ground Monitoring Soil Moisture Networks

The important role of ground monitoring networks in further understanding the processes governing the transfer of heat, mass, and energy between the terrestrial ecosystems and the atmosphere and to improve predictions of parameters characterizing surface–atmosphere exchange from remote sensing data was already underlined in Chapter 1. As has already been discussed, the increasing number of Earth Observation (EO) missions with the launch of more and more sophisticated remote sensing radiometers that are continuously becoming available has made developing the necessary infrastructure for validating the remote sensing–derived surface heat fluxes and/or surface SMC data a key priority. In the remainder of this section the currently active global *in situ* monitoring networks providing appropriate validated ground measurements of soil moisture are briefly reviewed. Knowing this is particularly important from a remote sensing point of view, as availability of validated *in situ* SMC data is needed to evaluate the performance of EO-based algorithms and operational products that are being developed or offered to the community. Conducting such detailed and thorough benchmarking studies is particularly important for the development and distribution of operational products before those are made available for use to the wider community.

2.4.1 International Soil Moisture Network

The International Soil Moisture Network (ISMN; http://ismn.geo.tuwien.ac.at/) is an international cooperation initiated by the Global Energy and Water Exchanges (GEWEX) project and European Space Agency (ESA) with the purpose of establishing and maintaining a database of harmonized global *in situ* soil moisture and promoting scientific studies on calibration and validation of satellite based and modeled soil moisture products (Dorigo et al. 2011a,b). The ISMN soil moisture measurement hosting facility is coordinated by GEWEX in collaboration with the Group of Earth Observation (GEO) and the Committee on Earth Observation Satellites (CEOS). The data portal has been implemented by the Vienna University of Technology, who also hosts it.

Within a fully automated process chain, collected data are harmonized in terms of measurement unit, sampling interval, and metadata, and after a basic quality check, they are stored in a database. The ISMN is being made possible through the voluntary contributions of scientists and networks from around the world. The soil moisture and meteorological data sets contained in the ISMN are shared by the different network operating organizations on a voluntary basis and free of cost. In June 2012, the ISMN contained the data of 40 networks, which together contain more than 1600 stations. Available data sets include historical observations as well as near-real-time measurements. Examples of the networks include the US Climate Reference network (United States, 114 stations), SCAN (United States, 182 stations), SNOTEL (United States, 381 stations), UMBRIA (Italy, 7 stations), OZNET (Australia, 64 stations), REMEDHUS (Spain, 18 stations), and SMOSMANIA (France, 21 stations). Apart from several recently established operational networks that share their data with the ISMN, the Global Soil Moisture Data Bank (Robock et al. 2000) merged its historical data collection with the ISMN and has now been closed. A complete list of networks that have already shared their soil moisture measurements with the ISMN are summarized in http://ismn.geo.tuwien.ac.at/networks. A Web interface of the ISMN with a map showing the spatial distribution of the stations as of October 2012 is shown in Figure 2.3. Users are able to access the harmonized data sets easily through this Web portal. A full description of the ISMN was recently given by Dorigo et al. (2011b). Currently, an enhanced quality control is being implemented, which should support the user in filtering the data sets for spurious observations (Gruber et al. 2013).

2.4.2 FLUXNET Network

FLUXNET, which was also mentioned in Chapter 1, is a global "network of networks" measuring a number of parameters, including soil moisture. It provides the entire required infrastructure for compiling, archiving, and distributing continuous measurements of soil moisture in different ecosystem conditions, acquired simultaneously to a number of other parameters characterizing land surface interaction processes. An overview of this network was provided in Chapter 1 and will not be included here for brevity. In respect to soil moisture measurement, in particular, at each FLUXNET site, soil moisture is measured at least in two depths (surface and root zone) at half-hourly time intervals. Soil moisture is a core parameter estimated in most, if not all, FLUXNET sites using largely the same type of ground instrumentation. As is also done with all the other parameters, the half-hourly soil moisture data are then passed from each site to their regional networks and then on to FLUXNET.

FIGURE 2.3

(See color insert.) Geographical distribution of the ISMN sites. Place marks indicate the station coordinates of the different networks contained in the ISMN database. The different colors refer to the various networks that provide data.

2.5 Conclusions

In view of the importance of information on the spatial distribution of SMC, its measurement has attracted the attention of scientists from many disciplines and decades of efforts have been dedicated to its estimation. This has led to a relatively mature understanding of the relationships between topographic, climatological/meteorological, hydrological, biotic, abiotic, and anthropogenic factors and SMC at localized scales.

This chapter has provided an overview of the wide variety of approaches available for measuring those parameters directly using ground instrumentation. As has been documented, there are many different options that can be considered, each having distinct practical advantages and disadvantages. Generally, the use of ground instrumentation has certain advantages, such as a relatively direct measurement, instrument portability, easy installation operation and maintenance, the ability to provide measurement at different depths, and also the relative maturity of the methods. However, use of ground measurement techniques has proven very difficult to implement practically over large areas. This is mainly because they can be complex, labor intensive, sometimes destructive to the study site, and not always reliable. Even if dense ground measurements are available, the heterogeneity of the observation site has to be taken into account. Different measurement methods, sensors, calibrations, and installation depths have to be considered as well when comparing measurements from different sites. In addition, the use of ground-based methods often requires the deployment of extensive equipment in the field in order to provide only localized estimates of SMC, making them unsuitable for measuring SMC over large spatial scales. Gravimetric sampling and networks of impedance probes based on dielectric methods appear to be two of the most reliable methods of estimating surface SMC at an accuracy level of ~4% vol vol^{-1} or better, although inaccurate installation can lead to much larger errors than this.

With regard to the availability of ground measurements, an overview of the currently available operational networks showed that the FLUXNET and ISMN networks are currently the densest monitoring networks globally providing *in situ* observations of SMC and other ancillary parameters. One of the key advantages of FLUXNET in comparison to ISMN is that it provides simultaneously to soil moisture archived and new data of a large number of other *in situ* variables characterizing land surface interactions over well-organized ground station networks established in different ecosystems all over the world. One of the key advantages of ISMN includes the fact that it provides easy and rapid access to harmonized and quality-checked SMC ground measurements from a large number of locations distributed all over the world, amalgamating the efforts of different groups attempting to provide long-term measurements of SMC. On both networks the ground data provided are largely based on uniformly adopted and well-established measurement techniques between the sites, allowing data comparability between sites and studies belonging to each network, or even between these two networks. Thus, although FLUXNET and ISMN have two different purposes, they can work complementarily. FLUXNET coordinates the networks and makes decisions of where and which measurements should be taken, whereas the ISMN acts only as a hosting and harmonization facility with the key objective to collect and fuse everything that is available. Therefore it just holds data from important ground monitoring networks.

Clearly, there is a need for efforts such as those of FLUXNET and ISMN to be further supported by the scientific and user community in the future in order to be able to more accurately measure SMC spatially over different ecosystem conditions and to better understand how the latter interplays with other parameters characterizing land surface interactions of the Earth system. Also, the increasing number of satellite missions related to soil moisture retrieval being placed in orbit has made it indispensable to develop the necessary infrastructure for validating remote sensing–derived estimates of SMC (an overview of methods is provided in Chapter 4). Thus a continuation of the initiatives supporting and expanding such ground observational networks in the future would be a very valuable investment from multiple perspectives, and its importance cannot be overstated.

Acknowledgments

This work was produced in the framework of the PROgRESSIon (Prototyping the Retrievals of Energy Fluxes and Soil Moisture Content) project implementation, funded by the European Space Agency (ESA) Support to Science Element (STSE) under contract STSE-TEBM-EOPG-TN-08-0005. Dr. Petropoulos gratefully acknowledges the financial support provided by the Agency. Wouter Dorigo, Angelika Xaver, and Alexander Gruber are funded through ESA's SMOS Soil Moisture Network Study—Operational Phase (ESA ESTEC Contract no. 4000102722/10). Dr Petropoulos and Griffiths wish to also thank Anthony Smith for his assistance in the preparation of some of the figures used in the present chapter.

References

Atchley, A. L., and R. M. Maxwell (2011): Influences of subsurface heterogeneity and vegetation cover on soil moisture, surface temperature and evapotranspiration at hillslope scales, *Hydrogeol. J.*, 19(2), 289–305.

Baker, J. M., and R. R. Allmaras (1990): System for automating and multiplexing soil moisture measurement by time-domain reflectometry, *J. Soil Sci. Soc. Am.*, 54(1), 1–6.

Bell, K. R., B. J. Blanchard, T. J. Schmugge, and M. W. Witzak. (1980): Analysis of surface soil moisture variations within large field sites, *Water Resour. Res.*, 16, 796–810.

Bissonnais, Y., O. Cerdan, V. Lecomte, H. Benkhadra, V. Souchere, and P. Martin (2005): Variability of soil surface characteristics influencing runoff and interrill erosion, *Catena*, 62, 111–124.

Brocca, L., F. Melone, T. Moramarco, and R. Morbidelli (2010a): Antecedent wetness conditions estimation based on ERS scatterometer data, *J. Hydrol.*, 364(1–2), 73–87.

Brocca, L., F. Melone, T. Moramarco, and R. Morbidelli (2010b): Spatial-temporal variability of soil moisture and its estimation across scales, *Water Resour. Res.*, 46, W02516, doi:10.1029/2009wr008016.

Calvet J. C., and J. Noilhan (2000): From near-surface to root-zone soil moisture using year-round data, *J. Hydrometeorol.*, 1(5), 393–411.

Choi, M., W. P. Kustas, M. C. Anderson, R. G. Allen, F. Li, and J. H. Kjaersgaard (2009): An intercomparison of three remote sensing-based surface energy balance algorithms over a corn and soybean production region (Iowa, U.S.) during SMACEX, *Agric. For. Meteorol.*, 149(12), 2082–2097.

Crave, A., and C. Gascuel-Odoux (1997): The influence of topography on time and space distribution of soil surface water content, *Hydrol. Processes*, 11, 203–210.

Crow, W. T., A. A. Berg, M. H. Cosh, A. Loew, B. P. Mohanty, R. Panciera, P. de Rosnay et al. (2012): Upscaling sparse ground-based soil moisture observations for the validation of coarse-resolution satellite soil moisture products, *Rev. Geophys.*, 50(2), RG2002.

Dean, T. J., J. P. Bell, and A. J. B. Baty (1987): Soil moisture measurement by an improved capacitance technique, Part I. Sensor design and performance, *J. Hydrol.*, 93, 67–78.

Dobriyal, P., A. Qureshi, R. Badola, and S. A. Hussain (2012): A review of the methods available for estimating soil moisture and its implications for water resource management, *J. Hydrol.*, 458–459, 110–117.

Dorigo, W. A., W. Wagner, R. Hohensinn, S. Hahn, C. Paulik, M. Drusch, S. Mecklenburg et al. (2011a): The International Soil Moisture Network: A data hosting facility for global in situ soil moisture measurements, *Hydrol. Earth Syst. Sci.*, 15, 1675–1698, doi:10.5194/hess-15-1675-2011.

Dorigo, W., P. Van Oevelen, W. Wagner, M. Drusch, S. Mecklenburg, A. Robock, and T. Jackson (2011b): A new international network for in situ soil moisture data, *Eos Trans., AGU*, 92, 141–142.

Dukes, M. D., L. Zotarelli, and K. T. Morgan (2010): Use of irrigation technologies for vegetable crops in Florida, *Horticultural Technol.*, 20(1), 133–142.

Engman, E. T. (1992): Soil moisture needs in earth sciences. In *Proceedings of International Geoscience and Remote Sensing Symposium (IGARSS)*, IEEE, New Brunswick, NJ, pp. 477–479.

Famiglietti, J., J. Rudnicki, and M. Rodell (1998): Variability in surface moisture content along a hillslope transect: Rattlesnake Hill, Texas, *J. Hydrol.*, 210, 259–281.

Francis, C. F., J. B. Thomes, A. Romero Diaz, F. Lopez Bermudez, and G. C. Fisher (1986): Topographic control of soil moisture, vegetation cover and land degradation in a moisture stressed Mediterranean environment, *Catena*, 13, 211–225.

Fu, B., and L. Chen (2000): Agricultural landscape spatial pattern analysis in the semiarid hill area of the Loess Plateau, China, *J. Arid Environ.*, 44, 291–303.

Fu, B., and H. Gulinck (1994): Land evaluation in area of severe erosion: The Loess Plateau of China, *Land Degrad. Rehabil.*, 5, 33–40.

Gao, X. D., P. T. Wu, X. N. Zhao, Y. G. Shi, J. W. Wang, and B. Q. Zhang (2011): Soil moisture variability along transects over a well-developed gully in the Loess Plateau, China, *Catena*, 87, 357–367.

Gardner, C. M. K., D. A. Robinson, K. Blyth, and J. D. Cooper (2001): Soil water content. In *Soil and Environmental Analysis: Physical Methods* (K. A. Smith and C. E. Mullins, Eds.), 2nd ed. Marcel Dekker, New York, pp. 1–64.

Gardner, W. H. (1986): Water content. In *Methods of Soil Analysis. Part 1. Physical and Mineralogical Methods* (A. Klute, Ed.), 2nd ed., *Agron. Ser.*, no. 9, American Society of Agronomy, Madison, WI, pp. 493–544.

Glenn, E., A. R. Huete, P. L. Nagler, K. K. Hirschboeck, and P. Brown (2007): Integrating remote sensing and ground methods to estimate evapotranspiration, *Crit. Rev. Plant Sci.*, 26, 139–168.

Gomez-Plaza, A., M. Martinez-Mena, J. Albaladejo, V. M. Castillo (2001): Factors regulating spatial distribution of soil water content in small semiarid catchments, *J. Hydrol.*, 253, 211–226.

Goodspeed, M. J. (1981): Neutron moisture meter theory. In *Soil Water Assessment by the Neutron Method*, CSIRO, Australia, p. 17.

Grayson, R. B., A. W. Western, F. H. S. Chiew, and G. Bloschl (1997): Preferred states in spatial soil moisture patterns: Local and non-local controls, *Water Resour. Res.*, 33, 2897–2908.

Gruber, A., W. A. Dorigo, S. Zwieback, A. Xaver, and W. Wagner (2013): Characterizing coarse-scale representativeness of in-situ soil moisture measurements from the International Soil Moisture Network, *Vadose Zone J.*, 12(2), doi:10.2136/vzj2012.0170

Hawley, M. E., T. J. Jackson, and R. H. McCuen (1983): Surface soil moisture variation on small agricultural watersheds, *J. Hydrol.*, 62, 179–200.

Herbrard, O., M. Voltz, P. Andrieux, and R. Moussa (2006): Spatiotemporal distribution of soil surface moisture in a heterogeneously farmed Mediterranean catchment, *J. Hydrol.*, 329, 110–121.

Hillel, D. (1998): *Environmental Soil Physics*, Academic Press, San Diego, CA.

Hills, R. C., and S. G. Reynolds (1969): Illustrations of soil moisture variability in selected areas and plots of different sizes, *J. Hydrol.*, 8, 27–47.

Huang, P. M., Y. Li, and M. E. Sumner (2011): *Handbook of Soil Sciences Properties and Processes*, 2nd ed., CRC Press, Boca Raton, FL.

Jones, J. A. A. (1997): *Global Hydrology: Processes, Resource and Environmental Management*, Addison Wesley, Reading, MA.

Kuruku, Y., F. B. Sanli, M. T. Esetlili, M. Bolca, and C. Goksel (2009): Contribution of SAR images to determination of surface moisture on the Menemen Plain, Turkey, *J. Remote Sens.*, 30(7), 1805–1817.

Liancourt, P., A. Sharkhuu, L. Ariuntsetseg, B. Boldgiv, B. R. Helliker, A. F. Plante, P. S. Petraitis et al. (2012): Temporal and spatial variation in how vegetation alters the soil moisture response to climate manipulation, *Plant Soil*, 351(1), 249–261.

Lull, H. W., and K. G. Reinhart (1955): Soil moisture measurement, *U.S. Forest Serv. Southern For. Exp. Sta. Occas. Pap.*, 140, 56 pp.

Mariotto, I., and V. P. Gutschick (2010): Non-Lambertian corrected albedo and vegetation index for estimating land evaporation in a heterogeneous semiarid landscape, *Remote Sens.*, 2, 926–938.

Moore, I. D., G. J. Burch, and D. H. Mackenzie (1988): Topographic effects on the distribution of soil surface water and the location of ephemeral gullies, *Trans. Am. Soc. Agric. Eng.*, 31, 1098–1107.

Ng, E., and P. C. Miller (1980): Soil moisture relations in the southern California chaparral, *Ecology*, 61, 98–107.

Niemann, K. O., and M. C. R. Edgell (1993): Preliminary analysis of spatial and temporal distribution of soil moisture on a deforested slope, *Phys. Geogr.*, 14, 449–464.

Nofziger, D. L. (1978): Errors in gamma-ray measurements of water content and bulk density in non-uniform soils, *Soil Sci. Soc. Am. J.*, 42(6), 845–850.

Nyberg, L. (1996): Spatial variability of soil water content in the covered catchment at Gårdsjön, Sweden, *Hydrol. Processes*, 10, 89–103.

Ochsner, T., A. Cosh, R. Cuenca, W. Doringo, C. Draper, Y. Hagimoto, Y. Kerr, K. Larson, E. Njoku, E. Small, and M. Zreda (in press). The state-of-the-art in large scale monitoring of soil moisture, *Soil Sci. Soc. Am. J.*

Panciera, R. (2009). Effect of land surface heterogeneity on satellite near- surface soil moisture observations, PhD thesis, University of Melbourne, Australia.

Patel, N. R., R. Anapashsha, S. Kiumar, S. K. Saha, V. K. Dadhwal (2009): Assessing potential of MODIS derived temperature/vegetation condition index (TVDI) to infer soil moisture status, *Int. J. Remote Sens.*, 30(1), 23–39.

Penna, D., H. van Meerveld, A. Gobbi, M. Borga, and G. Fontana (2011): The influence of soil moisture on threshold runoff generation processes in an alpine headwater catchment, *Hydrol. Earth Syst. Sci.*, 15(3), 689–702.

Pires, L., O. S. Bacchi, and K. Reichardt (2005): Soil water retention curve determined by gamma ray beam attenuation, *Soil Tillage Res.*, 82(1), 89–97.

Qiu, Y., B. Fu, J. Wang, and L. Chen (2001): Soil moisture variation in relation to topography and land use in a hillslope catchment of the Loess Plateau, China, *J. Hydrol.*, 240, 243–263.

Radatz, T. F., A. M. Thompson, and F. W. Madison (2012): Soil moisture and rainfall intensity thresholds for runoff generation in southwestern Wisconsin agricultural watersheds, *Hydrol. Processes*, doi:10.1002/hyp.9460.

Reynolds, S. G. (1970a): The gravimetric method of soil moisture determination, I: A study of equipment, and methodological problems, *J. Hydrol.*, 11, 258–273.

Reynolds, S. G. (1970b): The gravimetric method of soil moisture determination, II: Typical required sample sizes and methods of reducing variability, *J. Hydrol.*, 11, 274–287.

Reynolds, S. G. (1970c): The gravimetric method of soil moisture determination, III: An examination of factors influencing soil moisture variability, *J. Hydrol.*, 11, 288–300.

Robinson, D. A., C. S. Campbell, J. W. Hopmans, B. K. Hornbuckle, S. B. Jones, R. Knight, F. Ogden, J. Selker, and O. Wendroth (2008): Soil moisture measurement for ecological and hydrological watershed-scale observatories: A review, *Vadose Zone J.*, 7, 358–389.

Robinson, M., and T. J. Dean (1993): Measurement of near surface soil water content using a capacitance probe, *Hydrol. Processes*, 7, 77–86.

Robock, A., K. Y. Vinnikov, G. Srinivasan, J. K. Entin, S. E. Hollinger, N. A. Speranskaya, S. Liu, and A. Namkhai (2000): The global soil moisture data bank, *Bull. Am. Meteorol. Soc.*, 81(6), 1281–1299.

Salvucci, G. D. (2001): Estimating the moisture dependence of root zone water loss using conditionally averaged precipitation, *Water Resour. Res.*, 37(5), 1357–1365.

Seneviratne, S. I., T. Corti, E. L. Davin, M. Hirschi, E. B. Jaeger, I. Lehner, B. Orlowsky, and A. J. Teuling (2010): Investigating soil moisture-climate interactions in a changing climate: A review, *Earth Sci. Rev.*, 99, 125–161.

Sequin, B., and B. Itier (1983): Using midday surface temperature to estimate daily evaporation from satellite thermal IR data, *Int. J. Remote Sens.*, 4, 371–383.

Shaw, E. (1994): *Hydrology in Practice*, Spon Press, Oxford.

Sivapalan, M., K. J. Beven, and E. F. Wood (1987): On hydrological similarity: 2. A scaled model of storm runoff production, *Water Resour. Res.*, 23(12), 2266–2278.

Smithson, P., K. Addison, and K. Atkinson (2008): *Fundamentals of the Physical Environment*, Routledge, London.

Stafford, J. V. (1988): Remote, non-contact and *in situ* measurement of soil moisture content: A review, *J. Agric. Eng. Res.*, 41(3), 151–172.

Striegl, A. M., and S. P. Loheide II (2012): Heated distributed temperature sensing for field scale soil moisture monitoring, *Ground Water*, 50(3), 340–347.

Taylor, S. A. (1955): Field determinations of soil moisture, *Agric. Eng.*, 26, 654–659.

Verstraeten, W. W., F. Veroustraete, and J. Feyen (2008): Assessment of evapotranspiration and soil moisture content across different scales of observation, *Sensors*, 8, 70–117

Vinnikov, K. Y., A. Robock, N. A. Speranskaya, and C. A. Schlosser (1996): Scales of temporal and spatial variability of midlatitude soil moisture, *J. Geophys. Res.*, 101, 7163–7174.

Walker, J. P., and P. R. Houser (2004): Requirements of a global near-surface soil moisture satellite mission: Accuracy, repeat time and spatial resolution, *Adv. Water Resour.*, 27, 785–801.

Western, A. W., R. B. Grayson, G. Bloschl, G. R. Willgoose, and T. A. McMahon (1999): Observed spatial organization of soil moisture and its relation to terrain indices, *Water Resources Res.*, 35(3), 797–810.

Wetzel, R. G. (1983): *Limnology*. 2nd ed. Saunders College Publishing, New York. 767 pp.

Wiebe, H. H., R. W. Brown, and J. Barker (1977): Temperature gradient effects on in situ hygrometer measurements of water potential, *Agron. J.*, 69, 933–939.

Wu, C., J. M. Chen, J. Pumpanen, A. Cescatti, B. Marcolla, P. D. Blanken, J. Ardö et al. (2012): An underestimated role of precipitation frequency in regulating summer soil moisture, *Environ. Res. Lett.*, 7(2), 024011.

Yang, H., Z. Cong, Z. Liu, and Z. Lei (2010): Estimating sub-pixel temperatures using the triangle algorithm, *Int. J. Remote Sens.*, 31(23), 6-47-6060.

Zazueta, F. S., and J. Xin (1994): Soil moisture sensors, Bull. 292, Florida Cooperative Extension Service, Institute of Food and Agricultural Sciences, University of Florida. Available at http://www.p2pays.org/ref/08/07697.pdf (accessed on May 4, 2011).

Zhao, Y., S. Peth, A. Reszkowska, L. Gan, J. Krümmelbein, X. Peng, and R. Horn (2011): Response of soil moisture and temperature to grazing intensity in a Leymus chinensis steppe, Inner Mongolia, *Plant Soil*, 340, 89–102.

Zreda, M., W. J. Shuttleworth, X. Zeng, C. Zweck, D. Desilets, T. Franz, R. Rosolem et al. (2012): COSMOS: The COsmic-ray Soil Moisture Observing System, *Hydrol. Earth Syst. Sci. Discuss.*, 9, 4505–4551.

Zribi, M., N. Baghdadi, N. Holah, and O. Fafin (2005): New methodology for soil surface moisture estimation and its application to ENVISAT-ASAR multi-incidence data inversion, *Remote Sens. Environ.*, 96, 485–496.

3

Remote Sensing of Surface Turbulent Energy Fluxes

George P. Petropoulos

CONTENTS

3.1 Introduction

Understanding natural processes of the Earth system as well as the interactions of its different components with man-made activities—especially in the context of global climate change—has been recognized by the global scientific community as a very urgent and important research direction requiring attention for further investigation (e.g., Battrick et al. 2006). In this framework, being able to accurately estimate parameters such as the fluxes of latent (LE) and sensible (H) heat is of great importance, given their relevance to a number of physical processes of the Earth system, their central role to the global energy and water cycle, as well as their significant practical value in a large number of regional and global scale applications (see Chapter 1).

In view of the importance of information on the spatial distribution of surface heat fluxes, their measurement has attracted the attention of scientists from many disciplines, and decades of effort have been dedicated to their estimation. As seen in Chapter 1, a number of approaches have been developed for measuring directly heat and moisture fluxes using ground instrumentation. Clearly, use of ground instrumentation has certain advantages, including the ability to provide a relatively direct measurement, as well as the easy installation operation and maintenance of the equipment involved. However, those techniques are often rather complex and labor-intensive to operate and can be destructive to the area in which measurements are conducted. In addition, their use often requires extensive and at times expensive equipment to be deployed in the field to provide only localized measurements. Thus, although use of ground instrumentation can provide accurate estimates of the energy fluxes, it appears to be an impractical solution when information on the spatiotemporal variation of those parameters is required.

Nowadays, Earth Observation (EO) technology is recognized as the only viable solution for obtaining estimates of both LE and H fluxes at the spatiotemporal scales and accuracy levels required by many applications (Glenn et al. 2007; Li et al. 2009b). This chapter aims to provide the reader with an overview of the range of techniques available for estimating LE and H fluxes from EO data. In this framework, the key methods for temporally extrapolating the instantaneous flux estimates to daytime or daily average values are also considered. Part II of the book presents in more detail the workings of some of the algorithms reviewed herein providing also examples of case studies in which those have been implemented at different places of the world.

3.2 Remote Sensing of Turbulent Energy Fluxes

The prospect and capability of EO technology in estimating the turbulent fluxes of LE and H using initially handheld and airborne thermometers was recognized in the 1970s. The potential of spaceborne remote sensing technology in deriving regional maps of LE and H fluxes became available first in 1972 with the launch of the Landsat MSS, and later on, in 1978 with the HCMM (Heat Capacity Mapping Mission) and TIROS-N satellites.

In comparison to other approaches such as ground instrumentation or the use of simulation process models, the use of EO technology offers a number of advantages in estimating the LE and H fluxes. One of the most significant is that it makes possible to assess the spatial and temporal variability of surface fluxes over large areas. Besides, EO technology permits monitoring repeatedly various biophysical parameters at a variety of scales ranging from local to even continental. The latter is rather important as examining the same area of study at multiple scales has enabled the scientific community to investigate some of the previously insoluble problems, such as that of spatial variability and scale of observation of the energy fluxes. What is more, the use of remote sensing allows deriving directly the water consumed by the soil-water vegetation system without the need to quantify or model often very complex hydrological processes.

Given the advantages of EO technology, a plethora of methods have been proposed for estimating energy fluxes from space utilizing spectral information acquired in all regions of the electromagnetic radiation (EMR) spectrum. A nonexhaustive list of EO systems and

TABLE 3.1

Examples of Spaceborne Sensors Currently in Orbit Providing Observations Appropriate to Derive Surface Heat Fluxes

Sensor	Manufacturer	Platform	Spatial Resolution	Spectral Resolution	Temporal Resolution
ASTER	NASA/ ERSDAC	Terra	VNIR: 15 m SWIR: 30 m TIR: 90 m	VNIR: 4 SWIR: 6 TIR: 5	16 days
Landsat TM/ ETM+	NASA/U.S. Department of Defense	Landsat	VNIR: 30 m SWIR: 30 m TIR: 120 m (TM)/60 m (ETM+)	VNIR: 4 SWIR: 2 TIR: 1	16 days
MODIS	NASA	Terra and Aqua	VNIR: 250 m/500 m SWIR: 500 m TIR: 1 km	VNIR: 18 SWIR: 2 TIR: 16	2 daytime/ 2 nighttime
AVHRR/3	NASA	NOAA	VNIR: 1.1 km SWIR: 1.1 km TIR: 1.1 km	VNIR: 2 SWIR: 1 TIR: 2/3	2 daytime/ 2 nighttime
AATSR	ESA	ENVISAT	VNIR: 1 km SWIR: 1 km TIR: 1 km	VNIR: 3 SWIR: 1 TIR: 3	2 daytime/ 2 nighttime
SEVIRI	EUMETSAT/ ESA	Meteosat-2	VNIR: 1.1 and 3.0 km SWIR: 3 km TIR: 3 km	VNIR: 4 TIR: 8	96 scenes per day (every 15′)

of their technical specifications providing at present data suitable for the retrieval of surface energy fluxes and SMC is summarized in Table 3.1.

An overview of the remote sensing–based estimation of LE and *H* turbulent fluxes is provided next, following largely the classification of the techniques proposed by other investigators (e.g., Glenn et al. 2007; Gowda et al. 2008; Li et al. 2009b; Petropoulos et al. 2009a). A summary of the relative strengths and limitations of the techniques reviewed is also made available in Table 3.2.

3.2.1 Residual Methods of Surface Energy Balance

Development of residual methods is traced back to the early 1970s (Sone and Horton 1974; Verma et al. 1976). The vast majority of these approaches are fundamentally based on the principles of energy conservation. In those techniques, generally land surface characteristics such as albedo, leaf area index (LAI), vegetation indices (VIs), surface roughness, surface emissivity, and surface temperature (T_s) are derived from satellite observations. Those parameters are subsequently combined with ground observations to produce spatially distributed maps of fluxes of net radiation (R_n), soil heat (*G*), and *H* fluxes. Then, the LE flux is computed as a residual of the energy balance equation.

In mathematical terms, for a typical vegetated land surface the energy balance equation is expressed as

$$R_n - G - S - H - LE = 0 \qquad (3.1)$$

where R_n is the net radiation (net short and net longwave), *G* is the soil heat flux, and *S* is the rate of heat storage in the plant canopy due to photosynthesis (W m^{-2}). The quantity ($R_n - G$) is commonly called the "available energy" (e.g., Diak et al. 2004).

TABLE 3.2

Summary of the Different Types of Methods Employed in the Retrieval of Latent and Sensible Heat Fluxes by Remote Sensing

Group	Methods	Advantages	Disadvantages	Examples of Techniques
Energy balance residual methods	One-layer models	Good accuracies over homogeneous regions, simple and straightforward implementation, offer extrapolation of instantaneous to daily fluxes	Heat storage (S) term and energy for photosynthesis are omitted, applied only on cloud-free and daytime conditions, site-specific, overestimate H flux term over heterogeneous landscapes, sensitive to atmospheric correction	SEBS (Su, 2002) SEBAL (Bastiaanssen, 1995; Texeira et al. 2009) METRIC (Allen et al. 2007)
	Two-source models	More accurate fluxes retrievals over heterogeneous surfaces in comparison to one-layer models, better physical concept than one-layer models, offer extrapolation of instantaneous to daily fluxes	Applied only on cloud-free and daytime conditions, often more ground measurements are needed, component temperatures of soil and vegetation need to be specified, some methods require two satellite passes in a given day, sensitive to atmospheric correction	Norman TSM (Norman et al. 1995) ALEXI (Anderson et al. 1997) TSEB (French et al. 2002)
T_s/VI scatterplot methods	Various methods (grouped generally depending on the information required in scatterplot axes)	Often simple and minimal input requirements, ability to easily extrapolate the instantaneous LE/H flux estimate to daytime average value, not accurate atmospheric correction required	Subjectivity introduced in their implementation, linear relationship of T_s/VI to fluxes not realistic, T_s/VI domain not easily formed using coarse resolution data, often require whole range in SMC in image to be implemented, some methods empirical, some need site-specific calibration	See review of methods by Petropoulos et al. (2009a)
Assimilation methods	Various methods linking remote sensing data with simulation models, based on different approaches	Minimize the mismatch between the observations and models, able to assimilate more variables than just LE/H fluxes, easy interpolation, ability to produce estimates at very high temporal resolution, ability to merge information from different data sources	Large requirement in input parameters, high level of expertise is often required for their implementation, more computationally demanding, uncertainty in their predictions increases as a function of surface heterogeneity	Randall et al. (1996); Margulis and Entekhabi (2003); Capparini et al. (2003)
Microwave methods	Methods used mainly either to combine MW data with TSM or relate fluxes directly with MW measurements	Their implementation is not restricted by daytime/clouds conditions, able to capture seasonal and diurnal variations of LE/H fluxes, some methods require few or no ground measurements	Offer extrapolation of instantaneous to daily fluxes, spatial resolution much coarser from that of TIR instruments, not so widely used today, issues in validation need to be taken care of, some empirical	Kustas et al. (2003); Min and Lin (2006); Min et al. (2010)

The S term in Equation 3.1 generally is less than a few percent of R_n (Meyers and Hollinger 2004). It is also a parameter that quantitatively is not easy to be measured even by ground instrumentation (Wilson et al. 2001). As a result, in most remote sensing energy balance–based approaches for the computation of LE flux, this term is generally omitted. Nevertheless, it should be noted that in the case of vegetation types with a significant canopy amount (e.g., as forests), S can become high particularly so over instantaneous periods (Meyers and Hollinger 2004; Sanchez et al. 2008).

The R_n term represents the total heat energy partitioned into G, H, and LE fluxes. R_n is generally modeled as the sum of the incoming and outgoing radiation components of the shortwave and longwave portions. Previous research has demonstrated that R_n can be calculated from well-established approaches based on primarily remotely sensed data. Generally, comparisons between modeled and ground-based measurements of R_n have shown an uncertainty in the estimation of R_n by remote sensing in the range of 5%–10% in comparison to ground observations (e.g., Sanchez et al. 2008; Yang et al. 2010). An overview of the EO-based methods employed in deriving all the components of R_n including relevant operational products currently available can be found in Chapter 5.

The G flux is defined as the heat energy responsible for changing the temperature of the substrate soil volume. When Equation 3.1 above is evaluated over a 24-h time period, G flux is commonly assumed to be negligible. However, when Equation 3.1 is applied on an instantaneous basis as is the case when satellite observations are used, the assumption of neglecting G is not valid. Hence, in such cases, evaluation of the instantaneous LE flux as a residual term from the energy balance equation makes indispensable the computation of the G flux component. Remote sensing approaches for the G flux estimation have been largely based on assumptions developed around relationships between the ratio of G/R_n and satellite-derived vegetation indices (VIs). Comparison studies specifically of G flux predicted by such simplified techniques versus *in situ* observations have shown an uncertainty of 20%–30% in the estimation of G flux alone (Kustas and Norman 1996; Li et al. 2009b; Ma et al. 2011).

Yet, remote sensing–based estimation of H flux is the most difficult and also challenging to be derived. The main difficulties relate essentially to the uncertainty introduced in the estimation of the aerodynamic and surface resistances. This is because those vary considerably spatially, particularly as a function of surface heterogeneity (e.g., Schmugge et al. 2002; Verstraeten et al. 2008). Thus, as H flux essentially introduces the largest uncertainty in the LE flux estimation, the methods in which LE flux is computed as a residual from the energy balance equation differ essentially on the way the H flux term is modeled. Generally, H flux prediction accuracy has been found to be largely dependent on the complexity by which the soil and vegetation components are modeled (Norman et al. 2003; Courault et al. 2005; Li et al. 2009b). Thus the remote sensing–based methods employed are generally divided into two broad categories, depending on whether or not the land surface is modeled as a single land surface or is separated into bare soil and vegetation layers, so-called "one-layer" and "two-source (or dual layer)" models, respectively.

The remainder of this section provides an overview of the use of the most widely used one- and two-layer modeling schemes by the remote sensing community. A detailed account to some of these modeling schemes including examples from case studies is provided in Part II of the volume.

3.2.1.1 One-Layer Models

In the case of one-layer approaches, the most commonly applied method to estimate regionally H flux from EO data is by the use of remotely derived radiometric surface temperature

(T_s). The physical concept for the use of temperature measurements in the estimation of H flux originates from Monteith and Szeicz (1962) and Monteith (1963) and is based in the following expression:

$$H = \rho C_p \frac{T_o - T_A}{R_A}$$ (3.2)

where ρ is the air density (kg m^{-3}), C_p is the specific heat of air at constant pressure (J kg^{-1} K^{-1}) [the component ρC_p is called the volumetric heat capacity (J m^{-3} °C^{-1})], T_0 is the aerodynamic temperature (°C, this parameter relates to the efficiency of heat exchange between the land surface and overlying atmosphere), T_A is the air temperature (°C) at a reference height, and R_A is the aerodynamic resistance between canopy height and the ambient environment above the canopy (s m^{-1}).

Equation 3.2 is a one-layer bulk transfer equation and is based on the assumption that the radiometric surface temperature ($T_R(\theta)$) measured by a thermal radiometer is identical or very close to the aerodynamic temperature (T_o) (e.g., Asrar 1989; Bastiaanssen 2000). Yet, although the surface brightness temperature is approximately the same as aerodynamic temperature over non-vegetated surfaces, it is not the same over most naturally vegetated surfaces. Estimation errors in the H flux even of the order of 100 W m^{-2} have been reported when ($T_R(\theta)$) is simply used to replace T_o and if atmospheric effects and surface emissivity are not properly considered (Chavez and Neale 2003; Gowda et al. 2008). Thus, to account for this difference, in the one-layer modeling schemes an extra resistance (R_{EX}) is added to R_A in Equation 3.2, which takes the following form:

$$H = \rho C_p \frac{T_R(\theta) - T_A}{R_A + R_{EX}}$$ (3.3)

where R_{EX} is the so-called excess resistance, which accounts for the non-equivalence of T_o and $T_R(\theta)$.

R_A is usually estimated using local data on wind speed, stability conditions, and roughness length, even though area averaging of roughness lengths is considered to be highly nonlinear (Hasager and Jensen 1999). A number of one-layer models have been developed so far based on the above concept. Some of the most widely used ones by the remote sensing community are reviewed below.

3.2.1.1.1 Surface Energy Balance System

Surface Energy Balance System (SEBS) is a one-layer model developed by Su (2002) for the estimation of surface energy balance parameters including the LE and H fluxes. Briefly, SEBS consists of three components: (1) a set of algorithms used for the estimation of different components of energy balance and relevant land surface parameters; (2) an extended model to establish the roughness length for heat transfer (Su 2002); and (3) a formulation to extrapolate the instantaneous flux estimates to daytime average and daily totals based on the energy balance at limiting wet and dry cases. SEBS generally requires input parameters derived by both remote sensing (e.g., albedo, emissivity, temperature, vegetation cover, and LAI) and ground observations (air pressure, humidity, and wind speed at a reference height). A more detailed description of SEBS with results from its implementation in different case studies is furnished in Chapter 7, whereas reference to this model from an operational perspective is made as well in Chapter 10.

SEBS ability to compute LE and H fluxes has been generally evaluated over a range of land cover types using different types of EO data. For example, Su et al. (2005) evaluated SEBS with Landsat ETM+ imagery over an agricultural area in Iowa, United States, using ground observations from a SMACEX field experiment. Authors reported the SEBS-derived LE fluxes to be ~90% to the corresponding ground measurements of LE acquired over different crop types. In another study, McCabe and Wood (2006) showed the LE flux retrievals from Landsat exhibiting closer agreement to ground measurements (RMSD = 61.94 W m^{-2}) in comparison to when SEBS was combined with ASTER-derived data (RMSD = 82.04 W m^{-2}). Yang et al. (2010) evaluated SEBS using coarser resolution MODIS data acquired over a mainly cropland area in north China. Comparisons of the LE and H fluxes predicted by SEBS versus concurrent eddy covariance *in situ* measurements showed a significant underestimation of the H fluxes by SEBS and an overestimation of the LE fluxes with biases of –28 and 103 W m^{-2}, respectively. Ma et al. (2011) using ASTER data acquired over a region in China evaluated SEBS ability to predict different components of the energy balance. Energy flux components estimated from SEBS were predicted in their study with acceptable accuracy over partially vegetated areas such as their test region.

Some of SEBS advantages include the following: (1) the uncertainty from the surface temperature or meteorological variable estimation in SEBS can be limited with consideration of the energy balance at the limiting cases; (2) formulation of the roughness height for heat transfer in SEBS is not based on using fixed values; and (3) *a priori* knowledge of the actual turbulent heat fluxes is not required by the model. However, SEBS operation is possible only on clear sky days. Yet, perhaps the main limitation of SEBS is that it requires aerodynamic roughness height to be estimated, a parameter of which estimation by remote sensing actually remains a challenge until today (Gowda et al. 2008).

3.2.1.1.2 *Surface Energy Balance Algorithm for Land (SEBAL)*

Surface energy balance algorithm for land (SEBAL) is exploiting both empirical relationships and physical parameterizations for computing the energy partitioning at the regional scale based primarily on EO data and a small number of ground observations. A full description of the SEBAL can be found in Bastiaanssen (1995) and Bastiaanssen et al. (1998). Briefly, key remote sensing input parameters to the model include the incoming radiation, T_s, Normalised Difference Vegetation Index (NDVI, Rouse et al. 1973), and albedo maps, whereas different semiempirical relationships are used for estimating variables such as emissivity and roughness length. First, the R_n and G fluxes are computed based on a series of parameters (e.g., surface temperature T_s, reflectance-derived albedo values, VIs, Leaf Area Index (LAI), and surface emissivity). H flux is subsequently estimated from a bulk aerodynamic resistance formulation and the temperature difference between the land surface and air computed from extreme (i.e., wet and dry) image pixels. Those are used to develop a linear relationship between temperature difference and surface temperature. Then, LE flux is solved as a residual from the energy balance equation. Chapter 6 presents the workings of SEBAL implementation in more detail, providing results from case studies.

SEBAL ability to provide regional estimates of the different components of the energy balance equation has been extensively examined by many investigators using different types of EO data. For example, Bastiaanssen et al. (2005) evaluated SEBAL over different climatic conditions and spatial scales. An accuracy of 85% was reported by the authors in the estimation of LE fluxes at field scale, of 95% at seasonal scale, and of 96% at annual scale. Bashir et al. (2008) using both Landsat ETM+ and MODIS image data acquired over a region in Sudan showed a mean absolute error in LE fluxes of approximately 0.9 mm day^{-1} for MODIS and of 0.9 mm day^{-1} for the Landsat images. Kongo et al. (2011) examined the combined use of SEBAL with MODIS imagery for deriving spatially distributed LE fluxes for a region in Africa analyzing in total 28 MODIS images. LE fluxes predicted by SEBAL showed a difference ranging from –14% to +26% in comparison to corresponding measurements derived from a large scintillometer. Recently, Sun et al. (2011) evaluated SEBAL using Landsat ETM+ data acquired over a wetland area in China. Authors reported an overestimation of the predicted daily LE estimates on the order of 10% in comparison to ground measurements from evaporation pans.

A key advantage of SEBAL is that the LE and H fluxes derived from this technique are not sensitive to accurate retrievals of T_s. This is because SEBAL accommodates an automatic internal calibration, which can be done separately for each image analyzed. Also, similar to SEBS, SEBAL implementation is based largely on EO-based inputs and a small number only of ground observations, which potentially make its implementation easy and cost-effective. Yet, it is dependent on the selection of representative pixels in an image for dry and wet conditions, which might include some degree of user subjectivity. Also, as underlined by Li et al. (2009b), SEBAL application over nonflat terrain sites requires making adjustments to certain input variables of the model (e.g., T_s), as otherwise prediction of the H and LE fluxes can be dramatically affected. Last but not least, SEBAL also requires cloud-free conditions and heterogeneity in moisture conditions to be implemented.

3.2.1.1.3 METRIC

METRIC (Allen et al. 2007) is essentially based on the SEBAL with the main difference being in the way of determining the wet and dry pixels. In particular, in contrast to SEBAL, METRIC does not make assumptions of zero H flux, neither that $H = R_n - G$ at the wet image pixel; instead of that, a soil water budget is applied for the hot pixel to verify that LE flux is indeed zero. What is more, in METRIC the extreme pixels are selected in an agricultural setting in which cold pixels should have biophysical characteristics similar to the reference crop (alfalfa). Besides, another difference in comparison to SEBAL is that in METRIC the alfalfa reference evapotranspiration fraction (ETrF) mechanism is used to extrapolate instantaneous LE flux to daily flux rates. ETrF is defined as the ratio of the remotely sensed instantaneous LE flux (LE) to the reference LE flux (ETr, e.g., mm h^{-1}), where the latter is derived at the time of the satellite overpass from ground data.

A number of experiments have been conducted attempting to appraise the use of METRIC for predicting the LE and H fluxes. Allen et al. (2007) evaluated METRIC in two agricultural areas in the Idaho area, United States, using *in situ* estimates from lysimeters and reported differences in LE ranging between 1% and 4%. Gowda et al. (2008) applied METRIC with Landsat TM acquired over an agricultural area located in Texas High Plains, United States. Comparisons of the daily LE fluxes predicted by the model versus predictions from a soil moisture budget showed a low error for the case of well-irrigated and high biomass corn vegetation [2.0 mm day^{-1} (17.1%) and 0.5 mm day^{-1} (6.0%) for each day]. Chavez et al. (2009) combined METRIC with Landsat TM for the Texas High Plains, United States. Authors reported a generally close agreement of METRIC-predicted daily LE fluxes in comparison to those derived from the lysimeters, with a mean bias error ±RMSD of 0.4 ± 0.7 mm day^{-1}. Khan et al. (2010) compared LE fluxes predicted from METRIC using MODIS data acquired over several Ameriflux sites versus eddy covariance *in situ* reference measurements. A mean bias from the ground observations lower than 15% and a seasonal bias lower than 8% in the daily LE fluxes agreement was reported.

Generally, METRIC is largely characterized by similar types of advantages and disadvantages as the other one-layer models reviewed previously, specifically those of the SEBAL model. However, in comparison to SEBAL, various studies have indicated that METRIC appears to have an advantage over SEBAL in providing more accurate estimates of energy fluxes under advective conditions (e.g., Chavez et al. 2009).

3.2.1.2 Two-Source Models

Essentially, the development of the early two-source modeling (TSM) schemes can be traced back to the end of the 1980s. Shuttleworth and Wallace (1985) and Shuttleworth and Gurney (1990) were among the first to introduce the idea. Generally, in comparison to the one-layer models, in TSM schemes energy fluxes are partitioned between the soil and vegetation allowing the estimation of the LE and *H* fluxes over incomplete canopies. A number of TSMs have been proposed over the years. A review of some of the most widely used ones is provided in the remainder of this section, underlying their comparative advantages and disadvantages in respect to their practical implementation.

3.2.1.2.1 Norman et al. (1995) Model

A significant contribution to the TSM techniques development was made by Norman et al. (1995). Briefly, their TSM was based on the assumption that the contribution of the canopy and soil layers to *H* fluxes depends on the temperature differences between each layer and the atmosphere and on the coupling that is assumed to be between layers. Retrievals of the LE and *H* fluxes from the Normal et al. (1995) model have also been the subject of many studies conducted in different regions. Norman et al. (1995) first evaluated their TSM using data from FIFE (Sellers et al. 1992) and MONSOON 90 (Kustas et al. 1991) field campaigns. Authors reported Root Mean Square Difference (RMSD) between the model predictions and corresponding *in situ* measurements between 35 and 60 W m^{-2} for *G*, *H*, and LE fluxes, respectively. French et al. (2000) evaluated independently the Norman et al. (1995) model using airborne remote sensing data from the Thermal Infrared Multispectral Scanner (TIMS) acquired in the SMACEX and SMEX02 field experiments in the United States. Their results showed *H* fluxes predictions close to *in situ* eddy covariance with average differences of 43 W m^{-2}.

Gonzalez-Dugo et al. (2009) using Landsat TM imagery examined the model of Norman et al. (1995) and a one-layer model in terms of their ability to estimate the surface fluxes. The Norman et al. (1995) model returned slightly more accurate predictions of the *H* and LE fluxes, in comparison to the one-layer model, with an RMSD lower than 31 W m^{-2} in the estimation of the *H* fluxes. Li et al. (2008) evaluated the effect of pixel resolution to the energy fluxes computed from the model of Norman et al. (1995) using Landsat TM images and *in situ* data collected in 2004 during the Soil Moisture Field Experiment (SMEX) conducted in Arizona, United States. In their study, the Norman et al. (1995) TSM was implemented at three spatial resolutions, namely, 30, 120, and 960 m. A generally satisfactory agreement was found between predicted and *in situ* measurements in the heat fluxes for the case of the 30 and 120 m spatial resolution, with RMSDs between 20 and 35 W m^{-2} and a mean RMSD of 33 W m^{-2}. However, for the comparisons at 960 m, agreement between the compared LE fluxes was comparatively lower, with a mean RMSD of 40 W m^{-2}. The latter was attributed by the authors to the inferior representation of the land surface heterogeneity, which was not possible to be depicted at this coarse spatial resolution.

In contrast to the one-layer models reviewed in the previous section, the Norman et al. (1995) TSM has the advantage that it contains the difference between radiometric temperature and aerodynamic temperature. The latter overcomes the problem of an empirical resistance depending on the radiometer view angle. In addition, the model offers the advantage that can account for variation in surface resistances due to variation in vegetation cover and surface roughness, without requiring redefinition of "excess" resistance without any additional ground observations.

3.2.1.2.2 ALEXI/DISALEXI Model

Another TSM scheme was developed by Anderson et al. (1997). Their model was initially named Two-Source Time Integrated Model (TSTIM) and was later renamed by Mecikalski et al. (1999) as Atmosphere-Land Exchange Inverse (ALEXI). In comparison to the Norman et al. (1995) TSM, ALEXI's main difference is that it includes a scheme where the growth of the atmospheric boundary layer (ABL) is coupled to the temporal changes in surface radiometric temperature from the Geosynchronous Operational Environmental Satellite (GOES). Detailed description of the model can be found in the work of Anderson et al. (1997) and also in Chapter 8, which also presents case studies of practical applications of the model outputs in monitoring drought and crop condition.

Validation of ALEXI by Anderson et al. (1997) and later by Mecikalski et al. (1999) with ground data from the FIFE (Sellers et al. 1992) and MONSOON'90 (Kustas et al. 1991) field campaigns confirmed the model's ability to estimate LE and H fluxes with uncertainty at least comparable to other modeling schemes. Later, Kustas et al. (2003) and Norman et al. (2003) developed a scheme that they called the Disaggregated Atmosphere Land Exchange Inverse model (DisALEXI). That allowed deriving spatially distributed estimates of the energy fluxes by combining low- and high-resolution satellite observations without the need of local observations. Authors demonstrated the use of DisALEXI from Southern Great Plains, United States. Their comparisons showed an agreement between the predicted surface fluxes and corresponding ground-based measurements within 10%–12%. An overview of the ALEXI/DISALEXI model including results from the model use in various locations of the world was provided recently by Anderson et al. (2010).

All in all, a key advantage of ALEXI in comparison to other TSMs including the Norman et al. (1999) model reviewed earlier is that it allows taking into consideration the temporal changes of brightness temperatures and relates the rise in air temperature above the canopy and the growth of the Atmosphere Boundary Layer (ABL) to the time-integrated fluxes from the surface. This, as argued by the model developers, dramatically reduces the errors in the conversion to radiometric surface temperatures due to uncertainties associated to emissivity and atmospheric correction. Besides, ALEXI is able to take under consideration the viewing angle effects in the calculation of vegetation fraction. Also, in comparison to other one-layer and TSMs, ALEXI is able to offer a scheme for disaggregating the predicted fluxes at spatial resolutions of moderate and fine resolution thermal infrared imagery from polar-orbiting systems. The latter allows generating daily maps of surface heat fluxes at very high spatial resolution (Anderson et al. 2010).

3.2.1.2.3 TSEB Model

The Two-Source Energy Balance (TSEB) model developed by French et al. (2002) consisted essentially of a refinement of the original Norman et al. (1995) model that was conducted, where T_s was derived by employing the temperature emissivity separation (TES) algorithm (Gillespie et al. 1998) using the ASTER thermal bands. Thus TSEB was particularly suitable for computing LE fluxes from high spatial resolution multispectral data such as those from the ASTER sensor. French et al. (2002) compared surface heat fluxes by TSEB against corresponding *in situ* observations from the Bowen ratio. TSEB returned a nearly ideal estimate of LE fluxes in their study. However, significant discrepancies with respect to the ground measurements were reported in the remaining flux components, with the most significant one being the H flux overestimated by TSEB by 90 W m^{-2}. Anderson et al. (2008) investigated the use of a canopy light use efficiency based model within TSEB, using as a case study the SGP97 field experiment data from the El Reno experimental site in the United States. Authors showed that this replacement resulted in a reduction in the predicted error in LE fluxes estimation by TSEB from 15% to 9% for LE flux and from 16% to 12% for all the other energy balance components combined. Sanchez et al. (2008) introduced a simplified version of TSEB, called the Simplified Two-Source Energy Balance (STSEB) model. Their model was aimed for use specifically with the dual-angle instrument ATSR. Authors evaluated STSEB for a test site in Maryland, United States, using input data acquired from ground instrumentation rather than remote sensing radiometers. An RMSD ranging from 15 to 50 W m^{-2} in the prediction of the R_n, G, LE, and H fluxes by their model was reported in comparisons performed versus eddy covariance ground measurements.

3.2.2 Methods Based on the T_s/VI Scatterplot

A different group of techniques employed for the estimation of hydrometeorological fluxes from remote sensing data has its basis on the relationships between a satellite-derived VI and surface temperature (T_s), when these are plotted in a scatterplot. An overview of the theoretical basis of these methods including the biophysical properties encapsulated on a T_s/VI scatterplot can be found in Petropoulos et al. (2009a).

Briefly, assuming that cloud-contaminated pixels and those containing standing water surface have been masked out from a remotely sensed dataset, per pixel-level values of T_s and VI collected from any satellite imagery usually form a triangular (or trapezoidal) shape in the T_s/VI feature space, as shown in Figure 3.1. In this representation, each yellow

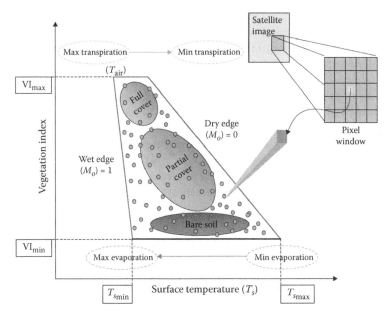

FIGURE 3.1
(See color insert.) Summary of the key descriptors and physical interpretations of the T_s/VI feature space "scatterplot." (Adapted from Petropoulos, G. et al., *Adv. Phys. Geogr.*, 33, 1–27, 2009.)

circle represents the measurements from a single image pixel. The emergence of the triangular (or trapezoid) shape in the T_s/VI feature space is the result of the low variability of T_s and its relative insensitivity to soil water content variations over areas covered by dense vegetation but its increased sensitivity (and thus larger spatial variation) over areas of bare soil. The right-hand-side border of the triangle (or trapezoid) (the so-called "dry edge" or "warm edge") shown in Figure 3.1 is defined by the locus of points of highest temperature but which contains differing amounts of bare soil and vegetation and is assumed to represent conditions of limited surface soil water content and zero evaporative flux from the soil. Likewise, the left-hand border (the so-called "wet edge" or "cold edge") corresponds to the set of cooler pixels that have varying amounts of vegetation, which represent those pixels at the limit of maximum surface soil water content. Variation along the lower edge (i.e., the "base") of the triangle (or trapezoid) is representing pixels of bare soil and is assumed to reflect the combined effects of soil water content variations and topography, while the triangle's (or trapezoid's) apex equates to full vegetation cover (as this expressed by the highest VI value). Points within the triangular space correspond to pixels with varying VI (i.e., fractional vegetation cover F_r) and surface soil water content between those with bare soil and those with dense vegetation. For data points having the same VI, T_s can range markedly. The triangle's (or trapezoid's) "dry edge" is considered to represent the lower limit of evapotranspiration for the different vegetation conditions found at that value of F_r within the scene, whereas the reverse is implied for the "wet edge."

Petropoulos et al. (2009a) classified all the available methods to five groups using as a catalogue the relationships between the thermal and optical spectral range to form the scatterplot. Following this notation. The remainder of this section aims to provide an overview of key T_s/VI techniques, discussing in this context their relative strengths and limitations.

3.2.2.1 Methods Based on the T_s and Simple VI Scatterplot

Price (1990) introduced for the first time the retrievals of spatially explicit maps of LE fluxes from the T_s/NDVI triangular space. Their method was based on a relatively simple conceptually mathematical description based on estimating first the LE fluxes at the points of full vegetation cover and for dry and wet bare soil, which were corresponding to the extremities of the T_s/NDVI envelope space. On a similar concept, Jiang and Islam (1999, 2001) proposed estimating regional LE fluxes based on the combination of the satellite T_s/NDVI scatterplot with a simplified form of the Priestley–Taylor equation for LE fluxes. Zhang et al. (2006) estimated the instantaneous and daily LE fluxes from the T_s/VI scatterplot, using the latter specifically for estimating the dry and wet edges of the scatterplot and for developing subsequently correlations to obtain the surface moisture availability, called temperature vegetation cover index. Tan et al. (2010) recently recommended a technique for determining quantitatively and robustly the dry and wet edges. Authors applied their approach for a region in China MODIS Terra images. The comparisons of their predicted LE fluxes versus corresponding ground measurements from a large aperture scintillometer showed an RMSD of ~25.1 W m^{-2} in the H fluxes estimation.

In summary, a key advantage of this group of approaches is their relative independence from site-specific tuning of model parameters. The latter is particularly important to be considered in case of an operational deployment. However, those methods are based on the assumption of a linear relationship between T_s and F_r for computing the fluxes for each image pixel, which might not necessarily represent the real-world case.

Last but not least, for a wider application of those methods, special attention should be paid to considering the impact of clouds, standing water, and sloping terrain, particularly on T_s estimation.

3.2.2.2 Methods Based on the T_s and Albedo Scatterplot

A number of studies have demonstrated a correlation between satellite-derived T_s and surface reflectance for areas with spatially invariant atmospheric conditions, suggesting that the derived relationships can be applied to determine the effective land surface properties (Menenti et al. 1989). Using this concept, Roerink et al. (2000) proposed the Simplified Surface Energy Balance Index (S-SEBI) for deriving spatially explicit maps of surface energy fluxes from remote sensing observations. This method has its basis on the spatial variation of T_s and broadband albedo (rather than a VI) and simple satellite-based estimates of R_n and G fluxes (Figure 3.2).

For a given albedo value and assuming Lambertian approximation in surface reflectance, the variation of T_s between the wet and dry edges of the T_s/VI scatterplot was related to the variations in land surface water availability. Then, the available energy $(R_n - G)$ is partitioned into H and LE fluxes according to the actual temperature of the surface (i.e., T_{kin}), assuming conditions of constant global radiation and T_{air}. Gomez et al. (2005) extended the S-SEBI concept to estimate the daily total LE fluxes via an integration of the instantaneous LE fluxes over the whole day and conversion into a daily value by accounting for the latent heat of vaporization (λ). Bhattacharya et al. (2010) based on the S-SEBI concept presented a new technique for the estimation of regional LE fluxes by relating the Evaporative Fraction (EF) to the available energy, bypassing the computation of H flux in the energy balance equation. Investigators, using data over an agricultural region in India from Indian geostationary meteorological satellite (Kalapana-1) sensor (VHRR), showed

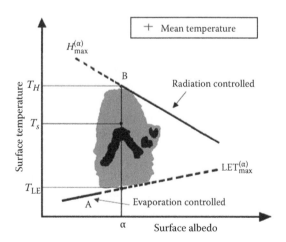

FIGURE 3.2
The concept of T_s/albedo scatterplot methods for the retrieval of LE and daily LE fluxes proposed by Roerink et al. (2000). $LET_{max}^{(\alpha)}$ represents the line of pixels showing the maximum LE (i.e., pixels from the wetter parts of the image), whereas $H_{max}^{(\alpha)}$ represents the line of pixels exhibiting the driest surfaces. (Adapted from *Physics and Chemistry of the Earth, Part B, Hydrology, Oceans and Atmosphere*, 25(2), G. Roerink, Z. Su, and M. Menenti, A simple remote sensing algorithm to estimate the surface energy balance, 147–157. Copyright 2000, with permission from Elsevier.)

that predicted daily ET fluxes by their method were within the range of 25%–32% of the *in situ* observations. Various studies have evaluated S-SEBI in deriving spatially distributed maps of energy fluxes at different environmental conditions (e.g., Roerink et al. 2000; Sobrino et al. 2007; Zahira et al. 2009). It generally appears that instantaneous and daily LE fluxes can be derived with accuracy of about 50 W m^{-2} and 1 mm day^{-1}, respectively.

A key advantage of this group of approaches is their independence from additional meteorological data, provided that surface hydrological extremes are present in the image. What is more, in contrast to other T_s/VI methods reviewed thus far that attempt to determine a fixed temperature for wet and dry conditions representative of the entire area of interest and/or for each land use class, this type of method assumes that the extreme temperatures for the wet and dry conditions vary with changing surface reflectance; the latter might appear a more realistic assumption. An important advantage of the Bhattacharya et al. (2010) technique, in particular, is that it can be implemented without the need of any ground observations. This is making potentially their method a very good choice for operational use. Also, their method avoids computation of *H* flux. Yet, generally speaking, between the key limitations of those approaches is their applicability on cloud-free days only and their requirement of identifying extreme points in the scatterplot domain. Also, the method proposed by Gomez et al. (2005) requires T_s measurements for both bare soil and full vegetation cover and the assumption of constant atmospheric conditions (mainly global radiation, wind speed, and T_{air}) over the entire studied region. Thus large prediction errors in LE flux can be returned when applied over areas of highly varying atmospheric conditions.

3.2.2.3 Methods Based on ($T_s - T_{air}$) and VI Scatterplot

Moran et al. (1994) introduced a concept termed the "vegetation index–temperature (VIT) trapezoid" for estimation of LE fluxes from the T_s/VI domain in areas of partial vegetation cover. Briefly, Moran et al. (1994) methodology was based on the use of crop water stress index (CWSI) of Jackson et al. (1981) and of a new index defined as the Water Deficit Index (WDI), linked to computed values of the four vertices of the T_s/VI trapezoid and was subsequently related to the retrieval of the LE fluxes at any point within the trapezoid (Figure 3.3). Validation of their method conducted by various researchers using different types of satellite data has showed that energy fluxes retrievals can be predicted with an accuracy on the order of 30–40 W m^{-2} (Moran et al. 1994, 1996; Li and Lyons 1998).

From a different perspective, Jiang and Islam (2003) proposed deriving spatially distributed estimates of LE fluxes by modifying the Jiang and Islam (1999) methodology presented earlier (in Section 3.2.2.1) using the ($T_s - T_{air}$)—also known as DT—in place of T_s and also by using the F_r parameter, as a proxy of vegetation amount, in place of the NDVI. Authors argued that locations where DT = 0 always represent the true cold edge of the triangular space, where *H* flux is negligible (near zero) and LE flux is equal to the available energy ($R_n - G$), that is, EF = 1.0. Stisen et al. (2008) introduced a modification of the Jiang and Islam (2003) technique replacing the ($T_s - T_{air}$) by either the T_s recorded at 1200 UTC or the surface temperature difference (dT_s) between the times 1200 and 0800 UTC. In addition, they included for the first time a nonlinear interpretation of the T_s/VI domain, based on a nonlinear interpolation of φ parameter with respect to NDVI (where φ represented the complex effects of the *A* and *B* parameters of the Priestley–Taylor equation). A number of studies have validated the Jiang and Islam (2003) modeling scheme and of its variants at dissimilar geographical regions and using different types of remote sensing data

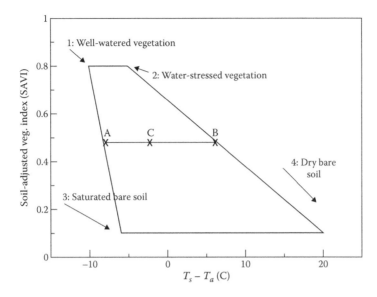

FIGURE 3.3

Illustration of the principles of the trapezoidal method of Moran et al. (1994) for the estimation of the instantaneous LE fluxes from the $(T_s - T_{air})$/VI domain. T_s is the land surface temperature and T_a is the surface air temperature. According to the authors, having a measurement of $(T_s - T_a)$ at any point C inside the trapezoid allows one to equate the ratio of actual to potential LE with the ratio of the distances CB and AB. (Adapted from *Remote Sensing Environment*, 49, M. S. Moran, T. R. Clarke, Y. Inoue, and A. Vidal, Estimating crop water deficit using the relation between surface-air temperature and spectral vegetation index, 246–263. Copyright 1994, with permission from Elsevier.)

(Jiang and Islam 2003; Venturini et al. 2004; Batra et al. 2006; Stisen et al. 2008; Shu et al. 2011). Such studies have shown that this group of methods are able to often provide estimates of instantaneous energy fluxes at an accuracy of about 50 W m^{-2} and daytime fluxes (expressed by the term evaporative fraction E which will be discussed later on) from 0.08 to 0.19, respectively.

Overall, a very important advantage of this group of methods is their independence from absolute accuracy of the T_s measures. This is because DT equal to zero always represented the true cold edge of the triangle/trapezoidal domain where EF equals zero. Also, all methods were dependent on a very small number of *in situ* observations, for example, the Moran et al. (1996) method was dependent only on R_n, vapor pressure deficit, wind speed, and T_{air}. Clearly, the most important advantage of the latter approach in particular was the use of the WDI, which allowed a relatively straightforward computation of the instantaneous LE fluxes within the trapezoidal domain for both heterogeneous and homogeneous surfaces, in contrast with the CWSI that was applicable only for homogeneous areas. Specifically the technique proposed by Stisen et al. (2008) has the advantage that included the utilization of the high temporal resolution geostationary MSG SEVIRI data, which allows the monitoring of the temperature diurnal variation. A further advantage of this is that it provides a nonlinear assumption of the triangular/trapezoid domain of (F_r, DT) feature space in solving for the energy fluxes, which might be a more realistic approximation of reality. Yet, the Stisen et al. (2008) technique does not allow for the presence of water stress over full F_r where EF is zero along the observed dry edge.

3.2.2.4 Methods Based on Day and Night T$_s$ Difference and VI Scatterplot

Briefly, implementation of this group of methods has been based on the existence of a strong relationship between the daytime and nighttime T_s and soil moisture and thermal inertia (van de Griend 1985; Jordan and Shih 2000). Tan (1998) and Chen et al. (2002) first proposed the idea of the LE fluxes retrieval from the difference between the day and night T_s versus the radiometric VI via a modeling scheme they called diurnal surface temperature variation (DSTV). DSTV has been based on implementing firstly a simple linear mixture model on the DSTV/VI domain to determine the fractional contributions of the values for each pixel from vegetation, dry soil, and wet soil surfaces. Then, an index is computed termed "vegetation and moisture coefficient," which for each image pixel was expressed as the sum of the weighted components from vegetation cover, dry soil, and wet soil. The latter is used to determine the actual LE flux, following the procedure detailed by Chen et al. (2002). Wang et al. (2006) proposed a variant of the Jiang and Islam (2003) method, in which the daytime T_s was replaced by the day – night T_s difference and NDVI (ΔT_s – NDVI).

Validation studies on techniques belonging to this group of approaches have generally showed errors in the prediction of daily evapotranspiration between 2.8% and 23.9% and RMSDs varying from 3.08 to 5.74 mm day^{-1} (Tan 1998; Chen et al. 2002; Wang et al. 2006). Although their implementation appears to be dependent of only a very small number of ground measurements, those techniques have certain shortcomings. First of all, those methods are based on the assumption of three dominant land cover types in the mixture modeling scheme, which cannot always be found in a satellite scene. Also, the method requires for its implementation two satellite derived T_s observations, one acquired during daytime and one during nighttime conditions. Evidently, this condition cannot be satisfied by all remote sensing sensors currently in orbit that are otherwise capable of being implemented by those methods.

3.2.2.5 Methods Based on Coupling the T$_s$/VI Scatterplot with a Soil Vegetation Atmosphere Transfer Model

An alternative approach for deriving spatially distributed maps of LE and H fluxes, termed the "triangle" method, is based on the coupling of the T_s/VI feature space with a land biosphere model, namely, a Soil Vegetation Atmosphere Transfer (SVAT) model. SVAT models are essentially vertical views of hydrological processes that consider the transport of water and energy just below the surface across the interface and within and through the vegetation canopy. This type of method aims to combine the horizontal coverage and spectral resolution of EO data with the vertical coverage and fine temporal continuity of SVAT models. At present, the "triangle" has been implemented with the SimSphere model (see Petropoulos et al. 2009b for a review of the model use); yet, any other SVAT model with similar functionalities can be used.

Overviews of the triangle method implementation can be found in the works of Carlson (2007) and Petropoulos and Carlson (2011). Briefly, this approach is based on a quantitative interpretation of the T_s/F_r scatterplot using a one-dimensional SVAT model the latest version of which is called SimSphere (distributed from Aberystwtyh University, UK, http:// www.aber.ac.uk/simsphere), with constraints imposed by the warm edge and the extreme values of the T_s/F_r scatterplot for bare soil and full vegetation, respectively. The SVAT model is first initialized using representative test site data. Then, at the time of the sensor overpass the model is iterated for all possible combinations of F_r and M_o and the simulated outputs of T_s and the surface energy fluxes are recorded at each iteration. As the next step, a third-order polynomial equation is derived linking the modeled soil surface moisture (M_o) to the values of T_s and F_r, along with similar equations linking each of the LE and H fluxes to M_o and F_r. Although these equations are empirically derived, they are based on physical representations of the biophysical processes operating within the SVAT model simulation. Thus the technique can be used to provide estimates of Mo simultaneous to the retrievals of the surface heat fluxes.

Gillies et al. (1997) extended the method by proposing and the computation of the ratios of LE or H to R_n (LE/R_n or H/R_n) inside the triangle domain, along with the instantaneous energy fluxes and M_o derived at the time of satellite overpass. Brunsell and Gillies (2003) using the SVAT model within the "triangle" demonstrated a procedure to interpolate the satellite-derived temperatures at different times from that of the satellite overpass time. This allowed the authors to perform comparisons of the energy fluxes acquired at different overpass times. Petropoulos et al. (2009b, 2010, 2013) performed detailed sensitivity analysis to SimSphere, providing for the first time a detailed insight into its architecture and discussing important implications of the model use within the "triangle" approach. A number of studies have evaluated the ability of this group of methods in deriving spatially distributed estimates of energy fluxes and Mo. It appears that the technique can predict instantaneous LE and H fluxes with a standard error in the order of ±10% and ±30%, respectively, and the daytime fluxes with an RMSD of around 0.15. Also, M_o prediction error can be about 16% (Gillies et al. 1997; Brunsell and Gillies 2003; Petropoulos and Carlson 2011).

In comparison to other T_s/VI techniques reviewed so far, the triangle group of methods has some noticeable advantages. First of all, contrary to all other T_s/VI techniques (except the recent study by Stisen et al. 2008), the "triangle" provides a nonlinear interpretation of the T_s/VI space and thus a solution for the computation of the spatially distributed estimates of the turbulent fluxes and M_o, which, in general, seems to be a more realistic assumption. Furthermore, the technique offers the potential of deriving additional parameters, namely, the soil surface moisture and the daytime average LE and H fluxes via a relatively simple and straightforward way. Furthermore, the triangle puts forward the prospect to construct similar T_s/VI triangles on successive days with a virtually identical configuration of isopleths over the entire T_s/VI space, allowing to monitor land surface processes that can be linked to other phenomena (such as urbanization; e.g., Owen et al. 1998; Arthur-Hartanft et al. 2003). Last but not least, it offers the possibility to extrapolate the instantaneous estimates of the energy fluxes from one time of day to another as was recently demonstrated by Brunsell and Gillies (2003). Yet, potential downsides related to this method of implementation include the requirement of a large number of input parameters in the SVAT model initialization and user expertise and familiarity that might be required with such type of model operation.

3.2.3 Data Assimilation Methods

Another approach in the regional estimation of surface heat fluxes includes the combined use of information derived from remote sensing with deterministic land surface process models, in some cases SVAT models. Models have some very important advantages, which justify their development and continuous use to contemporary modeling schemes together with EO data. Those are able to often provide access to a detailed description of soil and vegetation canopy processes and not only to a limited number of final variables such as evapotranspiration, soil moisture, or net primary production. Another important advantage is that their time resolution, which usually is less than 1 h, is in good agreement with the dynamic of atmospheric and surface processes. Because of their fine vertical coverage and fine temporal continuity, land surface models have become attractive for applications combined with EO data, acquired instantaneously (Olioso 1992). This has led to a number of studies attempting to combine those models with EO data for obtaining spatially distributed parameters characterizing land surface interaction processes, including surface heat fluxes. Overviews of assimilation approaches in remote sensing can be found for example in Kalma et al. (2008) and Li et al. (2009b). Generally, two types of approaches can be distinguished: (1) the so-called forcing methods and (2) the assimilation methods.

Forcing methods are based on forcing the model input with the remote sensing measurement. Briefly, those methods consist of setting some of the input quantities in the model at values estimated from remote sensing measurements. Models' parameterization usually requires an extensive amount of parameters, including information on vegetation structure (e.g., LAI and height), optical properties of soil and vegetation, physiological properties of vegetation (e.g., stomatal conductance description, water transfer from soil to plants), thermal and hydraulic properties of the soil, and atmospheric conditions (e.g., air temperature and humidity, wind speed, incident radiation). Some of these parameters are derived from remote sensing. Various relevant studies have shown that the most adequate variables to be estimated from remote sensing are vegetation fraction (F_r), LAI, albedo, and emissivity (Courault et al. 2005). For example, the multilayer canopy–surface–layer terrestrial biosphere–atmosphere model Advanced Canopy–Atmosphere–Soil Algorithm (ACASA; Pyles et al. 2000) is an example of a forcing type of model developed to calculate face energy, mass, and momentum exchanges, as well as the microclimatic conditions, trace gas exchanges, and associated turbulence statistics, over vegetated regimes.

Assimilation methods have to do with adjusting some of the model parameters aiming to minimize the difference between the information derived from remote sensing and that from the model simulations (Figure 3.4). By doing that, the reproduction of the

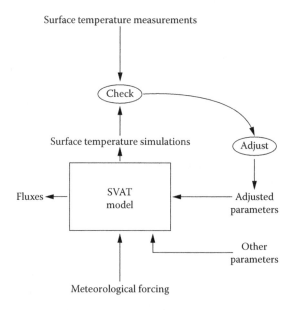

FIGURE 3.4
Schematic representation of the assimilation method based on recalibration of SVAT model surface temperature measurements (in Olioso et al. 1999). In this example, comparisons are made between surface temperatures estimated by the model with the surface temperature measured. Satellite-derived surface temperatures can equally be used for forcing.

diurnal courses of canopy fluxes from instantaneous measurements is then possible (e.g., Soer 1980; Ottle and Vidal-Madjar 1994). Assimilation methods are generally divided into two subcategories, namely, (1) the sequential methods (e.g., Ensemble Kalman Filter and optimal interpolation) (Caparrini et al. 2003; Crow and Kustas 2005; Margulis et al. 2005), in which the course of state variables in the model is corrected at each time remote sensing data are available; and (2) the variational methods (e.g., four-dimensional variational assimilation, e.g., Margulis and Entekhabi 2003), where re-initialization or change of unknown parameters in the model is performed using data sets acquired over temporal windows of several days/weeks. In sequential assimilation, each individual observation influences the estimated state of the flow only at later times and not at previous times, whereas variational assimilation aims to adjust the model solution globally to all the observations available over the assimilation period (Talagrand 1997).

The principle of any data assimilation scheme is essential to minimize the mismatch between the observations and models by adjusting components under the fundamental physical constraints. More often, SVAT models are used to estimate variables directly in relation with hydrological or meteorological models and drive these models from remote sensing, such as in the approaches of Ottle and Vidal-Madjar (1994) and Calvet et al. (1998). Then, a correction of possible temporal drift of the variable that is dynamically predicted by SVAT models may be done. This method is also found in the literature as "re-initialization" instead of "recalibration," as an initial value of a variable is adjusted instead of a model parameter (Moulin et al. 1998). It is perhaps worthwhile to note that Olioso et al. (1999) in an overview of the data assimilation methods in remote sensing included the "triangle" method discussed previously (Section 2.2.5) in this group of assimilation methods.

Important advantages of the data assimilation approaches to mapping surface energy fluxes over traditional retrieval methods include the following (see Li et al. 2009b): (1) the assimilation procedure estimates not only LE or *H* fluxes but also the various intermediate variables related to the turbulent heat fluxes in a numerical model; (2) estimates of the turbulent heat fluxes are continuous in time and space since the dynamic models used in the assimilation procedure interpolate the measurements taken at discrete sampling times; (3) the data assimilation procedure can produce estimates at a much finer resolution; (4) data assimilation schemes can merge spatially distributed information obtained from many data sources with different resolutions, coverage, and uncertainties (Margulis et al. 2002). A main downside of data assimilation techniques to retrieve regional LE/*H* fluxes is that they are relatively more computationally demanding in comparison to methods utilizing EO data alone. As noted, for example, by Courault et al. (2005), one of the main problems arising when using SVAT model is the spatial resolution of EO data. Indeed, the detailed process descriptions provided by these models are based on local parameters that are not systematically adequate with the information collected with several meter size pixels. These difficulties yield to the development of approaches aiming in defining "effective" parameters corresponding to these composite surfaces (Noilhan and Lacarrere 1995) or at disaggregating the pixel content into elementary responses for each land use class (Courault et al. 1998). Some parameters like LAI can, for example, be effortlessly averaged using arithmetic laws, in comparison to some other (e.g., T_s) in which aggregation schemes can be more complex.

3.2.4 Microwave-Based Methods

A different group of approaches for estimating the surface heat and moisture fluxes from EO data has been based on the exploitation of microwave (MW) remote sensing data. Such methods frequently work synergistically with other types of remote sensing radiometers operating at different parts of the EMR spectrum. Various techniques exploiting MW observations have been developed and successfully applied to a range of environmental conditions and in conditions of low to moderate vegetation cover. Such approaches can be divided mainly to (1) methods combining MW observations to TSM schemes and (2) methods relating energy fluxes with MW land surface emissivity measurements acquired at two different wavelengths via empirical or semiempirical relationships.

The first group of approaches has been based on combining MW observations within a TSM scheme. A key effort in this direction includes the technique proposed by Kustas et al. (1994). Authors offered a revised version of the Norman et al. (1995) TSM (presented in Section 3.2.1.2.1) that allowed estimating the energy fluxes via the combined use of optical and MW observations. Their proposed TSM scheme incorporated remote sensing surface temperature estimation derived from a passive MW radiometer. Apart from the availability of the MW measurements, implementation of their model was based on the availability of optical satellite data (for estimating F_r or LAI) and land cover map from which the vegetation height and the surface roughness characteristics were estimated. In addition, the model implementation was dependent on a small number of ground observations (i.e., wind speed, T_{air}, relative humidity, and incoming solar radiation), as well as soil type/ texture. The major advantage of this model over the traditionally used TSM of Norman et al. (1995) was the inclusion of MW observations, which was making estimation of the T_s independent of the atmospheric attenuation effects and clouds passing. Apart from the obvious advantages of the use of MW observations in a TSM scheme, a limitation of the models was that in contrast to the original TSM of Norman et al. (1995), the model did not include any mechanism for explicitly reducing canopy transpiration from its potential rate, in the case of moisture-induced vegetation stress. Kustas et al. (1994) evaluated their model over an area in Arizona, United States, using passive MW brightness temperature images from the Push Broom Microwave Radiometer (PBMR) and ground observations from selected stations belonging to the METFLUX network. A good agreement between the modeled and the ground LE and H fluxes was found, with RMSDs of 65 and 36 W m^{-2}, respectively, and a mean absolute percentage difference of 31% and 23%, respectively.

Kustas et al. (2003a) developed a two-source model that was able to use as a driving remote sensing input for the estimation of surface heat fluxes either microwave-derived near-surface soil moisture or radiometric surface temperature. Differences in the two TSM architectures are described by Li et al. (2006). Briefly, in comparison to the original TSM, that was using optical/TIR data, in the MW-based model, MW data are used to estimate soil moisture that is used along with the Priestley–Taylor formula (Priestley and Taylor 1972) to compute transpiration from vegetation, and the H flux is computed as a residual from the energy balance. Li et al. (2006) using Landsat ETM+ data, as well as airborne and ground observations acquired during the SMACEX and SMEX02 field experiments conducted in Iowa, provided an independent evaluation of the ability of these two two-source modeling schemes of Kustas et al. (2003) for deriving regional maps of surface heat fluxes. A close agreement in the LE and H fluxes predictions by the two models was reported by the authors with a mean RMSD of ~45 W m^{-2} for both fluxes.

The second group of MW-based approaches attempts to infer energy fluxes by relating via empirical or semiempirical relationships vegetation-related parameters with an MW land surface emissivity measurement acquired at two different wavelengths, namely, the Microwave Emissivity Difference Vegetation Index (EDVI) (defined by Min and Lin 2006). This type of method has been based on previous studies establishing the existence of semiempirical relation between the optical depth at MW wavelengths and vegetation water content, which varies systematically with both wavelength and canopy structure (Jackson and Schmugge 1991). As EDVI is directly linked to volumetric soil moisture content, the fast changes of EDVI represent canopy response to the changes of environmental conditions, such as water potential. EDVI values are derived from a combination of satellite MW measurements and optical as well as TIR observations. Min and Lin (2006) estimated the EDVI based on MW observations from the SSM/I sensor acquired over a forested region in the eastern United States and attempted to develop an empirical relationship relating EDVI to surface heat flux measurements acquired over the region concurrently to the satellite overpass. EDVI was found sensitive to EF fluxes, with a correlation coefficient (R) higher than 0.79 for cloudless conditions and higher than when NDVI was used in place of EDVI.

Li et al. (2009a) based on EDVI developed another algorithm for estimating EF and instantaneous LE fluxes from dense vegetation cover, by using the high temporal resolution of EDVI. Investigators linked the seasonal trend of EDVI to the variance of canopy resistance due to the interrelationship among leaf development, environmental condition, and MW radiation. Li et al. (2009b) evaluated their algorithm using data from the same test site used by Min and Lin (2006) using also SSM/I satellite data. They reported a correlation coefficient (R) of 0.83 between the predicted and observed LE fluxes and bias and standard deviation of 3.31 and 79.63 W m^{-2}, respectively. Authors pointed out as a key advantage of their approach that all key inputs required for its implementation could be replaced by satellite remote sensing and reanalysis data, which highlighted it as an important characteristic for a potential operational use of their method.

Min et al. (2010) developed and subsequently applied the EDVI for a region in the Amazon using satellite observations from the MW AMSR-E radiometers in combination with MODIS cloud products and reanalysis meteorological data from NCEP. Authors performed comparisons for cloud-free days and reported that EDVI was able to capture vegetation variation from dense vegetation of the Amazon rainforest to the short/sparse vegetation of the savannah, under all-weather conditions. Also, in agreement to the previous findings by Min and Lin (2006), good correlations were found in their study between EDVI computed from the MW data and the optical indices such as the DVI and the Enhanced Vegetation Index (EVI), with an important advantage of EDVI being that it did not saturate in comparison to other indices.

All in all, MW-based remote sensing approaches are able to be applied under all-weather conditions, including cloudy days, which is an important advantage in comparison to other modeling schemes. Nevertheless, it should be noted that at present available passive MW imaging systems suitable for retrievals of soil moisture have spatial resolutions in the range of 35–65 km, which is much coarser than that of TIR radiometers that varies from 60 m with ASTER and Landsat to 5 km with the Geostationary Operational Environmental Satellite (GOES). Perhaps this is the main reason that explains their not so wide use and growth today in comparison to other approaches for modeling the surface fluxes from remote sensing observations.

3.3 Extrapolation Methods of the Instantaneous Estimates

Retrieval of regional estimates of surface heat fluxes from remote sensing has been largely based on deriving instantaneous estimates of those parameters at the time of satellite overpass each time. However, for many practical applications (e.g., irrigation planning and water needs assessment), these instantaneous estimates can be indeed of very small use. A daytime or daily average LE or H flux estimate would be more interesting and simultaneously useful in practice for use in hydrological and meteorological studies. This section is providing an overview of some of these key approaches developed for this purpose, discussing their relative strengths and limitations.

3.3.1 Extrapolation to Daytime Average Fluxes

Most of the techniques employed in temporally extrapolating instantaneous remotely sensed estimates of LE and H fluxes to daytime averages have their basis on the use of flux ratios. Those ratios have been shown to remain fairly conservative (constant) during the daytime period, thus allowing the extrapolation of a single flux estimate to a daytime average flux. The working hypothesis underlying the computation of the daytime average fluxes from those ratios, supported by several field campaigns and studies, is that these ratios are approximately constant during the daytime hours under clear-sky conditions, although in absolute value those may vary from day to day (e.g., Shuttleworth and Wallace 1985; Shuttleworth et al. 1989; Crago 1996).

On this basis, some investigators (Brutsaert and Sugita 1992) proposed the use of simple ratios of instantaneous LE or H flux to instantaneous R_n (namely, LE/R_n, H/R_n). An example of the use of those ratios as expressions of daytime average fluxes was seen in the so-called "triangle" method (Carlson, 2007) reviewed earlier. Various authors have shown that generally reasonable estimates of daytime energy fluxes can be obtained from the use of those ratios (e.g., Brutsaert and Sugita 1992; Farah et al. 2004; Gentine et al. 2007; Abdelghani et al. 2008). However, wider use of those simple ratios as approximations of representative daytime average fluxes deserves further investigation (e.g., Crago 1996a,b). Evidently, most authors assume that the net daily soil heat flux (G) is zero (e.g., Kustas and Norman 1996); that is, the total downward flux in daytime and the total upward flux at night balance each other out. Thus, when compared to the other components of the surface energy balance, net G flux can be considered negligible when summed over a day (e.g., Kustas and Norman 1996; Cammalleri et al. 2012). Also, instantaneously G flux can be quite large and can show large spatiotemporal variations because it involves the thermal properties of the soil that vary largely with moisture content (e.g., Perez et al. 1999). What is more, in semiarid areas the contribution of G flux at a daily or daytime scale cannot be neglected as it is high instantaneously (Kustas et al. 1994). Moreover, particularly in forests (as high biomass systems), the storage term (S) calculated by measuring the change in soil temperature over the averaging period can be significant, representing up to 70%–80% of G flux.

Among the most widely employed alternative "flux ratio"–based techniques used in energy balance modeling as proxies of the daytime average fluxes is the evaporative fraction (EF) and nonevaporative fraction (NEF) (e.g., Shuttleworth and Wallace 1985; Shuttleworth et al. 1989; Sugita and Brutsaert 1991; Crago 1996a). Those ratios have generally received special interest because those have been linked with the available energy that is the primary driver of the LE and H fluxes. In particular, the EF and NEF ratios are defined as follows (e.g., Shuttleworth et al. 1985, 1989):

$$\mathrm{EF}_i = \frac{\mathrm{LE}_i}{(R_n - G)_i} = \frac{\mathrm{LE}_i}{(\mathrm{LE} + H)_i} \tag{3.4}$$

$$\mathrm{NEF}_i = \frac{H_i}{(Rn - G)_i} = \frac{H_i}{(\mathrm{LE} + H)_i} = 1 - \mathrm{EF}_i \tag{3.5}$$

where i is the time of the satellite overpass.

From the above equations it can be easily also observed that EF_i and NEF_i are also related to the Bowen ratio β (i.e., the ratio of H to LE) shown as follows:

$$EF_i = \frac{1}{1+\beta}, \ NEF_i = \frac{\beta}{1+\beta} \tag{3.6}$$

However, in contrast to the Bowen ratio, which has no upper limit (i.e., if LE approaches zero then the Bowen ratio tends to infinity), EF and NEF can vary between 0.0 and 1.0. EF equaling 0.0 essentially means that the surface is very dry and there is no evapotranspiration. On the other hand, EF equaling to 1.0 means that the land surface is very wet and there is maximum evapotranspiration. EF provides an indication of how much of the available energy is used for evapotranspiration, that is, for transpiration from vegetation and evaporation from the soil. Understandably, as long as moisture is available, energy will be used for its evaporation, whereas if no moisture is left, the available energy will be directed into H processes and EF will approach zero. In general, when there is higher near surface soil moisture, EF tends to increase since the LE flux is favored as a more efficient way for dissipating heat, while when the soil is dry LE is limited, and therefore, EF increases. As would be expected the reverse situation is applicable in the NEF flux ratio. Hence, it is clear why EF has been related to some measures of soil saturation state, and thus has the potential according to many researchers to be used as a drought indicator (e.g., Cammalleri et al. 2012; Mahfouf and Noilhan 1991; Billi and Caparini 2006; Niemeyer and Vogt 1999).

A number of investigators also evaluated the use of EF or NEF as satisfactory approximations of the daytime average surface heat flux (e.g., Kustas et al. 1994; Zhang and Lemuear 1995; Crago 1996a,b; Chavez et al. 2008; Petropoulos et al. 2010). Generally, among the most important advantages of the EF and NEF ratios as expressions of the daytime average fluxes are their simplicity and flexibility to ingest a small number of input data. What is more, this technique can operate regardless of climate and type of ground biomass and can be applicable in many different remote sensing systems acquiring data at a range of spatial resolutions. Amongst its disadvantages is that retrievals of daytime fluxes by this method are dependent on the accuracy of the technique used for predicting the instantaneous surface heat fluxes. Also, it should be noted that the role of vegetation heterogeneity and topography on daytime average EF may at times be significant (e.g., Smith et al. 1992), but their effects on the daytime variability of the energy budget are still not completely clear (Crago 1996a). It appears that compensatory relationships exist between the roles of radiation and wind speed (Gentine et al. 2007). Last but not least, this approach cannot be applied under non-clear-sky conditions, and in such cases when the weather conditions are variable such as under strong radiative forcing and humid conditions, the hypothesis of constant EF may not be satisfied (e.g., Gentine et al. 2011; Sugita and Brustaert 1991).

3.3.2 Extrapolation to Daily Energy Fluxes

3.3.2.1 Simplified Empirical Regression Methods (or Direct Simplified Methods)

In the late 1970s, Jackson et al. (1977) proposed an empirical method for estimating daily LE, based on the surface energy balance and the difference between instantaneous surface temperature (T_s) and the simultaneous air temperature T_{air}, according to

$$LE_{daily} = R_{n_{daily}} - B(T_s - T_{air})^n \tag{3.7}$$

where B and n are the site-specific regression coefficients dependent on surface roughness, wind speed, atmospheric stability, etc. (Li et al. 2009b), which are determined either by linear least squares fit to data or by simulations based on a SVAT model (Carlson and Buffum 1989; Carlson et al. 1995).

The principal assumptions on which the method of operation is based include the fact that G flux is negligible when considering the daily estimation of LE fluxes, and also that the ratio of H/R_n is also constant during daytime. Details on this method foundation can be found, for example, in the work of Seguin and Itier (1983), whereas the implications of the regression coefficients in the simplified equation are discussed by Carlson et al. (1995). In Equation 3.7, R_n and T_s are computed in a relatively straightforward way from optical and thermal data, whereas the estimation of T_{air} is perhaps the most problematic to be acquired. The latter can be estimated in different ways, including geostatistical interpolation methods of meteorological observations or using empirical relationships with VIs (see review by Courault et al. 2005).

Various studies have evaluated the above equation, confirming its potential use as a simple method for deriving coarse estimates of daily LE fluxes under a variety of atmospheric conditions and vegetation covers (Carlson and Buffum 1989; Carlson et al. 1995; Tsouni et al. 2008; Delogu et al. 2012). Generally, those studies have shown that error in the predicted daily LE fluxes by the Jackson et al. (1977) method is about 1 mm day^{-1}. Also, a number of refinements of the Jackson et al. (1977) equation have been proposed over the years. Some schemes have been based on parameterizations of the B and n coefficients as a function of wind speed, roughness, and criterions of atmospheric stability (Vidal and Perrier 1989; Lagouarde and McAneney 1992). For example, Carlson and Buffum (1989) used the simplified equation that might be more applicable to regional-scale ET estimations if the air temperature and wind speed were measured or evaluated at a level of 50 m, owing to the fact that at this height variation of the meteorological parameters, they are assumed to be insensitive to surface characteristics.

Generally, key advantages of the simplified empirical regression methods include their ease and straightforward implementation along with their low demand in data provision. Also, these methods are offering relatively sufficient accuracy in the estimation of daily evapotranspiration values from a very small number of ground-measured variables. What is more, those modeling approaches do not require estimates of instantaneous energy fluxes to be computed previously and thus are not affected by the accuracy of the instantaneous LE and H flux estimation (as is the case with other approaches discussed next). The need for local calibration of the empirical coefficients is perhaps the single most important disadvantage to be taken under consideration in using those practically. Parameters such as the surface roughness and wind speed cannot be directly provided by remote sensing, and this perhaps is the main reason limiting their use so far (see Li et al. 2009b). Besides, most of these methods are based on the assumption that the T_{air} and R_n are not spatially varying, which can be valid only if canopy albedo and surface temperature do not vary significantly, which is not always the case especially under very heterogeneous surfaces.

3.3.2.2 EF-Based Methods

A widely used approach employed for the temporal extrapolation of instantaneous LE fluxes to daily values has been related to the EF introduced earlier in Section 3.1. On the basis of the EF concept of Shuttleworth et al. (Shuttleworth and Wallace 1985; Shuttleworth et al. 1989), Sugita and Brutsaert (1991) introduced the idea of the use of this ratio for deriving daily totals of evapotranspiration rates from corresponding instantaneous estimates. In particular, the authors suggested that the daytime evapotranspiration could be calculated from

$$LE_d = EF_e Q_d \qquad (3.8)$$

where EF_e is the evaporative fraction determined from one or more instantaneous flux values and Q_d is the total daytime available energy, determined as

$$Q_d = \int_{t_2}^{t_1} (R_n - G) dt \qquad (3.9)$$

where R_n is the net radiation and G is the ground heat flux, $(t_2 - t_1)$ represents the time daylight period defined as the time of day when $(R_n - G)$ is greater than or equal to zero. EF is computed as

$$EF_e = \frac{\sum_{i}^{n} LE_i}{\sum_{i}^{n} (LE_i + H_i)} \qquad (3.10)$$

where the subscript i refers to the ith measurement and n is the total number of daytime measurements.

Therefore, to calculate daily total LE from Equation 3.7, it is necessary to determine the EF term, which can be obtained using remote sensing data and the daytime total available energy, as described earlier (Section 3.1). This method is very easy and straightforward to be applied, and these are its key advantages. The main downsides are the requirement for more than one instantaneous EF estimate and that of daytime total available energy. Also, the accuracy of the daily evapotranspiration fluxes will depend on the accuracy of the instantaneous LE and H fluxes and the method cannot be applied on cloudy days.

Others (e.g., Kustas et al. 1994; Sarwar and Bill 2007) have proposed the retrievals of daily LE fluxes from the midday EF based on the following equation:

$$LE_{daily} = 0.0345 \cdot EF_{midday} + R_{n_{daily}} \qquad (3.11)$$

where $R_{n_{daily}}$ is the daily net radiation (W m^{-2}) and 0.0345 is a conversion factor. The daily net radiation is computed, using standard meteorological data as follows:

$$R_{n_{daily}} = (1-\alpha)K\downarrow - 110\frac{K\downarrow}{K_{exo}\downarrow} \qquad (3.12)$$

where α is the surface albedo, $K\downarrow$ is the global radiation at the surface level, and $K_{exo}\downarrow$ is the theoretical extraterrestrial radiation.

Another variant of the EF-based method in temporally extrapolating the instantaneous energy fluxes to daily total energy fluxes is employed in the SEBS model reviewed earlier (see Section 3.2.1.1), in which the daily LE flux can be determined as

$$LE_{daily} = 8.64 \cdot 10^7 \cdot \overline{EF} \cdot \frac{\overline{R_n} - \overline{G}}{\lambda \rho_w} \tag{3.13}$$

where LE_{daily} is the actual LE flux on a daily basis (mm day^{-1}), $\overline{EF}$ is the daily average EF, which is approximated by SEBS, R_n and G are the daily net radiation and soil heat flux, respectively, λ is the latent heat of vaporization (J kg^{-1}), and ρ_w is the density of water (kg m^{-3}). In SEBS, since the daily G flux is close to zero because the downward flux in the daytime and the upward flux at night balance each other approximately, the daily LE flux depends only on the total net radiation flux, which in SEBS is computed as

$$\overline{R_n} = (1 - \alpha) \cdot K_{24}^{\downarrow} + \varepsilon \cdot L_{24} \tag{3.14}$$

where $K_{24}^{\downarrow}$ is the daily incoming global radiation, L_{24} is the daily net longwave radiation, α is the daily average albedo, and ε is the emissivity.

In SEBAL, estimation of the daily LE fluxes is also based on the EF concept, where the daily LE flux is also linked to EF, as follows:

$$LE_{daily} = \frac{86400 \cdot 10^3}{\lambda \rho_w} \cdot EF \cdot R_{n_{daily}} \tag{3.15}$$

where $R_{n_{daily}}$ is the 24 h averaged daily R_n, λ is the latent heat of vaporization (J kg^{-1}), ρ_w is the density of water (kg m^{-3}), and LE_{daily} is expressed in mm day^{-1}.

3.3.2.3 Sinusoid Relationship

On the basis of the assumption that the diurnal course of LE generally follows the course of solar radiation throughout the daylight period, Jackson et al. (1983) demonstrated that the ratio of total solar radiation to an instantaneous value taken during midday could be approximated by a sine function. On the basis of this finding, the authors proposed the computation of the daily total LE flux from an instantaneous LE estimate and a conversion coefficient given as a function of latitude, day of year, and time of day. More specifically, according to Jackson et al. (1983), the instantaneous solar irradiance (Si) on a clear day can be approximated as

$$S_i = S_m \sin\left(\frac{\pi t}{N}\right) \tag{3.16}$$

where S_m is the maximum irradiance at solar noon, t is the time beginning at sunrise, and N is the time period between sunrise and sunset in units of t. Integration of Equation 3.16 yields the daily total irradiance S_d:

$$S_d = \int_0^N S_m \sin\left(\frac{\pi t}{N}\right) dt = \left(2N\big/\pi\right) S_m \tag{3.17}$$

Therefore the ratio of total daily irradiance to the instantaneous irradiance at time t is given by

$$J = S_d\big/S_i = 2N\big/[\pi \sin(\pi t/N)] \tag{3.18}$$

where t and N in the argument of the sine function must have the same units. The daylight period N can be expressed by

$$N = 0.945 \cdot [\alpha + b \sin^2[\pi(D + 10)/365]] \tag{3.19}$$

where a and b are latitude-dependent constants; the constant a represents the shortest daylight period of the year and the constant b is the amount that must be added to a in order to obtain the longest daylight of the year; and D is the day of the year. The constants a and b can be calculated as a function of latitude (Jackson et al. 1983):

$$\alpha = 12.0 - 5.69 \cdot 10^{-2} L - 2.02 \cdot 10^{-4} L^2 + 8.25 \cdot 10^{-6} L^3 - 3.15 \cdot 10^{-7} L^4 \tag{3.20}$$

$$b = 0.123 \cdot L - 3.10 \cdot 10^{-4} L^2 + 8.00 \cdot 10^{-7} L^3 + 4.99 \cdot 10^{-7} L^4 \tag{3.21}$$

where L is the latitude in degrees.

Using the similarity between the diurnal course of evapotranspiration and the one of solar irradiance, daytime LE can be estimated as

$$\text{LE}_d = E_i \left(S_d\big/S_i\right) = \text{LE}_i \left[2N\big/\pi \sin(\pi t/N)\right] \tag{3.22}$$

where LE_d is the daytime evapotranspiration in MJ m^2 day^{-1} and LE_i is the instantaneous evapotranspiration in W m^{-2}, which can be obtained from the energy balance equation using EO data. The method also requires the conversion coefficient (J), which is a simple function of latitude, day of year, and time of day.

Jackson et al. (1983) evaluated the above approach and showed a strong correlation between measured and estimated daily LE fluxes. For example, Zhang and Lemeur (1995), in a comparison of different methods for the daily estimation of LE fluxes for cloud free days using data from the HAPEX-MOBIHLY experiment in France, reported comparable results in the predicted daily LE fluxes between the method of Jackson et al. (1983) and the EF method of Sugita and Brutsaert (1991). An important advantage of the Jackson et al. (1983) method for potential operational implementation is that it requires only one instantaneous value of evapotranspiration and the conversion coefficient, which is a function of latitude and the day of year. In contrast, the Sugita and Brutsaert (1991) method requires more than one instantaneous EF fraction and daytime total available energy values in deriving the daily LE fluxes.

On the basis of the same concept, Xie (1999) suggested the transformation of the instantaneous LE flux to daily LE fluxes based on the sinusoid relationship between daily and instantaneous LE fluxes, as follows:

$$\frac{\text{LE}_d}{\text{LE}} = \frac{2N_E}{\pi \sin(\pi t / N_E)} \tag{3.23}$$

where LE_d is the daily LE flux, LE is the instantaneous LE flux, t is the time of satellite sensor overpass, and N_E is the daily hours, which can be calculated by minus two from the daily sunshine hours.

Various investigators have applied the above method in arid and semiarid regions using different types of remote sensing data, reporting generally satisfactory results in daily LE predictions. For example, Yunhao et al. (2003) compared daily LE fluxes derived by the Xie (1999) method for a region in China derived from Advanced Very High Resolution Radiometer (AVHRR) observations and reported a generally satisfactory agreement with the corresponding ground measurements with an average relative error of 16% and R^2 of 0.77. The latter scheme or variants of it has been widely applied today by many in deriving daily LE flux estimates (e.g., Zhang et al. 2006; Liu et al. 2010).

3.4 Conclusions and Future Outlook

This chapter presented an overview of the most widely approaches employed key methodologies employed in deriving spatiotemporal estimates of LE and H fluxes from remote sensing. It can be concluded that remote sensing–based estimation of energy fluxes is performed in quite diverse, unconnected ways. Our review showed that an average RMSD value of about 50 W m^{-2} and relative error of 15%–30% in the estimation of LE and H fluxes by remote sensing is expected to be found by many techniques. Taking into account that an absolute accuracy of ~50 W m^{-2} in energy fluxes is generally considered satisfactory for a large range of applications, it is clear that EO technology is showing a promising potential in this direction.

From an algorithmic point of view, nearly all available methods in deriving estimates of LE and H fluxes have been developed essentially for cloud-free conditions. What is more, for the majority of the techniques employed, thermal remote sensing data consist of one of the most crucial inputs required, together with optical data. Also, interestingly, a survey of the different methods showed that more multifaceted physical and analytical modeling approaches do not necessarily provide more accurate energy fluxes retrievals than, for example, more empirically based techniques. Indeed, the different groups of methods reviewed are characterized by different strengths and limitations. Evidently, more work is required to address some of the key limitations imposed in obtaining more accurate retrievals of energy fluxes from space and even more to put forward an operational implementation scenario in the retrievals of those parameters at a global scale. Actually, several approaches at present investigate the perspective of global operational estimation of energy fluxes by remote sensing. Along these lines are, for example, the ESA-funded WACMOS and PROgRESSIon (European Space Agency 2012) projects, which explore the development of a series of prototype products from the synergy of remote sensing observations acquired from different ESA-funded missions.

Given the limitations inherent to both sensing systems and modeling approaches, it is clear that it will be necessary to direct efforts toward the investigation of the potential added value from the exploitation of the synergistic use of remote sensing systems and/or modeling approaches. The data and technique complementarity (independent information) and interchangeability (similar information) will assist in developing approaches that better address aspects currently unsolved, such as the spatial and temporal resolution or land surface heterogeneity. In addition, intercomparison studies between different modeling approaches at dissimilar ecosystem conditions, such as those currently implemented in the framework of the LandFLux GEOWEX project, should be further fostered. Such initiatives will allow one to not only appreciate the relative strengths and limitations between the models but also appreciate their prediction accuracy under different ecosystem conditions globally. To this end, the availability of relevant ground observations from global networks such as the FLUXNET and CarboEurope, which are able to provide systematically long-term measurements of relevant land surface parameters, should be continuously supported. This is important, as such data assist not only in validating satellite-derived and modeling results but also in better understanding the relevant physical processes that occurred at different scales of observation using EO data.

Last but not least, the recent launch of remote sensing radiometers such as that of Landsat 8 and the forthcoming launch of SENTINELS 2/3 will continue the heritage of thermal infrared observations from space. It is also expected to foster the development of new modeling techniques for improving the retrieval accuracy of energy fluxes from EO technology. Yet, this prospect remains to be seen in years to come.

Acknowledgments

This work was produced in the framework of the PROgRESSIon (Prototyping the Retrievals of Energy Fluxes and Soil Moisture Content) project implementation, funded by the European Space Agency (ESA) Support to Science Element (STSE) under contract STSE-TEBM-EOPG-TN-08-0005. Dr. Petropoulos gratefully acknowledges the financial support provided by the Agency.

References

Abdelghani, C., J. C. B. Hoedjes, J.-C. Rodriquez, C. J. Watts, J. Gatatuza, F. Jacob, and Y. H. Kerr (2008): Using remotely sensed data to estimate area-averaged daily surface fluxes over a semi-arid mixed agricultural land, *Agric. For. Meteorol.*, 148, 330–342.

Allen, R. G., M. Tasumi, A. Morse, R. Trezza, J. L. Wright, W. Bastiaanssen, W. Kramber, I. J. Lorite, and C.W. Robison (2007): Satellite-based energy balance for mapping evapotranspiration with internalized calibration (METRIC)—Applications, *J. Irrig. Drain. Eng.*, 133(4), 395–406.

Anderson, M. C., J. M. Norman, G. R. Diak, and W. P. Kustas (1997): A two-source time Integrated model for estimating surface fluxes for thermal infrared satellite observations, *Remote Sens. Environ.*, 60, 195–216.

Anderson, M. C., J. M. Norman, W. P. Kustas, R. Houborg, P. J. Starks, and N. Agam (2008): A thermal-based remote sensing technique for routine mapping of land surface carbon, water and energy fluxes from field to regional scales, *Remote Sens. Environ.*, 112, 4227–4241.

Anderson, M. C., W. P. Kustas, J. M. Noorman, C. R. Hain, J. R. Mecikalski, L. Schultz, M. P. Gonzalez-Dugo et al. (2010): Mapping daily evapotranspiration at field to global scales using geostationary and polar orbiting satellite imagery, *Hydrol. Earth Syst. Sci. Discuss.*, 7, 5957–5990.

Arthur-Hartanft, T., T. N. Carlson, and K. C. Clarke (2003): Satellite and ground-based microclimate and hydrologic analyses coupled with a regional urban growth model, *Remote Sens. Environ.*, 86, 385–400.

Asrar, G. (1989): *Theory and Applications of Optical Remote Sensing*, John Wiley, New York.

Bashir, M. A., T. Hata, H. Tanakamaru, A. W. Abdelhadi, and A. Tada (2008): Satellite-based energy balance model to estimate seasonal evapotranspiration for irrigated sorgum: A case study from the Gezira scheme, Sudan, *Hydrol. Earth Syst. Sci.*, 12, 1129–1139.

Bastiaanssen, M. D. (2000): SEBAL-land based sensible and latent heat fluxes in the irrigated Gediz Basin, Turkey, *J. Hydrol.*, 229, 87–100.

Bastiaanssen, W. G. M. (1995): Regionalisation of surface flux densities and moisture indicators in composite terrain. PhD Thesis, Wageningen, Agricultural University, the Netherlands: 273 pp.

Bastiaanssen, W. G. M., E. J. M. Noordman, H. Pelgrum, G. Davids, B. P. Thoreson, and R. G. Allen (2005): SEBAL model with remotely sensed data to improve water-resources management under actual field conditions, *J. Irrig. Drain. Eng.*, 131(1), 85–93.

Bastiaanssen, W. G. M., M. Menenti, R. A. Feddes, and A. A. M. Holstag (1998): A remote sensing surface energy balance algorithm for land (SEBAL): 1. Formulation, *J. Hydrol.*, 198–212.

Batra, N., S. Islam, V. Venturini, G. Bisht, and L. Jiang (2006): Estimation and comparison of evapotranspiration from MODIS and AVHRR for clear sky days over the Southern Great Plains, *Remote Sens. Environ.*, 103, 1–15.

Battrick, B. (2006): The changing Earth: New scientific challenges for ESA's Living Planet Programme, *Eur. Space Agency Spec. Publ.*, ESA SP-1304.

Bhattacharya, B. K., K. Mallick, and J. S. Parihar (2010): Regional clear sky evapotranspiration over agricultural land using remote sensing data from Indian geostationary meteorological satellite, *J. Hydrol.*, 387, 65–80.

Billi, P., and F. Caparrini (2006): Estimating land cover effects on evapotranspiration with remote sensing: a case study in Ethiopian Rift valley, *Hydrol. Sci.*, 51(4), 655–668.

Brunsell, N. A., and R. R. Gillies (2003): Scale issues in land-atmosphere interactions: Implications for remote sensing of the surface energy balance, *Agric. For. Meteorol.*, 117, 203–221.

Brutsaert, W., and M. Sugita (1992): Application of self-preservation in the diurnal evolution of the surface energy budget to determine daily evaporation, *J. Geophys. Res.*, 97, D17, 18377–18382.

Calvet, J. C., J. Noilhan, and P. Bessemoulin (1998): Retrieving the root zone soil moisture from surface soil moisture or temperature estimates: A feasibility study based on field measurements, *J. Appl. Meteorol.*, 37, 371–386.

Cammalleri, C., G. Ciraolo, G. La Loggia, and A. Maltese (2012): Daily evapotranspiration assessment by means of residual surface energy balance modeling: A critical analysis under a wide range of water availability, *J. Hydrol.*, 452–453(C), 119–129.

Caparrini, F., F. Castelli, and D. Entekhabi (2003): Mapping of land-atmosphere heat fluxes and surface parameters with remote sensing data, *J. Hydrometeorol.*, 5, 145–159.

Carlson, T. N. (2007): An overview of the "triangle method" for estimating surface evapotranspiration and soil moisture from satellite imagery, *Sensors*, 7, 1612–1629.

Carlson, T. N., and M. J. Buffum (1989): On estimating total daily evapotranspiration from remote surface temperature measurements, *Remote Sens. Environ.*, 29, 197–207.

Carlson, T. N., W. J. Capehart, and R. R. Gilies (1995): A new look at the simplified method for remote sensing of daily evapotranspiration, *Remote Sens. Environ.*, 54, 161–167.

Chavez, J. L., and C. M. U. Neale (2003): Validating airborne multispectral remotely sensed heat fluxes with ground energy balance tower and heat flux source area (footprint) functions. ASAE Paper No. 033128. St. Joseph, Michigan.

Chavez, J. L., C. M. U. Neale, L. E. Hipps, J. H. Prueger, and W. P. Kustas (2005): Comparing aircraft-based remotely sensed energy balance fluxes with eddy covariance tower data using heat flux source area functions, *J. Hydrometeorol.*, 6, 923–940.

Chavez, J. L., C. M. U. Neale, J. H. Prueger, and W. P. Kustas (2008): Daily evapotranspiration estimates from extrapolating instantaneous airborne remote sensing ET values, *Irrig. Sci.*, 27, 67–81.

Chavez, J. L., P. H. Gowda, T. A. Howell, and K. Copeland (2009): Radiometric surface temperature calibration effects on satellite based evapotranspiration estimation, *Int. J. Remote Sens.*, 30(9), 2337–2354.

Chen, J.-H., C.-E. Kan, C.-H. Tan, and S.-F. Shih (2002): Use of spectral information for wetland evapotranspiration assessment, *Agric. Water Manag.*, 55, 239–248.

Courault, D., P. Clastre, P. Cauchi, and R. Del´ecolle (1998): Analysis of spatial variability of air temperature at regional scale using remote sensing data and a SVAT model, In *Proceedings of the First International Conference on Geospatial Information in Agriculture and Forestry*, vol. II, ERIM International, Ann Arbor, Mich.

Courault, D., B. Sequin, and A. Olioso (2005): Review on estimation of evapotranspiration from remote sensing data: From empirical to numerical modeling approaches, *Irrig. Drain. Syst.*, 19, 223–249.

Courtier, P., E. Andersson, W. Heckley, J. Pailleux, D. Vasiljevic, M. Hamrud, A. Hollingsworth et al. (1998): The ECMWF implementation of three-dimensional variational assimilation (3D-Var). I: Formulation, *Q. J. R. Meteorol. Soc.*, 124, 1783–1807.

Crago, R. D. (1996a): Conservation and variability of the evaporative fraction during the daytime, *J. Hydrol.*, 180, 173–194.

Crago, R. D. (1996b): Daytime evaporation from conservation of surface flux ratios. In *Scaling Up in Hydrology Using Remote Sensing*, Chapter 16. John Wiley, New York.

Crow, W. T., and W. P. Kustas (2005): Utility of assimilating surface radiometric temperature observations for evaporative fraction and heat transfer coefficient retrieval, *Boundary Layer Meteorol.*, 115, 105–130.

Delogu, E., G. Boulet, A. Olioso, B. Coudert, J. Chirouze, E. Ceschia, V. Le Dantec et al. (2012): Reconstruction of temporal variations of evapotranspiration using instantaneous estimates at the time of satellite overpass, *Hydrol. Earth Syst. Sci.*, 16(8), 2995–3010.

Diak, G., J. R. Mecikalski, M. C. Anderson, J. M. Norman, W. P. Kustas, R. D. Torn, and R. L. Dewolf (2004): Estimating land surface energy budgets from space: Review and current efforts at the University of Wisconsin-Madison and USDA-ARS, *Bull. Am. Meteorol. Soc.*, 65–78, doi: 10.1175/BAMS-85-1-65.

European Space Agency (2012): Prototyping the retrievals of energy fluxes and soil surface moisture. Available at http://due.esrin.esa.int/stse/projects/stse_project.php?id=148 (Accessed on November 24, 2012).

Farah, H. O., W. G. M. Bastiaanssen, and R. A. Feddes (2004): Evaluation of the temporal variability of the evaporative fraction in a tropical watershed, *Int. J. Appl. Earth Obs. Geoinf.*, 5, 129–140.

French, A., T. Schmugge, and W. P. Kustas (2002): Estimating evapotranspiration over El-Reno, Oklahoma with ASTER imagery, *Agronomie*, 22, 105–106.

Gentine, P., D. Entekhabi, A. Chehbouni, G. Boulet, and B. Duchemin (2007): Analysis of evaporative fraction diurnal behaviour, *Agric. For. Meteorol.*, 143, 13–29.

Gentine, P., D. Entekhabi, and J. Polcher (2011), The diurnal behavior of evaporative fraction in the soil-vegetation-atmospheric boundary layer continuum, *J. Hydrometeorol.*, 12(6), 1530–1546, doi:10.1175/2011JHM1261.1.

Gillespie, A., S. Rokugawa, T. Mantsunaga, J. S. Cothern, S. Hook, and A. Kahle (1998): A temperature and emissivity separation algorithm for Advanced Spaceborne Thermal Emission and Reflection Radiometer (ASTER) images, *IEEE Trans. Geosci. Remote Sens.*, 36, 1113–1126.

Gillies, R. R., T. N. Carlson, J. Cui, W. P. Kustas, and K. S. Humes, (1997): Verification of the "triangle" method for obtaining surface soil water content and energy fluxes from remote measurements of the Normalized Difference Vegetation Index NDVI and surface radiant temperature, *Int. J. Remote Sens.*, 18, 3145–3166.

Glenn, E., A. R. Huete, P. L. Nagler, K. K. Hirschboeck, and P. Brown (2007): Integrating remote sensing and ground methods to estimate evapotranspiration, *Crit. Rev. Plant Sci.*, 26, 139–168.

Gomez, M., A. Olioso, J. A. Sobrino, and F. Jacob (2005): Retrieval of evapotranspiration over the Aplilles/ReSeDA experimental site using airborne POLDER sensor and a thermal camera, *Remote Sens. Environ.*, 96, 399–408.

Gonzalez-Dugo, M. P., C. M. U. Neale, L. Mateos, W. P. Kustas, J. H. Prueger, M. C. Anderson, and F. Li (2009): A comparison of operational remote sensing-based models for estimating crop evapotranspiration, *Agric. Forest. Meteorol.*, 149, 1843–1853.

Gowda, P. H., J. L. Chavez, T. A. Howell, T. H. Marek, and L. L. New (2008): Surface energy balance based evapotranspiration mapping in the Texas High plains, *Sensors*, 8, 5186–5201.

Hasager, C., and N. O. Jensen (1999): Surface-flux aggregation in heterogeneous terrain, *Q. J. R. Meteorol. Soc.*, 125, 267–284.

Jackson A. C., J. P. Butler, E. J. Millet, F. G. Hoppin, and S. V. Dawson (1977): Airway geometry by analysis of acoustic pulse response measurements, *J. Appl. Physiol.*, 43(3), 523–536.

Jackson, R. D., J. L. Hatfield, R. J. Reginato, S. B. Idso, and P. J. Pinter (1983): Estimation of daily evapotranspiration from one time-of-day measurements, *Agric. Water Manage.*, 7, 351–362.

Jackson, R. D., D. B. Idso, R. J. Reginato, and P. J. Pinter (1981): Canopy temperature as a crop water stress indicator, *Water Resour. Res.*, 17, 1133–1138.

Jackson, T., and T. Schmugge (1991): Vegetation effects on the microwave emission of soils, *Remote Sens. Environ.*, 36, 203–212.

Jiang, L., and S. Islam (1999): A methodology for estimation of surface evapotranspiration over large areas using remote sensing observations, *Geophys. Res. Lett.*, 26, 2773–2776.

Jiang, L., and S. Islam (2001): Estimation of surface evaporation map over Southern Great Plains using remote sensing data, *Water Resour. Res.*, 37, 329–340.

Jiang, L., and S. Islam (2003): An intercomparison of regional heat flux estimation using remote sensing data, *Int. J. Remote Sens.*, 24, 2221–2236.

Jordan, J. D. and S. F. Shih (2000): Satellite-based diurnal and seasonal thermal patterns of natural, agricultural, and urban land-cover vs. soil type in Florida. In *Proceedings of the 2nd International Conference on Geospatial Information in Agriculture and Forestry (ICGIAF)*, vol. I. ERIM International, Ann Arbor, Mich., pp. 489–495.

Kalma, J. D., T. R. McVicar, and M. F. McCabe (2008): Estimating land surface evaporation: A review of methods using remotely sensed surface temperature data, *Surv. Geophys.*, 29(4–5), 421–469.

Khan, S. I., Y. Hong, B. Vieux, and W. Liu (2010): Development and evaluation of an actual evapotranspiration estimation algorithm using satellite remote sensing and meteorological observational network in Oklahoma, *Int. J. Remote Sens.*, 31(14), 3799–3819.

Kongo, M. V., G. W. P. Gewitt, and S. A. Lorentz (2011): Evaporative water use of different land uses in the upper-Thukela river basin assessed from satellite imagery, *Agric. Water Manage.*, 98(11), 1727–1739.

Kustas, W. P., A. N. French, J. L. Hatfield, T. J. Jackson, M. S. Moran, A. Rango, J. C. Ritchie et al. (2003) Remote sensing research in hydrometeorology, *Photogramm. Eng. Remote Sens.*, 69(6), 631–646.

Kustas, W. P., D. C. Goodrich, M. S. Moran, S. A. Amer, L. B. Bach, J. H. Blanford, A. Chehbouni et al. (1991): An interdisciplinary field study of the energy and water fluxes in the atmosphere-biosphere system over semi-arid rangelands: Description and some preliminary results, *Bull. Am. Meteorol. Soc.*, 72, 1683–1705.

Kustas, W. P., J. Hatfield, and J. H. Prueger (2005): The Soil Moisture–Atmosphere Coupling Experiment (SMACEX): Background, hydrometerological conditions, and preliminary findings, *J. Hydrometeorol.*, 6, 791–804.

Kustas, W. P., M. S. Moran, K. S. Humes, D. I. Stannard, P. J. Pinter, Jr., L. E. Hipps, E. Swiatek, and D. C. Goodrich (1994): Surface energy balance estimates at local and regional scales using optical remote sensing from an aircraft platform and atmospheric data collected over semiarid rangelands, *Water Resour. Res.*, 30(5), 1241–1259.

Kustas, W., J. M. Norman, M. C. Anderson, and A. N. French (2003): Estimating subpixel surface temperatures and energy fluxes from the vegetation index–radiometric temperature relationship, *Remote Sens. Environ.*, 85, 429–440.

Kustas, W. P., and J. M. Norman (1996): Use of remote sensing for evapotranspiration monitoring over land surfaces, *Hydrol. Sci. J.*, 41, 495–516.

Lagouarde, J. P., and K. J. McAneney (1992): Daily sensible heat flux estimation from a single measurement of surface temperature and maximum air temperature, *Boundary Layer Meteorol.*, 59(4), 341–362.

Li, F., and T. J. Lyons (1998): Estimation of regional evapotranspiration through remote sensing, *J. Appl. Meteorol.*, 38, 1644–1654.

Li, F., W. P. Kustas, M. C. Anderson, T. J. Jackson, R. Bindlish, and J. H. Prueger (2006): Comparing the utility of microwave and thermal remote sensing constrains in two-source energy balance modelling over an agricultural landscape, *Remote Sens. Environ.*, 101, 315–328.

Li, F., W. P. Kustas, M. C. Anderson, J. H. Prueger, and R. L. Scott (2008): Effect of remote sensing spatial resolution on interpreting tower-based flux observations, *Remote Sens. Environ.*, 112, 337–349.

Li, R., Q. Min, and B. Lin (2009a): Estimation of evapotranspiration in a mid-latitude forest using the Microwave Emissivity Difference Vegetation Index (EDVI), *Remote Sens. Environ.*, 113, 2011–2018.

Li, Z.-L., R. Tang, Z. Wan, Y. Bi, C. Zou, B. Tang, G. Yan, and X. Zhang (2009b): A review of current methodologies for regional evapotranspiration estimation from remotely sensed data, *Sensors*, 9, 3801–3852.

Liu, R., J. Wen, X. Wang, L. Wang, H. Tian, T. T. Zhang, X. K. Shi., J. H. Zhang, and S. H. N. Lv (2010): Actual daily evapotranspiration estimated from MERIS and AATSR data over the Chinese Loess Plateau, *Hydrol. Earth Syst. Sci.*, 14, 47–58.

Ma, W., Y. Ma, Z. Su, J. Wang, and H. Ishikawa (2011): Estimating surface fluxes over middle and upper streams of the Heihe river basin with ASTER imagery, *Hydrol. Earth Sci.*, 15, 1403–1413.

Mahfouf, J. F., and J. Noilhan (1991): Comparative study of various formulations of evaporation from bare soil using in situ data, *J. Appl. Meteorol.*, 30, 1354–1365.

Margulis, S. A., and D. Entekhabi (2003): Variational assimilation of radiometric surface temperature and reference-level micrometeorology into a model of the atmospheric boundary layer and land surface, *Mon. Weather Rev.*, 131, 1272–1288.

Margulis, S., J. Kim, and T. Hogue (2005): A comparison of the Triangle retrieval and variational data assimilation methods for surface turbulent flux estimation, *J. Hydrometeorol.*, 6, 1063–1072.

Margulis, S. A., D. McLaughlin, D. Entekhabi, and S. Dunne (2002): Land data assimilation and estimation of soil moisture using measurements from the Southern Great Plains 1997 Field Experiment, *Water Resour. Res.*, 38, 1–18.

McCabe, M., and E. F. Wood (2006): Scale influences on the remote estimation of evapotranspiration using multiple satellite sensors, *Remote Sens. Environ.*, 105, 271–285.

Mecikalski, J. R., G. R. Diak, M. C. Anderson, and J. M. Norman (1999): Estimating fluxes on continental scales using remotely-sensed data in an atmospheric-land exchange model, *J. Appl. Meteorol.*, 38, 1352–1369.

Menenti, M., W. Bastiaanssen, D. van Eick, and M. A. Karim (1989): Linear relationships between surface reflectance and temperature and their application to map evaporation of groundwater, *Adv. Space Res.*, 9, 165–176.

Meyers, T. P., and S. E. Hollinger (2004): An assessment of storage terms in the surface energy balance of maize and soybean, *Agric. For. Meteorol.*, 125, 105–115.

Min, Q., and B. Lin (2006): Remote sensing of evapotranspiration and carbon update at Harvard Forest, *Remote Sens. Environ.*, 100, 379–387.

Min, Q., B. Lin, and R. Li (2010): Remote sensing vegetation hydrological states using passive microwave measurements, *IEEE J. Selected Topics Appl. Earth Obs. Remote Sens.*, 3(1), 124–131.

Monteith, J. L. (1963): Gas exchange in plant communities. In Environmental Control of Plant Growth (L. T. Evans, Ed.), Academic Press, New York, pp. 95–112.

Monteith, J. L., and G. Szeicz (1962): Radiative temperature in the heat balance of natural surfaces, *Q. J. R. Meteorol. Soc.*, 88, 496–507.

Moran, M. S., T. R. Clarke, Y. Inoue, and A. Vidal (1994): Estimating crop water deficit using the relation between surface-air temperature and spectral vegetation index, *Remote Sens. Environ.*, 49, 246–263.

Moran, M. S., A. F. Rahman, J. C. Washburne, D. C. Goodrich, M. A. Weltz, and W. P. Kustas (1996): Combining the Penman-Monteith equation with measurements of surface temperature and reflectance to estimate evaporation rates of semi-arid grassland, *Agric. For. Meteorol.*, 80, 87–109.

Moulin, S., A. Bondeau, and R. Delecolle (1998): Combining agricultural crop models and satellite observations: from field to regional scales, *Int. J. Remote Sens.*, 19, 1021–1036.

Noilhan, J., and P. Lacarrere (1995): GCM gridscale evaporation from mesoscale modelling, *J. Climate*, 8, 206–223.

Norman, J. M., M. C. Anderson, W. P. Kustas, A. N. French, J. Mecikalski, R. Torn, G. R. Diak et al. (2003): Remote sensing of surface energy fluxes at 101-m pixel resolutions, *Water Resour. Res.*, 39(8), 1221.

Norman, J. M., W. P. Kustas, and K. S. Humes (1995): A two-source approach for estimating soil and vegetation energy fluxes in observations of directional radiometric surface temperature, *Agric. For. Meteorol.*, 77, 263–293.

Olioso, A. (1992): Simulation des changes d'6nergie et de masse d'un convert v6g&al, dans le but de relier ia transpiration et la photosynthese anx mesures de reflectance et de temp6rature de surface, Doctorate thesis, Universite de Montpellier II, 260 pp.

Olioso, A., H. Chauki, D. Courault, and J.-P. Wigneron (1999): Estimation of evapotranspiration and photosynthesis by assimilation of remote sensing data into SVAT models, *Remote Sens. Environ.*, 68, 341–356.

Olioso, A., Y. Inoue, J. Demarty, J. P. Wigneron, I. Braud, S. Ortega-Farias, P. Lecharpentier et al. (2002): Assimilation of remote sensing data into crop simulation models and SVAT models. In: Sobrino (Ed.), *Proceedings of the 1st International Symposium on Recent Advances in Quantitative Remote Sensing*, Universitat de València, Valencia, France, September 16–18, 2002, pp. 329–338.

Ottle, C., and D. Vidal-Madjar (1994): Assimilation of soil moisture inferred from infrared remote sensing in a hydrological model over the HAPEX-MOBILHY region, *J. Hydrol.*, 158, 241–264.

Owen, T. W., T. N. Carlson, and R. R. Gillies (1998): Remotely sensed surface parameters governing urban climate change, *Int. J. Remote Sens.*, 19, 1663–1681.

Perez, P. J., F. Castellvi, M. Ibanez, and J. I. Rosell (1999): Assessment of reliability of Bowen ratio method for partitioning fluxes, *Agric. For. Meteorol.*, 97, 141–150.

Petropoulos G., and T. N. Carlson (2011): Retrievals of turbulent heat fluxes and soil moisture content by Remote Sensing. In *Advances in Environmental Remote Sensing: Sensors, Algorithms, and Applications*, CRC Press, Boca Raton, FL, 469–501.

Petropoulos, G., T. N. Carlson, M. J. Wooster, and S. Islam (2009a): A review of T_s/VI remote sensing based methods for the retrieval of land surface fluxes and soil surface moisture content, *Adv. Phys. Geogr.*, 33(2), 1–27.

Petropoulos, G., M. J. Wooster, M. Kennedy, T. N. Carlson, and M. Scholze (2009b): A global sensitivity analysis study of the 1d SimSphere SVAT model using the GEM SA software, *Ecol. Modell.*, 220(19), 2427–2440.

Petropoulos, G., M. J. Wooster, T. N. Carlson, and N. Drake (2010): Synergy of the SimSphere land surface process model with ASTER imagery for the retrieval of spatially distributed estimates of surface turbulent heat fluxes and soil moisture content, paper presented at European Geosciences Union 2010, May 2–7, Vienna, Austria.

Petropoulos, G., M. Ratto, and S. Tarantola (2010): A comparative analysis of emulators for the sensitivity analysis of a land surface process model. *6th International Conference on Sensitivity Analysis of Model Output*, July 19–22nd, 2010, Milan, Italy, on Procedia-Social and Behavioral Sciences, vol. 2 (6), 7716–7717, doi:10.1016/j.sbspro.2010.05.194.

Petropoulos G., H. Griffiths, and P. Ioannou-Katidis (2013): Sensitivity Exploration of SimSphere Land Surface Model Towards its Use for Operational Products Development from Earth Observation Data, Chapter 14, 21 pages. To appear in book titled as "Advancement in Remote Sensing for Environmental Applications", edited by S. Mukherjee, M. Gupta, P. K. Srivastava and T. Islam, Springer [accepted].

Price, J. C. (1990): Using spatial context in satellite data to infer regional scale evapotranspiration, *IEEE Trans. Geosci. Remote Sens.*, 28, 940–948.

Priestley, C. H. B., and R. J. Taylor (1972): On the assessment of surface heat flux and evaporation using large scale parameters, *Mon. Weather Rev.*, 100, 81–92.

Pyles, R. D., B. C. Weare, and K. T. Pawu (2000): The UCD Advanced-Canopy-Atmosphere-Soil Algorithm (ACASA): Comparisons with observations from different climate and vegetation regimes, *Q. J. R. Meteorol. Soc.*, 126, 2951–2980.

Roerink, G., Z. Su, and M. Menenti (2000): S-SEBI: A simple remote sensing algorithm to estimate the surface energy balance, *Phys. Chem. Earth, Part B*, 25(2), 147–157.

Rouse, J. W., R. H. Haas, J. A. Schell, and D. W. Deering (1973): Monitoring vegetation systems in the Great Plains with ERTS. In 3rd ERTS Symposium, NASA SP-351 I, pp. 309–317.

Sanchez, J. M., W. P. Kustas, V. Caselles, and M. C. Anderson (2008): Modelling surface energy fluxes over maize using a two-source patch model and radiometric soil and canopy temperature observations, *Remote Sens. Environ.*, 112, 1130–1143.

Sarwar, A., and R. Bill (2007), Mapping evapotranspiration in the Indus Basin using ASTER data, *Int. J. Remote Sens.*, 28(22), 5037–5046.

Schmugge, T. J., W. P. Kustas, J. C. Ritchie, T. J. Jackson, and A. Rango (2002): Remote sensing in hydrology, *Adv. Water Resour.*, 25, 1367–1385.

Sellers, P. J., F. Hall, G. Asrar, D. E. Strebel, and R. E. Murphy (1992): An overview of the First International Satellite Land Surface Climatology project (ISLSCP) Field Experiment (FIFE), *J. Geophys. Res.*, 97, 18,345–18,371.

Sequin, B., and B. Itier (1983): Using midday surface temperature to estimate daily evaporation from satellite thermal IR data, *Intern. J. Rem. Sens.*, 4, 371–383

Shu, Y., S. Stisen, K. H. Jensen, and I. Sandholt (2011): Estimation of regional evapotranspiration over the North China Plain using geostationary data, *Int. J. Appl. Earth Obs. Geoinf.*, 13, 192–206.

Shuttleworth, W., and R. Gurney (1990): The theoretical relationship between foliage temperature and canopy resistance in sparse crops, *Q. J. R. Meteorol. Soc.*, 116, 497–519.

Shuttleworth, W., and J. Wallace (1985): Evaporation from sparse crops: an energy combination theory, *Q. J. R. Meteorol. Soc.*, 111, 1143–1162.

Shuttleworth, W. J., R. J. Gurney, A. Y. Hsu, and J. P. Ormsby (1989): FIFE: The variation in energy partition at surface flux sites. IAHS Publ., 186, 67–74.

Smith, E. A., A. Y. Hsu, W. L. Crosson, R. T. Field, L. J. Fritchen, R. J. Gurney, and E. Kanemasu et al. (1992): Area-averaged surface fluxes and their time-space variability over the FIFE experimental domain, *J. Geophys. Res.*, 97, 18,599–18,622.

Sobrino, J. A., M. Gómez, J. C. Jiménez-Muñoz, and A. Olioso (2007): Application of a simple algorithm to estimate daily evapotranspiration from NOAA–AVHRR images for the Iberian Peninsula, *Remote Sens. Environ.*, 110(2), 139–148.

Soer, G. J. R. (1980): Estimation of regional evapotranspiration and soil moisture conditions using remotely sensed crop surface temperature, *Remote Sens. Environ.*, 9, 27–45.

Sone, L., and M. Horton (1974): Estimating evapotranspiration using canopy temperatures: Field evaluation, *Agron J.*, 66, 450–454.

Stisen, S., I. Sandholt, A. Norgaard, R. Fensholt, and K. H. Jensen (2008): Combining the triangle method with thermal inertia to estimate regional evapotranspiration—Applied to MSG SEVIRI data in the Senegal River basin, *Remote Sens. Environ.*, 112, 1242–1255.

Su, H., M. F. McCabe, E. F. Wood, Z. Su, and J. H. Prueger (2005): Modeling evapotranspiration during SMACEX: Comparing two approaches local- and regional-scale prediction, *J. Hydrometeorol.*, 6(6), 910–922.

Su, Z. (2002): The Surface Energy Balance System (SEBS) for estimation of turbulent heat fluxes, *Hydrol. Earth Syst. Sci.*, 6(1), 85–89.

Sugita, M., and M. Brutsaert (1991): Daily evaporation over a region from lower boundary layer profiles, *Water. Resour. Res.*, 27, 747–752.

Sun, Z., B. Wei, W. Su, W. Shen, C. Wang, D. You, and Z. Liu (2011): Evapotranspiration estimation based on the SEBAL model in the Nansi Lake Wetland of China, *Math. Comput. Modell.*, 54(3–4), 1086–1092.

Talagrand, O. (1997): Assimilation of observations, an introduction, *J. Meteorol. Soc.*, 75, 191–209.

Tan, C.-H. (1998): Regional scale evapotranspiration estimation using vegetation index and surface temperature from NOAA AVHRR satellite data, PhD thesis, University of Florida, Tallahassee, p. 244.

Tan, R., Z.-L. Li, and B. Tang (2010): An application of the T_s/VI triangle method with enhanced edges determination for evapotranspiration estimation from MODIS data in arid and semi-arid regions: implementation and validation, *Remote Sens. Environ.*, 114, 540–551.

Tsouni, A., C. Kontoes, D. Koutsoyiannis, P. Elias, and N. Mamassis (2008): Estimation of actual evapotranspiration by remote sensing: Application in Thessaly Plain, Greece, *Sensors*, 8, 3586–3600.

Van de Griend, A., P. J. Camil]o, and R. J. Gurney (1985): Discrimination of soil physical parameters, thermal inertia and soil moisture from diurnal surface temperature fluctuations, *Water Resour. Res.*, 21(7), 997–1009.

Venturini, V., G. Bisht, S. Islam, and L. Jiang (2004): Comparison of evaporative fractions estimated from AVHRR and MODIS sensors over South Florida, *Remote Sens. Environ.*, 93, 77–86.

Verma, S., N. Rosenburg, B. Blad, and M. Baradas (1976): Resistance energy balance method for predicting evapotranspiration, determination of boundary layer resistance, and evaluation of error effects, *Agric J.*, 68, 776–782.

Verstraeten, W. W., F. Veroustraete, and J. Feyen (2008): Assessment of evapotranspiration and soil moisture content across different scales of observation, *Sensors*, 8(1), 70–117.

Vidal, A., and A. Perrier (1989): Analysis of a simplified relation for estimating daily evapotranspiration from satellite thermal IR data, *Int. J. Remote Sens.*, 10(8), 1327–1337.

Wang, K., Z. Li, and M. Cribb (2006): Estimation of evaporative fraction form a combination of day and night land surface temperatures and NDVI: A new method to determine the Priestley-Taylor parameter, *Remote Sens. Environ.*, 102, 293–305.

Wilson, K. B., P. J. Hanson, P. J. Mulholland, D. D. Baldocchi, and S. D. Wullschleger (2001): A comparison of methods for determining forest evapotranspiration and its components: Sapflow, soil water budget, eddy covariance and catchment water balance, *Agric. For. Meteorol.*, 106, 153–168.

Xie, X. Q. (1999): Estimation of daily evapotranspiration from one-time of day remotely sensed canopy temperature, *Remote Sens. Environ. China*, 6, 253–259.

Yang, H., Z. Cong, Z. Liu, and Z. Lei (2010): Estimating sub-pixel temperatures using the triangle algorithm, *Int. J. Remote Sens.*, 31(23), 6-47-6060.

Yunhao, C., L. Xiaobing, J. Guiffei, and S. Peijun (2003): An estimation model for daily regional evapotranspiration, *Int. J. Remote Sens.*, 24(1), 199–205.

Zahira, S., H. Abderrahmane, K. Mederbal, and D. Frederic (2009): Mapping latent heat flux in the western forest covered regions of Algeria using remote sensing data and a spatialised model, *Remote Sens.*, 1, 795–817.

Zhang, L., and R. Lemeur (1995): Evaluation of daily evapotranspiration estimates from instantaneous measurements, *Agric. For. Meteorol.*, 74, 139–154.

Zhang, Y., C. Liu, Y. Lei, Y. Tang, Q. Yu, Y. Shen, and H. Sun (2006): An integrated algorithm for estimating regional latent heat flux and daily evapotranspiration, *Int. J. Remote Sens.*, 27(1), 129–152.

4

Satellite Remote Sensing of Surface Soil Moisture

Brian W. Barrett and George P. Petropoulos

CONTENTS

4.1 Introduction

Soil is arguably the Earth's most valuable nonrenewable resource and undoubtedly the most biologically diverse part of the biosphere. Roughly half of a soil's volume is composed of mineral and organic content, while the other half consists of pores. Soil moisture (or water) content (SMC) refers to the amount of water in these pores and generally refers to the water contained in the unsaturated soil zone (e.g., Hillel 1998). SMC is affected by the soil texture (determines water holding capacity), topography (affects runoff and infiltration), land cover (influences evapotranspiration), and climate (precipitation, wind, humidity, and solar illumination), and, as a result, SMC is highly variable both in space and time.

Although soil moisture comprises only a tiny percentage (~0.001%) of the total global water budget, its importance and influence in the hydrological cycle cannot be understated. It is a key parameter in the exchange of mass and energy at the land surface–atmosphere

TABLE 4.1

Summary of Different Methods of SMC Retrieval by Remote Sensing

Group	Methods	Advantages	Disadvantages	Example Studies
Optical	Reflectance-based methods	Good spatial resolution, many satellites available to use, hyperspectral sensors promising, based on mature technology	Weak relationship to SMC when > vegetation cover, not applied during cloudy conditions and nighttime, poor temporal resolution	Liu et al. 2002, 2003; Lobell and Asner 2002
	TIR-based methods	Good spatial resolution, many satellites available, methods relating SMC to thermal inertia show promise	Weak relationship to SMC when > vegetation cover, not able to be applied during cloudy conditions, poor temporal resolution, SMC retrievals sensitive to Earth's atmosphere	Verstraeten et al. 2006; Minacapilli et al. 2009; Lu et al. 2009; Carlson 1986
MW passive	Various methods proposed (e.g., statistical, neural networks, and model inversion)	Promising results in SMC estimation particularly over bare soil surfaces, use not limited by clouds, high temporal resolution	Coarse spatial resolution, SMC retrieval influenced by vegetation cover and surface roughness	Njoku et al. 2003; Draper et al. 2009; Albergel et al. 2011, 2012; Champagne et al. 2011; Li et al. 2010; Sanchez et al. 2012; Bircher et al. 2012
MW active	SAR-based methods (empirical, semiempirical, physically based)	Fine spatial resolution, use not limited by clouds and/or nighttime conditions	SMC accuracy influenced by surface roughness and vegetation cover amount, poor temporal resolution (but this will improve with the greater number of sensors to come available)	Oh 2004; Holah et al. 2005; Baghdadi et al. 2006; Paloscia et al. 2008; Thoma et al. 2008
	Scatterometer	High temporal resolution, high sensitivity to SMC	Coarse spatial resolution, signal also influenced by surface roughness and vegetation	Bartalis et al. 2007; Naeimi et al. 2009b; Albergel et al. 2009; Brocca et al. 2010, 2011
	Altimeter	Attenuation by the vegetation layer is minimized by the nadir-looking configuration	Fatras et al. (2012) is the first demonstration of the potential of radar altimetry to derive surface soil moisture. Further studies are needed	Fatras et al. 2012

(continued)

TABLE 4.1 (Continued)

Summary of Different Methods of SMC Retrieval by Remote Sensing

Group	Methods	Advantages	Disadvantages	Example Studies
Synergistic methods	Optical and TIR	High spatial resolution, wide selection of satellite sensors to choose from, simple and straightforward implementation, based on mature technology	Mostly methods are of empirical nature (transferability difficult), limited to cloud-free and daytime conditions, poor temporal resolution, low penetration depth	Ghulam et al. 2007; Wang et al. 2007; Petropoulos et al. 2010; Petropoulos and Carlson 2011; Sobrino et al. 2012
	Active and passive MW	Link the high spatial resolution of active MW systems to the coarser resolution passive MW sensors, better temporal resolution and improved SMC retrieval	SMC scaling and validation issues	Chauhan 1997; Lee and Anagnostou 2004; Narayan et al. 2006; Li et al. 2011; Liu et al. 2011, 2012
	MW and optical	Vegetation and surface roughness effects can be minimized	SMC scaling and validation issues, different SMC measurement depths	Chauhan et al. 2003; Wang et al. 2004; Kuruku et al. 2009; Choi and Hur 2012

boundary, controlling the partitioning of incoming radiant energy into latent and sensible heat fluxes through evaporation and transpiration by plants (Delworth and Manabe 1989; Entekhabi et al. 1996). Moreover, it determines the fate of precipitation as surface runoff or infiltration, and, as a result, it has an important control in the geochemical cycling of nutrients (Raich and Schlesinger 1992), availability of transpirable water for plants, and groundwater recharge. In addition, SMC has a strong influence on the feedback between the land surface and climate, affecting the dynamics of the atmosphere boundary layer and thereby having a direct relationship to weather and global climate (Patel et al. 2009; Seneviratne et al. 2010).

More recently, the prominence of climate change issues in societal thinking and policy decision making has heightened awareness of our carbon footprint. As soil can act as both a major source and sink for atmospheric carbon dioxide (CO_2), soil moisture along with temperature and the organic matter content in the soil are the major limiting factors in the amount and transport of this carbon within the environment. A very warm dry soil will have decreased microbial activity, whereas a moist soil will have increased activity, thereby increasing respiration in the soil and emission of CO_2 (Orchard and Cook 1983). Within this context, the influence of soil moisture in the hydrological, carbon, and energy cycles is gaining increasing attention. Consequently, accurate estimation of SMC is crucial for planning and management of water resources, particularly in ecosystems in which water is a limiting resource, such as in many arid and semiarid regions (Mariotto and Gutschick 2010).

The large spatial and temporal heterogeneity of soil moisture, however, makes it a difficult parameter to measure on a routine basis over large areas. Many techniques and instruments have been developed—and continue to evolve and improve—to measure the moisture content of soils. A comprehensive overview of the methods and advances in soil moisture measurement using conventional approaches can be found in Chapter 2 and elsewhere (e.g., Gardner 1986; Topp and Ferré 2002; Robinson et al. 2003; Evett and Parkin 2005; Robinson et al. 2008). Spaceborne remote sensing has the capability to monitor soil moisture over large areas at regular time intervals, and several approaches for soil moisture retrieval have been developed using optical, thermal infrared (TIR), and microwave (MW) sensors over the last three decades. The principal advantage of remote sensing, as opposed to conventional measurement methods, is that spatially distributed and frequent (ranging from bi-daily to monthly) observations of soil moisture can be made rapidly over a large area. The latter is rather important as examining the same area of study at multiple scales enables the investigation of some of the previously insoluble problems, such as that of spatial variability and scale of observation.

This chapter provides a synthesis of the efforts made toward the retrieval of soil moisture from remote sensing data acquired from all possible regions of the electromagnetic radiation (EMR) spectrum. This is followed by an overview of the different types of remote sensing operational products providing estimates of SMC at a global scale. A summary of the methods reviewed herein, along with their relative strengths and limitations, is made available in Table 4.1.

4.2 Optical Remote Sensing of Soil Moisture

Remote sensing of soil moisture in the optical domain utilizes wavelengths between 0.4 and 2.5 μm and measures the reflected radiation of the Sun from the Earth's surface. Examples of commonly used spaceborne optical sensors used for estimating SMC can be found in Chapter 3 (Table 3.1). A soil's reflectance is determined by the inherent scattering and absorption properties of its various components (mineral, air, water, and organic content) and the way they are arranged within the soil (Liu et al. 2002). Given that water absorbs energy, a soil with higher SMC will, in theory, have less reflective intensity than soils containing less moisture. Various studies have explored the relationship between spectral reflectance and SMC (e.g., Bowers and Hanks 1965; Stoner and Baumgardner 1981; Lobell and Asner 2002; Liu et al. 2003) and have demonstrated that soil reflectance decreases with increasing SMC, although Liu et al. (2002) found that after a critical point, soil reflectance can actually increase with soil moisture.

Most of the approaches proposed have been based around the rationale of developing an empirical spectral vegetation index (VI) that can indicate vegetation spectral properties (e.g., growth, amount, stress) and degree of vegetation moisture stress, through measures of vegetation health and photosynthesis and thereby allow indirect estimates of SMC even when the soil surface is not visible. Commonly used VIs include the normalized difference vegetation index (NDVI), enhanced vegetation index (EVI) (Huete et al. 1994), normalized difference water index (NDWI) (Gao 1996), and the normalized multiband drought index (Wang and Qu 2007). Various studies have evaluated the effects of soil moisture on the spectral characteristics of vegetation as well as the relationships between soil moisture and a variety of VIs at multiple spatial and spectral scales (Gu et al. 2008; Wang et al. 2008; Liu et al. 2011; Farrar et al. 1994; Liu et al. 2012). Recent work has explored the development of hyperspectral sensors in surface SMC retrieval. The narrowband spectral information acquired in the visible, near-infrared (NIR) and shortwave infrared (SWIR) wavelengths permits material identification as a function of their spectral absorption features. The development of hyperspectral sensors has resulted in the evolution of VIs that are specifically designed to take advantage of the finer spectral resolution and more detailed spectral information of the imaged vegetation [e.g., moisture stress index (MSI; Hunt and Rock 1989), chlorophyll absorption in reflectance index (MCARI; Daughtry et al. 2000), disease-water stress index (DWSI; Apan et al. 2004)].

Generally, reflectance-based approaches are capable of providing estimates of SMC at high spatial and temporal resolutions exploiting satellite observations of a mature technology. Yet, they are unable to penetrate cloud cover and vegetation canopies and are also highly attenuated by the Earth's atmosphere. Moreover, the influence of various "noise" elements (e.g., organic matter, texture, and color) can confuse the interpretation of the measured soil reflectance (Ben-Dor et al. 1999) and limit the effectiveness of reflectance-based approaches for SMC determination. Because of these noise controls, efforts to directly relate soil reflectance to moisture have achieved success only when models are fit for specific soil types in the absence of vegetation cover (e.g., Muller and Décamps 2000).

4.3 TIR Remote Sensing of Soil Moisture

Remote sensing of soil moisture in the TIR domain (wavelengths ranging between 3.5 and 14 μm) is based on the relationship between the thermal properties of soil, namely, the heat capacity and thermal conductivity, and the SMC (see Table 3.1 in Chapter 3 for examples of TIR sensors). In summary, these methods involve the use of the thermal inertia P (J m^{-2} K^{-1} s$^{-0.5}$), which describes the impedance of soil to temperature variations and is defined as (Minacapilli et al. 2009):

$$P = \sqrt{\lambda \rho C} \tag{4.1}$$

where λ (W m^{-1} K^{-1}) is the soil thermal conductivity, ρ (kg m^{-3}) is the soil bulk density, and C (J kg^{-1} K^{-1}) is the soil heat capacity that can be expressed as:

$$C = \frac{\rho_b}{\rho_s} C_s + m_v C_w \tag{4.2}$$

where ρ_b (kg m^{-3}) is the dry bulk density of soil, ρ_s is the density of the solid phase (~2650 kg m^{-3}), m_v is the volumetric soil moisture (m^3 m^{-3}), and C_s and C_w (J kg^{-1} K^{-1}) are the heat capacities of the solid and liquid phases, respectively.

The thermal inertia approach (Price 1977) is the most commonly used method for soil moisture retrieval using TIR observations (Pratt and Ellyett 1979; Verstraeten et al. 2006). This approach generally involves two steps: measuring thermal inertia from thermal remote sensing observations and subsequently retrieving SMC from models describing thermal inertia as a function of soil moisture (e.g., Kahle 1977; Xue and Cracknell 1995; Cai et al. 2007; Lu et al. 2009). It is based on the fact that water bodies have a higher thermal inertia than dry soils and exhibit a lower diurnal temperature fluctuation. As SMC increases, thermal inertia proportionally increases and the diurnal temperature fluctuation range is reduced. Therefore, by measuring the amplitude of the diurnal temperature change, one can develop a relationship between the temperature change and SMC. However, the relationship between diurnal temperature and SMC is also a function of soil type and is largely limited to bare soil conditions, given the low penetrating ability of the thermal region of the EMR spectrum.

Provided that atmospheric and meteorological conditions have been accounted for, surface temperature (T_s) is primarily dependent on the thermal inertia of the soil (i.e., soil surface temperatures are used to infer the spatial variability in the thermal inertia of the soil). However, if the vegetation cover obscures more than ~20% of the soil surface, the imaged scene may have little relation to the radiation properties of the actual soil surface and instead provide measurements of the thermal properties of the vegetation. As a result, the thermal inertia approach for estimating SMC is difficult to apply to large-scale soil moisture monitoring.

A different approach to deriving estimates of surface soil SMC is based on measurements in the thermal bands of the EMR spectrum being used in combination with VIs derived from visible and infrared wavebands (Friedl and Davis 1994). The strong negative relationship between land surface temperature and the vegetation cover fraction (represented by VIs) has been widely used to study the moisture content of land surfaces (Goward et al. 1985; Nemani and Running 1989; Sandholt et al. 2002). Generally speaking, such methods suffer from all the characteristics of an empirically derived methodology (e.g., luck of transferability to other regions, fine-tuning, and weakness to describe physical processes). Nevertheless, some of the techniques proposed have been based particularly on the combination of T_s and VI and have shown promise in being able to provide significant information in SMC distribution, particularly over partially or fully vegetated areas. More detailed reference to this group of approaches is provided later herein (Section 4.5.1).

4.4 MW Remote Sensing of Soil Moisture

By using the MW portion of the EMR spectrum, distinct capabilities can be gained over the optical and TIR regions and greater insight into surface processes can be achieved. MW remote sensing uses the MW and radio part of the EMR spectrum, with frequencies between 0.3 and 300 GHz—corresponding to wavelengths between 1 m and 1 mm. Owing to their longer wavelengths, compared to visible and infrared radiation, MWs are largely unaffected by cloud cover, haze, rainfall, and aerosols and so are not as susceptible to atmospheric scattering, which affects the shorter wavelengths (Engman 1990).

MW remote sensing encompasses both active and passive forms, depending on the mode of operation. Active techniques provide their own source of illumination to measure the intensity of EMR that is scattered back from the surface to the sensor. In contrast, passive MW sensors measure the naturally emitting radiation from the land surface. The theory behind (both active and passive) MW remote sensing of soil moisture is based on the large contrast between the dielectric properties of liquid water ($\varepsilon \approx 80$) and dry soil ($\varepsilon \approx 4$), which results in a high dependency of the complex dielectric constant on volumetric soil moisture (m_v). By measuring the strength of the MW signal, the amount of water in the soil can be determined through the soil dielectric properties. However, interpreting the MW signal received from a soil surface and subsequently determining the signal part that is actually from the soil water content is a difficult and generally challenging task. This is due to the

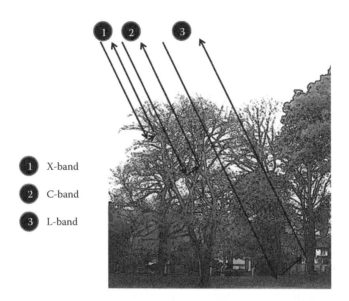

FIGURE 4.1
(See color insert.) Vegetation influence on backscatter at three different wavelengths.

number of sensor (wavelength, polarization, and incidence angle) and surface characteristics that influence the received signal. The dominant surface parameters are the vegetation cover and surface roughness. For example, in the case of radar (an active sensor) it can be seen from Figure 4.1 that the effect of vegetation on backscatter generally decreases with increasing wavelength. The shorter wavelength MW signal (X-band, ~3 cm) interacts mainly with the top of the canopy cover, while longer wavelengths (L-band, ~24 cm) are able to penetrate further into the canopy and reflect from the soil surface. Similarly, the influence of surface roughness is another major limiting factor in the retrieval of soil moisture and is discussed in more detail in Chapter 13.

4.4.1 Passive MW Techniques

All physical objects with a temperature above absolute zero (0 K or –273°C) emit (MW) energy of some magnitude. Passive MW sensors (radiometers) detect this naturally emitting MW radiation within their field of view and operate in a similar manner to thermal sensors. (Recall that a thermal sensor measures the emanating EMR [wavelength $(\lambda) \approx$ 3–14 μm] from the Earth's surface, whereas conventional optical remote sensors are concerned with the reflective properties of the surface or objects.) The emitted MW radiation is usually expressed in terms of brightness temperature, where the brightness temperature emitted by the Earth's surface (T_B) is determined by its physical temperature T and its emissivity e_p according to:

$$T_B = e_p T \tag{4.3}$$

where the subscript p denotes the signal polarization (vertical or horizontal). For smooth soil surfaces, the soil emissivity can be approximated from the reflectivity $\Gamma_{s,p}$ of the surface:

$$e_p = 1 - \Gamma_{s,p} \tag{4.4}$$

where the surface reflectivity coefficients at vertical and horizontal polarizations can be predicted from the Fresnel reflection equations ($\Gamma_{s,p} \approx \Gamma_{0,p}$) as:

$$\Gamma_{ov} = \left| \frac{\varepsilon_s \cos\theta - \sqrt{\varepsilon_s - \sin^2\theta}}{\varepsilon_s \cos\theta + \sqrt{\varepsilon_s - \sin^2\theta}} \right|^2 \tag{4.5}$$

$$\Gamma_{oh} = \left| \frac{\cos\theta - \sqrt{\varepsilon_s - \sin^2\theta}}{\cos\theta + \sqrt{\varepsilon_s - \sin^2\theta}} \right|^2 \tag{4.6}$$

where θ is the incidence angle and ε_s is the complex dielectric constant of soil. See Wigneron et al. (2003) for a comprehensive overview of the physical basis of the MW emission from bare soil and vegetated surfaces.

The first spaceborne low-frequency (L-band, 1.4 GHz) radiometer utilized for soil moisture observations was the Skylab S-194 instrument launched in 1973 and in operation until 1977 (Lerner and Hollinger 1977). Since then, a range of sensors (at mostly C-band) have been in operation to provide soil moisture measurements on a global scale (see Table 4.2 for a list of the major spaceborne radiometers). All of these instruments are typically characterized by broad spatial coverage and high temporal resolution but coarse spatial resolutions (25–50 km) (in order to detect the low quantities of emitted MW radiation, the field of view for passive sensors must be large enough to detect sufficient energy to record a signal, resulting in a low spatial resolution). As a result, the use of passive MW observations is more suitable for global-scale studies rather than regional or watershed analysis.

Although theoretical and experimental studies have shown that L-band radiometers are optimal for soil moisture sensing, it was not until recently (with the launch of SMOS in November 2009) that an L-band radiometer on an orbiting platform has been available since the Skylab mission in the 1970s. This was due to the significant challenge of building an L-band antenna large enough to provide reasonable spatial resolution. SMOS uses a novel two-dimensional L-band interferometric radiometer (Microwave Imaging Radiometer with Aperture Synthesis, MIRAS) and can achieve global coverage every 3 days. Figure 4.2 illustrates an example of a global monthly composite of SMOS retrieved soil moisture values for September 2012, where lowest values are shown in blue and highest values in red.

The fundamental relationship between emissivity and surface soil moisture was demonstrated through a number of large-scale field experiments using ground-based or aircraft-mounted radiometers [e.g., FIFE 1987 (Wang et al. 1990), Monsoon 90, Washita'92 (Jackson et al. 1995), Southern Great Plains 1997 (Jackson et al. 1999)]. Although the early spaceborne MW radiometers had poor resolution (e.g., Skylab S-194) or short wavelengths (e.g., SMMR and SSM/I), their capability to retrieve soil moisture from a satellite platform was demonstrated (Eagleman and Lin 1976; Jackson and Schmugge 1989). However, difficulties arise in the inversion of soil moisture from brightness temperatures owing to the influence of surface roughness, vegetation cover, and soil texture. In order to retrieve soil moisture from brightness observations, corrections must be made for these perturbing effects. The degree to which these corrections must be made can be reduced significantly by using low-frequency observations in the 1–5 GHz range, owing to the greater vegetation penetration and reduced atmospheric attenuation at longer wavelengths (Njoku and Entekhabi 1996; Jackson and Schmugge 1989). Models of varying complexity were developed, for

TABLE 4.2

Characteristics of Major Spaceborne Scatterometers and Radiometers

Platform	Sensor	Frequency (GHz)	Polarization	Highest Spatial Resolution (km)	Swath Width (km)	Mission
Scatterometers						
ERS-1	SCAT	5.3	VV	25	500	Jul 1991 to Mar 2000
ERS-2	SCAT	5.3	VV			Apr 1995 to Sep 2011
MetOp-A	ASCAT	5.3	VV	25	1100	Oct 2006–present
MetOp-B	ASCAT	5.3	VV			Sept 2012–present
MetOp-C	ASCAT	5.3	VV			
Radiometers						
Nimbus-7	SMMR	6.6	H, V	148 × 95	780	Nov 1978 to Aug 1987
		10.7		91 × 59		
		18		55 × 41		
		37		27 × 18		
DMSP	SSM/I	19.3	H, V	69 × 43	1394	Jul 1987–present
		36.5		37 × 28		
TRMM	TMI	10.7	H, V	63 × 39	878	Dec 1997–present
		19.4		30 × 18		
		37		16 × 10		
AQUA	AMSR-E	6.9	H, V	74 × 43	1445	May 2002 to Oct 2011
		10.7		51 × 30		
		18.7		27 × 16		
		36.5		14 × 8		
Coriolis	Windsat	6.8	H, V	40 × 60	1025	Jan 2003–present
		10.7		25 × 38		
		37		8 × 13		
SMOS	MIRAS	1.4	H, V	~35	1000	Nov 2009–present
GCOM	AMSR2	6.9	H, V	35 × 62	1450	May 2012–present
		10.7		24 × 42		
		18.7		14 × 22		
		36.5		7 × 12		
SMAP	Radiometer	1.26	H, V	~36	1000	2014
		1.41				

example, Eagleman and Lin (1976) used simple linear regression to establish the relationship between Skylab-S194 brightness temperature and soil moisture, while later, models of increasing intricacy (e.g., Choudhury et al. 1979; Mo et al. 1982; Owe et al. 2001; Jackson et al. 1982) were developed to account for the surface roughness and vegetation effects on the MW signal. For example, the MW radiation emitted from a vegetation canopy* can be confused with that from the desired soil emission, increasing the emissivity of the surface and hence the MW brightness temperature (i.e., making the surface appear dryer than it really is). Many of these models are based on radiative transfer theory, deriving volumetric soil moisture from the soil emissivity by inverting the Fresnel relationships and a dielectric mixing model (Wigneron et al. 2003). For example, many studies have used the Land Parameter Retrieval Model (LPRM) (Owe et al. 2008) to derive soil moisture from the brightness temperatures (e.g., de Jeu et al. 2008; Champagne et al. 2011). For more detailed

---

* Vegetation will also absorb or scatter radiation emanating from the soil so the effects of vegetation on the microwave emission may be considered as two-fold.

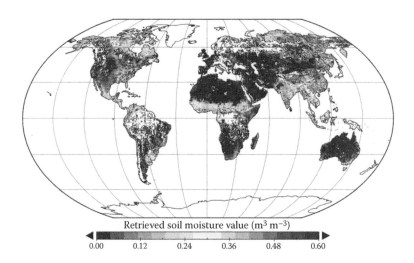

FIGURE 4.2
(See color insert.) Global monthly average of the SMOS (ascending orbit) soil moisture values for September 2012. Data obtained from the Centre Aval de Traitement des Données SMOS (CATDS), operated for the Centre National d'Etudes Spatiales (CNES, France) by IFREMER (Brest, France).

information on the different algorithms used in passive MW techniques for soil moisture retrieval, the reader is referred to Wigneron et al. (2003).

As noted, inversion of soil moisture from passive MW observations is extremely complex, as the MW emission from soils depends not only on its moisture content but also on other surface characteristics such as soil type, temperature, and vegetation cover. Developing global passive MW soil moisture products therefore is a significant challenge. Among the products, problems persist with, for example, the unknown impacts of undetected radio frequency interference (RFI) (Oliva et al. 2012) and land cover heterogeneity (Loew 2008). RFI can be a serious problem for passive (and active) MW sensing of the Earth. Anthropogenic RFI generally originates from TV transmitters, radio links, and terrestrial radars and badly contaminates satellite observations. The problem is more serious for passive sensors, because the relatively weak signal emitted by the Earth can be easily obscured by strong RFI signals that add unpredictable noise to the measurements (Njoku et al. 2005; Castro et al. 2012). Consequently, many passive MW sensors are designed to operate within protected frequency bands allocated for Earth Exploration Satellites, space research, and radio astronomy. For example, SMOS operates within the protected 1400–1427 MHz band. However, Figure 4.3a displays SMOS data for September 2012 where there are many incidences of illegal RFI signals, particularly in southern and eastern Europe, Asia, and the Middle East. Compare this with Figure 4.3b depicting suspected RFI in January 2010 and it can be seen that the influence of RFI has not been drastically reduced within this time period despite the efforts of the European Space Agency and National Frequency Protection entities to reduce or switch off these emitters. The consequence of these illegal emissions is that no reliable estimates of soil moisture are possible in the affected areas.

4.4.2 Active MW Techniques

Active MW sensors provide their own source of illumination and measure the difference in power between the transmitted and received EMR. Active sensors can be divided

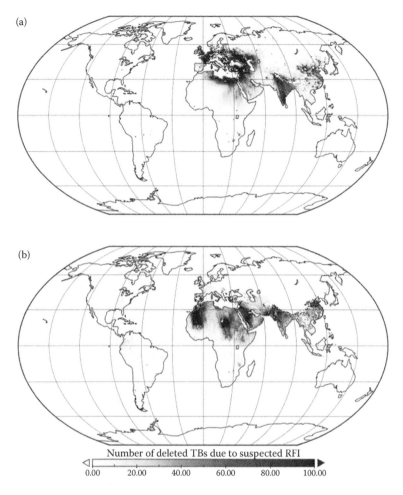

FIGURE 4.3
(See color insert.) Global monthly count of deleted SMOS brightness temperatures (ascending orbit) due to suspected RFI for (a) September 2012 and (b) January 2010. Areas of highest RFI pollution are depicted in dark red. Data obtained from the Centre Aval de Traitement des Données SMOS (CATDS), operated for the Centre National d'Etudes Spatiales (CNES, France) by IFREMER (Brest, France).

into two distinct categories: imaging radars [rear aperture radars and synthetic aperture radars (SARs)] and nonimaging radars (altimeters and scatterometers). Scatterometers are generally used to obtain information on wind speed and direction over ocean surfaces, although numerous applications to soil moisture measurement have been used in the past. Altimeters are primarily used to determine height measurements, traditionally over the oceans and cryosphere, although applications in geodesy, hydrology, and the atmospheric sciences have also been explored.

4.4.2.1 Scatterometers

Scatterometers measure the normalized radar cross section with high radiometric accuracy and relatively coarse spatial resolution (~50 km). Over bare or sparsely vegetated land surfaces, much of the variability of the received backscatter measurements is the result of changes in the soil dielectric constant. Table 4.2 provides a list of major spaceborne scatterometers. Despite being designed to measure wind speed and direction over the sea surface, scatterometers onboard the ERS-1, ERS-2, and the METeorological OPerational (METOP) satellites have been shown to be useful for soil moisture retrieval (e.g., Pulliainen et al. 1998; Zine et al. 2005). These approaches typically used inversion methods based on physical backscattering models and limitations therefore exist in their validity over natural surfaces. The TU-Wien change detection model (Wagner et al. 1999a,b) derives relative changes in surface soil moisture and indirectly accounts for surface roughness and land cover. Several studies have shown this technique to overcome some of the limitations of the physical-based inversion models when sufficient long-term measurements are available (Bartalis et al. 2007; Naeimi et al. 2009a). Wagner et al. (2003) used this method to produce the first global soil moisture dataset using ERS 1/2 scatterometer measurements. This dataset is available since 2002 from the Vienna University of Technology Web site (http://www.ipf.tuwien.ac.at/radar), and further information on how it is computed is given in Section 4.6.2 herein.

4.4.2.2 Synthetic Aperture Radar

The increasing number of SAR satellites now available (at multiple frequencies and polarizations) with higher spatial resolutions and revisit times offers greater potential than ever to improve the quality with which surface soil moisture can be retrieved from SAR data (see Table 4.3 for a list of major spaceborne SAR sensors). All (satellite) SAR sensors previous to 2002 were capable only of single configuration observations where the influence of surface roughness and vegetation on the backscatter could not be easily differentiated from that of soil moisture. The launch of ENVISAT marked the beginning of the trend toward multiconfiguration sensors, with the Advanced SAR (ASAR) featuring greater capability in terms of coverage, range of incidence angles, polarization, and modes of operation than any of its predecessors. Many different approaches have been developed to retrieve surface soil moisture using various configurations of SAR measurements (see Barrett et al. (2009) for a comprehensive overview). In order to take account of the various sensor configurations and surface parameters, many backscattering models have been developed over the past 30 years to help define the relationship between the radar signal and certain biophysical parameters. These models are generally categorized into three groups: theoretical (or physical), empirical, and semiempirical models and are discussed in more detail next.

4.4.2.2.1 Soil Moisture Retrieval Using Theoretical Scattering Models

Theoretical or physical models simulate the radar backscatter in terms of soil attributes such as dielectric constant and surface roughness [root-mean-square (rms) height and correlation length] for an area with known characteristics. The standard theoretical backscattering models are the Kirchhoff approximation (KA), which consists of the geometrical optics model (GOM), the physical optics model (POM), and the small perturbation model (SPM). Generally, these models can be applied only in the case of specific roughness conditions where typically GOM is best suited for very rough surfaces, POM for surfaces with intermediate roughness, and SPM for very smooth surfaces. One widely used such model is the integral equation model (IEM), originally developed by Fung et al. (1992). This is a physical-based radiative transfer model that unites the Kirchhoff models and the SPM to create a method that was applicable to a wider range of roughness conditions, and in theory, not limited to any one location. Owing to the complexity of the IEM, many approximate solutions and improvements to the original model have been developed (Chen et al.

TABLE 4.3

Characteristics of Major Spaceborne SAR Systems

Platform	Sensor	Frequency (GHz)	Polarization	Highest Spatial Resolution	Swath Width (km)	Mission
SEASAT	SAR	1.27	HH	25 m	100	Jun–Oct 1978
SIR-A	SAR	1.27	HH	40 m	50	Nov 1981
SIR-B	SAR	1.27	HH	25 m	30	Oct 1984
Almaz-1	SAR	3.1	HH	13 m	172	Mar 1991 to Oct 1992
ERS-1	AMI	5.3	VV	30 m	100	Jul 1991 to Mar 2000
JERS-1	SAR	1.27	HH	18 m	75	Feb 1992 to Oct 1998
SIR-C/X-SAR	SIR-C X-SAR	1.27, 5.3, 9.6	VV, HH, HV, VH, HH	30 m	10–200	Apr 1994 Oct 1994
ERS-2	AMI	5.3	VV	30 m	100	Apr 1995 to Sep 2011
RADARSAT-1	SAR	5.3	HH	10 m	100–170	Nov 1995–
SRTM	C-SAR X-SAR	5.3, 9.6	VV, HH HH	30 m	50	Feb 2000
ENVISAT	ASAR	5.3	VV, HH, HH/VV HV/HH, VH/VV	30 m	100–400	Mar 2002 to Apr 2012
ALOS	PALSAR	1.27	Quad-pol	10 m	70	Jan 2006–May 2011
TerraSAR-X	X-SAR	9.6	Quad-pol	1 m	10–100	Jun 2007–
RADARSAT-2	SAR	5.3	Quad-pol	3 m	10–500	Dec 2007–
COSMO/ SkyMed Series	SAR-2000	9.6	Quad-pol	1 m	10–200	Jun and Dec 2007, Oct 2008, Nov 2010
TecSAR	SAR	9.6	HH, HV, VH, VV	1 m	40–100	Jan 2008
SAR-Lupe	SAR	9.6	–	<1 m	–	Dec 2006, Jul and Nov 2007, Mar and Jul 2008, Jul 2008–
Kondor-5	SAR	3.3	HH, VV	1 m	–	2009
TanDEM-X	SAR	9.6	Quad-pol	1 m	10–150	Jun 2010–
RISAT-1	SAR	5.3	Quad-pol	3 m	30–240	Apr 2012–
RISAT-2	SAR-X	9.6		3 m	10–650	Apr 2009–
MapSAR	SAR	1.27	Quad-pol	3 m	20–55	2011
KompSAT-5	SAR	9.6	HH, HV, VH, VV	20 m	100	2011
HJ-1C	SAR	3.3	HH, VV	20 m	–	2012
ARKON-2	SAR	9.6, 1.27, 0.3	–	2 m	–	2013
Sentinel-1A	C-SAR	5.3	Quad-pol	9 m	80–400	2013
ALOS-2	PALSAR-2	1.27	Quad-pol	1 m	25–350	2013
NovaSAR-S	SAR	3.3	Quad-pol	6 m	15–750	2013
SAOCOM-1A	SAR-L	1.27	Quad-pol	10 m	20–350	2014
SMAP	SAR	1.27	HH, HV, VV	3 km	30–1000	2014
Sentinel-1B	C-SAR	5.3	Quad-pol	9 m	80–400	2015

(continued)

TABLE 4.3 (Continued)

Characteristics of Major Spaceborne SAR Systems

Platform	Sensor	Frequency (GHz)	Polarization	Highest Spatial Resolution	Swath Width (km)	Mission
RISAT-1A	SAR	5.3	Quad-pol	3 m	30–240	2015
DESDynI	SAR	1.27	Quad-pol	25 m	>340	2016
CSG-1	SAR 2000 SG	9.6	HH, HV, VH, VV	<1 m	10–320	2015
CSG-2	SAR 2000 SG	9.6	HH, HV, VH, VV	<1 m	10–320	2016
RADARSAT Constellation Mission	SAR	5.3	Quad-pol	3 m	20–500	2016–2017

2000; Thoma et al. 2008; Song et al. 2009) and been used in various studies with mixed results (e.g., Baghdadi and Zribi 2006; Álvarez-Mozos et al. 2007; Paloscia et al. 2008).

Since *a priori* fine resolution information on the surface roughness is required, in order to be able to accurately invert soil moisture, the use of theoretical models for soil moisture retrieval over large areas is challenging because of the difficulty of describing the natural surface roughness and topography over such large areas (such as those typically covered in radar image swaths). IEM also neglects scattering from the subsurface soil volume, which may be important for dry soil conditions and long wavelengths. Therefore the complexity and the restrictive assumptions made when deriving the models as well as the requirement of a very detailed knowledge of the surface roughness (which is only achievable through intensive roughness measurement campaigns) means that they can seldom be used to invert soil moisture to a high degree of accuracy over natural surfaces.

4.4.2.2.2 *Soil Moisture Retrieval Using Empirical Scattering Models*

The difficulty encountered in the application of theoretical models has led to the development of empirical and semiempirical models. Empirical backscattering models, generally derived from experimental measurements, gain insight into the interaction of MWs with natural surfaces through simple retrieval algorithms (e.g., linear regression). Examples of studies where empirical models have been used with varying degrees of success include those by Mathieu et al. (2003), Holah et al. (2005), Álvarez-Mozos et al. (2007), and Shoshany et al. (2000). Many of those empirically based studies have generally shown that a linear relationship between the backscattering coefficient and SMC is a reliable approximation for one study site, assuming that roughness does not change between successive radar measurements (Zribi et al. 2005). Various authors have proposed calibration approaches for adjusting those empirical models to other implementation conditions (e.g., Baghdadi et al. 2008; Zribi et al. 2005).

While the use of empirical backscattering models is a simple and straightforward option for relating radar backscatter to SMC, there are several limitations that must be considered. Empirical models, by their nature, are generally only valid to the surface conditions and radar parameters at the time of the experiment, and therefore because of limitations in observation frequency, incidence angles, and surface roughness, empirical models may not be applicable for datasets other than those used in their development

(Chen et al. 1995). Another limitation is that they require many *in situ* soil moisture measurements, obtained over time, for their implementation. Consequently, large databases over a variety of study sites are essential to ensure that developed (and proposed) models are robust and transferable to other datasets, irrespective of surface conditions and sensor configuration.

4.4.2.3 *Soil Moisture Retrieval Using Semiempirical Scattering Models*

Semiempirical backscattering models offer a compromise between the complexity of the theoretical backscattering models and the simplicity of the empirical models. They start from a physical background and incorporate experimental or simulated datasets to simplify the theoretical backscattering model. Their main advantage is that they are not site dependent and may be applied when little or no information on surface roughness is available. The most widely used semiempirical models include those developed by Oh et al. (1992) and Dubois et al. (1995) and, to a lesser extent, the model by Shi et al. (1997). Details of their operation and information on their use by the interested users' community can be found, for example, in the works of Verhoest et al. (2008) and Barrett et al. (2009).

The Oh model relates the ratios of radar backscatter in separate polarizations (HH, HV, and VV) to volumetric soil moisture and soil surface roughness. The model addresses both the co- and cross-polarized backscatter coefficient but does not account for multiple or secondary scattering processes. Although the model is based on truck-mounted scatterometer measurements (at three frequencies: C-, L-, and X-band), it has been applied successfully to spaceborne SAR measurements. The Oh model has been subject to a number of improvements since its original development, aimed largely at incorporating the effects of incidence angle (Oh et al. 1992) and to model cross-polarized backscatter coefficients (Oh et al. 2002). Oh (2004) ultimately introduced a new formulation in the model such that the correlation length could be ignored. The main advantage of the Oh model is that only one surface parameter (namely, the rms height) is required, and when multipolarized data are available, both the dielectric constant and surface roughness can be inverted without the need for field measurements (Álvarez-Mozos et al. 2007). However, the model has restrictive validity ranges of $9 \leq SMC \leq 31\%$ and $0.1 \leq ks \leq 6$, where ks is the normalized rms surface roughness.

The Dubois model accounts for copolarized backscatter only (as they are less sensitive to system noise and generally easier to calibrate than cross-polarized backscatter) and has a validity range of $SMC \leq 35\%$ and $ks \leq 2.5$ for incidence angles greater than 30°. Similar to the Oh model, the Dubois model was formulated using truck-mounted scatterometer data, but at six different frequencies between 2.5 and 11 GHz. A number of studies have used the Dubois model, reporting generally satisfactory results in deriving SMC from radar observations (e.g., Baghdadi and Zribi 2006; Leconte et al. 2004), with best results achieved over bare to sparsely vegetated surfaces (Neusch and Sties 1999). However, various other investigators have reported not so satisfactory or ambiguous results (Baghdadi et al. 2006; Álvarez-Mozos et al. 2007). The less commonly used model by Shi is a simplification of the IEM, derived using L-band measurements and valid only for copolarized terms. All three of these models (Oh, Dubois, and Shi) are valid only over bare soil surface conditions, although some studies have shown good results over sparsely vegetated soil surfaces. The semiempirical water cloud model developed by Attema and Ulaby (1978) has been shown to adequately represent the backscatter from vegetated soil surfaces and has found widespread use among the radar modeling community.

4.5 Synergistic Methods in Soil Moisture Retrieval

Data fusion or synthesis studies have come about as a direct result of the difficulties encountered in discriminating between the multiple influences on the MW signal and that from soil moisture using a single data source. Most studies in the past have dealt with synergies between (1) optical and thermal observations, (2) active and passive MW observations, and (3) MW and optical/thermal observations. These approaches aim to exploit the increased information content that can be extracted by combining multiple datasets originating from different sensors. An overview of these three types of data fusion approaches is provided in the following sections.

4.5.1 Optical and Thermal Data Synergies

A large number of investigations have focused on examining the potential for deriving surface SMC over land surfaces from the synergistic use of remote sensing observations acquired simultaneously in the optical and TIR parts of the EMR spectrum. A brief overview of the available T_s/VI techniques for the estimation of SMC is provided here and the interested reader is referred to the review by Petropoulos et al. (2009a) for more detailed information. To date, the majority of research has documented the potential of obtaining information about surface SMC from heterogeneous land surfaces when remotely sensed T_s and VI measures are plotted in two-dimensional feature space that has been shown to form a T_s/VI scatterplot that is largely bounded by a triangular domain, provided that a full range of fractional vegetation cover and soil moisture is represented in the data. Figure 4.4 represents a schematic description of the soil temperature and vegetation dependence on soil moisture. In the T_s/VI scatterplot, the top of the triangle corresponds to the case of

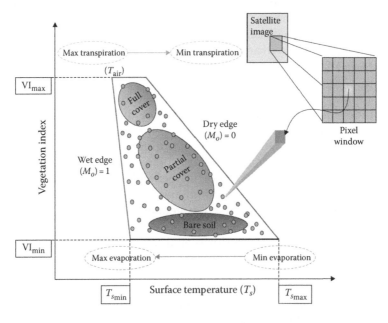

FIGURE 4.4
Summary of the key descriptors and physical interpretations of the T_s/VI feature space "scatterplot." (Adapted from Petropoulos, G. et al., *Progress in Physical Geography* 33(2):224–250, 2009.)

an area with full vegetation coverage, whereas the base of the triangle corresponds to bare soil. Between the triangle top and its base, the majority of the triangle corresponds to land with varying vegetation cover. High-temperature pixels correspond to dry soil conditions, while lower variability is observed in the case of wet soil conditions. This T_s/VI domain is related to the SMC, based on the assumption that there is an increased variability in T_s over bare soil (particularly over dry surfaces) than over a fully vegetated surface, which is due to the variations in surface SMC. Surface soil moisture varies from left (high) to right (low) in the triangle and scaled temperature and VI as the abscissa and ordinate, respectively, such that T_s and VI can be obtained using the following formulae:

$$T_s = \frac{T - T_{s\min}}{T_{s\max} - T_{s\min}} \tag{4.7}$$

$$VI = \frac{VI - VI_{\min}}{VI_{\max} - VI_{\min}} \tag{4.8}$$

Several studies have attempted to relate the satellite-derived T_s/VI feature space either directly to surface SMC, or indirectly, by linking the physical properties of this feature space to environmental variables mainly related to drought conditions. Sandholt et al. (2002) first proposed a technique for deriving direct estimates of surface SMC by linking the T_s/VI scatterplot with an index, which they termed the temperature vegetation dryness index (TVDI). Validation of their method using Advanced Very High Resolution Radiometer (AVHRR) data conducted against simulated SMC maps produced by the MIKE SHE distributed hydrological model (Abbott et al. 1986) at a test site in West Africa indicated spatially similar patterns in the distributed estimates of soil moisture ($R^2 = 0.70$). Vincente-Serrano et al. (2004) also examined the spatial relationships of soil surface moisture derived from the Sandholt et al. (2002) methodology applied to both AVHRR and Landsat ETM+ images acquired over a study site in northeast Spain. In this study, the authors used the fractional vegetation cover (F_r) instead of the NDVI and showed a close correlation of the SMC estimates derived from the two datasets ($R \sim 0.66$), similar to correlations between T_s and NDVI maps ($R \sim 0.74$ and $R \sim 0.76$, respectively). Hassan et al. (2007) proposed a scheme for improving the SMC estimates in high elevation regions, derived by the TVDI method of Sandholt et al. (2002). Authors demonstrated the use of their technique for a region in eastern Canada using 8-day composite images of NDVI and T_s products from MODIS. Authors reported a strong relationship in spatial distribution of land surface wetness predicted by their scheme against estimates of SMC from a process-based model ($R^2 = 95.7\%$). More recently, Patel et al. (2009) investigated the potential of the original TVDI method of Sandholt et al. (2002) for mapping SMC in a region in India. The authors reported a high correlation between the T_s and NDVI parameters ($R^2 = 0.8$). A high correlation was also found between the TVDI and *in situ* soil moisture at different depths (0–15, 15–30, and 30–45 cm).

Various studies have also attempted to relate the satellite-derived T_s/VI feature space to drought conditions, and thus indirectly to surface SMC distribution. Such studies have been largely based on the computation of spectral indices combining information from both the reflected and TIR emitted parts of the EMR. For example, Wang et al. (2004) proposed the use of the vegetation temperature condition index (VTCI) for retrieving information on the spatial variation in drought condition. Authors suggested that the VTCI isolines can be drawn in the NDVI versus T_s scatterplot with the upper limit of the triangle representing the NDVI maximum and the lower limit representing T_s at maximum NDVI. Katou and Yamaguchi (2005) used the NDWI (Gao 1996), NDVI, and T_s to calculate the vegetation water temperature index 1 (VWTI1) and the VWTI2 to measure the strength of stress and influence of stress on vegetation, respectively. Ghulam et al. (2007) developed the perpendicular drought index (PDI) using reflectance of NIR and red bands of the ETM+ image based on the spectral patterns of soil moisture variations in NIR–Red space. More recently, Shakya and Yamaguchi (2010), based on the Katou and Yamaguchi (2005) indices, proposed a new index named vegetation water temperature condition index (VWTCI) for monitoring drought on a regional scale. The VWTCI included the NDWI, which measures the water status in vegetation, the NDVI, and T_s data.

An alternative approach to the estimation of SMC from T_s/VI feature space has been based on coupling the T_s/VI domain with a soil vegetation atmosphere transfer (SVAT) model. In this method, estimated surface SMC is obtained indirectly from a parameter called the "moisture availability (M_o)," which is loosely equated with the fraction of field capacity for the surface SMC. A recent overview of the method workings, so-called triangle method, has been made available by Carlson (2007) and Petropoulos and Carlson (2011). This method was originally proposed by Carlson et al. (1990) and later extended by Carlson et al. (1995b) and Gillies et al. (1997). In this approach, the outputs from a land surface model, specifically a SVAT model, are coupled with the T_s and the F_r via empirically derived correlations developed between the relevant input (e.g., F_r, M_o) and output (e.g., LE, T_s) parameters of the physically derived model, parameterized for the time of satellite overpass. This method has been implemented so far using the SimSphere SVAT model (currently distributed globally from Aberystwyth University, UK, http://www.aber.ac.uk/simsphere), an overview of which was provided by Petropoulos et al. (2009b), but equally any SVAT model can be used. Various studies have evaluated the ability of the triangle method and its variants in deriving spatially distributed estimates of SMC using different types of satellite datasets (Carlson et al. 1995a; Capehart and Carlson 1997; Petropoulos et al. 2010; Petropoulos and Carlson 2011; Sobrino et al. 2012). Results from those studies have generally indicated an ability of these methods to provide estimates of SMC with a standard error of about 10%, an accuracy acceptable for many practical applications. It is worthwhile to note at this point that a variant of the triangle method is at present being explored for prototyping the operational estimation of SMC from the synergy of satellite observations from ESA-funded/cofunded missions in the framework of the PROgRESSIon (ESA 2012a) research project.

The key advantages of synergistic techniques involving optical and thermal remote sensing data include their simplicity in terms of their implementation, as well as the requirement of easily obtained parameters from remote sensing data (namely, a VI and T_s). These techniques also include all the advantages of both the optical and TIR methods reviewed previously (i.e., provide fine spatial and temporal resolution for SMC estimation, use of mature technology, data easily accessible from operational satellites, and long historical data). Specifically, the triangle method of Gillies et al. (1997) has some distinct advantages over other analogous synergistic approaches. Although its implementation requires some degree of expertise by the user and also a large number of input parameters due to the use of the SVAT model, it is able to provide a nonlinear interpretation of the T_s/VI space and thus a solution for the computation of the spatially distributed SMC estimates. Furthermore, the method has already demonstrated a potential to construct similar T_s/VI triangles on successive days [see review by Carlson (2007)], which allows monitoring changes of SMC and its relationship with respect to other surface processes or phenomena (e.g., Owen et al. 1998; Arthuer-Hartanft et al. 2003). However, the use of synergistic techniques between optical and TIR remote sensing data also comes with the inherent limitations that are commonly found in all optical and TIR techniques (i.e., shallow soil penetration, cloud-free conditions, and infrequent coverage at spatial resolutions suitable for watershed management). In addition, many of the synergistic approaches require a full or at least very wide range of both NDVI and SMC conditions within a study region for their practical implementation, a condition that generally cannot always be satisfied, particularly over large homogeneous areas.

4.5.2 MW and Optical Data Synergies

A number of studies have explored the synergy of MW and optical/TIR data for surface SMC retrieval and a range of methodologies along these lines have been proposed. Following the evolution of satellite technologies, researchers working in this direction have mostly investigated the contribution of mainly SAR images to optical and/or of SAR to optical/ TIR images in order to determine SMC more accurately. A key advantage, especially of SAR with optical/TIR data, is that it allows minimizing the vegetation biomass and surface roughness effects to the radar backscatter. For example, Wang et al. (2004) proposed an approach for minimizing surface roughness and vegetation effects and extracted soil moisture in sparsely to moderately vegetated areas using an ERS-2/Landsat TM synergy. The ratio between two different SAR images (wet and dry seasons) was used to reduce the effect of surface roughness, while Landsat data were used to calculate the NDVI to account for the influence of vegetation. Similarly, Yang et al. (2006) proposed a technique to estimate the change in the surface SMC using a synergy of C-band Radarsat ScanSAR data and optical data from Landsat TM and AVHRR. Kuruku et al. (2009) also proposed a pixel-based image fusion technique for determining soil moisture regionally based on the synergistic use of observations from Radarsat-1 Fine Beam mode with SPOT-2 imagery. In their scheme, the image fusion approach allowed the backscatter value of Radarsat-1 to be added to the intensity band. Hence the fused image was able to provide a better result regarding the water/moisture content of the planted fields.

Concerning the synergy of passive MW data with optical/TIR data, a variation of the T_s/VI triangle method of Carlson et al. (1995b) and Gillies et al. (1997) was proposed by Chauhan et al. (2003) for the operational retrieval of 1-km-resolution surface SMC maps from the National Polar-orbiting Operational Environmental Satellite System (NPOESS). Their technique was based on the synergy between the MW-derived soil moisture estimates from the Conical Scanning Microwave Imager/Sounder (CMIS) with the optical/ infrared instrument Visible/Infrared Imager Radiometer Sensor Suite (VIIRS). An initial verification of their proposed scheme was carried out by the authors for a grassland site in Oklahoma, United States, using data from the Special Sensor Microwave Imager (SSM/I) (operating at frequency of 19.4 GHz and having a spatial resolution of 25 km) and the AVHRR (1 km spatial resolution). Predicted surface SMC at high resolution agreed reasonably well with the low-resolution results in both their magnitude and spatiotemporal pattern, with an RMSD of the order of 5% in surface SMC estimation. In another study, a methodology for deriving regional estimates of soil SMC based on the synergy of passive MW data from AMSR-E instrument with visible wavelengths of MODIS and also topographic attributes derived from the SRTM Digital Elevation Model was proposed by Temimi et al. (2010). In their method, the MODIS images were used to compute the F_r required for computing their modified wetness index and to assess the vegetation dynamics and growth. More recently, Piles et al. (2011) demonstrated the use of brightness temperatures measured by SMOS with VIS/IR satellite data from Landsat and MODIS sensor using a variant of the triangle method to derive soil moisture estimates for selected sites belonging to the Oznet soil moisture Australian network. In another study, Choi and Hur (2012) used a synergistic approach between AMSR-E soil moisture operational products and MODIS data to perform a spatial downscaling to 1-km spatial resolution from 25 km.

4.5.3 Active and Passive MW Data Synergies

Active systems (SAR in particular) are likely to be more sensitive to surface features such as surface roughness and vegetation biomass and structure, whereas a passive MW radiometer is likely to be more sensitive to the near surface SMC (Njoku et al. 2000; Lee and Anagnostou 2004). Furthermore, SAR sensors have a clear and very important advantage over passive sensors, that being their finer spatial resolution. The higher spatial resolution SAR data, when combined with passive MW observations that are less sensitive to backscattering contributing factors (such as surface roughness and vegetation amount), allow resolving more accurately the effects of the contributing factors to the received backscattering signal and potentially determining more accurately the subpixel variability of passive-derived soil moisture. This has been confirmed by many studies examining the synergy between active and passive MW data. These studies generally determine the vegetation amount and surface roughness parameters from active (SAR) observations and combine these with coarser spatial resolution passive MW observations of brightness temperature to retrieve regional estimates of surface SMC. Studies by Njoku et al. (2002) and Narayan et al. (2004, 2006) have all demonstrated the synergistic use of active and passive MW observations for deriving spatially distributed estimates of near surface SMC.

Zribi et al. (2003) developed an algorithm using a high temporal resolution scatterometer combined with high spatial resolution SAR and observed high correlations ($R^2 > 0.8$) when soil moisture estimates were compared with ground measurements. Lee and Anagnostou (2004) proposed a method combining observations from the Tropical Rainfall Measuring Mission (TRMM) MW imager (TMI) channel and the precipitation radar to estimate near surface SMC and vegetation properties. Evaluation of their technique was performed for a region of Oklahoma, United States, using satellite observations from three consecutive years. Their study demonstrated that the use of coincident passive/active MW observations has the potential to increase the number of estimated geophysical variables, especially, in cases of moderate to low vegetation. In a recent study, Li et al. (2011) evaluated two SMC retrieval methods that are based on the synergy of passive and active observations. The two retrieval methods were applied to Advanced Microwave Scanning Radiometer for EOS (AMSR-E) and QuikSCAT/Seawinds (The Seawinds scatterometer on NASA's Quick Scatterometer) observations that have been carried out for the SMEX03 (Soil Moisture Experiment 2003) region, north of Oklahoma. The two methods compared the three-parameter retrieval (THRA) and the two-parameter retrieval (TWRA) approaches. These two methods were principally different in the way the roughness parameter was estimated. Results from this study showed the TWRA achieved a higher accuracy than THRA in dealing with the active and passive MW observations at different overpass times, with an RMSD of 0.037 and 0.089, respectively.

More recently, studies have focused on the evaluation and harmonization of soil moisture datasets from coarse resolution active (ERS-1/2, MetOp) and passive (SMMR, SSM/I, TMI, AMSR-E, Windsat) MW sensors. ESA, in collaboration with the Global Energy and Water Cycle Mission Observation Strategy (GEWEX) of the World Climate Research Program, launched the Water Cycle Multi-Mission Observation Strategy (WACMOS) project in 2008. The WACMOS project (duration: 2009–2012) produced a global long-term (1978–2008) soil moisture product based on merged active and passive coarse resolution single sensor MW

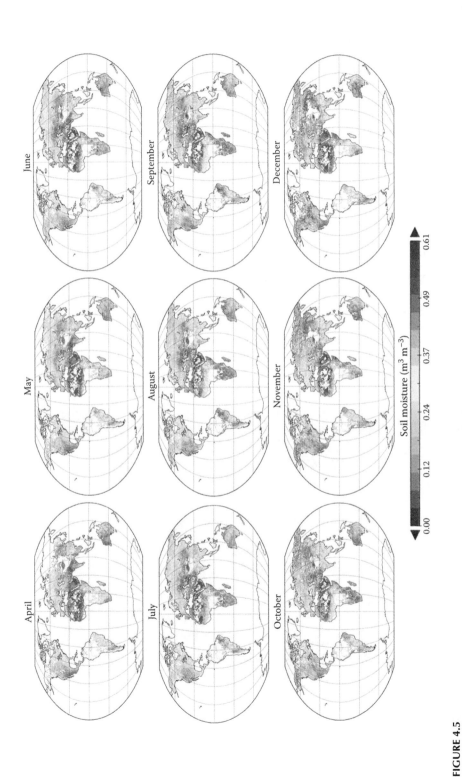

FIGURE 4.5

(See color insert.) Global mean monthly volumetric soil moisture values for 2010 from the ECV product dataset. (Data source: The ESA CCI SM project. Liu, Y. Y. et al., *Remote Sensing of Environment* 123:280–297, 2012; Wagner, W. et al., Fusion of active and passive microwave observations to create an Essential Climate Variable data record on soil moisture. Paper presented at XXII ISPRS Congress, Melbourne, Australia, 2012.)

data products (see http://wacmos.itc.nl/). The Essential Climate Variable Soil Moisture (ECV SM) product (Liu et al. 2012; Wagner et al. 2012) was initiated within the WACMOS project and is being continued and refined in the context of the Climate Change Initiative (CCI) programme. See Figure 4.5 for an example of global mean monthly soil moisture values for 2010 generated using the ECV SM products. The CCI project uses multiple active and passive sensors; however, the future NASA Soil Moisture Active Passive (SMAP) mission (Entekhabi et al. 2010), a successor to the cancelled Hydrosphere State (HYDROS) mission (Entekhabi et al. 2004), will integrate L-band radar and radiometer measurements from a single platform to provide high-resolution and high-accuracy global maps of surface SMC every 2 to 3 days and may thereby overcome some of the limitations of using individual active or passive MW observations to determine SMC.

4.6 Operational Global Soil Moisture Products from Remote Sensing

Global satellite-based SMC observations started to become available in 2003 when the first global multiannual soil moisture dataset (1992–2000) derived from ERS-1 and ERS-2 SCAT observations was published (Wagner et al. 2003). Since then, efforts of the scientific community have continued in this direction resulting in a variety of products being developed using a variety of approaches and exploiting information acquired from different satellite sensors. This section provides an overview of the range of global soil moisture products that are or where until recently available from the different sensors (Figure 4.6). Table 4.4 displays the main characteristics of the different SMC products included in this overview.

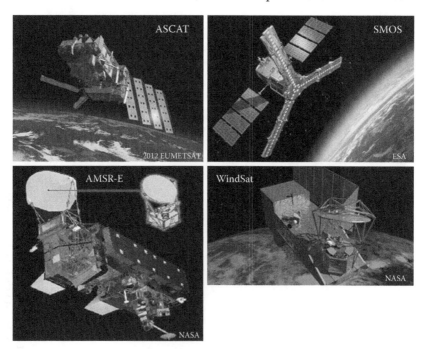

FIGURE 4.6
(See color insert.) Examples of the satellite sensors providing operational estimates of SMC.

TABLE 4.4

Remote Sensing Operational SMC Products

	AMSR-E[a]	ASCAT	SMOS	WINDSAT
Instrument type	C-band (six frequencies) (MW radiometer)	C-band (5.255 GHz) (MW scatterometer)	L-band (1.4 GHz) (MW radiometer)	C-band (five frequencies) (MW radiometer)
Spatial resolution	25 km	12.5 km, 25 km	35 km	25 km
Temporal resolution	Daily	3 days	1–3 days (2 passes)	Daily
Temporal coverage	2002–2011	Since 2007	Since 2010	Since 2003
Related validation studies	Rüdiger et al. (2009); Draper et al. (2009)	Brocca et al. (2011); Albergel et al. (2012)	Albergel et al. (2011); Bircher et al. (2012)[b]	Li et al. (2010); Parinussa et al. (2012)
Distributed by	JAXA (Japan)[c]	EUMETSAT	ESA	US Navy

[a] Since October 4, 2011, the AMSR-E instrument has stopped producing data due to an antenna problem.
[b] See special issue on the ESA SMOS mission published in *IEEE Transactions on Geoscience and Remote Sensing*, vol. 50, issue 5, part 1, 2012.
[c] There are different AMSR-E data products produced by four different agencies; JAXA, US National Snow and Ice Data Center (NSIDC), NASA and the US Department of Agriculture (USDA).

4.6.1 AMSR-E

Researchers at the US National Snow and Ice Data Center (NSIDC) produced the first operational global passive MW soil moisture product using brightness temperatures from the Advanced Microwave Scanning Radiometer (AMSR-E) (Njoku et al. 2003). This approach used six MW channels (three frequencies at two polarizations) to solve for soil moisture, vegetation water content, and surface temperature. Two versions of a global soil moisture product are available; the Level-2B surface soil moisture and the L3 Daily surface moisture products. Spatial coverage in both products is fixed at 25-km resolution and is global except for snow-covered and densely vegetated areas.

The Level-2B land surface product includes daily measurements of surface soil moisture (g cm^{-3}), vegetation/roughness water content (kg m^{-2}), and quality control variables. Input brightness temperature data at 10.7 GHz, corresponding to a 38-km mean spatial resolution, are resampled to a global cylindrical 25-km Equal-Area Scalable Earth Grid (EASE-Grid) cell spacing. The gridded Level 3 land surface product includes daily measurements of surface soil moisture (g cm^{-3}), vegetation/roughness water content interpretive information (kg m^{-2}), as well as brightness temperatures (in K) and quality control variables. Input brightness temperature data from the Level 2B product, corresponding to a 56-km mean spatial resolution for frequencies 6.9–36.5 GHz, and a 12-km mean spatial resolution for frequencies 36.5 and 89 GHz, are resampled to EASE-Grid cell spacing. Soil moisture in the top ~1 cm of soil is averaged over the retrieval footprint, and in both products, ancillary data are included that contain parameters such as time, geolocation, and quality assessment. Products are available from June 2002 to October 2011, when a failure in the instrument was reported and the sensor stopped providing data.*

The operational estimation of SMC by AMSR-E is based on the solution of a two-parameter inverse model that is exploiting a MW radiative transfer model to solve simultaneously

---

* Continuity of data will be provided by AMSR-2 [launched in May 2012 on the Global Change Observation Mission for Water research (GCOM-W1) satellite] and data will become publicly available in early 2013.

from the multifrequency channel brightness temperature, data surface SMC, and vegetation optical depth without *a priori* information. The flexible approach allows, in general, for the retrieval of soil moisture from a variety of frequencies. AMSR-E SMC retrieval algorithm does not explicitly model effects of topography, snow cover, clouds, and precipitation. A number of validation experiments have been carried out evaluating the accuracy of SMC AMRS-E products at different ecosystem conditions (e.g., Gruhier et al. 2008; Rüdiger et al. 2009; Draper et al. 2009). Generally high correlations with *in situ* observations have been reported in semiarid regions with R and RMSD values around 0.80 and 0.05 m^3/m^3, respectively (Wagner et al. 2007b; Draper et al. 2009; Rüdiger et al. 2009). Various verification studies conducted on the product accuracy have generally also shown that high error in the SMC by the algorithm is observed in areas of high vegetation cover. Further products produced by the Japanese Aerospace Exploration Agency (JAXA) (Koike et al. 2004), US Department of Agriculture (Jackson 1993), US Navy (Li et al. 2010), and the NASA Goddard Space Flight Center with the Vrije Universiteit Amsterdam (Owe et al. 2008) have become available and span the period from 1979 to present. These data products make use of a variety of passive MW sensors: Scanning Multichannel Microwave Radiometer (SMMR) on Nimbus-7, the Defense Meteorological Satellite Program (DMSP) Special Sensor Microwave Imager (SSM/I), and Advanced Microwave Scanning Radiometer on Earth Observing System (EOS) Aqua satellite (AMSR-E).

4.6.2 ASCAT

The Advanced Scatterometer (ASCAT) is a radar instrument onboard the MetOp-A satellite launched by ESA in October 2006 and EUMETSAT currently distributes soil moisture products at two spatial resolutions: 25 and 12.5 km. Operational SMC estimation from ASCAT is achieved by converting the backscatter measurements to soil moisture by applying the soil moisture retrieval algorithm developed by the Technical University of Vienna (TU Wien) (Naeimi et al. 2009a,b). This is essentially a modified version of the change detection algorithm, originally developed for soil moisture retrievals from ERS-1 and 2 (Wagner et al. 1999a,b) after some minor modifications. Briefly stated, instantaneous backscatter measurements are extrapolated to a reference incidence angle (taken at 40°) and compared to dry and wet backscatter references. By knowing the typical yearly vegetation cycle and how it influences the backscatter-incidence angle relationship for each location on Earth, the vegetation effects can be removed, revealing the soil moisture variations. The historically lowest and highest values of observed soil moisture are then assigned to the 0% (dry) and 100% (wet) references, respectively, and the soil moisture index is then computed. Considering that there is a linear relationship between SMC and radar backscatter, this results in a relative measure of surface (<2 cm) soil moisture that ranges between 0% (wilting point) and 100% (saturation) and direct estimates of soil moisture can be derived by multiplying it with the soil porosity. Areas covered by snow/ice and also any missing data are masked out from the algorithm and product delivery. Data continuity will be provided by the ASCAT instrument on the recently launched MetOp-B satellite (launched September 2012) and future launch of MetOp-C in 2017.

4.6.3 SMOS

Another SMC operational product is provided by the Soil Moisture and Ocean Salinity (SMOS) satellite, launched by the European Space Agency (ESA) in November 2009 in the framework of their Earth Explorer Opportunity Missions. SMOS was designed to measure

SMC over continental surfaces as well as ocean salinity using a low MW frequency—L-band (1.4 GHz) on the Microwave Imaging Radiometer with Aperture Synthesis (MIRAS) instrument. As of beginning of October 2010, SMOS soil moisture products are being distributed to all science users. The reprocessing of the SMOS data from the commissioning phase and beyond (January 2010 to January 2011) has now been completed and reprocessed data are now available. The reprocessed dataset comprises all Level 1 and the Level 2 Soil Moisture data products. The intention of this intermediate reprocessing was to provide a consistent dataset for the first year of mission operations, because there were several updates of the SMOS instrument settings and processor versions throughout commissioning. The SMC product from SMOS is provided as a Level 2 data product, with measurements of SMC at a spatial resolution of approximately 35 km at nadir and at a revisit period 1–3 days with equatorial crossing times 0600 and 1800. The SMC product provides not only the SMC retrievals but also a series of ancillary data derived from the processing, namely, nadir optical thickness, surface temperature, and roughness parameter (namely, dielectric constant and brightness temperature retrieved at the top of atmosphere and on the surface) with their corresponding uncertainties. A series of auxiliary data related to soil moisture are also provided. These auxiliary files contain information needed for generating soil moisture products, such as ECOCLIMAP surface cover information, ECMWF forecast geophysical fields, and vegetation optical thickness.

The SMOS SMC retrieval algorithm is based on L-band multiangular dual-polarized (or fully polarized) brightness temperatures. Previous studies have clearly underscored the potential of L-band in soil moisture estimation. At L-band, the sensitivity to SMC is high, whereas the sensitivity to tropospheric disturbances and surface roughness is minimal. To separate soil and vegetation contributions to land emissivity, SMOS algorithm makes full use of the dual-polarized multiincidence angle acquisitions. The retrieval concept involves using all the angular and polarized acquisitions to retrieve separately the soil moisture and vegetation water content. A full description of the retrieval algorithm can be found in the work of Kerr et al. (2012). SMOS product Calibration/Validation is implemented by ESA using the Valencia Anchor Station in eastern Spain and the Upper Danube Catchment in southern Germany as the two main test sites in Europe (e.g., see dall'Amico et al. 2012; Schlenz et al. 2012; Sanchez et al. 2012). The SMOS mission aims to provide global maps of SMC with accuracy better than 0.04 $m^3 m^{-3}$ as well as vegetation water content with an accuracy of less than 0.5 kg m^{-2} (Kerr et al. 2001).

4.6.4 WindSat

WindSat is the world's first satellite-based polarimetric MW radiometer that measures the partially polarized energy emitted, scattered, and reflected from the Earth's atmosphere and surfaces. The instrument is onboard the Coriolis mission launched in January 2003 by the Naval Research Laboratory Remote Sensing Division and the Naval Center for Space Technology for the US Navy and the National Polar-orbiting Operational Environmental Satellite System (NPOESS) Integrated Program Office. The SMC retrieval algorithm used for WindSat data is a physically based, multichannel, maximum-likelihood estimator using 10-, 18-, and 37-GHz data. The surface parameters considered by the algorithm include SMC, vegetation water content, land surface temperature, surface type, precipitation, and snow cover. The SMC data product is generated by iteratively minimizing the difference between WindSat measured brightness temperatures and those obtained from radiative transfer simulations. The first release of WindSat land data products includes daily global surface soil moisture, land classification, and the observation time. WindSat soil moisture retrievals have generally been validated using different types of multiscale data including soil mois-ture climatology, ground *in situ* network data, precipitation patterns, and vegetation data from AVHRR sensors. Overall, the SMC retrievals have been found to be very consistent with global dry/wet patterns of climate regimes (Jackson et al. 2005; Li et al. 2011; Parinussa et al. 2012). Authors reported a close agreement with ground SMC estimates, within the requirements for most science and operational applications (standard error of 0.04 $m^3 m^{-3}$).

All original sensor data (brightness temperatures) and Level 2 data products can be downloaded from their respective public source archives [SMMR, SSM/I, and AMSR-E are available from NSIDC (http://nsidc.org/data/), TMI from the Goddard Earth Sciences Data and Information Services Center (DES DISC; http://disc.sci.gsfc.nasa.gov/), WindSat from the US Naval Research Laboratory (http://www.nrl.navy.mil/WindSat/index.php), and SMOS data are available after registration with ESA (https://earth.esa.int/web) or from the Centre Aval de Traitement des Données SMOS (CATDS; Level-3 data products only) at http://catds.ifremer.fr/Data/Official-Products-from-CPDC]. ASCAT data can be accessed from EUMETCast after registration on the Earth Observation Portal (http://www.eumetsat.int/Home/Main/DataAccess/EUMETCast/index.htm?l=en.%20).

4.7 Conclusions

A large number of methods exist for the estimation of SMC using remote sensing data acquired at different parts of the EMR spectrum. Optical, TIR, MW, and even Global Positioning System (GPS) reflected signals (Katzberg et al. 2006; Larson et al. 2008) have been explored and shown to be sensitive to the soil dielectric constant. Despite the pre-dominance of optical sensors currently in orbit, limited success has been achieved in the exploitation of visible, near-infrared, shortwave infrared, and/or hyperspectral remote sensing observations on the retrieval of surface SMC. This is due partly to the fact that the optical signal has limited ability to penetrate clouds and vegetation canopy and is highly attenuated by the Earth's atmosphere. Active and passive MW remote sensing methods have had the greatest success in retrieving soil moisture in a temporally and spatially consistent manner. Most progress has been made with passive MW sensors, and while it may be argued that only active sensors meet the spatial resolution and coverage required for many of the applications of consistent soil moisture data, passive sensors have too found many applications in small-scale studies (e.g., Brocca et al. 2012; Matgen et al. 2012). In addition, fine-scale soil moisture products are still some time away from becoming operational, limited by spatial and temporal sampling characteristics, accuracy, surface roughness, and vegetation cover problems (Wagner et al. 2007a; Thoma et al. 2008; Barrett et al. 2009).

The majority of international efforts are directed toward the creation of (operational) multisource global soil moisture datasets, making best use of existing active (ERS-1/2, METOP A/B), passive (SMMR, SSM/I, TMI, AMSR-E, Windsat, SMOS), and future (SMAP) MW datasets. In addition, some efforts of the scientific community are directed toward the exploration of "synergistic" approaches, which exploit spectral information acquired from sensors operating at different parts of the EMR spectrum. The current availability of satel-lite observations from such a large array of satellite systems necessitates further investiga-tion of their potential synergies, aiming to ultimately develop and improve the estimation of surface SMC on an operational level. Yet, the creation of long-term, multisource soil moisture data products on an operational basis poses significant scientific and technical challenges (Wagner et al. 2009). These challenges include the development of a global *in situ* network for validation and in accurately merging satellite data from multiple sources. See Dorigo et al. (2011) for more information on the International Soil Moisture Network, maintained by TU Wien (http://www.ipf.tuwien.ac.at/insitu), and Liu et al. (2011, 2012) for more details on the methodology for blending active and passive datasets. In addition, the contrast between the point-scale nature of current ground-based soil moisture sensors and the spatial resolution of satellites used to retrieve soil moisture also poses a significant challenge for the validation of data products from current and upcoming soil moisture satellite missions. See the review by Crow et al. (2012) as well as Chapter 17 of this book for a more comprehensive description of the soil moisture upscaling problem, commonly used upscaling strategies, and the measurement density requirements for ground-based soil moisture networks.

Despite these limitations, the launch of the Soil Moisture Ocean Salinity (SMOS) satellite and the launch preparations for the Soil Moisture Active Passive (SMAP) mission demonstrate the real need and true potential of spaceborne sensors to provide spatially and temporally consistent and reliable soil moisture information. These dedicated soil moisture missions along with the availability of a greater number of satellites with enhanced spatial and temporal resolutions, in addition to improved retrieval algorithms and *in situ* validation networks, offer increased opportunities for deriving operational SMC data products. In addition, projects exploring the synergies between different satellite sensors such as the ESA-funded WACMOS and PROgRESSIon as well as the Soil Moisture CCI (ESA 2012b) is anticipated to further contribute to providing the most complete and consistent global soil moisture data record based on satellite observations.

Acknowledgments

Dr. Barrett's contribution to this work was produced as part of the Science, Technology, Research and Innovation for the Environment (STRIVE) Programme 2007–2013, financed by the Irish Government under the National Development Plan 2007–2013 and administered on behalf of the Department of the Environment, Heritage and Local Government by the Environmental Protection Agency (EPA). Dr. Petropoulos's contribution was supported by the PROgRESSIon (Prototyping the Retrievals of Energy Fluxes and Soil Moisture Content) project, funded by the European Space Agency (ESA) Support to Science Element (STSE) under contract STSE-TEBM-EOPG-TN-08-0005. The authors wish to acknowledge the reviewer for his valuable comments that helped improve this chapter's content.

References

Abbott, M. B., J. C. Bathurst, J. A. Cunge, P. E. O'Connell, and J. Rasmussen. 1986. An introduction to the European Hydrological System—Systeme Hydrologique Europeen,"SHE," 1: History and philosophy of a physically-based, distributed modelling system. *Journal of Hydrology* 87(1):45–59.

Albergel, C., P. de Rosnay, C. Gruhier, J. Muñoz-Sabater, S. Hasenauer, L. Isaksen, Y. Kerr, and W. Wagner. 2012. Evaluation of remotely sensed and modelled soil moisture products using global ground-based in situ observations. *Remote Sensing of Environment* 118:215–226.

Albergel, C., C. Rüdiger, D. Carrer, J. C. Calvet, N. Fritz, V. Naeimi, Z. Bartalis, and S. Hasenauer. 2009. An evaluation of ASCAT surface soil moisture products with in-situ observations in southwestern France. *Hydrology and Earth System Sciences* 13(2):115.

Albergel, C., E. Zakharova, J. C. Calvet, M. Zribi, M. Pardé, J. P. Wigneron, N. Novello, Y. Kerr, A. Mialon, and N. D. Fritz. 2011. A first assessment of the SMOS data in southwestern France using in situ and airborne soil moisture estimates: The CAROLS airborne campaign. *Remote Sensing of Environment* 115(10):2718–2728.

Álvarez-Mozos, J., M. Gonzalez-Audícana, and J. Casalí. 2007. Evaluation of empirical and semi-empirical backscattering models for surface soil moisture estimation. *Canadian Journal of Remote Sensing* 33:176–188.

Apan, A., A. Held, S. Phinn, and J. Markley. 2004. Detecting sugarcane 'orange rust' disease using EO-1 Hyperion hyperspectral imagery. *International Journal of Remote Sensing* 25(2):489–498.

Arthuer-Hartanft, T., T. N. Carlson, and K. C. Clarke. 2003. Satellite and ground-based microclimate and hydrologic analyses coupled with a regional urban growth model. *Remote Sensing of Environment* 86:385–400.

Attema, E. P. W., and F. T. Ulaby. 1978. Vegetation modelled as a water cloud. *Radio Science* 13(2):357–364.

Baghdadi, N., N. Holah, and M. Zribi. 2006. Calibration of Integral Equation Model for SAR data in C band and HH and VV polarizations. *International Journal of Remote Sensing* 27:805–816.

Baghdadi, N., and M. Zribi. 2006. Evaluation of radar backscatter models IEM, Oh and Dubois using experimental observations. *International Journal of Remote Sensing* 27:3831–3852.

Baghdadi, N., M. Zribi, C. Loumagne, P. Ansart, and T. P. Anguela. 2008. Analysis of TerraSAR-X data and their sensitivity to soil surface parameters over bare agricultural fields. *Remote Sensing of Environment* 112:4370–4379.

Barrett, B. W., E. Dwyer, and P. Whelan. 2009. Soil moisture retrieval from active spaceborne micro-wave observations: An evaluation of current techniques. *Remote Sensing* 1(3):210–242.

Bartalis, Z., W. Wagner, V. Naeimi, S. Hasenauer, K. Scipal, H. Bonekamp, J. Figa, and C. Anderson. 2007. Initial soil moisture retrievals from the METOP-A Advanced Scatterometer (ASCAT). *Geophysical Research Letters* 34(20):L20401.

Ben-Dor, E., J. R. Irons, and G. F. Epema. 1999. Soil reflectance. In *Remote Sensing for the Earth Sciences*, edited by A. N. Rencz and R. A. Ryerson. John Wiley, New York, pp. 111–188.

Bircher, S., J. E. Balling, N. Skou, and Y. H. Kerr. 2012. Validation of SMOS brightness temperatures during the HOBE airborne campaign, western Denmark. *IEEE Transactions on Geoscience and Remote Sensing* 50(5):1468–1482.

Bowers, S. A., and R. J. Hanks. 1965. Reflection of radiant energy from soils. *Soil Science* 100(2):130–138.

Brocca, L., S. Hasenauer, T. Lacava, F. Melone, T. Moramarco, W. Wagner, W. Dorigo, P. Matgen, J. Martínez-Fernández, and P. Llorens. 2011. Soil moisture estimation through ASCAT and AMSR-E sensors: An intercomparison and validation study across Europe. *Remote Sensing of Environment* 115(12):3390–3408.

Brocca, L., F. Melone, T. Moramarco, W. Wagner, and S. Hasenauer. 2010. ASCAT soil wetness index validation through in situ and modeled soil moisture data in central Italy. *Remote Sensing of Environment* 114(11):2745–2755.

Brocca, L., T. Moramarco, F. Melone, W. Wagner, S. Hasenauer, and S. Hahn. 2012. Assimilation of surface-and root-zone ASCAT soil moisture products into rainfall–runoff modeling. *IEEE Transactions on Geoscience and Remote Sensing* 50(7):2542–2555.

Cai, G., Y. Xue, Y. Hu, Y. Wang, J. Guo, Y. Luo, C. Wu, S. Zhong, and S. Qi. 2007. Soil moisture retrieval from MODIS data in Northern China Plain using thermal inertia model. *International Journal of Remote Sensing* 28(16):3567–3581.

Capehart, W. J., and T. N. Carlson. 1997. Decoupling of surface and near-surface soil water content: A remote sensing perspective. *Water Resources Research* 33(6):1383–1395.

Carlson, T. N. 1986. Regional-scale estimates of surface moisture availability and thermal inertia using remote thermal measurements. *Remote Sensing Reviews* 1(2):197–247.

Carlson, T. N. 2007. An overview of the "triangle method" for estimating surface evapotranspiration and soil moisture from satellite imagery. *Sensors* 7(8):1612–1629.

Carlson, T. N., W. J. Capehart, and R. R. Gillies. 1995a. A new look at the simplified method for remote sensing of daily evapotranspiration. *Remote Sensing of Environment* 54(2):161–167.

Carlson, T. N., R. R. Gillies, and T. J. Schmugge. 1995b. An interpretation of methodologies for indirect measurement of soil water content. *Agricultural and Forest Meteorology* 77(3–4):191–205.

Carlson, T. N., E. M. Perry, and T. J. Schmugge. 1990. Remote estimation of soil moisture availability and fractional vegetation cover for agricultural fields. *Agriculture and Forest Meteorology* 52:45–69.

Castro, R., A. Gutierrez, and J. Barbosa. 2012. A first set of techniques to detect radio frequency interferences and mitigate their impact on SMOS data. *IEEE Transactions on Geoscience and Remote Sensing* 50(5):1440–1447.

Champagne, C., H. McNairn, and A. A. Berg. 2011. Monitoring agricultural soil moisture extremes in Canada using passive microwave remote sensing. *Remote Sensing of Environment* 115(10):2434–2444.

Chauhan, N. S. 1997. Soil moisture estimation under a vegetation cover: Combined active and passive microwave remote sensing approach. *International Journal of Remote Sensing* 18(5):1079–1097.

Chauhan, N. S., S. Miller, and P. Ardanuy. 2003. Spaceborne soil moisture estimation at high resolution: A microwave-optical/IR synergistic approach. *International Journal of Remote Sensing* 24(22):4599–4622.

Chen, K. S., Wu Tzong-Dar, T. Mu-King, and A. K. Fung. 2000. Note on the multiple scattering in an IEM model. *IEEE Transactions on Geoscience and Remote Sensing* 38(1):249–256.

Chen, K. S., S. K. Yen, and W. P. Huang. 1995. A simple model for retrieving bare soil moisture from radar-scattering coefficients. *Remote Sensing of Environment* 54:121–126.

Choi, M., and Y. Hur. 2012. A microwave-optical/infrared disaggregation for improving spatial representation of soil moisture using AMSR-E and MODIS products. *Remote Sensing of Environment* 124:259–269.

Choudhury, B. J, T. J. Schmugge, A. Chang, and R. W. Newton. 1979. Effect of surface roughness on the microwave emission from soils. *Journal of Geophysical Research* 84(C9):5699–5706.

Crow, W. T., A. A. Berg, M. H. Cosh, A. Loew, B. P. Mohanty, R. Panciera, P. de Rosnay, D. Ryu, and J. P. Walker. 2012. Upscaling sparse ground-based soil moisture observations for the validation of coarse-resolution satellite soil moisture products. *Reviews of Geophysics* 50(2):RG2002.

dall'Amico, J. T., F. Schlenz, A. Loew, and W. Mauser. 2012. First results of SMOS soil moisture validation in the Upper Danube Catchment. *IEEE Transactions on Geoscience and Remote Sensing* 50(5):1507–1516.

Daughtry, C. S. T., C. L. Walthall, M. S. Kim, E. B. De Colstoun, and J. E. McMurtrey. 2000. Estimating corn leaf chlorophyll concentration from leaf and canopy reflectance. *Remote Sensing of Environment* 74(2):229–239.

de Jeu, R. A. M., W. Wagner, T. R. H. Holmes, A. J. Dolman, N. C. Van De Giesen, and J. Friesen. 2008. Global soil moisture patterns observed by space borne microwave radiometers and scatterometers. *Surveys in Geophysics* 29(4):399–420.

Delworth, T., and S. Manabe. 1989. The influence of soil wetness on near-surface atmospheric variability. *Journal of Climate* 2(12):1447–1462.

Dorigo, W. A, W. Wagner, R. Hohensinn, S. Hahn, C. Paulik, A. Xaver, A. Gruber, M. Drusch, S. Mecklenburg, and P. Van Oevelen. 2011. The International Soil Moisture Network: A data hosting facility for global in situ soil moisture measurements. *Hydrology and Earth System Sciences* 15(5):1675–1698.

Draper, C. S., J. P. Walker, P. J. Steinle, R. A. M. de Jeu, and T. R. H. Holmes. 2009. An evaluation of AMSR–E derived soil moisture over Australia. *Remote Sensing of Environment* 113(4):703–710.

Dubois, P. C., J. Van Zyl, and T. Engman. 1995. Measuring soil moisture with imaging radars. *IEEE Transactions on Geoscience and Remote Sensing* 33(4):915–926.

Eagleman, J. R., and W. C. Lin. 1976. Remote sensing of soil moisture by a 21-cm passive radiometer. *Journal of Geophysical Research* 81(21):3660–3666.

Engman, E. 1990. Progress in microwave remote sensing of soil moisture. *Canadian Journal of Remote Sensing* 16:6–14.

Entekhabi, D., E. G. Njoku, P. Houser, M. Spencer, T. Doiron, Kim Yunjin, J. Smith et al. 2004. The hydrosphere State (hydros) Satellite mission: An Earth system pathfinder for global mapping of soil moisture and land freeze/thaw. *IEEE Transactions on Geoscience and Remote Sensing* 42(10):2184–2195.

Entekhabi, D., E. G. Njoku, P. E. O'Neill, K. H. Kellogg, W. T. Crow, W. N. Edelstein, J. K. Entin et al. 2010. The Soil Moisture Active Passive (SMAP) Mission. *Proceedings of the IEEE* 98(5):704–716.

Entekhabi, D., I. Rodriguez-Iturbe, and F. Castelli. 1996. Mutual interaction of soil moisture state and atmospheric processes. *Journal of Hydrology* 184(1–2):3–17.

European Space Agency. 2012a. Prototyping the retrievals of energy fluxes and soil surface moisture. Available at http://due.esrin.esa.int/stse/projects/stse_project.php?id=148 (accessed on November 24, 2012).

European Space Agency. 2012b. Soil moisture essential climate variable. Available at http://www.esa-soilmoisture-cci.org/ (accessed on October 24, 2012).

Evett, S. R., and G. W. Parkin. 2005. Advances in soil water content sensing: The continuing maturation of technology and theory. *Vadose Zone Journal* 4(4):986–991.

Farrar, T. J., S. E. Nicholson, and A. R. Lare. 1994. The influence of soil type on the relationships between NDVI, rainfall, and soil moisture in semiarid Botswana. II. NDVI response to soil moisture. *Remote Sensing of Environment* 50(2):121–133.

Fatras, C., F. Frappart, E. Mougin, M. Grippa, and P. Hiernaux. 2012. Estimating surface soil moisture over Sahel using ENVISAT radar altimetry. *Remote Sensing of Environment* 123:496–507.

Friedl, M. A., and F. W. Davis. 1994. Sources of variation in radiometric surface temperature over a tallgrass prairie. *Remote Sensing of Environment* 48(1):1–17.

Fung, A. K., Z. Li, and K. S. Chen. 1992. Backscattering from a randomly rough dielectric surface. *IEEE Transactions on Geoscience and Remote Sensing* 30(2):356–369.

Gao, B. C. 1996. NDWI—A normalized difference water index for remote sensing of vegetation liquid water from space. *Remote Sensing of Environment* 58(3):257–266.

Gardner, W. 1986. Water content. In *Methods of Soil Analysis. Part 1. Physical and Mineralogical Methods*, edited by A. Klutem. American Society of Agronomy, Soil Science Society of America, Madison, WI. p. 493–544.

Ghulam, A., Q. Qin, and Z. Zhan. 2007. Designing of the perpendicular drought index. *Environmental Geology* 52(6):1045–1052.

Gillies, R. R., W. P. Kustas, and K. S. Humes. 1997. A verification of the 'triangle' method for obtaining surface soil water content and energy fluxes from remote measurements of the Normalized Difference Vegetation Index (NDVI) and surface radiant temperature. *International Journal of Remote Sensing* 18(15):3145–3166.

Goward, S. N., G. D. Cruickshanks, and A. S. Hope. 1985. Observed relation between thermal emission and reflected spectral radiance of a complex vegetated landscape. *Remote Sensing of Environment* 18(2):137–146.

Gu, Y., E. Hunt, B. Wardlow, J. B. Basara, J. F. Brown, and J. P. Verdin. 2008. Evaluation of MODIS NDVI and NDWI for vegetation drought monitoring using Oklahoma Mesonet soil moisture data. *Geophysical Research Letters* 35(22):L22401.

Gruhier, C., P. De Rosnay, Y. H. Kerr, E. Mougin, E. Ceschia, J. C. Calvet, and P. Richaume. 2008. Evaluation of AMSR-E soil moisture product based on ground measurements over temperate and semiarid regions. *Geophysical Research Letters* 35:L10405.

Hassan, Q. K., C. P.-A. Bourque, F.-R. Meng, and R. M. Cox. 2007. A wetness index using terrain-corrected surface temperature and normalised difference vegetation index derived from standard MODIS products: An evaluation of its use in a Humid forest-dominated region of eastern Canada. *Sensors* 7(10):2028–2048.

Hillel, D. 1998. *Environmental Soil Physics*. Academic Press, San Diego, CA.

Holah, N., N. Baghdadi, M. Zribi, A. Bruand, and C. King. 2005. Potential of ASAR/ENVISAT for the characterization of soil surface parameters over bare agricultural fields. *Remote Sensing of Environment* 96(1):78–86.

Huete, A., C. Justice, and H. Liu. 1994. Development of vegetation and soil indices for MODIS-EOS. *Remote Sensing of Environment* 49(3):224–234.

Hunt, E. R., and B. N. Rock. 1989. Detection of changes in leaf water content using near-and middle-infrared reflectances. *Remote Sensing of Environment* 30(1):43–54.

Jackson, T. J., T. J. Schmugge, and J. R. Wang. 1982. Passive microwave sensing of soil moisture under vegetation canopies. *Water Resources Research* 18(4):1137–1142.

Jackson, T. J. 1993. Measuring surface soil moisture using passive microwave remote sensing. *Hydrological Processes* 7(2):139–152.

Jackson, T. J., and T. J. Schmugge. 1989. Passive microwave remote sensing system for soil moisture: some supporting research. *IEEE Transactions on Geoscience and Remote Sensing* 27(2):225–235.

Jackson, T. J., R. Bindlish, A. J. Gasiewski, B. Stankov, M. Klein, E. G. Njoku, D. Bosch, T. L. Coleman, C. A. Laymon, and P. Starks. 2005. Polarimetric scanning radiometer C- and X-band microwave observations during SMEX03. *IEEE Transactions on Geoscience and Remote Sensing* 43(11):2418–2430.

Jackson, T. J., D. M. Le Vine, C. T. Swift, T. J. Schmugge, and F. R. Schiebe. 1995. Large area mapping of soil moisture using the ESTAR passive microwave radiometer in Washita'92. *Remote Sensing of Environment* 54(1):27–37.

Jackson, T. J., D. M. Le Vine, A. Y. Hsu, A. Oldak, P. J. Starks, C. T. Swift, J. D. Isham, and M. Haken. 1999. Soil moisture mapping at regional scales using microwave radiometry: The Southern Great Plains Hydrology Experiment. *IEEE Transactions on Geoscience and Remote Sensing* 37(5):2136–2151.

Kahle, A. B. 1977. A simple thermal model of the Earth's surface for geologic mapping by remote sensing. *Journal of Geophysical Research* 82(11):1673–1680.

Katou, A., and Y. Yamaguchi. 2005. Assessment of water and temperature stresses of vegetation in urban areas by the VWTI index. *Proceedings of the Conference of the Remote Sensing Society of Japan* 39:201–202.

Katzberg, S. J., O. Torres, M. S. Grant, and D. Masters. 2006. Utilizing calibrated GPS reflected signals to estimate soil reflectivity and dielectric constant: Results from SMEX02. *Remote Sensing of Environment* 100(1):17–28.

Kerr, Y. H., P. Waldteufel, P. Richaume, J. P. Wigneron, P. Ferrazzoli, A. Mahmoodi, A. Al Bitar, F. Cabot, C. Gruhier, and S. E. Juglea. 2012. The SMOS soil moisture retrieval algorithm. *IEEE Transactions on Geoscience and Remote Sensing* 50(5):1384–1403.

Kerr, Y. H., P. Waldteufel, J. P. Wigneron, J. Martinuzzi, J. Font, and M. Berger. 2001. Soil moisture retrieval from space: The Soil Moisture and Ocean Salinity (SMOS) mission. *IEEE Transactions on Geoscience and Remote Sensing* 39(8):1729–1735.

Koike, T., I. Nakamura, I. Kaihotsu, N. Davva, N. Matsurra, K. Tamagawa, and H. Fujii. 2004. Development of an advanced microwave scanning radiometer (AMSR-E) algorithm of soil moisture and vegetation water content. *Annual Journal of Hydraulic Engineering* 48:217–222.

Kuruku, Y., F. B. Sanli, M. T. Esetlili, M. Bolca, and C. Goksel. 2009. Contribution of SAR images to determination of surface moisture on the Menemen Plain, Turkey. *Journal of Remote Sensing* 30(7):1805–1817.

Larson, K. M., E. E. Small, E. D. Gutmann, A. L. Bilich, J. J. Braun, and V. U. Zavorotny. 2008. Use of GPS receivers as a soil moisture network for water cycle studies. *Geophysical Research Letters* 35:L24405.

Leconte, R., F. Brissette, M. Garlneau, and J. Rousselle. 2004. Mapping near-surface soil moisture with RADARSAT-1 synthetic aperture radar data. *Water Resources Research* 40:1–13.

Lee, K. H., and E. N. Anagnostou. 2004. A combined passive/active microwave remote sensing approach for surface variable retrieval using Tropical Rainfall Measuring Mission observations. *Remote Sensing of Environment* 92(1):112–125.

Lerner, R. M., and J. P. Hollinger. 1977. Analysis of 1.4 GHz radiometric measurements from Skylab. *Remote Sensing of Environment* 6(4):251–269.

Li, L., P. W. Gaiser, B. C. Gao, R. M. Bevilacqua, T. J. Jackson, E. G. Njoku, C. Rüdiger, J. C. Calvet, and R. Bindlish. 2010. WindSat global soil moisture retrieval and validation. *IEEE Transactions on Geoscience and Remote Sensing* 48(5):2224–2241.

Li, Q., R. Zhong, J. Huang, and H. Gong. 2011. Comparison of two retrieval methods with combined passive and active microwave remote sensing observations for soil moisture. *Mathematical and Computer Modelling* 54(3):1181–1193.

Liu, W., F. Baret, Gu Xingfa, Tong Qingxi, Zheng Lanfen, and Zhang Bing. 2002. Relating soil surface moisture to reflectance. *Remote Sensing of Environment* 81(2–3):238–246.

Liu, W., F. Baret, Xingfa Gu, Bing Zhang, Qingxi Tong, and Lanfen Zheng. 2003. Evaluation of methods for soil surface moisture estimation from reflectance data. *International Journal of Remote Sensing* 24(10):2069–2083.

Liu, S., O. A. Chadwick, D. A. Roberts, and C. J. Still. 2011. Relationships between GPP, satellite measures of greenness and canopy water content with soil moisture in Mediterranean-climate grassland and oak savanna. *Applied and Environmental Soil Science* 2011:839028.

Liu, S., D. A. Roberts, O. A. Chadwick, and C. J. Still. 2012. Spectral responses to plant available soil moisture in a Californian grassland. *International Journal of Applied Earth Observation and Geoinformation* 19:31–44.

Liu, Y. Y., W. A. Dorigo, R. M. Parinussa, R. A. M. De Jeu, W. Wagner, M. F. McCabe, J. P. Evans, and A. Van Dijk. 2012. Trend-preserving blending of passive and active microwave soil moisture retrievals. *Remote Sensing of Environment* 123:280–297.

Liu, Y. Y., R. M. Parinussa, W. A. Dorigo, R. A. M. De Jeu, W. Wagner, A. Van Dijk, M. F. McCabe, and J. P. Evans. 2011. Developing an improved soil moisture dataset by blending passive and active microwave satellite-based retrievals. *Hydrology and Earth System Sciences* 15(2):425–436.

Lobell, D. B., and G. P. Asner. 2002. Moisture effects on soil reflectance. *Soil Science Society of America Journal* 66(3):722–727.

Loew, A. 2008. Impact of surface heterogeneity on surface soil moisture retrievals from passive microwave data at the regional scale: The Upper Danube case. *Remote Sensing of Environment* 112(1):231–248.

Lu, S., Z. Ju, T. Ren, and R. Horton. 2009. A general approach to estimate soil water content from thermal inertia. *Agricultural and Forest Meteorology* 149(10):1693–1698.

Mariotto, I., and V. P. Gutschick. 2010. Non-Lambertian corrected albedo and vegetation index for estimating land evaporation in a heterogeneous semiarid landscape. *Remote Sensing* 2:926–938.

Matgen, P., F. Fenicia, S. Heitz, D. Plaza, R. De Keyser, V. Pauwels, W. Wagner, and H. Savenije. 2012. Can ASCAT-derived soil wetness indices reduce predictive uncertainty in well-gauged areas? A comparison with in situ observed soil moisture in an assimilation application. *Advances in Water Resources* 44:49–65.

Mathieu, R., M. Sbih, A. A. Viau, F. Anctil, L. E. Parent, and J. Boisvert. 2003. Relationships between Radarsat SAR data and surface moisture content of agricultural organic soils. *International Journal of Remote Sensing* 24(24):5265–5281.

Minacapilli, M., M. Iovino, and F. Blanda. 2009. High resolution remote estimation of soil surface water content by a thermal inertia approach. *Journal of Hydrology* 379(3–4):229–238.

Mo, T., B. J. Choudhury, T. J. Schmugge, J. R. Wang, and T. J. Jackson. 1982. A model for microwave emission from vegetation-covered fields. *Journal of Geophysical Research* 87(C13):11,229–11,237.

Muller, E., and H. Décamps. 2000. Modeling soil moisture-reflectance. *Remote Sensing of Environment* 76:173–180.

Naeimi, V., Z. Bartalis, and W. Wagner. 2009a. ASCAT soil moisture: An assessment of the data quality and consistency with the ERS scatterometer heritage. *Journal of Hydrometeorology* 10(2):555–563.

Naeimi, V., K. Scipal, Z. Bartalis, S. Hasenauer, and W. Wagner. 2009b. An improved soil moisture retrieval algorithm for ERS and METOP scatterometer observations. *IEEE Transactions on Geoscience and Remote Sensing* 47(7):1999–2013.

Narayan, U., V. Lakshmi, and T. J. Jackson. 2006. High-resolution change estimation of soil moisture using L-band radiometer and radar observations made during the SMEX02 experiments. *IEEE Transactions on Geoscience and Remote Sensing* 44(6):1545–1554.

Narayan, U., V. Lakshmi, and E. G. Njoku. 2004. Retrieval of soil moisture from passive and active L/S band sensor (PALS) observations during the Soil Moisture Experiment in 2002 (SMEX02). *Remote Sensing of Environment* 92(4):483–496.

Nemani, R. R., and S. W. Running. 1989. Estimation of regional surface resistance to evapotranspiration from NDVI and thermal-IR AVHRR data. *Journal of Applied Meteorology* 28(4):276–284.

Neusch, T., and M. Sties. 1999. Application of the Dubois-model using experimental synthetic aperture radar data for the determination of soil moisture and surface roughness. *ISPRS Journal of Photogrammetry and Remote Sensing* 54:273–278.

Njoku, E. G., and D. Entekhabi. 1996. Passive microwave remote sensing of soil moisture. *Journal of Hydrology* 184(1–2):101–129.

Njoku, E. G., P. Ashcroft, T. K. Chan, and L. Li. 2005. Global survey and statistics of radio-frequency interference in AMSR-E land observations. *IEEE Transactions on Geoscience and Remote Sensing* 43(5):938–947.

Njoku, E. G., T. J. Jackson, V. Lakshmi, T. K. Chan, and S. V. Nghiem. 2003. Soil moisture retrieval from AMSR-E. *IEEE Transactions on Geoscience and Remote Sensing* 41(2):215–229.

Njoku, E. G., W. J. Wilson, S. H. Yueh, and Y. Rahmat-Samii. 2000. A large-antenna microwave radiometer-scatterometer concept for ocean salinity and soil moisture sensing. *IEEE Transactions on Geoscience and Remote Sensing* 38(6):2645–2655.

Njoku, E. G., W. J. Wilson, S. H. Yueh, S. J. Dinardo, F. K. Li, T. J. Jackson, V. Lakshmi, and J. Bolten. 2002. Observations of soil moisture using a passive and active low-frequency microwave airborne sensor during SGP99. *IEEE Transactions on Geoscience and Remote Sensing* 40(12):2659–2673.

Oh, Y. 2004. Quantitative retrieval of soil moisture content and surface roughness from multi-polarized radar observations of bare soil surfaces. *IEEE Transactions on Geoscience and Remote Sensing* 42:596–601.

Oh, Y., K. Sarabandi, and F. T. Ulaby. 1992. An empirical model and an inversion technique for radar scattering from bare soil surfaces. *IEEE Transactions on Geoscience and Remote Sensing* 30(2):370–381.

Oh, Y., K. Sarabandi, and F. T. Ulaby. 2002. Semiempirical model of the ensemble-averaged differential Mueller matrix for microwave backscattering from bare soil surfaces. *IEEE Transactions on Geoscience and Remote Sensing* 40(6):1348–1355.

Oliva, R., E. Daganzo, Y. H. Kerr, S. Mecklenburg, S. Nieto, P. Richaume, and C. Gruhier. 2012. SMOS radio frequency interference scenario: Status and actions taken to improve the RFI environment in the 1400–1427-MHz passive band. *IEEE Transactions on Geoscience and Remote Sensing* 50(5):1427–1439.

Orchard, V. A., and F. J. Cook. 1983. Relationship between soil respiration and soil moisture. *Soil Biology and Biochemistry* 15(4):447–453.

Owe, M., R. A. M. de Jeu, and T. Holmes. 2008. Multisensor historical climatology of satellite-derived global land surface moisture. *Journal of Geophysical Research* 113(F1):F01002.

Owe, M., R. A. M. de Jeu, and J. Walker. 2001. A methodology for surface soil moisture and vegetation optical depth retrieval using the microwave polarization difference index. *IEEE Transactions on Geoscience and Remote Sensing* 39(8):1643–1654.

Owen, T. W., T. N. Carlson, and R. R. Gillies. 1998. Remotely sensed surface parameters governing urban climate change. *International Journal of Remote Sensing* 19:1663–1681.

Paloscia, S., P. Pampaloni, S. Pettinato, and E. Santi. 2008. A comparison of algorithms for retrieving soil moisture from ENVISAT/ASAR images. *IEEE Transactions on Geoscience and Remote Sensing* 46(10):3274–3284.

Parinussa, R. M., T. R. H. Holmes, and R. A. M. de Jeu. 2012. Soil moisture retrievals from the WindSat spaceborne polarimetric microwave radiometer. *IEEE Transactions on Geoscience and Remote Sensing* 50(7):2683–2694.

Patel, N. R., R. Anapashsha, S. Kiumar, S. K. Saha, and V. K. Dadhwal. 2009. Assessing potential of MODIS derived temperature/vegetation condition index (TVDI) to infer soil moisture status. *International Journal of Remote Sensing* 30(1):23–39.

Petropoulos, G., and T. N. Carlson. 2011. Retrievals of turbulent heat fluxes and soil moisture content by Remote Sensing. In *Advances in Environmental Remote Sensing: Sensors, Algorithms, and Applications*. CRC Press, Boca Raton, FL, pp. 469–502.

Petropoulos, G., T. N. Carlson, M. J. Wooster, and S. Islam. 2009a. A review of Ts/VI remote sensing based methods for the retrieval of land surface energy fluxes and soil surface moisture. *Progress in Physical Geography* 33(2):224–250.

Petropoulos, G., T. N. Carlson, and M. J. Wooster. 2009b. An overview of the use of the SimSphere Soil Vegetation Atmosphere Transfer (SVAT) model for the study of land-atmosphere interactions. *Sensors* 9(6):4286–4308.

Petropoulos, G., M. J. Wooster, T. N. Carlson, and N. Drake. 2010. Synergy of the SimSphere land surface process model with ASTER imagery for the retrieval of spatially distributed estimates of surface turbulent heat fluxes and soil moisture content. Paper presented at European Geosciences Union 2010, May 2–7, Vienna, Austria.

Piles, M., A. Camps, M. Vall-llossera, I. Corbella, R. Panciera, C. Rudiger, Y. H. Kerr, and J. Walker. 2011. Downscaling SMOS-derived soil moisture using MODIS visible/infrared data. *IEEE Transactions on Geosciences and Remote Sensing* 49(9):3156–3166.

Pratt, D. A., and C. D. Ellyett. 1979. The thermal inertia approach to mapping of soil moisture and geology. *Remote Sensing of Environment* 8(2):151–168.

Price, J. C. 1977. Thermal inertia mapping: A new view of the earth. *Journal of Geophysical Research* 82(18):2582–2590.

Pulliainen, J. T., T. Manninen, and M. T. Hallikainen. 1998. Application of ERS-1 wind scatterometer data to soil frost and soil moisture monitoring in boreal forest zone. *IEEE Transactions on Geoscience and Remote Sensing* 36(3):849–863.

Raich, J. W., and W. H. Schlesinger. 1992. The global carbon dioxide flux in soil respiration and its relationship to vegetation and climate. *Tellus, Ser. B* 44(2):81–99.

Robinson, D. A., C. S. Campbell, J. W. Hopmans, B. K. Hornbuckle, S. B. Jones, R. Knight, F. Ogden, J. Selker, and O. Wendroth. 2008. Soil moisture measurement for ecological and hydrological watershed-scale observatories: A review. *Vadose Zone Journal* 7(1):358–389.

Robinson, D. A., S. B. Jones, J. M. Wraith, D. Or, and S. P. Friedman. 2003. A review of advances in dielectric and electrical conductivity measurement in soils using time domain reflectometry. *Vadose Zone Journal* 2(4):444–475.

Rüdiger, C., J. C. Calvet, C. Gruhier, T. R. H. Holmes, R. A. M. de Jeu, and W. Wagner. 2009. An intercomparison of ERS-Scat and AMSR-E soil moisture observations with model simulations over France. *Journal of Hydrometeorology* 10(2):431–447.

Sanchez, N., J. Martinez-Fernandez, A. Scaini, and C. Perez-Gutierrez. 2012. Validation of the SMOS L2 Soil Moisture Data in the REMEDHUS Network (Spain). *IEEE Transactions on Geoscience and Remote Sensing* 50(5):1602–1611.

Sandholt, I., K. Rasmussen, and J. Andersen. 2002. A simple interpretation of the surface temperature/vegetation index space for assessment of surface moisture status. *Remote Sensing of Environment* 79(2):213–224.

Seneviratne, S. I., T. Corti, E. L. Davin, M. Hirschi, E. B. Jaeger, I. Lehner, B. Orlowsky and A. J. Teuling. 2010. Investigating soil moisture–climate interactions in a changing climate: A review. *Earth Sciences Reviews* 99(3–4):125–161.

Schlenz, F., A. Loew, and W. Mauser. 2012. Uncertainty assessment of the SMOS validation in the upper Danube catchment. *IEEE Transactions on Geoscience and Remote Sensing* 50(5):1517–1529.

Shakya, N., and Y. Yamaguchi. 2010. Vegetation, water and thermal stress index for study of drought in Nepal and central northeastern India. *International Journal of Remote Sensing* 31(4):903–912.

Shi, J., J. Wang, A. Y. Hsu, P. E. O'Neill, and E. T. Engman. 1997. Estimation of bare surface soil moisture and surface roughness parameter using L-band SAR image data. *IEEE Transactions on Geoscience and Remote Sensing* 35(5):1254–1266.

Shoshany, M., T. Svoray, P. J. Curran, G. M. Foody, and A. Perevolotsky. 2000. The relationship between ERS-2 SAR backscatter and soil moisture: Generalization from a humid to semiarid transect. *International Journal of Remote Sensing* 21(11):2337.

Sobrino, J. A., B. Franch, C. Mattar, J. C. Jiménez-Muñoz, and C. Corbari. 2012. A method to estimate soil moisture from airborne hyperspectral scanner (AHS) and aster data: Application to
SEN2FLEX and SEN3EXP campaigns. *Remote Sensing of Environment* 17:415–428.

Song, K., X. Zhou, and Y. Fan. 2009. Empirically adopted IEM for retrieval of soil moisture
from radar backscattering coefficients. *IEEE Transactions on Geoscience and Remote Sensing*
47(6):1662–1672.

Stoner, E. R., and M. F. Baumgardner. 1981. Characteristic variations in reflectance of surface soils.
Soil Science Society of America Journal 45(6):1161–1165.

Temimi, M., R. Leconte, N. Chaouch, P. Sukumal, R. Khanbilvardi, and F. Brissette. 2010. A combination of remote sensing data and topographic attributes for the spatial and temporal monitoring
of soil wetness. *Journal of Hydrology* 388(1):28–40.

Thoma, D. P., M. S. Moran, R. Bryant, M. M. Rahman, C. D. Holifield Collins, T. O. Keefer, R. Noriega
et al. 2008. Appropriate scale of soil moisture retrieval from high resolution radar imagery for
bare and minimally vegetated soils. *Remote Sensing of Environment* 112(2):403–414.

Topp, G. C., and P. A. Ferré. 2002. Water content. In *Methods of Soil Analysis. Part 4, Physical Methods*,
edited by J. H. Dane and G. C. Topp. Soil Science Society of America, Madison, WI, pp. 417–422.

Verhoest, N. E., H. Lievens, W. Wagner, J. Álvarez-Mozos, M. S Moran, and F. Mattia. 2008. On the
soil roughness parameterization problem in soil moisture retrieval of bare surfaces from synthetic aperture radar. *Sensors* 8(7):4213–4248.

Verstraeten, W. W., F. Veroustraete, C. J. van der Sande, I. Grootaers, and J. Feyen. 2006. Soil moisture
retrieval using thermal inertia, determined with visible and thermal spaceborne data, validated
for European forests. *Remote Sensing of Environment* 101(3):299–314.

Vincente-Serrano, S. M., X. Pons-Fernandez, and J. M. Cuadrat-Prats. 2004. Mapping soil moisture
in the central Ebro river valley with Landsat and NOAA satellite imagery: A comparison with
meteorological data. *International Journal of Remote Sensing* 25(20):4325–4350.

Wagner, W., G. Blöschl, P. Pampaloni, and J. Calvet. 2007a. Operational readiness of microwave
remote sensing of soil moisture for hydrologic applications. *Hydrology Research* 38(1):1–20.

Wagner, W., R. De Jeu, and P. J. van Oevelen. 2009. Towards multisource global soil moisture datasets
for unraveling climate change impacts on water resources. Paper presented at 33rd International
Symposium on Remote Sensing of Environment (ISRSE), May 4–8, Joint Research Centre of the
European Commission. Stresa, Italy.

Wagner, W., W. Dorigo, R. De Jeu, D. Fernandez, J. Benveniste, E. Haas, and M. Ertl. 2012. Fusion of
active and passive microwave observations to create an Essential Climate Variable data record
on soil moisture. Paper presented at XXII ISPRS Congress, Melbourne, Australia.

Wagner, W., G. Lemoine, M. Borgeaud, and H. Rott. 1999a. A study of vegetation cover effects on ERS
scatterometer data. *IEEE Transactions on Geoscience and Remote Sensing* 37(2):938–948.

Wagner, W., G. Lemoine, and H. Rott. 1999b. A method for estimating soil moisture from ERS scatterometer and soil data. *Remote Sensing of Environment* 70(2):191–207.

Wagner, W., V. Naeimi, K. Scipal, R. de Jeu, and J. Martínez-Fernández. 2007b. Soil moisture from
operational meteorological satellites. *Hydrogeology Journal* 15(1):121–131.

Wagner, W., K. Scipal, C. Pathe, D. Gerten, W. Lucht, and B. Rudolf. 2003. Evaluation of the agreement between the first global remotely sensed soil moisture data with model and precipitation
data. *Journal of Geophysical Research* 108(D19):4611.

Wan, Z., P. Wang, and X. Li. 2004. Using MODIS land surface temperature and normalized difference vegetation index products for monitoring drought in the southern Great Plains, USA.
International Journal of Remote Sensing 25(1):61–72.

Wang, C., J. Qi, S. Moran, and R. Marsett. 2004. Soil moisture estimation in a semiarid rangeland
using ERS-2 and TM imagery. *Remote Sensing of Environment* 90(2):178–189.

Wang, J. R., J. C. Shiue, T. J. Schmugge, and E. T. Engman. 1990. The L-band PBMR measurements of surface soil moisture in FIFE. *IEEE Transactions on Geoscience and Remote Sensing*
28(5):906–914.

Wang, L., and J. J. Qu. 2007. NMDI: A normalized multi-band drought index for monitoring soil and
vegetation moisture with satellite remote sensing. *Geophysical Research Letters* 34(20):L20405.

Wang, L., J. J. Qu, X. Hao, and Q. Zhu. 2008. Sensitivity studies of the moisture effects on MODIS SWIR reflectance and vegetation water indices. *International Journal of Remote Sensing* 29(24):7065–7075.

Wigneron, J. P., J. C. Calvet, T. Pellarin, A. A. Van de Griend, M. Berger, and P. Ferrazzoli. 2003. Retrieving near-surface soil moisture from microwave radiometric observations: Current status and future plans. *Remote Sensing of Environment* 85(4):489–506.

Xue, Y., and A. P. Cracknell. 1995. Advanced thermal inertia modelling. *Remote Sensing* 16(3):431–446.

Yang, H., J. Shi, Z. Li, and H. Guo. 2006. Temporal and spatial soil moisture change pattern detection in an agricultural area using multitemporal Radarsat ScanSAR data. *International Journal of Remote Sensing* 27(19):4199–4212.

Zine, S., L. Jarlan, P. L. Frison, E. Mougin, P. Hiernaux, and J. P. Rudant. 2005. Land surface parameter monitoring with ERS scatterometer data over the Sahel: A comparison between agro-pastoral and pastoral areas. *Remote Sensing of Environment* 96(3):438–452.

Zribi, M., N. Baghdadi, N. Holah, and O. Fafin. 2005. New methodology for soil surface moisture estimation and its application to ENVISAT-ASAR multiincidence data inversion. *Remote Sensing of Environment* 96:485–496.

Zribi, M., S. Le Hégarat-Mascle, C. Ottlé, B. Kammoun, and C. Guerin. 2003. Surface soil moisture estimation from the synergistic use of the (multiincidence and multiresolution) active microwave ERS Wind Scatterometer and SAR data. *Remote Sensing of Environment* 86:30–41.

5

Remote Sensing of the Land Surface Radiation Budget

Shunlin Liang, Xiaotong Zhang, Tao He, Jie Cheng, and Dongdong Wang

CONTENTS

5.1 Introduction

Land surface energy balance is central to characterizing land surface ecological, hydrological, and biogeochemical processes. The land surface energy balance equation can be written as

$$R_n = G + H + \lambda\text{ET} \tag{5.1}$$

where R_n is the all-wave net radiation, G is the soil heat flux, H is the sensible heat flux, and λET is the latent heat flux in which λ is the latent heat of evaporation of water and ET is the rate of evaporation of water.

The land surface radiation budget (SRB), characterized by the net radiation (R_n), represents the balance between incoming radiation from the atmosphere and outgoing longwave and reflected shortwave radiation from the Earth's surfaces. All-wave net radiation is the sum of shortwave net radiation $\left(R_n^s\right)$ and longwave net radiation $\left(R_n^l\right)$ and can be expressed by

$$R_n = R_n^s + R_n^l = (1 - \alpha_{sw})F_d^s + F_d^l - F_u^l = (1 - \alpha_{sw})F_d^s + \varepsilon F_d^l - \sigma\varepsilon T_s^4 \tag{5.2}$$

where α_{sw} is the land surface shortwave broadband albedo that will be discussed in Section 5.3, F_d^s is the shortwave downward flux and often called insolation that will be discussed in Section 5.2, F_d^l and F_u^l are the longwave downward and upwelling radiation, respectively, ε is the longwave broadband emissivity, T_s is the surface skin temperature, and σ is the Stefan–Boltzmann's constant. All these four quantities will be discussed in Section 5.4. The all-wave net radiation (R_n) will be discussed in Section 5.5.

As illustrated in other chapters in this book, estimation of heat fluxes (H, ET) and soil moisture content from remotely sensed data largely depends on SRB components. For example, there are multiple methods for estimating ET, as summarized by Wang et al. (2012), almost all of which require shortwave incident radiation $\left(F_d^s\right)$ or net radiation $\left(R_n^s\right)$, and many methods also use land surface temperature (T_s). Liang et al. (2012b) recently reviewed various remote sensing methods for estimating soil moisture contents, and some of them that rely on optical data use land surface temperature (T_s) and albedo (α_{sw}). Therefore accurately estimating SRB and its individual components is highly relevant to the entire book.

Figure 5.1 is the updated global annual mean energy budget from various sources (Stephens et al. 2012). Of the total solar radiation entering the Earth system, about 30% is reflected back to space, about 22% absorbed by the atmosphere, and about 48% by the surface. From Figure 5.1, the Earth surface (including both ocean and land) albedo is 0.12. Land surface albedo is about 0.24. In the longwave spectrum, the surface emits more radiation (398 W m^{-2}) than the received downward radiation from the atmosphere (345 W m^{-2}). Thus the total net radiation at the Earth's surface is about 112 W m^{-2}, and about 79% becomes latent heat flux for evapotranspiration.

These numbers represent our current and best knowledge on the Earth energy budget, which is constantly improving. For example, the downward longwave radiation (345 W m^{-2}) is significantly higher than the previous number (333 W m^{-2}) (Trenberth and Fasullo 2012). In general, most estimates of these components are largely uncertain. Figure 5.2 shows the uncertainties of surface energy budget components from both observations and

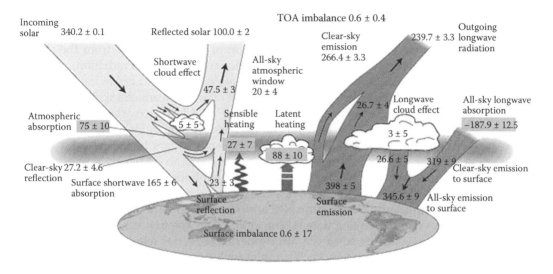

FIGURE 5.1
(See color insert.) The global annual mean energy budget of Earth for the approximate period 2000–2010. All fluxes are in W m^{-2}. Solar fluxes are in yellow and infrared fluxes in pink. The four flux quantities in purple-shaded boxes represent the principal components of the atmospheric energy balance. (Adapted from Stephens, G. L. et al., *Nat. Geosci.*, 5(10), 691–696, 2012.)

models. The observed fluxes (containing error estimates) are taken from Figure 5.1, and the climate model fluxes are from simulations archived under the World Climate Research Programme's Coupled Model Intercomparison Project phase 5 (CMIP5) twentieth-century experiments. The fluxes from a 16-model ensemble are summarized in terms of the range in model values (maximum and minimum fluxes) with the ensemble mean fluxes given in parentheses.

These numbers represent the SRB climatology on global scales, but there is a tremendous amount of variation in both space and time. In the temporal dimension, all SRB components are variable. For example, incident solar radiation has undergone significant variations on decadal timescales due to changes in clouds and air pollution. Surface albedo

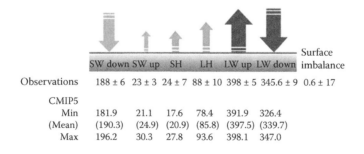

	SW down	SW up	SH	LH	LW up	LW down	Surface imbalance
Observations	188 ± 6	23 ± 3	24 ± 7	88 ± 10	398 ± 5	345.6 ± 9	0.6 ± 17
CMIP5							
Min	181.9	21.1	17.6	78.4	391.9	326.4	
(Mean)	(190.3)	(24.9)	(20.9)	(85.8)	(397.5)	(339.7)	
Max	196.2	30.3	27.8	93.6	398.1	347.0	

FIGURE 5.2
Observed and climate model deduced energy fluxes and uncertainties (all in W m^{-2}) at the surface (Stephens et al. 2012). "SW in" and "SW out" refer to the incoming and outgoing (reflected) solar fluxes at the top of atmosphere (TOA) and "LW out" is the outgoing longwave radiation. Similarly, "SW down" and "SW up" refer to downward and upward (reflected) solar fluxes at the surface, and "LW up" and "LW down" refer to the upward emitted flux of longwave radiation from the surface and the downward longwave flux emitted from the atmosphere to the surface, respectively. SH and LH refer to sensible and latent heat fluxes, respectively.

shows different increasing or decreasing trends at different spatial scales. With increasing radiative absorption due to the abundance of anthropogenic greenhouse gases in the atmosphere and consequent warming, the emission of thermal energy from the atmosphere toward the surface is increasing (known as downward thermal radiation).

Changes in individual radiation budget components result in changes of net radiation, which have great implications for the water cycle and other environmental changes. Wild et al. (2008) estimated that surface net radiation over land has increased by about 2 W m^{-2} per decade from 1986 to 2000, after several decades with no evidence of an increase. They attributed this increase to a combination of several factors including solar radiation "brightening" with more transparent atmospheres and an increased flux of downward longwave radiation, due to enhanced levels of greenhouse gases. Urbanization also increased the all-wave net radiation (Offerle et al. 2005).

Therefore it is important to determine SRB accurately. There are two methods for calculating SRB from remote sensing data (Liang et al. 2010). The first is based on radiative transfer calculations with the retrieved atmospheric properties (cloud and atmosphere parameters) and surface properties. The examples include SBR products from the Clouds and the Earth's Radiant Energy System (CERES) and the International Satellite Cloud Climatology Project (ISCCP). The second is to estimate SRB components directly from satellite observations, such as the Moderate Resolution Imaging Spectroradiometer (MODIS).

In the following four sections, we will document recent advances that have been made in estimating incident solar radiation, albedo, longwave net radiation, and all-wave net radiation. In addition to the methodology, existing satellite products and the spatial and temporal variations of these SRB components are also described. The conclusion provides a synthesis of the review main conclusions and suggests directions of future work.

5.2 Incident Solar Radiation

5.2.1 Background

The Sun is the only energy source for the climate system. Sunlight, which penetrates the atmosphere and reaches the Earth's surface, is crucial for life on our planet. It not only causes the formation of cloud, fog, snow, and rain but also heats the environment, induces pressure gradients, generates evaporation, and enables photosynthesis (Liang et al. 2012a; Wild 2011). The solar irradiance is mainly scattered by clouds, aerosols, or air molecules and absorbed by other gases. Thus only part of it can penetrate the atmosphere and reach the Earth's surface in the form of direct and diffuse radiation.

Most solar radiation is concentrated within the range of 250–2500 nm, which accounts for 99% of the total. The visible, infrared, and ultraviolet bands contain about 50%, 44%, and 6% of the total solar radiation, respectively. The incident shortwave solar radiation, also known as insolation, is referred to as total solar irradiance incident at Earth surface. The visible part (400–700 nm) of insolation is called photosynthetically active radiation (PAR). PAR constitutes the basic source of energy for biomass by controlling the photosynthetic rate of organisms on land, thus directly affecting plant growth. It is an indispensable variable in calculating gross primary production or net primary production (NPP) (Running et al. 1999, 2000).

Besides the spatial variations, there remain great temporal variations of incident solar radiation. The increasing and decreasing trends are often referred to as global brightening and dimming, caused mainly by changes in atmospheric constituents including cloud amount, cloud cover, and aerosols (Wang et al. 2009a; Wild 2011).

In this section, we will introduce the principles and methods for estimation of insolation and PAR from remotely sensed data. The advantages and disadvantages of the existing methods will be discussed, and the currently available ground measurements and satellite-derived radiation products will be presented.

5.2.2 Methodology

The insolation at surface has traditionally been estimated from surface meteorological observations using various empirical methods based on the relationship between ground measurements of radiation and some meteorological parameters, such as sunshine duration hours, surface pressure, and so on. Because of the long records of meteorological observations, this method is still widely used and therefore will be first introduced in Section 5.2.2.1. Given the atmospheric and surface properties, various radiative transfer (RT) codes, such as MODTRAN (Anderson et al. 1999), can be used for calculating incident solar radiation. Oreopoulos (2012) compared most of the radiative transfer codes for evaluation. The advantage of using rigorous RT codes is their strong physical basis. The disadvantage is that their calculations can be very time consuming. To improve the computational efficiency, various parameterization schemes have been developed and some of them will be described in Section 5.2.2. The lookup table (LUT) methods that do not rely on a radiative transfer code will be introduced in Section 5.2.3.

5.2.2.1 Empirical Method

The relative sunshine duration method is one of the classical empirical models to estimate surface shortwave radiation. There have been many models developed to estimate the surface insolation from sunshine duration or other meteorological parameters (El-Metwally 2004; Paulescu and Schlett 2003; Psiloglou and Kambezidis 2007). The most widely used method is that of Angstrom (1924), who presented a linear relationship between the ratio of average daily global radiation to the corresponding value on a completely clear day and the average daily sunshine duration to the possible sunshine duration. The relationship has achieved great success over lowland regions, but the coefficients of empirical methods are always site dependent. Yang et al. (2010) developed a hybrid empirical method to estimate downward shortwave radiation over high-altitude regions. The model is introduced as follows:

$$F_s = F_{s,\text{clr}} \tau_c \tag{5.3}$$

$$F_{s,\text{clr}} = F_b^s + F_d^s \tag{5.4}$$

$$\tau_c = 0.2505 + 1.1468 \left(\frac{n}{N} \right) - 0.3974 \left(\frac{n}{N} \right)^2 \tag{5.5}$$

where F_s is the all-sky daily-mean downward shortwave radiation, and $F_{s,\text{clr}}$ is the clear-sky daily-mean downward shortwave radiation. τ_c is the transmittance for daily-mean downward shortwave radiation due to cloud scattering and absorption, n is the actual sunshine duration, and N is the maximum possible sunshine duration. F_b^s and F_d^s are the daily-mean direct radiation and diffuse radiation at surface under clear-sky conditions. They can be estimated using

$$F_b^s = \frac{1}{t_2 - t_1} \int_{t_1}^{t_2} I_0 T_b \, dt \tag{5.6}$$

$$F_d^s = \frac{1}{t_2 - t_1} \int_{t_1}^{t_2} I_0 T_b \, dt \tag{5.7}$$

where I_0 is the solar irradiance on a horizontal level at top of atmosphere (TOA), and t_1 and t_2 are the sunrise and sunset, respectively. T_b and T_d are the broadband direct radiative transmittance and the diffuse radiative transmittance, respectively. The broadband direct and diffuse radiative transmittances are two functions of five damping processes in the atmospheric layer, namely, Rayleigh scattering, aerosol extinction, ozone absorption, water vapor absorption, and mixed gases absorption. The detailed calculation of these transmittances can be found in the work of Yang et al. (2006). The input parameters of this empirical method include the precipitable water, the Angstrom turbidity, and the thickness of the ozone layer as well as surface elevation and air pressure.

5.2.2.2 Parameterization Method

Solar radiation is partly absorbed by water vapor, gas, or ozone and also partly absorbed and scattered by aerosol. Parameterization methods estimate the surface solar radiation through the calculation of the amount of radiation absorbed, reflected, and scattered. In most of the transmittance parameterization methods, attenuation of direct solar radiation in the atmosphere is usually presented in terms of transmittance T_t,

$$F_b^s = I_0 \cos(\theta) T_t \tag{5.8}$$

where θ is the solar zenith angle and I_0 is the extraterrestrial incident radiation (W m^{-2}), varying slightly because of the distance between the Sun and the Earth:

$$I_0 = E_0 \left(1 + 0.033 \cos \left(2\pi d_n / 365\right)\right) T_t \tag{5.9}$$

E_0 is the extraterrestrial solar radiation (W m^{-2}), also known as the solar constant, and is the Julian day. The transmittance T_t is the composition of transmittances taking place in the atmosphere.

$$T_t = T_R T_A T_O T_W T_G T_N \tag{5.10}$$

where T_R, T_A, T_O, T_W, T_G, and T_N refer to the process of Rayleigh scattering, aerosol scattering, ozone absorption, water vapor absorption, uniformly mixed gas (carbon dioxide and oxygen) absorption, and nitrogen dioxide absorption, respectively. In some models, T_N is not considered in the transmittance formula because it does not influence the radiation significantly. The detailed formulae of these transmittance terms can be found in the work of Zhang et al. (2012a), Ryu et al. (2008), Gueymard (2003), and other relevant literatures. The parameterization methods need many atmospheric parameters including water vapor content, ozone content, aerosol optical thickness, the Angstrom turbidity coefficient, and so on. All these parameters are provided at meteorological sites or through satellite products, such as the MODIS land surface and atmosphere products.

In the following two subsections, we will introduce the parameterization schemes of the diffuse radiation in both clear- and cloudy-sky conditions, respectively.

5.2.2.2.1 Clear Sky

Parameterization methods for surface insolation estimation under clear-sky conditions have been widely reported and evaluated (Annear and Wells 2007; Houborg et al. 2007; Houborg and Soegaard 2004; Ryu et al. 2008). The methods for insolation estimation can be divided into the spectral and broadband models. Both the spectral and broadband models estimate surface insolation by calculation of the amount of solar radiation that is absorbed, reflected, and scattered. In spectral models, the amount of solar radiation in each waveband is calculated and the total solar radiation is estimated through spectral integration. Bird's model, which is one of the broadband models, was developed based on the comparison of SOLTRAN3 and SOLTRAN4. This model has been accurately evaluated in several studies (Annear and Wells 2007; Houborg et al. 2007; Houborg and Soegaard 2004; Ryu et al. 2008). With Bird's model, direct normal irradiance $\left(F_b^s \right)$ and diffuse irradiance $\left(F_d^s \right)$ can be obtained using

$$F_b^s = I_0 \cos(\theta) \cdot 0.9662 \cdot T_A T_R T_G T_O T_W \tag{5.11}$$

$$F_d^s = I_0 \cos(\theta) 0.79 T_O T_W T_G T_{AA} \frac{0.5(1-T_R) + B_a(1-T_{AS})}{1 - M + M^{1.02}} \tag{5.12}$$

where T_{AA} is the transmittance due to aerosol absorption, T_{AS} is the transmittance due to aerosol scattering, B_a is the constant aerosol forward scattering ratio (equal to 0.84), z is the solar zenith angle, and M is the optical air mass. The insolation is the sum of the direct irradiance, diffuse irradiance, and irradiance from multiple scattering between the ground and atmosphere that is calculated using

$$F_r^s = r_g r_s \left(F_b^s + F_d^s \right) / (1 - r_g r_s) \tag{5.13}$$

where r_s and r_g are the atmospheric and ground albedo, respectively.

5.2.2.2.2 Cloudy Sky

The broadband model used for cloudy-sky conditions, proposed by Choudhury (1982), is simple with readily accessible parameters. The surface incident solar radiation under cloudy conditions can be calculated if the cloud coverage, solar zenith angle, cloud optical thickness, and land surface albedo are given by

$$F_{cld}^s = I_0(1 - f_{cld} + f_{cld} T_{cld}) / (1 - f_{cld} r_g x_{cld}) \tag{5.14}$$

$$T_{cld} = (0.97(2 + 3 \cos \theta)) / (4 + 0.6\tau_{cld}) \tag{5.15}$$

$$x_{cld} = 0.6\tau_{cld} / (4 + 0.6\tau_{cld}) \tag{5.16}$$

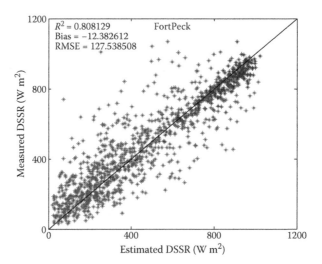

FIGURE 5.3
(See color insert.) Validation results of the clear-sky and broadband cloudy-sky models at the SURFRAD station based on the MODIS products from Terra 2003–2005 (blue represents a clear sky; red represents a cloudy sky). (Reprinted from Zhang, X. et al., *Advanced Remote Sensing: Terrestrial Information Extraction and Applications*, Copyright 2012, with permission from Academic Press.)

where I_0 is the extraterrestrial solar irradiance, f_{cld} is the fraction of cloud cover, τ_{cld} is the cloud optical thickness, θ is the solar zenith angle, and r_g is the surface albedo, all of which can be derived from MODIS products. Figure 5.3 shows the verification of the combined Bird clear-sky model and broadband cloudy-sky model at a Surface Radiation Budget Network (SURFRAD) site (Fort Peck, Montana).

5.2.2.3 LUT Method

If a complete set of the retrieved cloud and atmosphere parameters from other sources is available, radiative transfer codes can be used for calculating incident solar radiation. This approach has been used for estimating insolation from CERES (Wielicki et al. 1998), ISCCP (Zhang et al. 2004), the Global Energy and Water Cycle Experiment (GEWEX) (Pinker et al. 2003), and the Spinning Enhanced Visible and InfraRed Imager (SEVIRI) (Deneke et al. 2008). This approach has a clear physical basis, but use of multiple atmospheric and surface products does not enable us to generate the high spatial resolutions that are needed to meet the requirements of land applications. The uncertainties from these atmospheric properties can also affect the accuracy of the incident solar radiation products.

The alternative approach is to establish the relationship between the TOA radiance and surface incident insolation based on extensive radiative transfer simulations. To improve the computational efficiency, a LUT is used to determine incident solar radiation at surface based on TOA observations. Liang et al. (2006) presented a LUT method to estimate PAR under different illumination geometries and atmospheric conditions based on MODTRAN4 simulations. In the follow-up studies, a series of refinements and improvements have been made. For example, MODIS surface reflectance product (MOD09) was used to map PAR over China from MODIS data (Liu et al. 2008). This algorithm has been extended to estimate PAR from GOES data (Zheng et al. 2008) by taking into account topographic effects. The algorithm was also extended to estimate insolation over China from

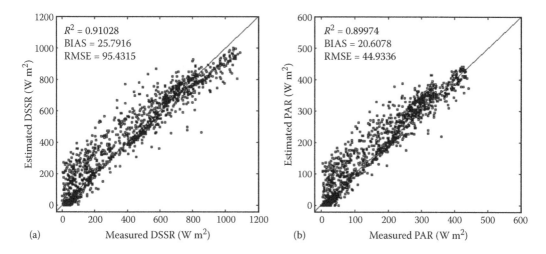

FIGURE 5.4
(a) Insolation and (b) PAR validation results for the SURFRAD Goodwin Creek site of 2008. (From Zhang, X. et al., *J. Geophys. Res.*, 2012.)

Geostationary Meteorological Satellite (GMS) 5 imagery by considering water vapor and the surface elevation (Lu et al. 2010). Huang et al. (2011) further extended the LUT scheme to estimate insolation by combining the Multifunctional Transport Satellite (MTSAT) data and MODIS data products.

Zhang et al. (2012b) applied this LUT method to both polar-orbiting and geostationary satellite data to generate global insolation and PAR products with 5-km spatial resolution and 3-h temporal resolution. The satellite data include the MODIS, the SEVIRI onboard the Meteosat Second Generation (MSG) satellite series, the MTSAT-1R, and the Geostationary Operational Environmental Satellite (GOES) Imager. Figure 5.4 shows the validation results of the insolation and PAR estimations based on the LUT method using both polar-orbiting and geostationary satellite data at SURFRAD Goodwin Creek site, Mississippi, in 2008.

5.2.3 Temporal Scaling of Solar Radiation

Because the satellite imagery is a snapshot of the Earth, the direct estimation of solar radiation from the imagery is usually the instantaneous value. However, the temporally integrated or averaged solar radiation during a period of time carries more importance for climatic and ecological applications. The integrated daily PAR is often used in the estimation of ecosystem productivity. For the geostationary satellites with frequent revisiting time (e.g., every 30 min for GOES), the temporal scaling from instantaneous values to daily value is straightforward. It will not result in large errors by assuming solar radiation changes linearly between two adjacent observations. The simple average of instantaneous values can be used as the temporal mean. Nevertheless, it is very challenging to estimate accurate daily mean solar radiation from few instantaneous values retrieved from polar-orbiting satellite sensors (e.g., MODIS) because the changes between two adjacent observations may be highly nonlinear.

One solution is to first interpolate the atmospheric parameters. For example, the ISCCP solar radiations were calculated at 3-h intervals from the interpolated inputs of atmosphere and surface parameters and averaged to obtain the daily mean (Zhang et al. 1995). It is also possible to calculate the daily value directly from the instantaneous values. For example, the daily means can be calculated from individual instantaneous values through an empirical relationship (Huang et al. 2012; Liang et al. 2006). However, such a relationship is dependent on the observation time of the instantaneous retrievals and the length of daytime. Wang et al. (2010) introduced an adjusted sinusoidal interpolation method to predict daily PAR from MODIS instantaneous PAR retrievals. Zheng and Liang (2011) developed a Bayesian method of ration (MOR) to combine sparse instantaneous PAR observations with prior knowledge to predict the daily value. By empirically linking the TOA daily flux with the cosine of solar zenith angle and using retrievals from Terra and Aqua to represent the atmospheric conditions in the morning and afternoon, Wang and Pinker (2009) estimated the daily mean solar radiation with the daily atmosphere data retrieved from the MODIS twin sensors.

5.2.4 Ground Measurements and Satellite-Derived Shortwave Radiation Products

5.2.4.1 Ground Measurements

Ground measurements play an important role in estimating land SRB for training/calibrating the algorithms and validating the satellite products. Currently, several global radiation networks are available (Liang et al. 2010), including the Global Energy Balance Archive (GEBA) (Gilgen and Ohmura 1999), the Baseline Surface Radiation network (BSRN) (Ohmura et al. 1998), the SURFRAD (Augustine et al. 2000, 2005), and FLUXNET (Baldocchi et al. 2001). The data obtained by these observation networks can be used to verify the empirical models or the accuracy of atmospheric radiative transfer simulation models, thus enabling the prediction of radiation values at other stations. The GEBA database, which is maintained by the Eidgenössische Technische Hochschule (ETH) Zurich, collects radiation measurements from over 2000 stations made since 1950. There are about 59 observation stations in the BSRN database, at least 40 of which are now still in operation. The observation accuracy of global irradiance, direct shortwave radiation, and scattered shortwave radiation are 5, 2, and 5 W m^{-2}, respectively. SURFRAD network has seven stations and provides precise, continuous, long-term observation data regarding the surface solar radiation for the entire United States with the relative error of the measurements between ±2% and ±5%. FLUXNET is composed of a series of regional networks, including AmeriFlux, CarboEuropeIP, AsiaFlux, KoFlux, OzFlux, Flux-Canada, and ChinaFlux. From 1996 to July 2009, 500 observation towers measured the fluxes and related parameters on a long-term basis. Additionally, other existing networks, including the Atmospheric Radiation Measurement (ARM), the GEWEX Asia Monsoon Experiment (GAME/AAN), and the Greenland Climate Network (GC-Net) also provide surface radiation measurements.

5.2.4.2 Existing Surface Radiation Products

There are two types of surface shortwave radiation products: (1) the satellite-retrieved data including the ISCCP C1 data (Pinker and Laszlo 1992), which have a 280-km spatial resolution (1983–2000), the global GEWEX SRB data, which have a 1° spatial resolution, and the CERES radiation budget data, which have a 140-km spatial resolution; and (2) the calculated results by general circulation models (GCMs). Of all the global PAR products, two are based on ISCCP C1 data: ISCCP-BR, obtained by multiplying the shortwave radiation product by a conversion factor of 0.5, and ISCCP-PL, obtained by the algorithm proposed

TABLE 5.1

Basic Information Regarding Major Radiation Products

Type of Data	Product Name	Spatial Resolution	Temporal Resolution
PAR	ISCCP BR	280 km	3 h
	ISCCP PL	280 km	3 h
	TOMAS PAR	1°	Monthly average
	GOES PAR	1 km	Daily
	MODIS PAR	4 km	Daily
	GLASS PAR	5 km	3 h
Insolation	ISCCP C1	280 km	3 h
	GEWEX	1°	3 h
	CERES	140 km	–
	GCMs	>1°	6 h
	GLASS	5 km	3 h

Source: Zhang, X. et al., Incident solar radiation. In *Advanced Remote Sensing: Terrestrial Information Extraction and Applications* (S. Liang, X. Li, and J. Wang, Eds.). Academic Press, San Diego, CA, pp. 127–173, 2012.

by Pinker and Laszlo (1992). The Total Ozone Mapping Spectrometer (TOMAS) PAR product is based on the data acquired by TOMAS at 370 nm. All of these products have a 2.5° spatial resolution. It has been reported that the spatial resolutions of the ISCCP-BR and ISCCP-PL products are coarser than those of the TOMAS-PAR product by 12%–16% and 8%–12%, respectively. The insolation and PAR products are generated at 5-km spatial resolution every 3 h from 2008 to 2010, which have been recently released globally via the Global Land Surface Satellite (GLASS) distribution system at the Center for Global Change Data Processing and Analysis, Beijing Normal University (Liang et al. 2013). The University of Maryland at College Park, with NASA support, also has generated PAR products using both MODIS and GOES data that are currently being distributed by the Oak Ridge National Laboratory DAAC. Table 5.1 summarizes the basic information of current available radiation products.

5.3 Surface Shortwave Albedo

5.3.1 Background

Land surface albedo is defined as the ratio of the surface reflected radiation to the incident radiation that reaches the Earth's surface and is a key geophysical parameter controlling the energy budget in land–atmosphere interactions (Dickinson 1983). Modulated by the surface shortwave albedo, the surface absorbs about twice as much solar shortwave radiation as the atmosphere does. Surface shortwave albedo plays a vital role in calculating the surface energy balance in global and regional models. The land surface albedo varies spatially and evolves seasonally, based on solar illumination changes, vegetation growth, and human activities such as cutting/planting forests and slash-and-burn agricultural practices.

Satellite remote sensing is an essential technique in estimating land surface albedo globally at various spectral, spatial, temporal, and angular resolutions. During the last decade, with the emergence of onboard remote sensors, many global satellite albedo products have been derived, such as the Advanced Very High Resolution Radiometer (AVHRR) (Csiszar and Gutman 1999), the MODIS onboard Terra and Aqua satellites (Lucht et al. 2000; Schaaf et al. 2002), the Multiangle Imaging SpectroRadiometer (MISR) on board Terra (Martonchik et al. 1998), the POLarization and Directionality of the Earth's Reflectances (POLDER) (Breon et al. 2002; Leroy et al. 1997), and the Visible/Infrared Imaging Radiometer Suite (VIIRS) (Liang 2003). As geostationary satellites observe the Earth's surface more frequently (every 15 to 30 min), albedo datasets with spatial resolution of 500 m to 5 km have been derived from Meteosat (Pinty et al. 2000a), the SEVIRI onboard the MSG satellite series (Geiger et al. 2008). Most of these satellite products are based on the atmospheric correction (e.g., Liang et al. 2000; Vermote et al. 1997) and surface anisotropy modeling (e.g., Maignan et al. 2004; Pinty et al. 2000a; Rahman et al. 1993; Roujean et al. 1992; Wanner et al. 1995) to retrieve surface properties from satellite observations. In the following sections, we will briefly introduce the different albedo retrieving methods, existing albedo products, and temporal and spatial characteristics of albedo.

5.3.2 Methods to Estimate Surface Albedo from Satellite Data

5.3.2.1 Traditional Methods

Satellite instruments do not directly measure surface albedo. Most satellite albedo estimation algorithms consist of three steps (Liang et al. 2010): atmospheric correction, surface directional reflectance modeling, and narrowband to broadband conversion. Atmospheric correction is needed to separate the contribution of surface properties (e.g., surface reflectance/albedo) from that of atmospheric properties (e.g., aerosol, water vapor, and ozone) in the signal observed by satellite. In the following subsections, we will briefly introduce the traditional method of estimating surface broadband albedo from surface reflectance.

5.3.2.1.1 Surface Anisotropy Modeling

Albedo is the integration of directional reflectance over the semihemispheric space. If the surface is not a Lambertian, the bidirectional reflectance distribution function (BRDF) is used to describe the anisotropic reflectivity characteristics.

BRDF models quantify the angular distribution of radiance reflected by an illuminated surface. Various models have been proposed to simulate or capture the anisotropic characteristics of the land surface (Lacaze et al. 1999; Liang 2007), including computer simulation models (Gastellu-Etchegorry et al. 2004), physical models using the canopy radiative transfer process (Kuusk 1995a,b; Pinty et al. 2006), and (semi-)empirical models based on various approximations of the radiative transfer process (Li and Strahler 1992; Rahman et al. 1993; Roujean et al. 1992). The quality of these models can be evaluated either through a comparison with simulations by other models of higher complexity or through comparison with measurements. In order to expedite the inversion procedure, complex computer simulation and physical models are not considered to be the optimal BRDF model herein.

Kernel-driven models are the most widely used BRDF models because of their easy implementation and time efficiency for the operational satellite products. Pokrovsky and Roujean (2003) made comparisons based on different kernel-based BRDF models and found that the Li-Sparse and Roujean models perform best when fitting the bidirectional reflectances. Maignan et al. (2004) evaluated a set of analytical models based on POLDER measurements and proposed an improved Ross-Li kernel model by adding an angular factor based on Breon's finding (2002) to better explain the "hot spot" effect, which occurs when the viewing and illumination directions coincide. By introducing the multiple scattering between the canopy and the soil and the relationship between the soil moisture and the soil reflectance into the Ross-Li kernel models, a recent method was proposed to build an angular and spectral kernel model (Liu et al. 2010). However, this method requires prior knowledge of soil moisture, which is difficult to obtain and therefore limits its operational application.

Polar-orbiting satellites can only observe the same location every few days, but derivation of the BRDF kernel parameters requires at least three cloud-free observations. In many satellite albedo algorithms, the surface anisotropy is assumed to be invariant within a certain time period (say, 10 or 16 days). In this way, enough clear-sky atmospherically corrected surface reflectances are collected as a temporal composite to estimate the BRDF kernel parameters using a least squares approximation approach.

5.3.2.1.2 Angular Integration

Albedo is the angular integration of bidirectional reflectance calculated from the BRDF models. There are two major albedo quantities, directional–hemispherical reflectance (DHR) and bihemispherical (BHR), defined according to the direction of the incoming radiation.

DHR, also known as the black-sky albedo (BSA) in MODIS product, specifies the fraction of direct radiation incident upon a surface that is diffusely reflected toward the upper hemisphere. DHR only considers the radiation from the solar incident direction. Therefore DHR is solar angle dependent:

$$\text{DHR}(\theta_s) = \frac{1}{\pi} \int_0^{2\pi} d\varphi \int_0^1 R(\theta_s, \theta_v, \varphi)\mu_v \cdot d\mu_v \qquad (5.17)$$

where $\mu = \cos(\theta)$.

BHR, also known as the white-sky albedo (WSA) in MODIS product, is the ratio of the diffusely reflected radiation to the diffuse incident radiation. BHR accounts for the diffuse downward radiation instead of the direct solar radiation. In theory, it does not rely on the solar zenith angle:

$$\text{BHR} = \frac{1}{\pi} \int_0^{2\pi} d\varphi \int_0^1 \text{DHR}(\theta_s)\mu_s \cdot d\mu_s \qquad (5.18)$$

Instead of directly calculating the integral, MODIS albedo products use the precalculated polynomial function to expedite the online computation. Both DHR and BHR are the ideal forms of albedo under simplified atmospheric conditions. DHR is close to the actual albedo if the atmosphere is completely clear, whereas BHR is for the situation that the sky is completely hazy. Usually, the actual albedo is the average of these two terms weighted by the direct/diffuse sky light ratio.

5.3.2.1.3 Narrowband-to-Broadband Conversion

Shortwave broadband albedo, defined as the averaged albedo in the solar range (0.3–3.0 μm), is required by many land surface models and weather forecast applications to quantify the overall solar shortwave net radiation. From the BRDF models, albedo from spectral bands can be easily estimated through the polynomial equation. Because the broadband albedo quantifies the ratio of the total reflected and incident radiation for a wide range of wavelengths, it can be expressed as the linear combination of spectral albedos weighted by distribution of the incident energy. Variation of atmospheric aerosol loading, water vapor content, and ozone will have impact on the downward radiation for different spectral ranges. Therefore surface broadband albedo is not the sole measure of surface reflective properties. Distribution of downward radiation at the surface needs to be considered in the broadband albedo calculation. To simplify the estimation of broadband albedo, general equations need to be developed to express the relationship between the spectral albedo and the integrated broadband shortwave albedo that shall account for various atmospheric conditions and illumination geometries over different surfaces for different satellite sensors (Liang 2001; Liang et al. 1999; Stroeve et al. 2005). The relationship can be established using the following linear equation:

$$\alpha_{sw} = \sum_{\lambda=1}^{n} Reg_{\lambda}\alpha_{\lambda} + Reg_{0} \tag{5.19}$$

where α_{sw} is the total shortwave albedo, α_{λ} ($\lambda \in [1, n]$) is the albedo of a spectral band, and Reg_{i} ($i \in [0, n]$) are the regression coefficients.

5.3.2.2 Direct Estimation Methods

In the conventional albedo algorithms discussed above, each step has clear physical foundation, but the errors from each step may accumulate and affect the final accuracy of the albedo product. An alternative approach, so-called direct estimation algorithm, is to estimate surface albedo directly from TOA observations, which combines all procedures together in one step through regression analysis aiming only to make a best estimation of broadband albedo.

The direct estimation method consists primarily of two steps (Liang 2003; Liang et al. 2005). The first produces a large database of TOA directional reflectance and surface albedo for a variety of surface and atmospheric conditions using radiative transfer model simulations. The second links statistically the simulated TOA reflectance with surface broadband albedo. The direct estimation method does not require surface reflectance resulting from atmospheric correction. Neither does it require an accumulation of observations during a certain period of time, which cannot capture the rapid surface changes. In an earlier study, Liang et al. (1999) developed such a direct retrieval algorithm using a feed-forward neural network. This was later improved upon by using linear regression analysis in each angular bin and applied to MODIS data (Liang 2003) and also improved to produce highly accurate daily snow/ice albedo more efficiently with mean bias of less than 0.02 and residual standard error of 0.04 (Liang et al. 2005). This algorithm has been adopted as a default algorithm for operationally mapping the land surface broadband albedo from VIIRS in the NPP program and the future Joint Polar Satellite System (JPSS) program. Qu et al. (2013) further extended this method for the GLASS albedo production.

If the surface reflectance product (after atmospheric correction of the TOA observations) is available, the last two steps can be combined to convert the directional reflectance to broadband albedo. Cui et al. (2009) developed such an empirical formula based on POLDER-1 multiangle imagery data.

5.3.2.3 Optimization Methods

On the basis of the "dark object" atmospheric correction algorithm in products derived from satellite sensors such as MODIS and SEVIRI, the retrievals of aerosol distribution and properties over land have shown valuable results. However, the use of this algorithm is restricted to land surface with low reflectance (e.g., water and dense vegetation), while over bright surfaces (snow, desert, and urban areas), it often fails to estimate the aerosol information accurately. Another problem of separating atmospheric correction and surface BRDF fitting exists in the Lambertian approximation in the radiative transfer procedure (He et al. 2012a). Therefore biases will emerge with "atmospheric corrected surface reflectances" and will further deteriorate the BRDF fitting process.

An alternative albedo estimation approach was proposed by retrieving the surface reflectance and aerosol optical depth (AOD) jointly using the optimization method based on SEVIRI data (Govaerts et al. 2010; Wagner et al. 2010). He et al. (2012a,b) further refined the algorithm by incorporating surface albedo climatology to improve the surface BRDF and instantaneous AOD estimations. They used available prior information on albedo and satellite observations to simultaneously derive the surface BRDF kernel parameters and instantaneous AOD in the context of a least squares approach through the minimization of the cost function:

$$J(X) = (A(X) - A^{Clm})B^{-1}(A(X) - A^{Clm}) + (R^{Est}(X) - R^{Obs})O^{-1}(R^{Est}(X) - R^{Obs}) + J_c \quad (5.20)$$

Here, X denotes the unknown variables to be estimated in one sliding window and it includes the surface BRDF model parameters and AOD. Two general assumptions are made here to reduce the complexity of the retrieving procedure and to generate the stable estimates as well: (1) the surface BRDF shape is stable within the sliding window and (2) the aerosol type and its properties (e.g., Angström exponent) do not change within the sliding window, but AOD varies from time to time. Because predefined aerosol types are used in this study, the intrinsic properties for each are not part of the unknown variables to be estimated. Then, X can be written in the following form:

$$X = [P_1, P_2, \ldots, P_{NB}, AOD_1, AOD_2, \ldots, AOD_{NO}]^T \quad (5.21)$$

where NB is the number of spectral bands from a certain satellite sensor, NO is the number of cloud-free observations involved in the inversion, P_i ($i \in [1, NB]$) is a set of BRDF model parameters (e.g., for kernel models, one set of P_i refers to three parameters: $f_{iso}, f_{vol},$ and f_{geo}), AOD_j ($j \in [1, NO]$) is the AOD value for the corresponding observation j, and $R_{i,j}^{Obs}$ and $R_{i,j}^{Est}$ refer to the observed and modeled TOA reflectance for a band and a given set of geometries (e.g., solar angle and viewing angle), respectively. The kernel models with considerations of hot spot effects (Maignan et al. 2004) and an atmospheric radiative transfer that accounts for the interaction between atmosphere and non-Lambertian surface are used in the forward simulation of the TOA observations. Extensive validation of surface albedo and directional reflectance from this algorithm shows reasonable high accuracies and the reduced data gaps. It is also found that AOD retrievals from this algorithm have similar accuracy with that of MODIS aerosol products (He et al. 2012a).

5.3.3 Ground Measurements and Satellite Products

Ground measurements of surface albedo can be directly calculated from the upward and downward shortwave radiation. Those measurements are available from the ground observation networks mentioned in Section 5.2.4, including SURFRAD, FLUXNET, BSRN, GC-Net, and so on. Table 5.2 summarizes the current global broadband albedo products.

TABLE 5.2

List of the Current Satellite Albedo Products

Albedo Products	Spatial Resolution	Temporal Resolution (Day)	Temporal Range
MODIS	0.5–1 km, 0.05°	8–16	2000–present
MISR	1.1 km	16	2002–present
MFG (Meteosat First Generation)	3 km	10	1982–2006
MSG	3 km	1	2005–present
GlobAlbedo	0.05°	16	1998–2011
POLDER	1/12°	10	Nov 1996 to Jun 1997, Apr 2003 to Oct 2003, Jul 2005–present
Vegetation	1 km, 0.5°	10	2002–2003 2009–present
Glass	1–5 km, 0.05°	8	1981–2010

On the basis of previous validation studies, the listed global products can achieve the accuracy requirements of 0.02 to 0.05 (in absolute values) for most climate and ecosystem applications. In this subsection, we will introduce some global products.

The MODIS sensors on Terra and Aqua provide measurements on a global basis every 1–2 days with seven spectral bands in the shortwave range for land applications. Current MODIS albedo products are generated on every 8 days from 2000 (Gao et al. 2005; Lucht et al. 2000; Schaaf et al. 2002). The atmospherically corrected surface reflectance data in a 16-day period are collected as the input of the kernel models. BRDF/albedo and the corresponding quality data are available at 500 m, 1-km resolution in sinusoidal projection, and 0.05° in latitude/longitude projection. In addition, the Nadir BRDF-Adjusted Reflectances are generated to help correct for the view angle effects that can be observed in the MODIS surface reflectance products.

VIIRS is a new generation of moderate resolution radiometer, serving as a substitute of MODIS. Land surface albedo is produced as a standard Environmental Data Record from VIIRS (NASA Goddard Space Flight Center 2011). Nine spectral bands (at 412, 445, 488, 555, 672, 865, 1240, 1610, and 2250 nm) from VIIRS with an approximately 750-m spatial resolution are used for surface albedo estimation. Two algorithms were designed to retrieve land surface albedo for dark (e.g., vegetation) and bright surfaces (e.g., snow), respectively. Dark Pixel Sub Algorithm is a heritage algorithm from the MODIS albedo algorithm. It contains three typical steps: atmospheric correction, BRDF modeling, and narrowband-to-broadband conversion. Details of such algorithms can be found in the previous part of this section. Bright Pixel Sub Algorithm directly estimates broadband albedo from the VIIRS TOA reflectance by using the regression models established through extensive radiative transfer simulation. Currently, the algorithm and product are being tested and validated.

The MISR onboard Terra is a multiangle sensor, which consists of nine cameras with nominal view zenith angles of 0.0°, 26.1°, 45.6°, 60.0°, and 70.5° in the along-track direction. Four spectral bands (446, 558, 672, and 866 nm) are available for each camera. Data are then averaged to a resolution of 1.1 km. Atmospheric correction is applied to the TOA observations, in which aerosol retrievals are averaged by a 17.6 × 17.6 km window (Martonchik et al. 1998). Unlike the MODIS one, the MISR albedo algorithm uses the three-parameter modified Rahman–Pinty–Verstraete BRDF model (Rahman et al. 1993) to fit the atmospherically corrected surface reflectance. The MISR 1.1-km data are derived using 1 day's data and available every 16 days. The products are also available for global maps at 0.5° resolution on a monthly basis.

The albedo retrieval algorithm for the Meteosat Visible and InfraRed Imager onboard MFG satellite series relies on the frequent temporal samplings of the TOA observations to estimate the surface anisotropy. After the atmospheric correction for aerosol, water vapor, and ozone, the RPV model is used to characterize the surface reflectances (Martonchik et al. 1998; Pinty et al. 2000a,b; Rahman et al. 1993). The MFG surface albedo product, recorded from 1982 (Meteosat 2) to 2006 (Meteosat 7), is derived in the visible band every 10 days at a resolution of 3 km. With the three-band SEVIRI instrument on board the MSG satellite series, albedo products are produced on a daily basis (Geiger et al. 2008). A simplified method for the atmospheric correction is applied. Both daily and 10-day averaged albedo products are available from the MSG satellite from 2002 in the Land Surface Analysis Satellite Applications Facility.

The GLASS albedo product from 1981 through 2010 is produced by Beijing Normal University, China, mainly from AVHRR and MODIS data. The albedo product from MODIS data is based on two direct albedo estimation algorithms from surface reflectance, TOA radiance (Qu et al. 2013), and the statistics-based temporal filtering fusion algorithm that integrates these two intermediate albedo products (Liu et al. 2013). The albedo product from AVHRR data is based on the similar direct estimation algorithm from the TOA observations (Liu et al. 2013). Global albedo maps derived from AVHRR and MODIS are available at 5 km and 0.05° for every 8 days. In addition, 1-km albedo data in sinusoidal projection derived from MODIS observations are provided at the same temporal resolution.

The GlobAlbedo project aims to provide global surface albedo data from 1998 to 2011 based on European satellites at three different spatial resolutions from 1 km, 0.05° to 0.5°. Data derived from SPOT4-VEGETATION, SPOT5-VEGETATION2, and the Medium Resolution Imaging Spectrometer are integrated using an optimal estimation approach (Lewis et al. 2012) and a gap-filling technique based on MODIS BRDF dataset (Lewis et al. 2012; Muller et al. 2012).

5.3.4 Spatial and Temporal Variation

Albedo is highly variable both spatially and temporally. Variations in surface conditions, such as snow cover, vegetation phenology, and soil moisture, are all accompanied by significant changes in land albedo. Drought or forest fires can lead to changes in surface albedo (Govaerts and Lattanzio 2008; Lyons et al. 2008). Human activities, such as deforestation (Loarie et al. 2011), irrigation (Zhu et al. 2011), and urbanization (Offerle et al. 2005), can greatly impact and alter the surface albedo. Steyaert and Knox (2008) reconstructed the albedo changes in the eastern half of the United States associated with extensive land cover changes over the past 350 years. The recession of snow cover associated with warm periods in Earth's history has led to more absorption of solar energy (and hence has amplified warming). Aerosols like dust and soot may also contaminate snow and greatly reduce its albedo (Hansen and Nazarenko 2004; Xu et al. 2009).

MODIS's 10-year-old archives of data have been widely investigated and analyzed. For example, Gao et al. (2005) found that the interannual variation of shortwave BHR is less than 0.01 over snow-free surfaces and the averaged values over 20° latitude vary little between consecutive years. Fang et al. (2007) calculated the variation of albedo over North America for different plant function types (PFTs) using MODIS data from multiple years and generated the shortwave albedo climatology for each of the PFTs. They found that the coefficient of variation shows a strong seasonal character, which increases in winter and

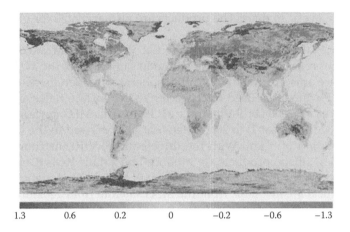

1.3   0.6   0.2   0   −0.2   −0.6   −1.3

FIGURE 5.5
(See color insert.) Global BHR variations from MODIS albedo product (2003–2008). (Zhang X. T. et al., Analysis of global land surface shortwave broadband albedo from multiple data sources, *IEEE J. Selected Topics Appl. Earth Obs. Remote Sens.*, 3(3), 296–305, © 2012 IEEE.)

spring and decreases in the growing season. They also found that the largest variation is strongly linked with winter snow and spring thaw. Zhang et al. (2010) analyzed 9 years of MODIS albedo data (2000–2008) for mapping the albedo variation globally (Figure 5.5) and found that although there was no significant global surface albedo change from 2000 to 2008, a decrease of about 0.01 for the Northern Hemisphere and an increase of about 0.01 for the Southern Hemisphere were identified during the same time period.

5.4 Longwave Downward and Upwelling Radiation

5.4.1 Longwave Downward Radiation

5.4.1.1 Backgrounds

The surface longwave radiation (4–100 μm) is of great importance in the Earth's climate system. The downward longwave radiation at surface $\left(F_d^l\right)$, the direct measure of radiative heating of atmosphere to surface, plays an important role in numerous applications, which require surface radiation and energy balance as input to predict evapotranspiration, snowmelt, surface temperature, and fog occurrence (Alados et al. 2012; Flerchinger et al. 2009; Liang et al. 2010). Knowledge of the spatial and temporal variability of the surface downward longwave radiation on local, regional, and global scales is therefore valuable.

F_d^l is the result of atmospheric absorbing, emission, and scattering. F_d^l is related to the downward specific intensity I_λ ($z = 0$, $-\mu$) at surface in an axial atmosphere as

$$F_d^l = 2\pi \int_{\lambda_1}^{\lambda_2} \int_0^1 I_\lambda(z=0,-\mu)\mu\,d\mu\,d\lambda \qquad (5.22)$$

where λ_1 and λ_2 describe the spectral range of the surface downward longwave radiation; λ is the wavelength; z is the altitude; and $\mu = \cos(\theta)$, where θ is the local zenith angle.

Under clear-sky conditions, the surface spectral radiance, $I_{\lambda,\text{clear}}(z = 0, -\mu)$, can be written as

$$I_{\lambda,\text{clear}}(z = 0, -\mu) = -\int_0^{z_s} B_\lambda(z') \frac{\partial T_\lambda(0, z'; -\mu)}{\partial z'} \, dz' \tag{5.23}$$

where $B(z')$ is the Planck function evaluated with the temperature at the altitude z', z_s is the altitude of the satellite, and T_λ is the transmittance from the surface to the altitude z'.

For overcast black cloud conditions, the surface spectral radiance, $I_{\lambda,\text{cloud}}(z = 0, -\mu)$, can be written as

$$I_{\lambda,\text{cloud}}(z = 0, -\mu) = B_\lambda(z_{cb}) T_\lambda(0, z_{cb}; -\mu)$$
$$- \int_0^{z_s} B_\lambda(z') \frac{\partial T_\lambda(0, z'; -\mu)}{\partial z'} \, dz' \tag{5.24}$$

where z_{cb} is the cloud base height.

Under clear-sky conditions, F_d^l can be estimated in the form of the Stefan–Boltzmann equation

$$F_d^l = \sigma \varepsilon_a T_a^4 \tag{5.25}$$

where σ is the Stefan-Boltzmann constant. ε_a and T_a are the clear-sky air emissivity and the screen level (about 2 m above the surface) temperature, respectively.

The dominant emitters of longwave radiation in the atmosphere are water vapor and, to a lesser extent, carbon dioxide and trace gases including ozone, methane, nitrogen dioxide, and aerosols (Kjaersgaard et al. 2007b). Moreover, the emissivity of atmosphere, which depends both on the vertical temperature profile and on the vertical distribution of relatively active constituents, is also an important factor to estimate F_d^l. Cloud emission also needs to be taken into account if cloud is present.

F_d^l is more difficult to estimate because only atmospheric window radiances at the TOA contain information on the near-surface radiation field (Schmetz 1989; Tang and Li 2008). The atmosphere below 500 m from the surface accounts for 80% of the total surface downward longwave radiation (Schmetz 1989). Under cloudy-sky conditions, the radiation fields at the surface and at the top of the atmosphere are entirely decoupled.

During the past several decades, much effort has been made on estimating F_d^l (Alados et al. 2012; Ellingson 1995; Nussbaumer and Pinker 2012; Prata 1996; Tang and Li 2008; Wang and Liang 2009d; Zhang et al. 2010; Zhou et al. 2007). Comprehensive reviews of these studies are also available from the literatures (Ellingson 1995; Flerchinger et al. 2009; Kjaersgaard et al. 2007b; Niemelä et al. 2001). These methods can be classified in three categories: profile-based, physical, and hybrid methods. The profile-based method calculates the downward longwave radiation using a radiative transfer model and satellite-derived atmospheric profiles. The advantage of the profile-based method is its basis in physics. The disadvantages are that using complex radiative transfer models and atmospheric profiles is often very time consuming. Moreover, the atmospheric profiles are not always available.

Physical methods always utilize parameterized radiative transfer model with quantities of ground or satellite observed parameters, such as surface temperature, relative humidity, and land surface broadband emissivity. The physical methods are usually very simple to implement. The hybrid methods usually estimate the F_d^l by simulating the surface downward longwave radiation using complex radiative transfer models and atmospheric profiles or using machine-learning approaches. Representatives of each algorithm for estimating F_d^l will be given in the following subsections.

5.4.1.2 Methodology

5.4.1.2.1 Profile-Based Methods

The profile-based methods calculating the F_d^l always use a radiative transfer model, and atmospheric temperature and moisture profiles. Frouin (1988) proposed a method for estimating downward longwave irradiance at the ocean surface from satellite radiance data. The downward longwave irradiance is computed with a fast and accurate radiative transfer model as a function of temperature, water vapor, ozone and carbon dioxide mixing ratios, fractional cloud coverage, emissivity of clouds, and cloud top and cloud base altitudes obtained from the GOES Visible and Infrared Spin Scan Radiometer (VISSR) data. Four methods were introduced in the paper (Frouin et al. 1988), and the results indicate that the proposed satellite methods perform similarly, with standard errors of estimate ranging from 21 to 27 W m^{-2} on a half-hourly timescale and from 16 to 22 W m^{-2} on a daily timescale, which correspond to 6%–8% and 4%–6% of the average measured values.

The CERES project is an investigation of cloud–radiation interactions in the Earth's climate system (Wielicki et al. 1996). Three algorithms are adopted to estimate the F_d^l including Model A (Inamdar and Ramanathan 1997b), Model B (Gupta 1989; Gupta et al. 1993), and Model C (Zhou and Cess 2001; Zhou et al. 2007). Models A and C are treated as the hybrid methods and will be briefly introduced in the following subsections. Model B proposed by Gupta (1989) [see also Gupta et al. (1993) and Gupta (2009)] is a profile-based method. The input parameters of this method are the surface temperature and emissivity, and the atmospheric temperature and humidity profiles. Clear-sky longwave downward radiation, denoted as $F_{d,\mathrm{clr}}^l$, is estimated from a parameterization radiation model and atmospheric parameters as

$$F_{d,\mathrm{clr}}^l = f(w)T_e^{3.7} \tag{5.26}$$

where $f(w)$ is a function of column water vapor and T_e is the effective emitting temperature of the lower atmosphere at different layers, which can be obtained using

$$T_e = k_s T_s + k_1 T_1 + k_2 T_2 \tag{5.27}$$

where T_s is the surface skin temperature, and T_1 and T_2 are the mean temperature of the surface to 800 and 800–600 hPa layers, respectively. And the all-sky longwave downward radiation $F_{d,\mathrm{all}}^l$, is derived by

$$F_{d,\mathrm{all}}^l = F_{d,\mathrm{clr}}^l + F_{d,\mathrm{cld}}^l \tag{5.28}$$

where $F_{d,\mathrm{cld}}^l$ is the radiation emitted by the cloud, which can be derived by

$$F_{d,\mathrm{cld}}^l = f(T_{cb}, w_c)A_c \tag{5.29}$$

where T_{cb} is the cloud base temperature, w_c is the water vapor below the cloud base, and A_c is the cloud fraction.

5.4.1.2.2 Parameterization Methods

It is generally believed that F_d^l is more difficult to estimate than the surface longwave upward radiation F_u^l, which is always estimated using the surface temperature and emissivity because F_d^l comes mainly from the near-surface layers of atmosphere. Many physical methods have been reported to estimate F_d^l (Alados et al. 2012; Flerchinger et al. 2009; Kjaersgaard et al. 2007b; Niemelä et al. 2001; Nussbaumer and Pinker 2012; Prata 1996).

5.4.1.2.2.1 Clear-Sky Physical Methods F_d^l can be estimated using the near-surface temperature and humidity based on the Stefan–Boltzmann equation (Crawford and Duchon 1999; Dilley and O'Brien 1998; Idso 1981; Niemelä et al. 2001; Prata 1996). As shown in Equation 5.25, the primary input parameters for physical methods are the screen level air temperature T_a and the atmospheric emissivity ε_a. The atmospheric emissivity is always estimated by screen level vapor pressure e. Table 5.3 summarizes formulae for calculating the atmospheric emissivity ε_a, proposed in previous literatures.

Kjaersgaard et al. (2007b) evaluated 20 simple models using long-term field measurements at two sites in Demark with mean bias errors under all-sky conditions ranging from –23 to 12 W m^{-2} and –18 to 15 W m^{-2} and root-mean-square errors (RMSEs) from 39 to 45 W m^{-2} and 30 to 36 W m^{-2}, respectively. The validation results of models in Brutsaet (1975), Prata (1996), and Swinbank (1963), and the FAO parameterization of the Brunt model (Brunt 1932) are better than the other models. Flerchinger et al. (2009) validated the accuracy of 13 algorithms for predicting F_d^l under clear skies at 21 sites across North American and China. The results indicated that the models proposed by Dilley and O'Brien (1998) and Prata (1996), with averaged RMSEs of approximately 23 W m^{-2} under clear-sky conditions, are adequate for predicting F_d^l.

Although the physical methods for estimating F_d^l have been widely reported, almost all of these models used the meteorological parameters as the input data. Physical models with satellite-derived meteorological parameters as input are rarely validated and compared. Wu et al. (2012) compared eight physical F_d^l estimation methods using MODIS products. The validated results showed that the uncertainty of the physical parameterization models is the key factor influencing the accuracy of F_d^l estimates rather than the uncertainty of

TABLE 5.3

Parameterizations for Estimating ε_a

Model Source	Clear-Sky Models
Dilley and O'Brien (1998)	$F_d^l = 59.38 + 113.7\left(\dfrac{T_a}{273.3}\right)^6 + 96.96\left(\dfrac{w}{25}\right)^{0.5}$ $w = 46.5/T_a$
Idso (1981)	$\varepsilon_a = 0.7 + 5.95 \times 10^{-5} e \exp\left(\dfrac{1500}{T_a}\right)$
Prata (1996)	$\varepsilon_a = 1 - (1+\eta)\exp(-(1.2+3\eta)^{0.5})$ $\eta = 46.5 e/T_a$

MODIS products. Moreover, the Bayesian model averaging (BMA) method is incorporated in this study to combine the predictive distribution of these models to achieve better accuracy (Wu et al. 2012). The integrated estimates for F_d^l based on the BMA method have lower RMSEs and higher coefficients of determination (R^2) than the best individual model. The RMSEs decreased by approximately 10 W m^{-2} at two forest sites and by approximately 4 W m^{-2} at other sites. The R^2 value increased at each site by more than 0.05.

5.4.1.2.2.2 All-Sky Physical Methods   Under the all-sky conditions, F_d^l is usually corrected by modifying the atmospheric emissivity. The atmospheric emissivity for all-sky is always corrected using the proposed clear-sky atmospheric emissivity and cloud cover. Table 5.4 summarizes the cloudy-sky atmospheric emissivity reported in some of the previous literatures.

Moreover, Aubinet (1994) proposed a method to directly estimate F_d^l using the vapor pressure e_0, temperature T_o, and clearness index (k) rather than using the clear-sky atmospheric emissivity correction. Four schemes were introduced in this research. Each method utilized a different combination of temperature, vapor pressure, and clearness index based on the data availability.

5.4.1.2.3 Hybrid Methods

Hybrid methods are always based on extensive radiative transfer simulations, statistical analysis, or machine-learning algorithms. In this section, we will introduce some representatives of estimating F_d^l based on the hybrid method, including the models for MODIS and CERES, and the model for estimating F_d^l based on machine-learning methods, for instance, artificial neural networks (ANNs).

5.4.1.2.3.1 Longwave Downward Radiation Model for MODIS   Wang and Liang (2009d) developed a nonlinear model for estimating $F_{d,clr}^l$ from MODIS data. They simulated the thermal infrared (TIR) radiance and F_d^l with a large number of atmospheric profiles from the MODIS Terra atmosphere product and MODTRAN4. The nonlinear relationship between the MODIS TIR and F_d^l was established by statistical analysis (Equation 5.30). MODIS bands 27–29 and 31–34 were used in the model; bands 27, 28, and 29 are the water vapor channels; bands 33 and 34 are related to the near-surface air temperature; and bands 29, 31, and 32 can be used to retrieve the skin temperature.

TABLE 5.4

Parameterization for Estimating the Atmospheric Emissivity under Cloudy-Sky Conditions (c is the Cloud Fraction)

Model Source	All-Sky Models
Maykut and Church (1973)	$\varepsilon_{a,\,cloudy} = (1 + 0.22c^{2.75})\varepsilon_{a,\,clear}$
Sugita and Brutsaert (1993)	$\varepsilon_{a,\,cloudy} = (1 + uc^v)\varepsilon_{a,\,clear}$ u and v are coefficients
Unsworth and Monteith (1975)	$\varepsilon_{a,\,cloudy} = (1 - 0.84c)\varepsilon_{a,\,clear} + 0.84c$
Crawford and Duchon (1999)	$\varepsilon_{a,\,cloudy} = (1 - c) + c\varepsilon_{a,\,clear}$
Lhomme et al. (2007)	$\varepsilon_{a,\,cloudy} = 1.18 \times (1.37 - 0.34c)\left(\dfrac{e_0}{T_a}\right)^{1/7}$

$$F_d^l = L_{\text{Tair}}\left(a_0 + a_1 L_{27} + a_2 L_{29} + a_3 L_{33} + a_4 L_{34} \right.$$

$$\left. + b_1 \frac{L_{32}}{L_{31}} + b_2 \frac{L_{33}}{L_{32}} + b_3 \frac{L_{28}}{L_{31}} + c_1 H \right) \tag{5.30}$$

where L_i are MODIS TOA radiance of band i (W m^{-2} μm^{-1} sr^{-1}), and L_{Tair} represents the air temperature, which is equal to MODIS L_{31} in the nighttime models and equal to L_{32} in the daytime models. a_i, b_i, and c_i are regression coefficients corresponding to five view zenith angles (0_i° 15_i° 30°, 45°, and 60°) and two observation times (daytime and nighttime). H is the surface elevation (km). The surface F_d^l at the given view zenith angles was calculated by linear interpolation from the two surface F_d^l values with the closest of the five view zenith angles to the given view zenith angle.

5.4.1.2.3.2 Longwave Downward Radiation Model for CERES   The CERES products adopted three models for estimating the surface F_d^l as mentioned previously. Model B is a profile-based method. Model A was developed by Inamdar and Ramanathan (1997b) and is based on the assumption that the surface F_d^l is correlated with outgoing longwave radiation at the TOA component in the atmospheric window (8–12 μm) region. Consequently, the surface F_d^l is estimated in terms of outgoing longwave radiation components inside and outside the window region radiation components. The window region radiation component is a function of the surface temperature, air temperature 50 hPa above the surface, and the surface and TOA upward radiation in the window region. The nonwindow radiation component is a function of the surface temperature, air temperature 50 hPa above the surface, and the surface and TOA upward radiation in the nonwindow region.

The CERES Model C was developed by Zhou (2007) [see also Zhou and Cess (2001)]. Model C estimated $F_{d,\text{clr}}^l$ and $F_{d,\text{all}}^l$ separately. $F_{d,\text{all}}^l$ is a combination of clear- and cloudy-sky components, as shown in the following equations:

$$F_{d,\text{all}}^l = F_{d,\text{clr}}^l f_{\text{clr}} + F_{d,\text{cld}}^l (1 - f_{\text{clr}}) \tag{5.31}$$

$$F_{d,\text{cld}}^l = a_0 + a_1 F_u^l + a_2 \ln(1 + PWV) + a_3 [\ln(1 + PWV)]^2 \tag{5.32}$$

$$F_{d,\text{cld}}^l = b_0 + b_1 F_u^l + b_2 \ln(1 + PWV) + b_3 [\ln(1 + PWV)]^2$$
$$+ b_4 (1 + LWP) + b_5 (1 + IWP) \tag{5.33}$$

where F_u^l is the surface upwelling longwave radiation; PWV is the column precipitable water vapor; LWP is the cloud liquid water path; IWP is the ice water path; and $a_0 - a_3$ and $b_0 - b_5$ are the regression coefficients.

5.4.1.2.3.3 Machine-Learning Model for Estimating Longwave Downward Radiation   Machine-learning methods have been widely used in atmospheric science, such as decision tree learning, Bayesian networks, ANN, and so on. The ANN method is one of the most commonly used methods for estimating F_d^l. Chevallier (1998) proposed an ANN-based long-wave parameterization by splitting the contribution to longwave flux from clear-sky and from clouds. Krasnopolsky et al. (2005) developed an ANN, which calculates LW fluxes from various profiles of absorbing gases, temperature, cloudiness, and the surface upward flux.

A novel approach for estimating surface F_d^l under all-sky conditions is recently presented utilizing the ANN method by Nussbaumer and Pinker (2012). The model is driven with a combination of MODIS level 3 cloud parameters and from the European Center for Medium-Range Weather Forecast (ECMWF) ERA-Interim model. ANNs for $F_{d,\text{clr}}^l$ and $F_{d,\text{cld}}^l$ are trained separately. The $F_{d,\text{clr}}^l$ component is computed with a two-layer feed-forward ANN. The cloud contribution to F_d^l is estimated by using an independent ANN approach, in which the cloud base temperature is trained using the MODIS, CloudSat Cloud Profiling Radar, and the Cloud-Aerosol Lidar and Infrared Pathfinder Satellite Observation (CALIPSO) Cloud-Aerosol Lidar with Orthogonal Polarization (CALIOP) observations. The contribution of clouds to F_d^l is parameterized Stefan–Boltzmann law with a modified emissivity and the screen level temperature by following the work by Schmetz et al. (1986). It can be expressed as

$$F_{d,\text{cld}}^l = \varepsilon^* \, T_o^4 \tag{5.34}$$

where T_o is the screen level temperature and ε^* is the effective emittance related cloud contribution and is given as

$$\varepsilon^* = (1-\varepsilon)\varepsilon_{\text{cld}} C \exp\left(\frac{-(T_o - T_{cb})}{\alpha}\right) \tag{5.35}$$

where ε is the clear-sky effective emittance, ε_{cld} is the cloud emittance, C is the cloud fraction, T_{cb} is the temperature of the cloud base, and α is a fitting coefficient. Tables 5.5 and 5.6 summarize the input and output of $F_{d,\text{clr}}^l$ and cloud base temperature ANN. The R^2 of the daily average estimates of F_d^l at the BSRN sites is 0.98, and the bias and RMSE are –0.39 and 15.84 W m^{-2}, respectively.

TABLE 5.5

Input and Output Parameter for the Clear-Sky ANN

	Unit	Input Type
Input Parameters		
Total column water vapor	kg m^{-2}	Reanalysis
Screen level temperature	K	Reanalysis
Air temperature	K	Reanalysis
Surface elevation	m	Satellite
Land sea mask	0/1	Satellite
Output Parameter		
Clear-sky LWDN	W m^{-2}	

Source: Nussbaumer, E. A. and P. T. Pinker, *J. Geophys. Res.*, 117(D7): D07209, 2012.

TABLE 5.6

Input and Output Parameters for the Cloud Base Temperature ANN

	Unit	Input Type
Input Parameters		
Cloud top height	km	Satellite
Cloud top temperature	K	Satellite/Reanalysis
Cloud emissivity	0–1	Satellite
Screen level temperature	K	Reanalysis
Air temperature (15 levels)	K	Reanalysis
Output Parameter		
Clear base temperature	K	

Source: Nussbaumer, E. A. and P. T. Pinker, *J. Geophys. Res.*, 117(D7): D07209, 2012.

5.4.1.3 Variation and Its Characteristics in Climate Change

As mentioned before, the dominant emitter of longwave radiation in the atmosphere is water vapor. The air temperature is an important parameter to estimate F_d^l based on the Stefan–Boltzmann equation. An additional emitted radiation term is added when clouds are present. Thus longwave downward radiation variation is strongly correlated with water vapor, air temperature, and clouds. Generally, greenhouse gases are fundamental to our understanding of the Earth's climate and its change. Increases in greenhouse gases result in a warming. The increase in water vapor resulted from increased emission of radiation from the atmosphere to the surface causes more warming (Stephens et al. 2011). This is the general understanding of the positive water vapor feedback that exists in longwave downward radiation, water vapor, and temperature.

The changes in water vapor occurring near the Earth's surface are important. It is the lower-level water vapor that is responsible for the emission of infrared radiation to the surface and controls the surface warming (Garratt 2001; Santer et al. 2007). In addition, Stephens and Hu (2010) found that the changes of radiation balance in response to global warming control the change in global precipitation. These changes occur via changes of the absorbed solar radiation and the emitted longwave radiation from atmosphere to surface $\left(F_d^l\right)$ as a result of changes to lower-level water vapor.

Stephens et al. (2011) found that clouds increase the global, annual mean F_d^l in the range 24–34 W m^{-2}. This effect is strongly modulated by the underlying water vapor that reaches a maximum sensitivity of F_d^l to clouds occurring at high-latitude regions. The maximum effect of clouds at the bottom of the atmosphere appears in the drier regions, which is contrast to the TOA cloud effect that is at maximum in the moist tropics.

Garratt (2001) found that the sensitivity of F_d^l to water vapor and temperature is about 1.94 W m^{-2}(kg m^{-2})$^{-1}$ and 3.36 W m^{-2} K^{-1}, respectively, based on a simple F_d^l model proposed by Swinbank (1964). In a different context, the sensitivity of F_d^l to temperature is 7.2 W m^{-2} K^{-1} (Stephens and Hu 2010). Wang and Liang (2009c) applied two simple parameterization methods proposed by Brunt (1932) and Brutsaert (1975) to the globally available meteorological observations from 3200 stations, and they also compared the air temperature, relative humidity, and F_d^l trend with each other. The results showed that the increase in air temperature and atmospheric water vapor concentration are important factors controlling long-term F_d^l variations. In addition, they found that atmospheric water vapor concentration is the most important parameter controlling the long-term F_d^l variation trend. The relative humidity showed substantial spatial variations with the long-term F_d^l variation trend.

The longwave downward radiation trend has been widely reported (Prata 2008; Stephens et al. 2011; Wang and Liang 2009c; Wild et al. 2008). Prata (2008) concluded that over the period 1964–1990, there has been a global increase in the clear-sky longwave flux at the surface. The global trend is approximately +1.7 W m^{-2} per decade, and there is a strong latitudinal pattern, with greater increases occurring in the tropics and smaller increases at both poles. Moreover, the temperature increased +0.22 K per decade, and the precipitable water increased by +0.29 mm per decade. Wild et al. (2008) estimated an overall increase trend of +0.21 W m^{-2} year^{-1} for the period 1986–2000 using the reanalysis data ECMWF 40 year reanalysis (ERA-40) (Uppala et al. 2005) and ground measurements, and a trend of +0.26 W m^{-2} year^{-1} based on ground-measured longwave downward radiation for the period of 1992–2002 from 12 BSRN sites.

Wang and Liang (2009c) estimated the decadal variation in the longwave downward radiation. As a result of global warming, F_d^l increases almost everywhere and the global average trend is +1.9 W m^{-2} per decade from 1973 to 2008, as shown in Figure 5.6. Moreover, the trend of relative humidity is negative when correlated with the trend in air temperature. This indicates a trend toward drought where air temperature increases at a higher rate.

In conclusion, the longwave downward radiation variation is influenced by several factors. These include water vapor, cloud, air temperature, and so on, but the longwave downward radiation variation still cannot be quantitatively explained adequately and the contribution and significance of these factors remain controversial. Moreover, the long-term variation trend is not exactly the same using different data sources. Thus additional

Trend in downward longwave radiation (W m^{-2} year^{-1})

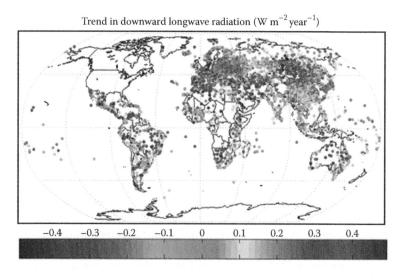

FIGURE 5.6

(See color insert.) Linear trend of daily downward longwave radiation at 3200 global stations (1973–2008). Each point in the figure represents one station, and the colors of the points indicate the trend values. (From Wang, K., and S. Liang., *J. Geophys. Res.*, 114(D19): D19101, 2009.)

research is still needed to clarify the causes and trends of longwave downward radiation variation as well as the character of F_d^l in climate change.

5.4.2 Longwave Upward Radiation

Theoretically, the surface upwelling longwave radiation F_u^l consists of two components: the surface longwave emission and the reflected surface downward longwave radiation F_d^l (Liang 2004).

$$F_u^l = \varepsilon \int_{\lambda_1}^{\lambda_2} \pi B(T_s) \, d\lambda + (1-\varepsilon) F_d^l \tag{5.36}$$

where ε is the surface broadband emissivity, T_s is the surface temperature, $B(T_s)$ is Planck's function, and λ_1 and λ_2 are the spectral range of surface upward longwave radiation (4–100 μm). Generally, we have two methods to evaluate F_u^l from remote sensing: temperature-emissivity method and hybrid method.

5.4.2.1 Temperature-Emissivity Method

There are three variables in Equation 5.36: ε, T_s, and F_d^l. Given that F_d^l has been retrieved from the methods mentioned in the previous subsection, only ε and T_s remain unknown. Satellite retrieval of ε and T_s from the radiometric measurements of thermal infrared and passive microwave sensors has been investigated for several decades. Land surface temperature (LST) detection by infrared thermal sensors provides finer spatial resolution, ranging from 90 m for the Advanced Spaceborne Thermal Emission and Reflection Radiometer (ASTER) and 1 km for MODIS to several dozen kilometers for meteorological satellites. Because of the limited penetration capacity of infrared radiation, infrared LST can be acquired only under clear-sky conditions. In contrast, microwave radiation is only slightly affected by atmospheric influences, which allows LST to be acquired in all weather conditions. Compared with infrared sensors, passive microwave sensors have coarser spatial resolution and lower accuracy.

Many algorithms for temperature and emissivity retrieval have been proposed and used to generate temperature and narrowband emissivity products (Liang et al. 2012a). The ASTER science team provides optional temperature and narrowband emissivity products at 90-m spatial resolution and 16-day temporal resolution (Gillespie et al. 1998). The MODIS science team provides multiple daily temperature and narrowband emissivity products operationally at 1- and 5-km spatial resolutions (Wan and Dozier 1996; Wan and Li 1997). The MODIS land surface temperature and emissivity calculated from MODIS narrowband emissivity have been widely used to calculate F_u^l (Bisht et al. 2005; Tang and Li 2008). The land surface temperature and emissivity derived from the AVHRR have also been applied to estimate F_u^l (Ma et al. 2012).

Estimating F_u^l with satellite-derived surface temperature and emissivity products is quite convenient. However, current satellite land surface temperature and emissivity products retain large uncertainties (Hulley and Hook 2009; Wang and Liang 2009b). For example, the MODIS global LST product is the most reliable LST product, with the accuracy of 1 K verified only for homogenous surfaces, such as water bodies and sandy land (Wan et al. 2002, 2004). For global land surfaces, homogeneity on a scale of 1 km is extremely rare. The retrieval accuracy for microwave sensors is far worse than 1 K. Thus the alternative solution is to estimate F_u^l from TOA longwave observations directly (Liang 2004).

5.4.2.2 Hybrid Method

Discounting the scattering of thermal infrared radiation, the clear-sky radiance measured by thermal infrared sensor can be approximated as

$$B_i(T_i) = \varepsilon_i B_i(T_s)\tau_i(\theta, \varphi, P_s \to 0) + \int_{P_s}^{0} B_i(T_p) \frac{\mathrm{d}\tau_i(\theta, \varphi, P_s \to 0)}{\mathrm{d}\ln P} \mathrm{d}\ln P$$

$$+ \frac{(1-\varepsilon_i)}{\pi} \int_{0}^{2\pi}\int_{0}^{\pi/2}\int_{P_s}^{0} B_i(T_p) \frac{\mathrm{d}\tau_i(\theta', \varphi', P \to P_s)}{\mathrm{d}\ln P} \cos\theta' \sin\theta' \,\mathrm{d}\ln P \,\mathrm{d}\theta' \,\mathrm{d}\varphi \cdot \tau_i(\theta, \varphi, P_s \to 0) \quad (5.37)$$

where $B_i(T_i)$ is the observed TOA radiance for band i, T_i represents the band brightness temperature for band i observed at TOA, ε_i is the surface emissivity in band i, T_s is the surface temperature, $\tau_i(\theta, \varphi, P_s \to 0)$ is the total atmospheric transmittance along the target to sensor in band i, θ and φ are the viewing zenith angle and azimuth angle, P_s is the atmospheric pressure at ground level, T_p is the air temperature at the level of atmospheric pressure P, $\tau_i(\theta, \varphi, P \to 0)$ is the channel transmittance of the atmosphere from the level of atmospheric pressure P to the TOA, and $\tau_i(\theta, \varphi, P \to P_s)$ is the band transmittance of the atmosphere from the level of atmosphere pressure P to the ground level P_s.

The first term on the right-hand side of Equation 5.37 is the surface radiation, the second term denotes the sum of the radiance contributions from all the atmospheric levels to the measured radiance, and the third term represents the hemispheric atmospheric downward longwave radiation reflected by the surface and then attenuated by the atmosphere along the path from the surface to the sensor. It is evident that the clear-sky TOA radiance contains information regarding the surface temperature, emissivity, and surface downward longwave radiation. The hybrid method derives the surface upwelling longwave radiation directly from the satellite TOA radiance or brightness temperature without separately estimating the three variables on the right-hand side of Equation 5.37. The advantage of this method is that the problem of separating the LST and emissivity is bypassed. As a result, a more accurate estimation of the surface upward longwave radiation may be achieved.

The hybrid method has mainly been used to estimate the surface downward longwave radiation (Liang et al. 2010). The general framework of hybrid methods (Wang and Liang 2009d; Wang et al. 2009b; Wang and Liang 2010) consists of two steps. The first is to generate simulated databases using extensive radiative transfer simulation. Surface upwelling longwave radiation and TOA radiance for a particular satellite instrument are simulated using a radiative transfer model and a large number of clear-sky atmospheric profiles. The physics governing the surface longwave radiation budget are embedded in the radiative transfer simulation processes. The second is to conduct a statistical analysis of the simulated databases to derive models.

Smith and Wolfe (1983) used a hybrid method to estimate the surface upward and downward longwave radiation at 1000-hPa pressure level using the TOA radiance obtained by the VISSR on board the National Oceanic and Atmospheric Administration (NOAA) GOES satellites. Meerkoetter and Grassl (1984) used the hybrid method to estimate the surface upward and net longwave radiation using split-window radiance data from the AVHRR.

However, these studies focused on estimating the surface longwave radiation over sea surfaces, and constant emissivity was usually assumed. Wang et al. (2009b,d, 2010) derived the clear-sky F_u^l hybrid models for MODIS and GOES over land surfaces, respectively. The emissivity effect is explicitly considered in the radiative transfer simulation process. In the MODIS hybrid method, the MODIS bands 29, 31, and 32 have the best prediction for the surface F_u^l, which is consistent with the physics that govern the surface upwelling long-wave radiation. All three bands are sensitive to the variations of the surface temperature. The column water vapor is also an important factor in estimating the surface F_u^l from the satellite TOA radiances. These three bands differ in their water vapor absorptions. Bands 31 and 32 are used to retrieve the column water vapor. Linear regression and neural network were used to establish the linear and nonlinear relationships between TOA radiance and F_u^l individually. The linear function was expressed as

$$F_u^l = a_0 + a_1 L_{29} + a_2 L_{31} + a_3 L_{32} \tag{5.38}$$

where a_0, a_1, a_2, and a_3 are regression coefficients and L_{29}, L_{31}, and L_{32} are the TOA radiances for MODIS channels 29, 31, and 32, respectively. Statistical analysis indicated that the linear model accounts for more than 99% of the variation in the simulated databases, with standard errors of 4.89 W m^{-2} (0° sensor viewing zenith angle) to 6.11 W m^{-2} (60° sensor viewing zenith angle). Table 5.7 summarizes the linear MODIS surface upwelling longwave radiation model fitting results.

ANN techniques have proven their ability in modeling nonlinear problems. Wang et al. (2009b) also developed a surface upward longwave radiation model using a single hidden-layer neural network provided by the S-Plus 7 statistical software package (Insightful 2005). The inputs to the neural networks were the MODIS TOA radiances from channels 29, 31, and 32, which are the same set of TOA radiances used to develop the linear model. The ANN model explains more than 99.6% of the variations in the simulated databases, which have standard errors that range from 3.07 to 3.70 W m^{-2} when the sensor view angles vary from 0° to 60°. Moreover, the nonlinear effect is significantly reduced. The radiative transfer simulations and statistical analyses for GOES sounders and GOES-R Advanced Baseline Imager (ABI) are similar to those for MODIS. Equation 5.39 shows the linear model developed to estimate the surface upwelling longwave radiation using the GOES Sounder channels 10, 8, and 7, which are comparable to MODIS channels 29, 31, and 32, respectively (Zhang et al. 2010):

$$F_u^l = b_0 + b_1 L_7 + b_2 L_8 + b_3 L_{10} \tag{5.39}$$

TABLE 5.7

Summary of MODIS Linear Surface Upwelling Longwave Radiation Model Fitting Results
(θ: Sensor View Zenith Angle; Unit of Standard Error: W m^{-2})

Θ	a_0	a_1	a_2	a_3	R^2	SE
0°	102.7589	121.3973	−100.4079	10.4963	0.993	4.89
15°	104.5829	123.4974	−103.0277	10.6894	0.993	4.94
30°	110.4514	129.9471	−111.2339	11.4267	0.992	5.10
45°	122.3125	141.1782	−126.4748	13.5455	0.991	5.41
60°	146.0408	157.2946	−152.6469	20.5749	0.990	6.11

Source: Wang, W. H., Surface longwave radiation budget. In *Advanced Remote Sensing: Terrestrial Information Extraction and Applications* (S. Liang, X. Li, and J. Wang, Eds.). Academic Press, San Diego, CA, pp. 273–300, 2012.

TABLE 5.8

GOES-12 Sounder Linear Surface Upward Longwave Radiation Model Fitting Results

VZA	0°	15°	30°	45°	60°
b_0	124.9927	125.9401	128.9878	135.2046	148.1727
b_1	−130.4156	−132.0319	−137.2860	−148.0604	−170.4925
b_2	153.7796	155.1242	159.4967	168.4509	187.0587
b_3	4.6375	4.8304	5.4884	6.9761	10.5636
SE(e)	2.99	3.04	3.21	3.64	4.95
R^2	0.999	0.999	0.999	0.998	0.997

Source: Wang, W. H., Surface longwave radiation budget. In *Advanced Remote Sensing: Terrestrial Information Extraction and Applications* (S. Liang, X. Li, and J. Wang, Eds.). Academic Press, San Diego, CA, pp. 273–300, 2012.

where b_j ($j = 0,3$) are regression coefficients. Table 5.8 summarizes the GOES-12 Sounder surface upward longwave radiation model fitting results. The model accounts for more than 99% of the variations in the simulated databases, with biases of zero and standard errors of less than 5.00 W m^{-2}.

Equation 5.40 shows the preliminary ABI (full system) surface upwelling longwave radiation model:

$$F_u^l = d_0 + d_1 L_{11} + d_2 L_{13} + d_3 L_{14} + d_4 L_{15} \tag{5.40}$$

where b_j ($j = 0,4$) are the regression coefficients. $L_{11} - L_{15}$ are the TOA radiance for ABI band 11–15. The ABI model fitting results are similar to those of the GOES-12 Sounder and MODIS. No concerns exist regarding the surface upward longwave radiation model because all channels used in the MODIS and GOES Sounder surface upwelling longwave radiation models are available in the ABI. Moreover, the ABI has an additional atmospheric window band at 10.35 µm that can also be used to estimate the surface F_u^l. All MODIS and GOES-12 Sounder models presented in this section were evaluated using ground measurements.

Wang and Liang (2009b,2010) validated the models using ground data from the SURFRAD sites. For MODIS, the RMSEs of the ANN model method are smaller than those of the other two methods at all sites. The average RMSEs of the ANN model method are 15.89 W m^{-2} (Terra) and 14.57 W m^{-2} (Aqua); the average biases are −8.67 W m^{-2} (Terra) and −7.21 W m^{-2} (Aqua). The biases and RMSEs for Aqua are about 1.3 W m^{-2} smaller than those of Terra. The biases and RMSEs of the ANN model method are 5 W m^{-2} smaller than those of the temperature-emissivity method and around 2.5 W m^{-2} smaller than those of the linear model method. For the GOES-12 Sounder linear model, the RMSEs are less than 21 W m^{-2} at 4 SURFRAD sites. Gui et al. (2010) validated the MODIS linear surface model using ground measurements from a 1-year period at 15 North American, Tibetan Plateau, Southeast Asian, and Japanese sites, with an average bias of −9.2 W m^{-2} and a standard deviation of 21.0 W m^{-2}.

5.4.3 Ground Measurements and Satellite Products of Longwave Radiation

Several surface radiation observation networks providing long-term surface longwave radiation measurements have been established, and these ground-based data are available publicly. The BSRN (Ohmura et al. 1998), the SURFRAD (Augustine et al. 2000), and the

TABLE 5.9

Summary of Existing Surface Longwave Radiation Budget Datasets

Available Products	CERES	WCRP/GEWEX	ISCCP
Spatial resolution	1°	1°	280 km
Temporal coverage	1998–present	1983–2007	1983–2007
Satellite	TRMM, Terra/Aqua	GOES	TOVS
Instrument footprint (km)	20	10 × 40	40
Stated accuracy of monthly averages (W m^{-2})	21	33.6	20–25

Source: Wang, W. H., Surface longwave radiation budget. In *Advanced Remote Sensing: Terrestrial Information Extraction and Applications* (S. Liang, X. Li, and J. Wang, Eds.). Academic Press, San Diego, CA, pp. 273–300, 2012.

FLUXNET (Baldocchi et al. 2001) are three major surface radiation observation networks. However, the existing ground-based observations are still geographically sparse for mapping radiation globally.

Three major long-term satellite surface longwave radiation budget datasets are currently available. The first dataset is derived from CERES, which is onboard the NASA Earth Observing System (EOS) Terra and Aqua satellites and the Tropical Rainfall Measuring Mission (TRMM) satellite (Gupta 1997; Inamdar and Ramanathan 1997a; Wielicki et al. 1998). The second long-term surface longwave radiation budget dataset is provided by the NASA World Climate Research Programme (WCRP)/GEWEX project using data from the GOES satellites (Stackhouse et al. 2011). The third dataset is an 18-year surface longwave radiation budget dataset derived from ISCCP data (Zhang and Rossow 2002; Zhang et al. 1995, 2004). Table 5.9 summarizes the spatial resolution, temporal coverage, satellite instrument, instrument footprint, and stated accuracy associated with each dataset. Gui et al. (2010) evaluated the CERES, WCRP/GEWEX, and ISCCP surface longwave radiation. The results show that the CERES datasets perform the best overall; however, all of the datasets are biased. Readers may consult this paper for more details.

Surface longwave radiation budget data are also available from GCM simulations. However, most GCMs have simplified land-surface emissivity, and considerable differences exist between model outputs and satellite-derived products (Liang et al. 2010).

5.5 All-Wave Net Radiation

5.5.1 Introduction

All-wave net radiation is the sum of shortwave and longwave net radiation. All-wave net radiation is closely correlated with shortwave net radiation and dominated by the shortwave net radiation except at the poles. It is generally greatest in the summer tropical and subtropical oceans with maxima near the Tropic of Cancer during July and the Tropic of Capricorn during January. The seasonal variability of all-wave net radiation is very similar to that of shortwave net radiation.

After we calculate shortwave and longwave net radiation using estimated individual components of surface radiation balance, we can directly calculate all-wave net radiation. Another method is to establish the empirical relationship between all-wave net radiation and solar shortwave radiation. These two methods are named as the component-based method and empirical method, respectively.

5.5.2 Methodology

5.5.2.1 Component-Based Method

In the previous subsections, we reviewed the method for estimating surface incident solar radiation, surface shortwave albedo, surface longwave downward radiation, and surface longwave upwelling radiation, from which we can calculate surface shortwave net radiation and surface longwave net radiation, and then the all-wave net radiation. The surface shortwave net radiation can also be directly estimated. For example, Kim and Liang (2010) proposed the hybrid method to estimate shortwave net radiation directly from MODIS TOA and surface reflectance data without separating cloudy- and clear-sky conditions. This method does not require ancillary data that typically have different spatial and spectral resolution. The shortwave net radiation estimated by this hybrid method is more accurate than that calculated by using surface incident solar radiation and surface shortwave albedo.

Bisht et al. (2005) proposed the scheme for deriving the components of surface energy budget over large heterogeneous areas under clear sky using only the MODIS products and calculated instantaneous net radiation for the first time. They also proposed the sinusoidal model that can estimate the diurnal cycle of net radiation with a single instantaneous net radiation estimated from the satellite. Bisht and Bras (2010) extended their method to cloudy-sky conditions and applied to estimate instantaneous and daily net radiation over the South Great Plains for a time period covering all seasons of 2006. The RMSE of instantaneous and daily average net radiation estimated under cloudy condition were 37 and 38 W m^{-2}, compared to ground measurements. Ryu et al. (2008) evaluated the land surface radiation balance derived from MODIS over a heterogeneous farmland area and a rugged deciduous forest in the Korea Flux Network. Their study suggested that the scale mismatch among the radiometer footprint, heterogeneity of the land surface properties, and MODIS extent should be considered in both the algorithm development and validation processes. The AVHRR data were also widely used to estimate the components of land surface radiation balance, including all-wave net radiation. Ma et al. (2012) derived surface shortwave albedo, land surface temperature, and emissivity from AVHRR data, land-surface, and aerological observation data, computed the downward shortwave and longwave radiation at surface, and determined the net radiation over the Tibetan Plateau by summing all the components. The validation results indicated that the derived surface shortwave albedo and land surface temperature for the Tibetan Plateau area were in good accordance with the land-surface status, and the absolute percentage difference between derived net radiation and the field observation was less than 10%. Combined AVHRR data and meteorological data, Hurtado and Sobrino (2012) calculated each component of surface radiation energy balance equation as well as the daily net radiation. The uncertainty of the estimated daily net radiation was about 1.5 W m^{-2}.

5.5.2.2 Empirical Method

Many researchers have proposed the statistical methods for calculating all-wave net radiation from other climatic parameters. A common procedure is to estimate all-wave net radiation using simple linear regression (Alados et al. 2003; Kaminsky and Dubayah 1997; Kjaersgaard et al. 2007a). The most commonly used equations are

$$R_n = a_1 F_d^s + b_1 \tag{5.41}$$

$$R_n = a_2(1-\alpha)F_d^s + b_2 \tag{5.42}$$

where a_1, a_2, b_1, and b_2 are the regression coefficients. Neither correction for longwave radiation nor factors affecting the longwave radiation components were included in Equations 5.41 and 5.42. Inclusion of cloud cover, subdivision of dataset by season, surface heating coefficient, and consideration of changes of surface albedo during single days and during a whole year in the regression have been suggested to improve the performance. Irmak et al. (2003) developed a model relating all-wave net radiation to a set of meteorological and geometrical variables. The general form of the model is

$$R_n = \beta_0 + \beta_1 T_{\max} + \beta_2 T_{\min} + \beta_3 F_d^s + \beta_4 d_r \tag{5.43}$$

where β_0 is the intercept, β_1, β_2, β_3, and β_4 are the coefficients for each independent variable, $T_{\max}$ is the daily maximum air temperature, $T_{\min}$ is the daily minimum air temperature, and d_r is the inverse relative Earth–Sun distance. Wang and Liang (2009a) proposed a method to estimate all-wave net radiation using satellite and conventional meteorological observation, which makes the daily all-sky all-wave net radiation global mapping become possible. The ratio of daily all-wave net radiation to daily shortwave net radiation S_n was formulated as follows:

$$(R_n/S_n) = a_0 + a_1 T_{\min} + a_2 DTR + a_3 VI + a_4 RH \tag{5.44}$$

where $T_{\min}$ is the daily minimum air temperature (or surface temperature), DTR is the range of daily air temperature (or surface temperature), VI is the MODIS global NDVI or EVI product with a spatial resolution of 1 km and 16-day temporal resolution, and RH is the relative humidity. This method was validated to accurately predict long-term variation in all-wave net radiation.

In the empirical method, all-wave net radiation is formulated as the linear function of incident solar radiation and other conventional meteorological observations. Generally, it is much easier to derive all-sky incident solar radiation than all-sky surface longwave net radiation. The empirical method is a promising way for the estimate of all-wave net radiation.

5.6 Conclusions

We have reviewed the recent advances in estimating land SRB components from satellite observations, including incident solar radiation, albedo, clear-sky longwave downward and upwelling radiation, and all-wave net radiation. For each component, we summarized the major recent advances in algorithm developments, introduced global satellite products and their characteristics, and also analyzed the spatial and temporal variations, and the attribution and consequences of their changes.

Although significant progress has been made, there are still many challenges. Remotely sensed products are usually generated from a series of processes, and the inversion process always introduces uncertainties. Many algorithms are available for one product, but each product is usually generated by one inversion algorithm. Because different algorithms have different strengths and weaknesses, it is almost impossible to identify the best algorithm. A better solution is a combination of multiple algorithms for producing the same product. Further efforts are needed to comprehensively and objectively evaluate different algorithms.

Another issue is that most satellite-generated products have been derived from individual sensors. For example, surface albedo products are produced from both MODIS and MISR data separately, although these two sensors are in the same satellite platform (Terra). More studies are needed to derive each product from data of multiple sensors effectively.

The retrieved satellite quantities have to be validated by ground measurements to quantify their uncertainties. The ground "point" measurements cannot match the size of the remote sensing pixel. Spatial upscaling always results in errors, but the use of remote sensing observations with different spatial resolutions will help minimize uncertainties. Overall, the number of current ground measurement networks is inadequate for a good spatial and temporal representation. Uncertainties of ground measurements in different networks, or even in the same network, vary considerably. Consistent processing and effective management of ground measurements from various sources to make them accessible to users remain a significant and ongoing challenge. In addition, different validation studies use the variable error measures, such as the bias, RMSE, RMS, percentage error, STD, and so on. This makes intercomparisons of the errors among ground measurements and satellite products difficult.

It is anticipated, in the near future, that we will be able to more accurately characterize the land SRB at different spatial and temporal scales from multiple satellite data, more mature inversion algorithms, and more and better ground measurements.

References

Alados, I., I. Foyo-Moreno, and L. Alados-Arboledas. 2012. Estimation of downwelling longwave irradiance under all-sky conditions. *Int. J. Climatol.*, 32(5): 781–793.

Alados, I., I. Foyo-Moreno, F. J. Olmo, and L. Alados-Arboledas. 2003. Relationship between net radiation and solar radiation for semi-arid shrub-land. *Agric. For. Meteorol.*, 116: 221–227.

Anderson, G. P. et al. 1999. MODTRAN4: Radiative transfer modeling for remote sensing. *Proc. SPIE*, 3866.

Angstrom, A. 1924. Solar and terrestrial radiation. Report to the international commission for solar research on actinometric investigations of solar and atmospheric radiation. *Q. J. R. Meteorol. Soc.*, 50(210): 121–126.

Annear, R. L., and S. A. Wells. 2007. A comparison of five models for estimating clear-sky solar radiation. *Water Resour. Res.*, 43(10): W10415.

Aubinet, M. 1994. Longwave sky radiation parameterizations. *Solar Energy*, 53: 147–154.

Augustine, J. A., J. Deluisi, and C. N. Long. 2000. SURFRAD-a national surface radiation budget network for atmospheric research. *Bull. Am. Meteorol. Soc.*, 81(10): 2341–2357.

Augustine, J. A., G. B. Hodges, C. R. Cornwall, J. J. Michalsky, and C. I. Medina. 2005. An update on SURFRAD: The GCOS surface radiation budget network for the continental United States. *J. Atmos. Oceanic Technol.*, 22(10): 1460–1472.

Baldocchi, D. et al. 2001. FLUXNET: A new tool to study the temporal and spatial variability of ecosystem-scale carbon dioxide, water vapor, and energy flux densities. *Bull. Am. Meteorol. Soc.,* 82: 2415–2434.

Bisht, G., and R. L. Bras. 2010. Estimation of net radiation from the MODIS data under all sky conditions: Southern Great Plains case study. *Remote Sens. Environ.,* 114: 1522–1534.

Bisht, G., V. Venturini, S. Islam, and L. Jiang. 2005. Estimation of the net radiation using MODIS (Moderate Resolution Imaging Spectrometer) data for clear sky days. *Remote Sens. Environ.,* 97: 52–67.

Breon, F. M., F. Maignan, M. Leroy, and I. Grant. 2002. Analysis of hot spot directional signatures measured from space. *J. Geophys. Res.,* 107(D16): 15.

Brunt, D. 1932. Notes on radiation in the atmosphere. I. *Q. J. R. Meteorol. Soc.,* 58(247): 389–420.

Brutsaert, W. 1975. On a derivable formula for long-wave radiation from clear skies. *Water Resour. Res.,* 11(5): 742–744.

Chevallier, F., F. Cheruy, N. A. Scott, and A. Chedin. 1998. A neural network approach for a fast and accurate computation of a longwave radiative budget. *J. Appl. Meteorol.,* 37(11): 1385–1397.

Choudhury, B. 1982. A parameterized model for global insolation under partially cloudy skies. *Solar Energy,* 29(6): 479–486.

Crawford, T. M., and C. E. Duchon. 1999. An improved parameterization for estimating effective atmospheric emissivity for use in calculating daytime downwelling longwave radiation. *J. Appl. Meteorol. Climatol.,* 38(4): 474–480.

Csiszar, I., and G. Gutman. 1999. Mapping global land surface albedo from NOAA AVHRR. *J. Geophys. Res.,* 104(D6): 6215–6228.

Cui, Y., Y. Mitomi, and T. Takamura. 2009. An empirical anisotropy correction model for estimating land surface albedo for radiation budget studies. *Remote Sens. Environ.,* 113(1): 24–39.

Deneke, H. M., A. J. Feijt, and R. A. Roebeling. 2008. Estimating surface solar irradiance from METEOSAT SEVIRI-derived cloud properties. *Remote Sens. Environ.,* 112(6): 3131–3141.

Dickinson, R. E. 1983. Land surface processes and climate surface albedos and energy balance. *Adv. Geophys.,* 25: 305–353.

Dilley, A. C., and D. M. O'Brien. 1998. Estimating downward clear sky long-wave irradiance at the surface from screen temperature and precipitable water. *Q. J. R. Meteorol. Soc.,* 124(549): 1391–1401.

El-Metwally, M. 2004. Simple new methods to estimate global solar radiation based on meteorological data in Egypt. *Atmos. Res.,* 69(3–4): 217–239.

Ellingson, R. G. 1995. Surface longwave fluxes from satellite observations: A critical review. *Remote Sens. Environ.,* 51(1): 89–97.

Fang, H. L. et al. 2007. Developing a spatially continuous 1 km surface albedo dataset over North America from Terra MODIS products. *J. Geophys. Res.,* 112(D20): D20206.

Flerchinger, G. N., W. Xaio, D. Marks, T. J. Sauer, and Q. Yu. 2009. Comparison of algorithms for incoming atmospheric long-wave radiation. *Water Resour. Res.,* 45: W03423.

Frouin, R., C. Gautier, and J.-J. Morcrette. 1988. Downward longwave irradiance at the ocean surface from satellite data: Methodology and in-situ validation. *J. Geophys. Res.,* 93(C1): 597–619.

Gao, F. et al. 2005. MODIS bidirectional reflectance distribution function and albedo Climate Modeling Grid products and the variability of albedo for major global vegetation types. *J. Geophys. Res.,* 110(D1): D01104.

Garratt, J. R. 2001. Clear-sky longwave irradiance at the Earth's surface—Evaluation of climate models. *J. Climate,* 14(7): 1647–1670.

Gastellu-Etchegorry, J. P., E. Martin, and F. Gascon. 2004. DART: A 3D model for simulating satellite images and studying surface radiation budget. *Int. J. Remote Sens.,* 25(1): 73–96.

Geiger, B., D. Carrer, L. Franchisteguy, J. L. Roujean, and C. Meurey. 2008. Land surface albedo derived on a daily basis from Meteosat Second Generation observations. *IEEE Trans. Geosci. Remote Sens.,* 46(11): 3841–3856.

Gilgen, H., and A. Ohmura. 1999. The global energy balance archive. *Bull. Am. Meteorol. Soc.,* 80(5): 831–850.

Gillespie, A. R. et al. 1998. A temperature and emissivity separation algorithm for Advanced Spaceborne Thermal Emission and Reflection Radiometer (ASTER) images. *IEEE Trans. Geosci. Remote Sens.*, 36: 1113–1126.

Govaerts, Y., and A. Lattanzio. 2008. Estimation of surface albedo increase during the eighties Sahel drought from Meteosat observations. *Global Planet. Change*, 64(3–4): 139–145.

Govaerts, Y. M., S. Wagner, A. Lattanzio, and P. Watts. 2010. Joint retrieval of surface reflectance and aerosol optical depth from MSG/SEVIRI observations with an optimal estimation approach: 1. Theory. *J. Geophys. Res.*, 115: D02203.

Gueymard, C. A. 2003. Direct solar transmittance and irradiance predictions with broadband models. Part I: detailed theoretical performance assessment. *Solar Energy*, 74(5): 355–379.

Gui, S., S. Liang, and L. Li. 2010. Evaluation of satellite-estimated surface longwave radiation using ground-based observations. *J. Geophys. Res.*, 115: D18214.

Gupta, S. K. 1989. A parameterization for longwave surface radiation from sun-synchronous satellite data. *J. Climate*, 2: 305–320.

Gupta, S. K. 1997. Clouds and the Earth's Radiant Energy System (CERES) algorithm theoretical basis document: An algorithm for longwave surface radiation budget for total skies (subsystem 4.6.3). Available at http://ceres.larc.nasa.gov/documents/ATBD/pdf/r2_2/ceres-atbd2.2-s4.6.3.pdf

Gupta, S. K., A. C. Wilber, W. L. Darnell, and J. T. Suttles. 1993. Longwave surface radiation over the globe from satellite data: An error analysis. *Int. J. Remote Sens.*, 14(1): 95–114.

Gupta, S. K. K., P. David, P. W. Stackhouse, Jr., A. C. Wilber, T. Zhang, and V. E. Sothcott. 2009. Improvement of Surface Longwave Flux Algorithms Used in CERES Processing. *J. Appl. Meteorol. Climatol.*, 49(7): 1579–1589.

Hansen, J., and L. Nazarenko. 2004. Soot climate forcing via snow and ice albedos. *Proc. Natl. Acad. Sci. USA*, 101(2): 423–428.

He, T. et al. 2012a. Estimation of surface albedo and directional reflectance from Moderate Resolution Imaging Spectroradiometer (MODIS) observations. *Remote Sens. Environ.*, 119: 286–300.

He, T. et al. 2012b. Daily-based estimation of surface albedo from Meteosat Second Generation (MSG) observations. Submitted to *J. Geophys. Res.* in revision.

Houborg, R. M., and H. Soegaard. 2004. Regional simulation of ecosystem CO_2 and water vapor exchange for agricultural land using NOAA AVHRR and Terra MODIS satellite data. Application to Zealand, Denmark. *Remote Sens. Environ.*, 93(1–2): 150–167.

Houborg, R., H. Soegaard, and E. Boegh. 2007. Combining vegetation index and model inversion methods for the extraction of key vegetation biophysical parameters using Terra and Aqua MODIS reflectance data. *Remote Sens. Environ.*, 106(1): 39–58.

Huang, G., M. Ma, S. Liang, S. Liu, and X. Li. 2011. A LUT-based approach to estimate surface solar irradiance by combining MODIS and MTSAT data. *J. Geophys. Res.*, 116(D22): D22201.

Huang, G. H., S. M. Liu, and S. L. Liang. 2012. Estimation of net surface shortwave radiation from MODIS data. *Int. J. Remote Sens.*, 33(3): 804–825.

Hulley, G. C., and S. J. Hook. 2009. Intercomparison of versions 4, 4.1 and 5 of the MODIS land surface temperature and emissivity products and validation with laboratory measurements of sand samples from the Namib desert, Namibia. *Remote Sens. Environ.*, 113: 1313–1318.

Hurtado, E., and J. A. Sobrino. 2012. Daily net radiation estimated from air temperature and NOAA-AVHRR data: A case study for the Iberian peninsula. *Int. J. Remote Sens.*, 22(8): 1521–1533.

Idso, S. B. 1981. A set of equations for full spectrum and 8- to 14- and 10.5- to 12.5-thermal radiation from cloudless skies. *Water Resour. Res.*, 17(2): 295–304.

Inamdar, A. K., and V. Ramanathan. 1997a. Clouds and the Earth's Radiant Energy System (CERES) algorithm theoretical basis document: estimation of longwave surface radiation budget from CERES (subsystem 4.6.2). Available at http://ceres.larc.nasa.gov/documents/ATBD/pdf/r2_2/ceres-atbd2.2-s4.6.2.pdf

Inamdar, A. K., and V. Ramanathan. 1997b. On monitoring the atmospheric greenhouse effect from space. *Tellus*, 49(2): 216–230.

Insightful. 2005. *S-PLUS 7 for UNIX User Guide*. Seattle, WA.

Irmak, S. et al. 2003. Predicting daily net radiation using minimum climatological data. *J. Irrig. Drain Eng.*, 129(4): 256–269.

Kaminsky, K. Z., and R. Dubayah. 1997. Estimation of surface net radiation in the boreal forest and northern prairie. *J. Geophys. Res.*, 102(D24): 29,707–29,716.

Kim, H. Y., and S. L. Liang. 2010. Development of a hybrid method for estimating land surface shortwave net radiation from MODIS data. *Remote Sens. Environ.*, 114(11): 2393–2402.

Kjaersgaard, J. H., R. H. Cuenca, and F. L. Plauborg. 2007a. Long-term comparison of net radiation calculation schemes. *Boundary Layer Meteorol.*, 123: 417–431.

Kjaersgaard, J. H., F. L. Plauborg, and S. Hansen. 2007b. Comparison of models for calculating daytime long-wave irradiance using long term dataset. *Agric. For. Meteorol.*, 143(1–2): 49–63.

Krasnopolsky, V. M., M. S. Fox-Rabinovitz, and D. V. Chalikov. 2005. New approach to calculation of atmospheric model physics: Accurate and fast neural network emulation of longwave radiation in a climate model. *Mon. Weather Rev.*, 133(5): 1370–1383.

Kuusk, A. 1995a. A fast, invertible canopy reflectance model. *Remote Sens. Environ.*, 51(3): 342–350.

Kuusk, A. 1995b. A Markov-chain model of canopy reflectance. *Agric. For. Meteorol.*, 76(3–4): 221–236.

Lacaze, R., J. L. Roujean, and J. P. Goutorbe. 1999. Spatial distribution of Sahelian land surface properties from airborne POLDER multiangular observations. *J. Geophys. Res.*, 104(D10): 12,131–12,146.

Leroy, M. et al. 1997. Retrieval of atmospheric properties and surface bidirectional reflectances over land from POLDER/ADEOS. *J. Geophys. Res.*, 102(D14): 17,023–17,037.

Lewis, P. et al. 2012. GlobAlbedo Algorithm Theoretical Basis Document V3.1. Available at http://www.globalbedo.org/docs/GlobAlbedo_Albedo_ATBD_V3.1.pdf

Lhomme, J. P., J. J. Vacher, and A. Rocheteau. 2007. Estimating downward long-wave radiation on the Andean Altiplano. *Agric. For. Meteorol.*, 145(3–4): 139–148.

Li, X. W., and A. H. Strahler. 1992. Geometric-optical bidirectional reflectance modeling of the DISCRETE crown vegetation canopy effect of crown shape and mutual shadowing. *IEEE Trans. Geosci. Remote Sens.*, 30(2): 276–292.

Liang, S., J. C. Stroeve, I. F. Grant, A. H. Strahler, and J. P. Duvel. 2000. Angular corrections to satellite data for estimating earth radiation budget. *Remote Sens. Rev.*, 18(2–4): 103–136.

Liang, S. L. 2001. Narrowband to broadband conversions of land surface albedo I Algorithms. *Remote Sens. Environ.*, 76(2): 213–238.

Liang, S. L. 2003. A direct algorithm for estimating land surface broadband albedos from MODIS imagery. *IEEE Trans. Geosci. Remote Sens.*, 41(1): 136–145.

Liang, S. L. 2007. Recent developments in estimating land surface biogeophysical variables from optical remote sensing. *Prog. Phys. Geogr.*, 31(5): 501–516.

Liang, S. L., B. Jiang, T. He, and X. F. Zhu. 2012b. Soil moisture content. In: S. Liang, X. Li and J. Wang (Editors), Advanced remote sensing: Terrestrial information extraction and applications. Academic Press, pp. 589–614.

Liang, S. L. 2004. Quantitative Remote Sensing of Land Surface. John Wiley, New York.

Liang, S. L. et al. 2006. Estimation of incident photosynthetically active radiation from Moderate Resolution Imaging Spectrometer data. *J. Geophys. Res.*, 111(D15): D15208.

Liang, S. L., B. Jiang, T. He, and X. F. Zhu. 2012b. Soil moisture content. In *Advanced Remote Sensing: Terrestrial Information Extraction and Applications*. S. Liang, X. Li, and J. Wang (Editors), Academic Press, San Diego, CA, pp. 589–614.

Liang, S. L., X. Li, and J. Wang (Ed.). 2012a. *Advanced Remote Sensing: Terrestrial Information Extraction and Applications*. Academic Press, San Diego, CA.

Liang, S. L., A. H. Strahler, and C. Walthall. 1999. Retrieval of land surface albedo from satellite observations: A simulation study. *J. Appl. Meteorol.*, 38(6): 712–725.

Liang, S. L., J. Stroeve, and J. E. Box. 2005. Mapping daily snow/ice shortwave broadband albedo from Moderate Resolution Imaging Spectroradiometer (MODIS): The improved direct retrieval algorithm and validation with Greenland in situ measurement. *J. Geophys. Res.*, 110(D10): D10109.

Liang, S. L., K. C. Wang, X. T. Zhang, and M. Wild. 2010. Review on estimation of land surface radiation and energy budgets from ground measurement, remote sensing and model simulations. *IEEE J. Selected Topics Appl. Earth Obs. Remote Sens.*, 3(3): 225–240.

Liang, S. et al. 2013. A long-term Global LAnd Surface Satellite (GLASS) dataset for environmental studies. *Int. J. Digital Earth*, doi:10.1080/17538947.17532013.17805262.

Liu, N. et al. 2013. Mapping spatially-temporally continuous shortwave albedo for global land surface from MODIS data. *Hydrol. Earth Syst. Sci.*, 17: 2121–2129, doi:2110.5194/hess-2117-2121-2013.

Liu, Q., L. Wang, Y. Qu, N. Liu, S. Liu, H. Tang, and S. L. Liang. 2013. Preliminary evaluation of the long-term GLASS albedo product. *Int. J. Digital Earth*, doi:10.1080/17538947.17532013.17804601.

Liu, R., S. Liang, H. He, J. Liu, and T. Zheng. 2008. Mapping incident photosynthetically active radiation from MODIS data over China. *Remote Sens. Environ.*, 112: 998–1009.

Liu, S. H., Q. A. Liu, Q. H. Liu, J. G. Wen, and X. W. Li. 2010. The angular and spectral kernel model for BRDF and albedo retrieval. *IEEE J. Selected Topics Appl. Earth Obs. Remote Sens.*, 3(3): 241–256.

Loarie, S. R., D. B. Lobell, G. P. Asner, Q. Mu, and C. B. Field. 2011. Direct impacts on local climate of sugar-cane expansion in Brazil. *Nature Climate Change*, 1(2): 105–109.

Lu, N., R. Liu, J. Liu, and S. Liang. 2010. An algorithm for estimating downward shortwave radiation from GMS-5 visible Imagery and its evaluation over China. *J. Geophys. Res.*, 115: D18102, doi:10.1029/2009JD013457.

Lucht, W., C. B. Schaaf, and A. H. Strahler. 2000. An algorithm for the retrieval of albedo from space using semiempirical BRDF models. *IEEE Trans. Geosci. Remote Sens.*, 38(2): 977–998.

Lyons, E. A., Y. F. Jin, and J. T. Randerson. 2008. Changes in surface albedo after fire in boreal forest ecosystems of interior Alaska assessed using MODIS satellite observations. *J. Geophys. Res.*, 113(G2): 15.

Ma, Y., L. Zhong, Y. Wang, and Z. Su. 2012. Using NOAA/AVHRR data to determine regional net radiation and soil heat fluxes over the heterogeneous landscape of the Tibetan Plateau. *Int. J. Remote Sens.*, 33(15): 4784–4795.

Maignan, F., F. M. Breon, and R. Lacaze. 2004. Bidirectional reflectance of Earth targets: Evaluation of analytical models using a large set of spaceborne measurements with emphasis on the Hot Spot. *Remote Sens. Environ.*, 90(2): 210–220.

Martonchik, J. V. et al. 1998. Determination of land and ocean reflective, radiative, and biophysical properties using multiangle imaging. *IEEE Trans. Geosci. Remote Sens.*, 36(4): 1266–1281.

Maykut, G. A., and P. E. Church. 1973. Radiation climate of Barrow Alaska, 1962–66. *J. Appl. Meteorol.*, 12(4): 620–628.

Meerkoetter, R., and H. Grassl. 1984. Longwave net flux at the ground from radiances at the top. IRS'84: Current Problems in Atmospheric Radiation, G. Fiocco, Ed., A. Deepak, 220–223.

Muller, J. P. et al. 2012. The ESA GlobAlbedo project for mapping the Earth's land surface albedo for 15 years from European sensors, IEEE Geoscience and Remote Sensing Symposium (IGARSS) 2012. IEEE, New Brunswick, NJ.

NASA Goddard Space Flight Center. 2011. Joint Polar Satellite System (JPSS) VIIRS Surface Albedo Algorithm theoretical basis document (for public release), Greenbelt, MD.

Niemelä, S., P. Räisänen, and H. Savijärvi. 2001. Comparison of surface radiative flux parameterizations: Part I: Longwave radiation. *Atmos. Res.*, 58(1): 1–18.

Nussbaumer, E. A., and P. T. Pinker. 2012. Estimating surface longwave radiative fluxes from satellites utilizing artificial neural networks. *J. Geophys. Res.*, 117(D7): D07209.

Offerle, B., P. Jonsson, I. Eliasson, and C. S. B. Grimmond. 2005. Urban modification of the surface energy balance in the West African Sahel: Ouagadougou, Burkina Faso. *J. Clim.*, 18(19): 3983–3995.

Ohmura, A. et al. 1998. Baseline Surface Radiation Network (BSRN/WCRP): New precision radiometry for climate research. *Bull. Am. Meterol. Soc.*, 79(10): 2115–2136.

Oreopoulos, L. et al. 2012. The continual intercomparison of radiation codes: Results from phase I. *J. Geophys. Res.*, 117(D6): D06118.

Paulescu, M., and Z. Schlett. 2003. A simplified but accurate spectral solar irradiance model. *Theoret. Appl. Climatol.*, 75(3): 203–212.

Pinker, R. T. et al. 2003. Surface radiation budgets in support of the GEWEX Continental-Scale International Project (GCIP) and the GEWEX Americas Prediction Project (GAPP), including the North American Land Data Assimilation System (NLDAS) Project. *J. Geophys. Res.*, 108(D22): 8844.

Pinker, R. T., and I. Laszlo. 1992. Modeling surface solar irradiance for satellite applications on a global scale. *J. Appl. Meteorol.*, 31(2): 194–211.

Pinty, B. et al. 2006. Simplifying the interaction of land surfaces with radiation for relating remote sensing products to climate models. *J. Geophys. Res.*, 111(D2): D02116.

Pinty, B. et al. 2000a. Surface albedo retrieval from Meteosat-1. Theory. *J. Geophys. Res.*, 105(D14): 18,099–18,112.

Pinty, B. et al. 2000b. Surface albedo retrieval from Meteosat-2. Applications. *J. Geophys. Res.*, 105(D14): 18,113–18,134.

Pokrovsky, O., and J. L. Roujean. 2003. Land surface albedo retrieval via kernel-based BRDF modeling: I. Statistical inversion method and model comparison. *Remote Sens. Environ.*, 84(1): 100–119.

Prata, A. J. 1996. A new long-wave formula for estimating downward clear-sky radiation at the surface. *Q. J. R. Meteorol. Soc.*, 122(533): 1127–1151.

Prata, F. 2008. The climatological record of clear-sky longwave radiation at the Earth's surface: evidence for water vapour feedback? *Int. J. Remote Sens.*, 29(17–18): 5247–5263.

Psiloglou, B. E., and H. D. Kambezidis. 2007. Performance of the meteorological radiation model during the solar eclipse of 29 March 2006. *Atmos. Chem. Phys. Discuss.*, 7: 6047–6059.

Qu, Y., Q. Liu, S. L. Liang, L. Wang, N. Liu, and S. Liu. 2013. Direct-estimation algorithm for mapping daily land-surface broadband albedo from MODIS data. *IEEE Trans. Geosci. Remote Sens.*, doi:10.1109/TGRS.2013.2245670.

Rahman, H., B. Pinty, and M. M. Verstraete. 1993. Coupled Surface-Atmosphere Reflectance (CSAR) model. 2. Semiempirical surface model usable with NOAA Advanced Very High-Resolution Radiometer data. *J. Geophys. Res.*, 98(D11): 20,791–20,801.

Roujean, J. L., M. Leroy, and P. Y. Deschamps. 1992. A bidirectional reflectance model of the Earth's surface for the correction of remote sensing data. *J. Geophys. Res.*, 97(D18): 20,455–20,468.

Running, S. W., R. Nemani, J. M. Glassy, and P. E. Thornton. 1999. MODIS daily Photosynthesis (PSN) and annual Net Primary Production (NPP) product (MOD17). Algorithm Theoretical Basis Document. Version 3.0. Available at http://ntsg.umt.edu/sites/ntsg.umt.edu/files/modis/ATBD/ATBD_MOD17_v21.pdf.

Running, S. W., P. E. Thornton, R. Nemani, and J. M. Glassy. 2000. Global terrestrial gross and net primary productivity from the earth observing system. In *Methods in Ecosystem Science*. Springer, New York, pp. 44–57.

Ryu, Y., S. Kang, S.-K. Moon, and J. Kim. 2008. Evaluation of land surface radiation balance derived from moderate resolution imaging spectroradiometer (MODIS) over complex terrain and heterogeneous landscape on clear sky days. *Agric. For. Meteorol.*, 148: 1538–1552.

Santer, B. D. et al. 2007. Identification of human-induced changes in atmospheric moisture content. *Proc. Natl. Acad. Sci.*, 104(39): 15,248–15,253.

Schaaf, C. B. et al. 2002. First operational BRDF, albedo nadir reflectance products from MODIS. *Remote Sens. Environ.*, 83(1–2): 135–148.

Schmetz, J. 1989. Towards a surface radiation climatology: Retrieval of downward irradiances from satellites. *Atmos. Res.*, 23(3–4): 287–321.

Schmetz, P., J. Schmetz, and E. Raschke. 1986. Estimation of daytime downward longwave radiation at the surface from satellite and grid point data. *Theoret. Appl. Climatol.*, 37(3): 136–149.

Smith, W. L., and H. M. Wolfe. 1983. Geostationary satellite sounder (VAS) observations of longwave radiation flux, The satellite systems to measure radiation budget parameters and climate change signal, Igls, Austria.

Stackhouse, Jr., P. W., S. K. Gupta, S. J. Cox, J. C. Mikovitz, T. Zhang, and L. M. Hinkelman. 2011. The NASA/GEWEX Surface Radiation Budget Release 3.0: 24.5-Year Dataset. GEWEX News, 21, No. 1, February, 10–12.

Stephens, G. L. et al. 2011. The global character of the flux of downward longwave radiation. *J. Clim.*, 25(7): 2329–2340.

Stephens, G. L. et al. 2012. An update on Earth's energy balance in light of the latest global observations. *Nature Geosci.*, 5(10): 691–696.

Stephens, G. L., and Y. Hu. 2010. Are climate-related changes to the character of global-mean precipitation predictable? *Environ. Res. Lett.*, 5(2): 025209.

Steyaert, L. T., and R. G. Knox. 2008. Reconstructed historical land cover and biophysical parameters for studies of land-atmosphere interactions within the eastern United States. *J. Geophys. Res.*, 113(D2): D02101.

Stroeve, J. et al. 2005. Accuracy assessment of the MODIS 16-day albedo product for snow: Comparisons with Greenland in situ measurements. *Remote Sens. Environ.*, 94(1): 46–60.

Sugita, M., and W. Brutsaert. 1993. Cloud effect in the estimation of instantaneous downward longwave radiation. *Water Resour. Res.*, 29(3): 599–605.

Swinbank, W. C. 1963. Long-wave radiation from clear skies. *Q. J. R. Meteorol. Soc.*, 89(381): 339–348.

Swinbank, W. C. 1964. Long-wave radiation from clear skies. *Q. J. R. Meteorol. Soc.*, 90(386): 488–493.

Tang, B., and Z. L. Li. 2008. Estimation of instantaneous net surface longwave radiation from MODIS cloud-free data. *Remote Sens. Environ.*, 112(9): 3482–3492.

Trenberth, K. E., and J. T. Fasullo. 2012. Tracking Earth's energy: From El Niño to global warming. *Surv. Geophys.*, doi: 10.1007/s10712-011-9150-2.

Unsworth, M. H., and J. L. Monteith. 1975. Long-wave radiation at the ground I. Angular distribution of incoming radiation. *Q. J. R. Meteorol. Soc.*, 101(427): 13–24.

Uppala, S. M. et al. 2005. The ERA-40 re-analysis. *Q. J. R. Meteorol. Soc.*, 131(612): 2961–3012.

Vermote, E. F., D. Tanre, J. L. Deuze, M. Herman, and J. J. Morcrette. 1997. Second simulation of the satellite signal in the solar spectrum, 6S: An overview. *IEEE Trans. Geosci. Remote Sens.*, 35(3): 675–686.

Wagner, S. C., Y. M. Govaerts, and A. Lattanzio. 2010. Joint retrieval of surface reflectance and aerosol optical depth from MSG/SEVIRI observations with an optimal estimation approach: 2. Implementation and evaluation. *J. Geophys. Res.*, 115: D02204.

Wan, Z., and J. Dozier. 1996. A generalized split-window algorithm for retrieving land-surface temperature form space. *IEEE Trans. Geosci. Remote Sens.*, 34: 892–905.

Wan, Z., and Z.-L. Li. 1997. A Physics-based algorithm for retrieving land-surface emissivity and temperature from EOS/MODIS data. *IEEE Trans. Geosci. Remote Sens.*, 35(4): 980–996.

Wan, Z., Y. L. Zhang, Q. C. Zhang, and Z.-L. Li. 2002. Validation of the land surface temperature products retrieved from Terra Moderate Resolution Imaging Spectrometer data. *Remote Sens. Environ.*, 83(12): 163–180.

Wan, Z., Y. Zhang, Q. C. Zhang, and Z.-L. Li. 2004. Quality assessment and validation of the MODIS global land surface temperature. *Int. J. Remote Sens.*, 25(1): 261–274.

Wang, D., S. Liang, R. Liu, and T. Zheng. 2010. Estimation of daily-integrated PAR from sparse satellite observations: comparison of temporal scaling methods. *Int. J. Remote Sens.*, 31(6): 1661–1677.

Wang, H., and R. T. Pinker. 2009. Shortwave radiative fluxes from MODIS: Model development and implementation. *J. Geophys. Res.*, 114: D20201.

Wang, K., R. E. Dickinson, and S. Liang. 2009a. Clear sky visibility has decreased over land globally from 1973 to 2007. *Science*, 323(5920): 1468–1470.

Wang, K., and S. Liang. 2009a. Estimation of daytime net radiation from shortwave radiation measurements and meteorological observations. *J. Appl. Meteorol. Climatol.*, 48: 634–643.

Wang, K., and S. Liang. 2009b. Evaluation of ASTER and MODIS land surface temperature and emissivity products using long-term surface longwave radiation observations at SURFRAD sites. *Remote Sens. Environ.*, 113(7): 1556–1565.

Wang, K., and S. Liang. 2009c. Global atmospheric downward longwave radiation over land surface under all-sky conditions from 1973 to 2008. *J. Geophys. Res.*, 114(D19): D19101.

Wang, K. C., R. E. Dickinson, and Q. Ma. 2012. Terrestrial evapotranspiration. In *Advanced Remote Sensing: Terrestrial Information Extraction and Applications* (S. Liang, X. Li, and J. Wang, Eds.). Academic Press, San Diego, CA, pp. 557–588.

Wang, W., and S. Liang. 2009d. Estimation of high-spatial resolution clear-sky longwave downward and net radiation over land surfaces from MODIS data. *Remote Sens. Environ.*, 113(4): 745–754.

Wang, W., S. Liang, and J. A. Augustine. 2009b. Estimating high spatial resolution clear-sky land surface upwelling longwave radiation from MODIS Data. *IEEE Trans. Geosci. Remote Sens.*, 47(5): 1559–1570.

Wang, W. H. 2012. Surface longwave radiation budget. In *Advanced Remote Sensing: Terrestrial Information Extraction and Applications* (S. Liang, X. Li, and J. Wang, Eds.). Academic Press, San Diego, CA, pp. 273–300.

Wang, W. H., and S. L. Liang. 2010. A method for estimating clear-sky instantaneous land surface longwave radiation with GOES sounder and GOES-R ABI data. *IEEE Geosci. Remote Sens. Lett.*, 7(4): 708–712.

Wanner, W., X. Li, and A. H. Strahler. 1995. On the derivation of kernels for kernel-driven models of bidirectional reflectance. *J. Geophys. Res.*, 100(D10): 21,077–21,089.

Wielicki, B. A. et al. 1996. Clouds and the Earth's Radiant Energy System (CERES): An Earth observing system experiment. *Bull. Am. Meteorol. Soc.*, 77(5): 853–868.

Wielicki, B. A. et al. 1998. Clouds and the Earth's Radiant Energy System (CERES): Algorithm overview. *IEEE Trans. Geosci. Remote Sens.*, 36(4): 1127–1141.

Wild, M. 2008. Short-wave and long-wave surface radiation budgets in GCMs: A review based on the IPCC-AR4/CMIP3 models. *Tellus Ser. A*, 60(5): 932–945.

Wild, M. 2011. Enlightening global dimming and brightening. *Bull. Am. Meteorol. Soc.*, 93(1): 27–37.

Wild, M., J. Grieser, and C. Schär. 2008. Combined surface solar brightening and increasing greenhouse effect support recent intensification of the global land-based hydrological cycle. *Geophys. Res. Lett.*, 35(17): L17706.

Wu, H., X. Zhang, S. Liang, H. Yang, and G. Zhou. 2012. Estimation of clear-sky land surface longwave radiation from MODIS data products by merging multiple models. *J. Geophys. Res.*, 117: D22107.

Xu, B. Q. et al. 2009. Black soot and the survival of Tibetan glaciers. *Proc. Natl. Acad. Sci. USA*, 106(52): 22,114–22,118.

Yang, K., J. He, W. Tang, J. Qin, and C. C. K. Cheng. 2010. On downward shortwave and longwave radiations over high altitude regions: Observation and modeling in the Tibetan Plateau. *Agric. For. Meteorol.*, 150(1): 38–46.

Yang, K., T. Koike, and B. Ye. 2006. Improving estimation of hourly, daily, and monthly solar radiation by importing global datasets. *Agric. For. Meteorol.*, 137(1–2): 43–55.

Zhang, X., S. Liang, G. Zhou, and H. Wu. 2012a. Incident solar radiation. In *Advanced Remote Sensing: Terrestrial Information Extraction and Applications* (S. Liang, X. Li, and J. Wang, Eds.). Academic Press, San Diego, CA, pp. 127–173.

Zhang, X., S. Liang, G. Zhou, and H. Wu. 2012b. Mapping global incident downward shortwave radiation and photosynthetically active radiation over land surfaces using multiple satellite data. *J. Geophys. Res.*, under review.

Zhang, X. T., S. L. Liang, K. C. Wang, L. Li, and S. Gui. 2010. Analysis of global land surface shortwave broadband albedo from multiple data sources. *IEEE J. Selected Topics Appl. Earth Obs. Remote Sens.*, 3(3): 296–305.

Zhang, Y.-C. and W. B. Rossow. 2002. New ISCCP global radiative flux data products. *GEWEX News*, 12(4): 7.

Zhang, Y., W. B. Rossow, A. A. Lacis, V. Oinas, and M. I. Mishchenko. 2004. Calculation of radiative fluxes from the surface to top of atmosphere based on ISCCP and other global datasets: Refinements of the radiative transfer model and the input data. *J. Geophys. Res.*, 109: D19105.

Zhang, Y. C., W. B. Rossow, and A. A. Lacis. 1995. Calculation of surface and top of atmosphere radiative fluxes from physical quantities based on ISCCP datasets. 1. method and sensitivity to input data uncertainties. *J. Geophys. Res.*, 100(D1): 1149–1165.

Zheng, T., S. Liang, and K. C. Wang. 2008. Estimation of incident PAR from GOES imagery. *J. Appl. Meteorol. Climatol.*, 47: 853–868.

Zheng, T., and S. L. Liang. 2011. A Bayesian approach to integrate satellite-estimated instantaneous photosynthetically active radiation product for daily value calculation. *J. Geophys. Res.*, 116: D15202.

Zhou, Y., and R. D. Cess. 2001. Algorithm development strategies for retrieving the downwelling longwave flux at the Earth's surface. *J. Geophys. Res.*, 106(D12): 12,477–12,488.

Zhou, Y., D. P. Kratz, A. C. Wilber, S. K. Gupta, and R. D. Cess. 2007. An improved algorithm for retrieving surface downwelling longwave radiation from satellite measurements. *J. Geophys. Res.*, 112(D15): D15102.

Zhu, X. F., S. L. Liang, Y. Z. Pan, and X. T. Zhang. 2011. Agricultural irrigation impacts on land surface characteristics detected from satellite data products in Jilin Province, China. *IEEE J. Selected Topics Appl. Earth Obs. Remote Sens.*, 4(3): 721–729.

Section II

Remote Sensing of Surface Energy Fluxes: Algorithms and Case Studies

6

A Comparative Study of Techniques for Modeling the Spatiotemporal Distribution of Heat and Moisture Fluxes at Different Agroecosystems in Brazil

Antônio Heriberto de Castro Teixeira, Fernando Braz Tangerino Hernandez, H. L. Lopes, Morris Scherer-Warren, and Luís Henrique Bassoi

CONTENTS

6.1 Introduction

The difficulties to measure the energy fluxes from mixed agroecosystems by field experiments make the use of remote sensing by satellite images a valuable application and its use for this purpose has already been demonstrated in different climate regions (Tang et al. 2009; Teixeira 2010; Miralles et al. 2011; Anderson et al. 2012; Pôças et al. 2013). Remote sensing is a suitable way for determining and mapping the spatial and temporal structure of the water and energy balance components. Hydrological models can be too complex and costly to be used for this purpose because of unavailability of datasets in different hydrological uniform subareas (Majumdar et al. 2007).

Several remote sensing algorithms have been developed, being based largely on the energy balance theories, highlighted by some advantages and shortcomings. The Surface Energy Balance Algorithm for Land (SEBAL) (Bastiaanssen et al. 1998), the Simplified Surface Energy Balance Index (S-SEBI) (Roerink et al. 2000), and the Surface Energy Balance System (SEBS) (Su 2002) are some examples. Those techniques can be applied to various agroecosystems without the need of crop classification, which is difficult to do in mixed agroecosystems.

Among others, Kustas et al. (2006) underline that the remote surface temperature and vegetation cover must be at high enough resolutions where different land surface conditions can be distinguished, with the validation of flux distributions predicted by modeling being important. Procedures for the validation of energy fluxes derived from remote sensing involve intercomparisons among methods (Liu et al. 2003; Su et al. 2005; Tasumi and Allen 2007; Teixeira 2010), with satellite and field energy balance measurements.

The disadvantage of many remote sensing energy balance methods, such as SEBAL and SEBS, is the need to identify hydrological extreme conditions, which is not required for the Two-Source Model (TSM) (Kustas and Norman 1999), dual-temperature difference (DTD) (Norman et al. 2000), and the Disaggregated Atmosphere Land Exchange Inverse (DisALEXI) model (Norman et al. 2003). The aerodynamic resistance-surface energy balance (RSEB) approach (Kalma and Jupp 1990) fails because small errors in the radiative temperature translate into large inaccuracies in the sensible heat flux (H) and then into estimates of actual evapotranspiration (ET).

The other problem in relation to the practical applicability of the remote sensing energy balance models, aiming at the end users, is the need of background knowledge in radiation physics involved in the description of these algorithms. For example, the suitability of applying the Penman–Monteith (PM) equation by the surface conductance algorithm has been shown by using remotely sensed vegetation indices, such as the normalized difference vegetation index (NDVI), together with weather data (Cleugh et al. 2007). Yet, perhaps the strongest advantages of the PM equation include the method applicability and the low sensitivity to input data and parameters. Its use is also highlighted by the model named Mapping Evapotranspiration with High Resolution and Internalized Calibration (METRIC) (Allen et al. 2007), which applies the ratio of ET to reference evapotranspiration (ET0) for extrapolating satellite overpass ET values to larger timescales.

Although the worldwide known SEBAL algorithm had been calibrated and validated with field radiation and energy balance measurements, showing satisfactory performances in the Brazilian semiarid region (Teixeira et al. 2009a,b), the major difficulty in its applicability for deriving estimates for the whole year is the assumption of zero latent heat flux (λE) for dry pixels. This is because during the rainy season, the mixed ecosystems of irrigated crops and natural vegetation are homogenously wet with the whole region presenting high ET rates.

Considering the simplicity of application and the absence of the need of neither crop classification nor extreme conditions, a model for ET acquirements based on the modeled ratio ET/ET0, called Simple Algorithm For Evapotranspiration Retrieving (SAFER), was recently developed in the semiarid conditions of Brazil, which has already been validated with field data from four flux stations involving irrigated crops and natural vegetation (Teixeira 2012a). A second biophysical model was also built for the same region allowing the acquirement of the surface resistance to the water fluxes (r_s), which, when establishing threshold values for r_s, is useful for classifying irrigated crops and natural vegetation (Teixeira et al. 2012b).

These two modeling approaches when used together with satellites having different spatial and temporal resolutions and agrometeorological stations can be appropriate for implementation allowing one to derive ET on large scales. In addition, together with the net radiation (R_n) acquired by the slob equation (Bruin and Stricker 2000; Teixeira et al. 2008, 2009a), the SAFER and r_s models allow the quantification of H daily values from different agroecosystems.

The SAFER algorithm has the additional advantage of the possibility for using daily weather data from either conventional or automatic agrometeorological stations. This is an important characteristic because it allows a historical evaluation of the energy balance components on a large scale, as data from automatic sensors are results from relatively recent advances on instrumental technology.

Given the recent development of SAFER, not many comparative studies have been performed so far, evaluating its performance, even more in a Brazilian environment. Hernandez et al. (in press) recently compared SEBAL and SAFER algorithms with the traditional Food and Agricultural Organization (FAO) crop coefficient (K_c) methodology (Allen et al. 1998). The remote sensing methods and water managements in corn, bean, and sugar cane growing under center pivot sprinkler irrigation systems were evaluated in the northwestern part of São Paulo State, Brazil. They concluded that ET results from both algorithms were higher than the water applied with these overestimations being mainly related to errors on crop stages evaluation by the irrigation manager. However, the authors pointed out that the SAFER algorithm performed better than SEBAL when these crops presented lower leaf areas, partially covering the soil beneath the canopies.

The objective of this chapter is two-fold: first, to perform a comparative study of SAFER and SEBAL evaluating their ability in modeling the spatiotemporal distribution of ET at different crops in Brazil using the FAO method as reference corresponding estimates; and second, to demonstrate SAFER's applicability in subsidizing the monitoring of land use effects on the energy exchange processes by using field and satellite measurements in the Brazilian semiarid environment.

As a result, this chapter is structured as follows: following the introduction, the study region, datasets, and the steps for modeling are described; one of the first comparisons between SAFER and SEBAL is then made available, comparing estimates by these two modeling approaches against corresponding estimates by the traditional FAO methodology; next, results from case studies are furnished, covering application of both SAFER and SEBAL methods with satellite images, in some irrigation crops and natural vegetation at different spatial and temporal scales in the Brazilian semiarid region.

6.2 Study Region

Figure 6.1 shows the locations of Petrolina (PE) and Juazeiro (BA) municipalities, the Nilo Coelho scheme, and the agrometeorological stations together with pictures of the main vegetation types involved in the semiarid region of the Brazilian Northeast.

In the semiarid region of Brazil, situated in the northeastern part of the country, disturbed currents of south, north, east, and west influence the climatology. Excluding the places of high altitude, all areas present annual averaged air temperatures higher than 24°C, even bigger than 26°C in the depressions at 200 to 250 m of altitude. The average maximum air temperature is 33°C and the average minimum is 19°C. The mean monthly values are in the range from 17°C to 29°C. The rainy period is concentrated between January and April, with the peak happening in March (Teixeira 2009). In the Nilo Coelho irrigation scheme the lowest irrigated area occurs in February, while the highest one is in July (Bastiaanssen et al. 2001).

The main commercial crops in the Brazilian semiarid region are mango orchards and vineyards surrounded by natural vegetation named Caatinga (Figure 6.1). Mango trees

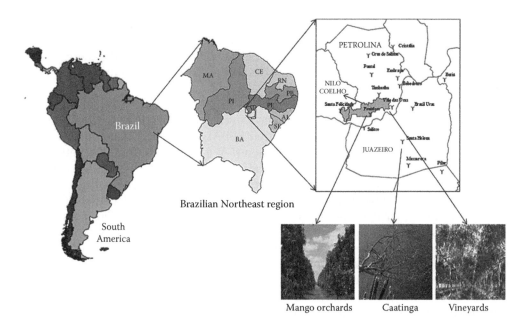

FIGURE 6.1
Location of the Brazilian semiarid study region with highlights for Petrolina (PE) and Juazeiro (BA) municipalities, Nilo Coelho irrigation scheme, the agrometeorological stations, and the main agroecosystems.

may have flowering and fruit set regularly every year. The vineyards growing under tropical warm conditions exhibit continuous and accelerated physiological processes, differently from those growing in temperate climates. Caatinga species are defined by bush aspects possessing small leaves and thorns, distributed in an irregular way, with soil patches of bare soil and clumps of vegetation interspersed (Teixeira 2009a,b).

The 15 agrometeorological stations showed in the right side of Figure 6.1 are inside irrigated and natural vegetation areas. They are equipped with radiometers to measure the incident solar radiation (RS↓); sensors for measuring air temperature (T_a) and relative humidity (RH); and wind speed (u), allowing the calculation of ET0 by the FAO Penman–Monteith method (Allen et al. 1998). Grids of RS↓, T_a, and ET_0 were used together with remote sensing retrieving parameters from MODIS and Landsat satellite images during the steps for modeling the energy balance components at different spatiotemporal scales.

6.3 Modeling the Energy Balance on a Large Scale

Among the satellite-based ET models developed during recent years the SEBAL algorithm has been widely used for heterogeneous surfaces. This model involves the spatial variability of the most agrometeorological variables and can be applied to various ecosystems. However, aiming simplicity of application, the SAFER algorithm was recently developed, having the advantages in terms of applicability in relation to SEBAL already commented in Section 6.1. In the following sections, the principles and descriptions of both models are done to facilitate understanding further comparisons and analyses.

6.3.1 Description and Validation of the SEBAL Model

In this chapter, applications of SEBAL algorithm are analyzed with Landsat 5 images only. SEBAL requires spatially distributed, visible, near-infrared, and thermal infrared data together with weather data. The algorithm computes R_n, H, and soil heat flux (G) for every pixel of a satellite image at the overpass time and λE is acquired as a residual in energy balance:

$$R_n - \lambda E - H - G = 0 \tag{6.1}$$

where all terms are expressed in W m^{-2}.

During the first step of SEBAL, the radiation balance is obtained by

$$R_n = \text{RS}{\downarrow} - \alpha_0\text{RS}{\downarrow} - \text{RL}{\uparrow} + \text{RL}{\downarrow} \tag{6.2}$$

where RS$\downarrow$ is in W m^{-2}, α_0 is the surface albedo, and RL$\uparrow$ and RL$\downarrow$ are the outgoing and the incoming longwave radiation (W m^{-2}), respectively.

For Landsat 5 the spectral radiance for each band (L_b) is computed as

$$L_b = \left[\frac{L_{\max} - L_{\min}}{\text{QCAL}_{\max} - \text{QCAL}_{\min}}\right](\text{DN} - \text{QCAL}_{\min}) + L_{\min} \tag{6.3}$$

where DN is the digital number of each pixel, L_{MAX} and L_{MIN} are calibration constants, QCAL$_{\max}$ and QCAL$_{\min}$ are the highest and lowest range of values for rescaled radiance in DN, respectively. The unit for L_b is W m^{-2} sr^{-1} μm^{-1} and QCAL$_{\max} = 255$ and QCAL$_{\min} = 0$.

In the radiation balance, the net shortwave radiation available at the Earth's surface depends on RS$\downarrow$ and α_0. The second parameter is calculated from satellite-measured spectral radiances for each narrowband, followed by mathematical expressions for spectral integration and atmospheric corrections.

The bands 1 to 5 and 7 with pixel size of 30 × 30 m from Landsat 5 provide data for the visible and near-infrared bands for the broadband planetary albedo (α_p) calculations. For each satellite band (α_{p_b}) it is calculated as

$$\alpha_{p_b} = \frac{L_b \pi d^2}{R_{a_b} \cos \varphi} \tag{6.4}$$

where d is the relative Earth–Sun distance, R_{a_b} is the mean solar irradiance at the top of the atmosphere for each band b (W m^{-2} μm^{-1}), and φ is the solar zenith angle. α_p is calculated as the total sum of the different narrowband α_{p_b} values according to weights for each band (w_b).

$$\alpha_p = \sum w_b \alpha_{p_b} \tag{6.5}$$

The weights for the different bands in Equation 6.5 are computed as the ratio of the amount of incoming shortwave radiation from the Sun in a particular band and the sum of incoming shortwave radiation for all the bands at the top of the atmosphere.

The atmosphere disturbs the signal reaching the satellite sensor. The satellite registered radiances are therefore affected by the atmospheric interaction in the radiative transfer path. Part of RS↓ is scattered back to the satellite before it reaches the Earth surface. Combination of field measurements of α_0 (Teixeira et al. 2008) with Landsat calculations of α_p resulted in a linear relationship for correcting all atmospheric disturbances:

$$\alpha_0 = a\alpha_p + b \tag{6.6}$$

where $a = 0.61$ and $b = 0.08$ are the satellite overpass regression coefficients found in the Brazilian semiarid conditions for the period of the images (Teixeira et al. 2009a; Teixeira 2010).

The Landsat overpass measurements yielded instantaneous values of surface albedo ($\alpha_{0_{inst}}$) that were systematically lower than those for 24 h ($\alpha_{0_{24}}$). Hence a second regression equation was applied to retrieve the daily from instantaneous values

$$\alpha_{0_{inst}} = a\alpha_{0_{24}} + b \tag{6.7}$$

where $a = 1.02$ and $b = 0.01$ are the regression coefficients found in the Brazilian semiarid region for the period of the satellite images (Teixeira et al. 2009a; Teixeira 2010).

Equation 6.7 has been considered in the computations of the 24-h values of R_n required for the determination of daily λE with the SEBAL model.

The spectral radiance of the thermal region of Landsat 5 (band 6), with a pixel size of 120×120 m, is converted into a radiation temperature applicable at the top of the atmosphere (T_{sat}) by inversion of Plank's law in the 10.4–12.5 µm bandwidth:

$$T_{sat} = \frac{K_2}{\ln\left(\dfrac{K_1}{L_{b(Thermal)}} + 1\right)} \tag{6.8}$$

where $L_{b(Thermal)}$ for Landsat 5 is the uncorrected spectral radiance for band 6 from the land surface and the conversion coefficients K_1 and K_2 are, respectively, 607.76 W m^{-2} sr^{-1} µm^{-1} and 1260.56 K for Landsat 5 (Schneider and Mauser 1996).

From field energy balance experiments (Teixeira et al. 2008), the vertical temperature difference (ΔT) between two heights was derived using data of H, T_a above the canopies, and the aerodynamic resistance r_a (Smith et al. 1989).

$$\Delta T = \frac{H r_a}{\rho_a c_p} \tag{6.9}$$

where ρ_a (kg m^{-3}) and c_p (J kg^{-1} K^{-1}) are the air density and air specific heat at constant pressure, respectively.

Experiments have demonstrated that, other than a thin surface such as bare soil or a short canopy, a difference of several degrees can be observed between radiometric and aerodynamic surface temperature (Troufleau et al. 1997). The thermal radiation measured by satellite sensors thus need to be corrected for both atmosphere emission and the difference between radiometric and aerodynamic surface temperature. To correct these conjugated effects, the field results from Equation 6.9 were used to fit a linear relationship between T_0 ($T_0 = T_a + \Delta T$) and T_{sat}:

$$T_0 = aT_{sat} + b \tag{6.10}$$

where the coefficients $a = 1.07$ and $b = -20.17$ were found for the Brazilian semiarid region for the period of the Landsat satellite images used (Teixeira et al. 2009a; Teixeira 2010).

It is realized that a physically based atmospheric correction would include transmittance and reflectance of spectral radiances (e.g., Schmugge et al. 1998), but for practical reasons, Equations 6.6 and 6.10 are doable.

The NDVI, which is an indicator related to the land cover, is obtained as

$$NDVI = \frac{\alpha_{p(NIR)} - \alpha_{p(RED)}}{\alpha_{p(NIR)} + \alpha_{p(RED)}} \tag{6.11}$$

where $\alpha_{p(NIR)}$ and $\alpha_{p(RED)}$ represent the planetary albedo over the ranges of wavelengths in the near-infrared (NIR) and red (RED) regions of the solar spectrum. For Landsat images these regions correspond to bands 4 and 3, respectively.

RS↓ was measured in the agrometeorological stations from Figure 6.1. RL↓ can be calculated by using the Stefan–Boltzmann equation with an empirically determined atmospheric emissivity (ε_a) and T_a data (Allen et al. 1998). RL↑ can also be obtained through the Stefan–Boltzmann equation with an empirically determined surface emissivity (ε_0) and T_0 acquired by the satellite after atmospheric correction (Teixeira et al. 2009a; Teixeira 2010).

The Stefan–Boltzmann equation applied for the satellite overpass time to calculate RL↓ is

$$RL\downarrow = \varepsilon_a \sigma T_a^4 \tag{6.12}$$

where σ (5.67×10^{-8} W m^{-2} K^{-4}) is the Stefan–Boltzmann constant and T_a is in K.

The atmospheric emissivity was calculated with the following relation:

$$\varepsilon_a = a \, (-\ln \tau_s)^b \tag{6.13}$$

where τ_s is the shortwave atmospheric transmissivity and the regression coefficients were $a = 0.942$ and $b = 0.103$ for the satellite overpass time in the semiarid region of Brazil (Teixeira et al. 2009a; Teixeira 2010). Equation 6.13 was thereafter applied to satellite images to obtain the large-scale incoming longwave radiation.

Field values of RL↑—jointly with estimates of T_0—gave the opportunity to quantify the thermal infrared emissivity (ε_0). The field values of ε_0 (Teixeira et al. 2008) were correlated with NDVI from satellite images:

$$\varepsilon_0 = a \, \ln NDVI + b \tag{6.14}$$

where a and b are regression coefficients being 0.06 and 1.00, respectively, for the Brazilian semiarid conditions (Teixeira et al. 2009a; Teixeira 2010).

The values of ε_0 were used for estimation of RL↑ in the regional radiation balance by applying Equation 6.12 replacing T_a and ε_a by T_0 and ε_0.

The second step of the SEBAL algorithm is to compute the large-scale values of G and H. Satellite overpass G values are calculated by modeling the ratio G/R_n, while those for H are obtained throughout the near-surface temperature gradients (ΔT). The first value for friction velocity (u_*) is computed for neutral atmospheric stability, using weather data from agrometeorological stations. The near-surface wind speed (u) is converted to a value at the blending height, where the effects from the land surface roughness can be neglected. A height of 200 m can be considered (u_{200}). The first estimate of u_* is used—conjunctively with surface roughness estimates—to infer the large-scale values of r_a. Corrections for atmosphere stability are obtained iteratively for each pixel. A series of iterations is required to determine new corrected values of u_* and r_a before obtaining numerical stability.

SEBAL computes ΔT by assuming its linear relationship with T_0, and the coefficients of this relationship are acquired by following an internal calibration procedure (Allen et al. 2007; Teixeira et al. 2009a). The algorithm considers two "anchor" pixels at which a value for H can be estimated on the basis of *a priori* knowledge of the fluxes over dryland (hot pixel) and wet terrain (cold pixel). The sensible heat fluxes in these sites (H_{hot} and H_{cold}) can be calculated applying Equation 6.1.

In the following the essential SEBAL equations are shown together with techniques for local calibrations and validations described by Teixeira et al. (2009a).

Field values of roughness length for momentum transfer (z_{0m}) in natural vegetation and irrigated crops (Teixeira et al. 2008) were used to estimate z_{0m} from remote sensing parameters:

$$z_{0m} = \exp\left[\left(a\frac{\text{NDVI}}{\alpha_0}\right) + b\right] \tag{6.15}$$

where z_{0m} is given in m and the regression coefficients for the semiarid conditions of Brazil were $a = 0.26$ m and $b = -2.21$ (Teixeira et al. 2009a).

Field measurements of G and R_n were used together with satellite measurements of T_0, α_0, and NDVI to calibrate the following multiple regression equation (Bastiaanssen et al. 1998):

$$\frac{G}{R_n} = T_0(a\alpha_0 + b)(1 - 0.98\,\text{NDVI}^4) \tag{6.16}$$

where T_0 is in degrees Celsius and a and b are regression coefficients, which were $a = -0.11°C^{-1}$, $b = 0.02$, respectively, for the Brazilian semiarid conditions (Teixeira et al. 2009a).

To compute H, the following equation is applied:

$$H = \frac{\rho_a c_p \Delta T}{r_a} \tag{6.17}$$

ΔT for each pixel is given as ($T_1 - T_2$), the temperature difference between two heights (z_1 and z_2), and its regionalization occurs by assuming a linear relationship with T_0:

$$\Delta T = aT_0 + b \tag{6.18}$$

where a and b are the fitting coefficients.

Originally, at the cold pixel H is neglected and λE is calculated as a residual in Equation 6.1, while in the hot pixel λE is assumed to be zero (Bastiaanssen et al. 1998). Teixeira et al. (2009a) did not neglect any energy fluxes in both wet pixel (mango orchard) and natural Caatinga (dry pixel). Mango orchard was chosen because it represents the largest irrigated area in the study region, while Caatinga is the driest natural vegetation, where in some periods of the year, ET is close to zero. Instead, they used the specific values for the ratio of ET to ET_0 in Equation 6.1 to estimate these fluxes in those pixels. The H values under these extreme conditions were used to infer ΔT applying Equation 6.17 in the iterative process.

The drag force between land and atmosphere is accounted by r_a, and this parameter requires the surface roughness conditions (z_{0h}) and friction velocities (u_*) to be specified:

$$r_a = \frac{\ln\left[\dfrac{(z-d)}{z_{0h}}\right] - \Psi_h[(z-d)/L]}{ku_*} \tag{6.19}$$

where z_{0h} (m) is the roughness length governing transfer of heat and vapor away from the land surface into the atmosphere, k is the von Karman's constant (0.41), z (m) is a reference height, d (m) is the displacement height taken as 4.67 z_{0m}, Ψ_h is the stability correction due to buoyancy, and L (m) is the Obukhov length.

The atmospheric surface-layer similarity theory was applied, using universal functions suggested by Businger et al. (1971) for solving the integrated stability functions of temperature (Ψ_h) and momentum (Ψ_m).

The following expressions are used for unstable situations:

$$\Psi_h = 2\ln\left(\frac{1+x^2}{2}\right) \quad x = \left(1-16\frac{z-d}{L}\right)^{\frac{1}{4}} \tag{6.20}$$

$$\Psi_m = 2\ln\left(\frac{1+x}{2}\right) + \ln\left(\frac{1+x^2}{2}\right) - 2\arctan(x) + \frac{\pi}{2} \tag{6.21}$$

For stable situations, the following is applied:

$$\Psi_h = \Psi_m = -5\frac{z-d}{L} \tag{6.22}$$

where L can be calculated as

$$L = -\frac{\rho_a c_p u_*^3 T_a}{kgH} \tag{6.23}$$

where g is the gravitational constant (9.81 m s^{-2}) and u_* expressed in m s^{-1} is given by

$$u_* = \frac{uk}{\ln\left[\dfrac{z-d}{z_{0m}}\right] - \Psi_m} \tag{6.24}$$

where u is the wind speed at the blending height (m s^{-1}) and Ψ_m is the stability correction for momentum transport at the blending height.

After applying the SEBAL calibrated equations in 10 Landsat images acquired in the Brazilian semiarid conditions spanning the years from 2001 to 2007, the results from Teixeira et al. (2009a) for the energy balance components showed excellent agreements on R_n and λE. The coefficient of determination (R^2) for R_n was 0.94, with a root-mean-square error (RMSE) of 17.5 W m^{-2}. For λE, R^2 was 0.93, with a RMSE of 33.8 W m^{-2}. Good agreements were achieved for H and G. For H, R^2 was 0.83. Part of the deviations could be attributed to the estimations of z_{0m}. The RMSE for H was 41.8 W m^{-2}. For G the local calibration yielded an R^2 of 0.81, with the RMSE of 13.3 W m^{-2}.

After calculating instantaneous values for λE, the satellite overpass time value of the evaporative fraction [EF = $\lambda E/(R_n - G)$] is multiplied by the daily values net radiation $\left(R_{n_{24}}\right)$ to acquire the latent heat flux for 24 h (λE_{24}):

$$\lambda E_{24} = \text{EF}_{\text{inst}} R_{n_{24}} \tag{6.25}$$

where EF_{inst} is the ratio of λE to the available energy at the satellite overpass time.

On the basis of field data (Teixeira 2008), Teixeira et al. (2009a) reported the need of adding a factor of 1.18 to extrapolate the instantaneous to daily values of EF, under the Brazilian semiarid region.

The daily values of R_n can be described by the 24-h values of net shortwave radiation, with a correction term for net longwave radiation for the same timescale throughout the Slob equation:

$$R_n = (1 - \alpha_0)\text{RS}\downarrow - a_1\tau_s \tag{6.26}$$

where a_1 is the regression coefficient of the relationship between net longwave radiation (R_{nl}) and atmospheric shortwave transmissivity (τ_s) on a daily scale (Bruin de and Stricker 2000; Teixeira et al. 2008). The term τ_s can be calculated by the ratio of RS$\downarrow$ measured by radiometers and the incident solar radiation at the top of the atmosphere (R_a).

After multiplying EF_{inst} by 1.18 to obtain the daily values, Teixeira et al. (2009a) reported that the relation between the satellite and field measurements of daily ET presented both better R^2 of 092 and a RMSE of 0.38 mm day^{-1} comparing field data and satellite measurements in irrigated crops and natural vegetation from 2001 to 2007 in the Brazilian semiarid region.

SEBAL has been validated in grapes, peaches, and almonds from Spain, Turkey, and California. These validations revealed that accumulated values of ET predicted by the model are within several percent from the measured values (Bastiaanssen et al. 2008).

Considering that the field measurements have their own sources of errors, the SEBAL accuracy in the Brazilian semiarid region after field calibrations is satisfactory. However, the difficulties of applying the model for different moisture conditions along the year and the lack of applicability, as discussed in the introduction section, make its implementation a difficult task in some situations.

6.3.2 Description and Validation of the SAFER Model

In the present study SAFER algorithm (Teixeira et al. 2012a) was applied and validated using both MODIS and Landsat images.

MODIS is an instrument onboard Terra platform with 36 spectral bands between 0.405 and 14.385 µm, acquiring data at three spatial resolutions (250, 500, and 1000 m). The land product used in the present work, included in this chapter, was Level-1B (L1B) dataset, which contains radiances for these bands. Only four (two reflectance and two thermal bands) were used, being the reflective solar bands (bands 1 and 2, red and near infrared) with a spatial resolution of 250 m and the thermal emissive bands (bands 31 and 32) with a spatial resolution of 1000 m.

The digital signals are written as scale integer (SI) representations, from which radiance for each band of the sensor L_b (W m^{-2} sr^{-1} µm^{-1}) can be calculated by using scale and offset terms:

$$L_b = R_{scale} \, (\text{SI} - R_{offset}) \tag{6.27}$$

$$R_{scale} = \frac{L_{max} - L_{min}}{32,767} \tag{6.28}$$

$$R_{offset} = -\frac{32,767 L_{min}}{L_{max} - L_{min}} \tag{6.29}$$

where SI is dimensionless, R_{scale} is the radiance rescaling gain factor in W m^{-2} sr^{-1} µm^{-1}, R_{offset} is a dimensionless radiance rescaling offset factor, L_{min} is the spectral radiance scaled to 0 in W m^{-2} sr^{-1} µm^{-1}, and L_{max} is the observed spectral radiance scaled to 32,767 in W m^{-2} sr^{-1} µm^{-1}.

Considering the MODIS red and infrared bands, α_{p_b} was calculated with Equation 6.4 by using, respectively, the bands 1 and 2. For α_0 calculations the reflectance values for these bands were used, according to the following regression equation (Valiente et al. 1995):

$$\alpha_0 = a + b\alpha_{p1} + c\alpha_{p2} \tag{6.30}$$

where α_{p1} and α_{p2} are the planetary albedo for bands 1 and 2 from MODIS satellite measurements, and a, b, and c are the regression coefficients obtained comparing these measurements with field data for the years of 2002 and 2004 (Teixeira et al. 2008), thus already including the atmospheric effects through the radiation path. The values found for the Brazilian semiarid conditions were 0.08, 0.41, and 0.14, respectively.

For the surface temperature (T_0), the MODIS bands 31 and 32 were used with Equation 6.8 considering L_b from these bands for conversion into T_{sat}. In this case, k_1 and k_2 were calculated with the following equations:

$$k_1 = 2hc^2 \, \lambda^{-5} \, 10^{-6} \tag{6.31}$$

$$k_2 = \frac{hc}{s\lambda} \tag{6.32}$$

where h is the Planck constant (6.62606896 × 10^{-34} J s^{-1}), c is the speed of light (2.99792458 × 10^8 m s^{-1}), s is the Boltzmann constant (1.3806504 × 10^{-23} J K^{-1}), and λ is the center wavelength in meter.

An initial implementation of the split window technique (Coll and Caselles 1997), having the aerodynamic T_0 data from the same energy balance experiments as for α_0 (Teixeira et al. 2008), showed that a simpler regression equation could be applied for deriving the T_0

with reasonable accuracy in relation to field experimental data in the Brazilian semiarid conditions:

$$T_0 = aT_{31} + bT_{32} \tag{6.33}$$

where T_{31} and T_{32} are the brightness temperature from bands 31 and 32, and the regression coefficients a and b were equally 0.50 for the Brazilian semiarid conditions, also including the atmospheric effects through the radiation path.

The SAFER model was also applied to Landsat images, as in the case of the SEBAL algorithm, using bands 1 to 5 and 7 for α_0 calculation and band 6 for acquiring T_0. Regression equations were again used with α_0 and T_0 obtained in the field and satellite measurements (Teixeira et al. 2009a; Teixeira 2010).

The images of NDVI, α_0, and T_0 are the only input parameters to model the ET/ET_0 instantaneous values. This ratio was then multiplied by the daily grids of ET_0 for estimating the ET large-scale values for 24 h.

Both the surface resistance (r_s) and ET/ET_0 are related to the soil moisture conditions and so are the remote sensing parameters. On the basis of this principle, a second model was applied to calculate r_s aiming a simplified vegetation classification. Studies examining relationships between vegetation indices with soil moisture and ET rates have been previously done in the semiarid region of Brazil (Teixeira et al. 2008; Teixeira et al. 2009a,b).

In our study, the following equation was used to acquire ET (Teixeira 2012a):

$$\frac{ET}{ET_0} = \exp\left[a + b\left(\frac{T_0}{\alpha_0 NDVI}\right)\right] \tag{6.34}$$

where a and b are the regression coefficients, being 1.9 and $-0.008°C^{-1}$, respectively, for the Brazilian semiarid conditions.

The Landsat satellite overpass time ET/ET_0 values and those for 24 h in irrigated mango orchards and Caatinga have been compared in a previous study, being the slope of the ET/ET_0 regression between satellite and field values close to 1.0 (Teixeira 2010). These results support the assumption that instantaneous and daily values for this ratio can be considered similar under cloud-free conditions (Allen et al. 2007).

To separate the energy balance components, including ET, from irrigated crops and natural vegetation, the following equation was applied with threshold values for r_s (Teixeira 2012b):

$$r_s = \exp\left[a\left(\frac{T_0}{\alpha_0}\right)(1 - NDVI) + b\right] \tag{6.35}$$

where r_s is given in s m^{-1}, and a and b, being -0.04 s m^{-1} °C^{-1} and 2.72, respectively, are the regression coefficients found for the semiarid conditions of Brazil.

To retrieve the daily values R_n, Equation 6.26 was also used. On average, the a_1 coefficient for the agroecosystems under the Brazilian semiarid conditions was found to be 143. Because of the T_a dependency on longwave radiation, it was investigated whether the variations of a_1 could be explained by those in the 24 h T_a values (Teixeira et al. 2008):

$$a_1 = bT_a - c \tag{6.36}$$

where T_a is in °C, and b and c were 7°C and 39.9, respectively, for the Brazilian semiarid conditions.

Transforming ET into energy units through the relation of 1 MJ m^{-2} day^{-1} = 0.408 mm day^{-1} (Allen et al. 1998), λE was quantified on a 24-h basis, and H was acquired as a residual in Equation 6.1, neglecting G for this timescale. Teixeira et al. (2008) has demonstrated that 24-h values for this last energy balance component is close to zero for irrigated crops and natural vegetation in the Brazilian semiarid region.

Equation 6.12 was also used for RL↓ but with daily values of 0.94 and 0.11 for the a and b regression coefficients in Equation 6.13, respectively (Teixeira 2008). Having the grids of RS↓ from the agrometeorological stations, the values for RL↑ on a large scale could be estimated as a residual in Equation 6.2.

To test the validity of Equation 6.34, Teixeira (2010) used field and remote sensing parameterizations, with 10 different Landsat images in the Brazilian semiarid region. The results showed good agreement, with $R^2 = 0.89$ and RMSE of 0.34 mm day^{-1}, fitting very well the line 1:1 when comparing the ET values in irrigated crops and natural vegetation.

6.4 Comparison between SEBAL and SAFER Methods

The challenges to make the remote sensing models applicable should be the implementation of agrometeorological stations, coupling weather data with satellite measurements and testing the algorithms over a diverse range of ecosystems. Although easier than SEBAL, before analyzing SAFER applications on a large scale in the Brazilian semiarid region, comparisons between the ET results from both models against water application by the FAO 56 approach (Allen et al. 1998) were first realized.

Different crops and soil cover conditions were considered in the northwestern side of São Paulo State, southeast Brazil. The studied commercial Bonança farm in São Paulo has central coordinates of 20°41′ 40″ S and 50°59′ 02″ W. One Landsat image acquired in July 12, 2010 (80 days after the last rain), was used together with weather data from an agrometeorological station close to the farm. The comparisons were done taking into account 20 center pivots covering a total irrigated area of 2110 ha with corn, bean, and sugar cane.

For using the SAFER model in São Paulo, the a coefficient of Equation 6.34 was first set to 1.0 considering the similarity between K_c values generally adopted by the farmer from FAO guidelines and ET/ET$_0$ results from the model in the wettest portions of the pivot irrigation areas for each specific crop.

The ET images obtained from the application of the two remote sensing algorithms are shown in Figure 6.2.

The center pivots of numbers 1–2 and 6–16 are corn, with the number 9 at the resting stage; 3 and 5 are bean; and 17–20 are sugar cane. The crops are at different days after planting (DAP) with pivot areas ranging from 70 to 160 ha.

One can clearly see the lower degree of uniformity of the ET values from SEBAL (Figure 6.2a) than for SAFER (Figure 6.2b) inside well-irrigated pivot areas. The average standard deviation (SD) for first method was 0.22 mm day^{-1}, while for the second one, the mean SD was 0.16 mm day^{-1}, indicating 38% more variation with SEBAL method than when applying the SAFER algorithm. As a first guess, the SEBAL results are less realistic, considering that inside the fully irrigated pivot areas, the water is, in general, uniformly applied.

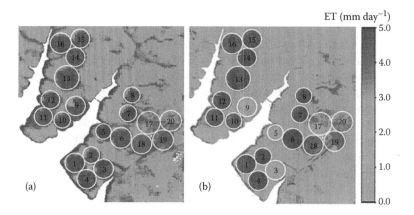

FIGURE 6.2
(See color insert.) Daily evapotranspiration (ET) from applications of (a) SEBAL and (b) SAFER models in a Landsat image acquired in July 12, 2010, for pivot center irrigated areas inside the Bonança farm, northwestern São Paulo, Brazilian southeast.

In fact, the generalized low SD values, around 0.20 mm day^{-1} for both algorithms, indicate reasonable uniformity of water application in the areas covered by the pivots, with the highest one for pivot 9. This last lack of uniformity is noted mainly for the SEBAL method, when the SD was 0.70 mm day^{-1}, while with SAFER application it was 0.30 mm day^{-1}. The reason for this as well as the difference between models is that the resting corn crop was with a very small pivot area being irrigated and low soil cover.

Comparing the ET rates from SEBAL and SAFER methods with the farm water management based on the FAO K_c methodology for all irrigation pivots, these rates were 45% and 29% higher, respectively. The RMSE was calculated taking the water applied as a reference according to DAP for bean (pivots 3 and 5 at 5 and 13 DAP, respectively) and corn crops (pivots 1–2 and 6–16, after 75 DAP). The results are shown in Figure 6.3.

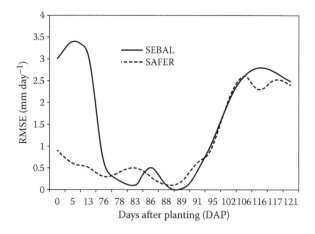

FIGURE 6.3
Relations between the root-mean-square errors (RMSE) from SEBAL and SAFER methods when comparing with water management based on FAO methodology taking as reference and according to the day after planting (DAP) for corn and bean crops.

In general, the ET results from both SAFER and SEBAL algorithms agreed well with water application from FAO methodology during the days with DAP from 75 to 95, with RMSE values around 0.4 mm day^{-1}. However, after this period, water application differed very much from ET results from both remote sensing methods, which presented good agreements between them, indicating a possibility of the use of low K_c values by the farmer after DAP = 95. The RMSE values for SEBAL algorithm at the initial stages are much larger than those for SAFER method.

Although sugar cane crops were not irrigated at the Landsat image acquirement day, the high ET rates in the pivots 18 and 19 (Figure 6.2) from both methods indicated sufficient soil moisture at the root zones. As the sugar cane crops were with high leaf areas (DAP = 120), and considering the similarity of the RMSE values at high DAP from Figure 6.3, the good agreements between SEBAL and SAFER in situations of large soil cover are reinforced.

Despite the disagreements between methods at low soil cover, satellite measurements can identify the crop stages, and the results indicated that in some situations these stages should be taken into account more accurately and objectively by the farmer, as for example, by applying relations between K_c and the accumulated degree days. According to Teixeira (2009), the advantage of using this relationship is the K_c values transfer to different thermal conditions during the crop stages.

One of the reasons for the differences between SEBAL and SAFER results in relation to water applied based on the FAO K_c methodology could be that the first algorithm gives much weight for the surface temperature that is calculated from thermal bands that have lower spatial resolution than the visible and near infrared ones and then H is obtained iteratively. At the crop initial stages the thermal conditions are more differentiated and the accuracy for obtaining G is also questionable for retrieving λE as residual in Equation 6.1 at the satellite overpass time. With application of the SAFER method, λE is acquired first than H and the ratio ET/ET_0 is related to α_0, NDVI, and ET_0, which, in turn, combine the soil–water–vegetation–atmosphere relations.

After having more confidence in SAFER algorithm, and considering also its applicability together with agrometeorological stations, it was applied in MODIS and Landsat satellite images to acquire ET on a municipality scale in the Brazilian semiarid region. Additional data of RS↓, T_a and ET_0 from automatic and conventional agrometeorological stations allowed retrieving the trend of the energy balance components at the irrigation scheme scale along the years.

6.5 Retrieving the Energy Balance Components at Different Spatial and Temporal Scales Using SAFER Modeling Scheme

The monthly ET values were analyzed for Petrolina (PE) municipality with MODIS images, while for Juazeiro (BA) municipality, this was done with Landsat images considering daily and annual timescales. For the Nilo Coelho irrigation scheme, Landsat images during the driest periods along the years were used for quantifying the land use change effects on the energy balance components as consequences of the replacement of natural vegetation by irrigated crops in the Brazilian semiarid conditions. The results and analyses are described throughout the following sections.

6.5.1 Evapotranspiration at the Municipality Scale

After calculating ET by applying the SAFER algorithm for the entire area showed in the right side of Figure 6.1 from MODIS and Landsat images, Petrolina (PE) and Juazeiro (BA) municipalities were extracted for water flux analyses at different spatial and temporal scales.

6.5.1.1 Monthly Evapotranspiration from the Petrolina Municipality

Fifteen ET/ET_0 MODIS images, with 6 days for 2010 and 9 for 2011, were used together with grids of the ET_0 monthly values during the year of 2011, representing different thermohydrological conditions along a year in the Brazilian semiarid region. The calculations involved the entire area involved by the agrometeorological stations showed in the right side of Figure 6.1. Petrolina municipality, Pernambuco (PE) state, northeast Brazil, was extracted for the monthly water fluxes analyses. Figure 6.4 presents the spatial distribution of the ET monthly values for 2011 after interpolation and averaging processes along the year.

Considering the Petrolina municipality as a whole, the spatial and temporal variation of ET along the year is evident, mainly when comparing the wettest period from February to April with the driest one, between July and September. During the rainy period the maximum ET values are verified in April, with an average of 61 mm month^{-1}. In November, there is another pick with a mean value of 58 mm month^{-1}. Intermediate water fluxes in natural vegetation occur just after the rainy season, from May to June, because the antecedent precipitation from January to April still keeps the Caatinga brushes wet and green.

As a consequence of the highest portions of the available energy used as *H* in Caatinga, during the driest period of the year from July to October, this natural vegetation presents the lowest ET values, while the irrigated fields show the highest ones. Stomata close in Caatinga species during the driest land conditions, limiting transpiration and photosynthesis, and, in general, irrigation intervals in agricultural crops are short (daily irrigation), with uniform water supplies, reducing the heat losses to the atmosphere.

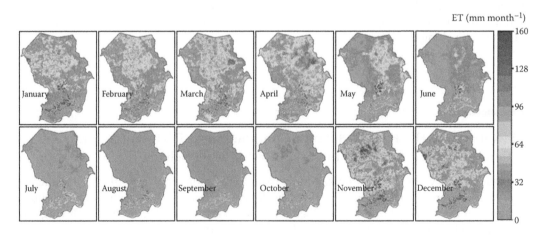

FIGURE 6.4
(See color insert.) Spatial distribution of the monthly values of evapotranspiration from MODIS images through SAFER model for the year 2011 in the Petrolina municipality, Pernambuco State, northeast Brazil.

According to Figure 6.4, moisture condition effects on the magnitude of the ET rates are strong. Pixels with values lower than 1.0 mm month^{-1} occur during the dry season representing very dry natural vegetation. The highest SD of 35 mm month^{-1} happened in November, as a consequence of the coupled effect of the start of the rains and the heterogeneity caused by differences of the crop stages, while the driest month of September presents the lowest SD of 17 mm month^{-1}.

Considering the whole year, the average annual ET rate for the mixture of agroecosystems was 438 ± 235 mm year^{-1}. Previous remote sensing studies in the semiarid conditions of the Brazilian northeast reported the highest annual ET values for table grapes and mango orchards with those for natural vegetation close to the amounts of annual rainfall (Teixeira et al. 2009b).

6.5.1.2 Daily and Annual ET from the Juazeiro Municipality

Still considering the municipality scale, three Landsat images for specific days on January 20, 2007, June 10, 2011, and November 17, 2009, were used to compute the daily ET rates also for the entire area involved by the agrometeorological stations showed on the right side of Figure 6.1. These dates represent different thermohydrological conditions along a year in the Brazilian semiarid region. The averaged ET/ET$_0$ image was multiplied by the grid of ET$_0$ for 2010 and the Juazeiro (BA) municipality extracted for quantification and analyses of the daily and annual ET rates together with the annual frequency histogram (Figure 6.5).

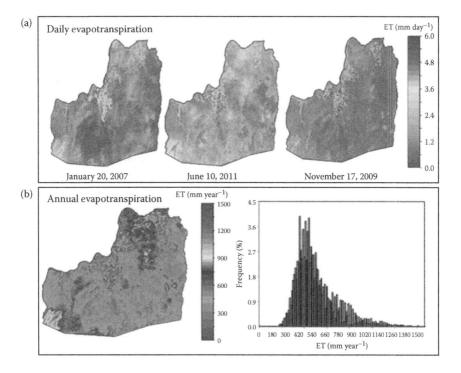

FIGURE 6.5
(See color insert.) Spatial distribution of evapotranspiration from Landsat images throughout the SAFER model in the Juazeiro municipality, Bahia (BA) State, northeast Brazil. (a) For specific days representative of different thermohydrological conditions along the year and (b) for the year 2010 together with annual ET frequency histogram.

According to Figure 6.5a, one can see the concentration of irrigated crops by the highest daily ET values in the superior portion of the Juazeiro municipality, closer to the São Francisco River.

At the start of the rainy period (represented by the image of January 20, 2007), the farmers, in general, do not irrigate, and some grape plots are under the resting stage, promoting ET daily values lower than 3.5 mm day^{-1} in irrigated areas. Considering natural vegetation, some Caatinga species are benefitted by the first rains and thus presenting high rates of biomass production. Under these conditions, considering the entire municipality, the average ET daily rate was 1.3 mm day^{-1}.

The daily ET rates rapidly increase after the rainy period (represented by the image of June 10, 2011) in both irrigated plots and natural vegetation areas. In irrigated areas, this is because the farmers reactivate the irrigation systems under continuous increases in the thermal conditions and there are much new vineyard growing seasons. Considering natural vegetation, the antecedent rains keep enough soil moisture at the root zone of the Caatinga species, leaving this ecosystem still actively consuming water. Under these circumstances the municipality ET daily values are around 1.9 mm day^{-1}.

During the driest period of the year (represented by the image of November 17, 2009), Caatinga converts large portion of the available energy into H, presenting the lowest ET daily rates. The commercial crops are with the highest values of the year, above 4.5 mm day^{-1} in well-irrigated plots, mainly from mango orchards and vineyards. In general, the irrigation intervals for these crops are short in absence of rains, and the water supply is uniform, lowering H while increasing λE. At the municipality scale, the daily ET rates are the lowest of the year, around 1.2 mm day^{-1}, owing to the largest area covered by dry natural vegetation.

From Figure 6.5b, the annual ET values representing Caatinga species are concentrated between 300 and 600 mm year^{-1}, while for irrigated crops they are in the range from 900 to 1500 mm year^{-1}. Pixels with the highest frequency present annual ET values around 500 mm year^{-1}; this low rate is a consequence of the predominance of the natural vegetation away from the bank of the São Francisco River.

Considering the annual average ET value for the whole Juazeiro municipal district, it was 40 mm year^{-1}; however, with a large standard deviation of 200 mm year^{-1}, evidencing the strong hydrological heterogeneity, as a consequence of the mixed natural and irrigated agroecosystems.

6.5.2 Energy Balance at the Irrigation Scheme Scale

For the irrigation perimeter scale, 10 Landsat images for the driest period of the year, from 1992 to 2011, were used to quantify the evolution of the daily available energy partition (R_n) into λE and H in the Nilo Coelho irrigation scheme. A conventional station, named Bebedouro, located at the Petrolina (PE) municipality (see Figure 6.1), together with an automatic one at the same place, was used in addition to apply regression equations for estimating the interpolated daily weather parameters before 2003.

After interpolations of RS↓, T_a, and ET$_0$ together with the remote sensing parameters α_0, T_0, and NDVI for all area showed in the right side of Figure 6.1, the irrigation perimeter was extracted for analyses of the energy balance components to see their trends along the years in the measure that the Caatinga species were replaced by irrigated crops.

The spatial and temporal variations for the daily values of the energy balance components resulted from Landsat images processing for each day/year from 1992 to 2011 in the Nilo Coelho irrigation scheme are presented in Figure 6.6.

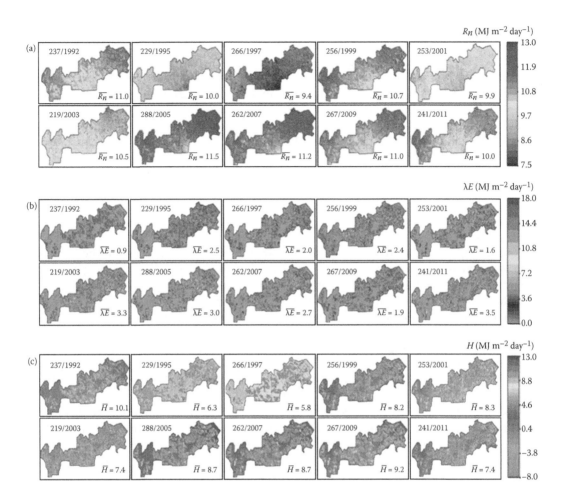

FIGURE 6.6
(See color insert.) Spatial and temporal variation of the daily energy balance components from Landsat images processing by the SAFER model and Slob equation, for the driest periods of the year, from 1992 to 2001, in the Nilo Coelho irrigation scheme, Brazil: (a) net radiation (R_n), (b) latent heat flux (λE), and (c) sensible heat flux (H). The bars mean average pixel values.

Figure 6.6a shows the spatial variation of the daily available energy (R_n), neglecting G for this timescale. The fraction of RS↓ transformed into R_n ranged from 42% to 50%, averaging 46%, confirming the assumption of this ratio being around 50% throughout field measurements in the Brazilian semiarid region (Teixeira et al. 2008). These agreements give more confidence for modeling the daily energy balance components by using Landsat satellite images and agrometeorological stations on a large scale.

In general, there is no clear distinction among the R_n pixel values from irrigated plots and natural vegetation areas along the years, as, for example, comparing the similar day of the year represented by the images for 1992 and 2011, with the largest difference in terms of land use change. Also, there are no large spatial distinctions in the SD values, which ranged from 0.40 to 0.55 MJ m⁻² day⁻¹. The strongest R_n dependence is on RS↓, with the largest R_n mean pixel value in 2005 (Figure 6.6a) corresponding to the highest level of RS↓ (25.4 MJ m⁻² d⁻¹), among the 10 days analyzed.

Clearly, one can distinguish irrigated areas from Caatinga species by the largest λE pixel values and the lowest ones for H under irrigation conditions (Figure 6.6b and c). H is even sometimes negative in the cropped areas, meaning heat horizontal advection from the warmer natural vegetation at the vicinities of irrigated plots. This happens mainly in crops with microsprinkler irrigation systems rather than those in drip irrigated ones (Teixeira 2009).

In relation to the daily λE pixel values (Figure 6.6b), their increments along the years are evident. Considering the entire irrigation perimeter, the average λE values increased 390% from 1992 to 2011. This means that the replacement of Caatinga species by irrigated crops raised almost four times the water vapor transferred to the lower atmosphere considering the period outside the rainy season. Declines in H along the years are also clear, with the average value in 1992 representing 136% of that for 2011 (Figure 6.6c). The extra water use by irrigated crops increases evapotranspiration and reduces air temperature close to the vegetated surfaces. This reduction together with increases in r_a contributes to declines in the H pixel values.

Considering the entire irrigation perimeter area, from 1992 to 1997, when the irrigated area was starting to increase, strong heat advection from the dry natural vegetation to irrigated areas is observed by the negative H values, making λE larger than R_n in very well irrigated parcels. The spatial variations in both fluxes are large, with an SD value around 3.5 MJ m^{-2} day^{-1}. Only at the start of the irrigation project in 1992 can one see a small spatial flux variation with an SD lower than 2.0 MJ m^{-2} day^{-1} for both λE and H (Figure 6.6b and c).

The trends of the average fractions of λE and H to R_n along the years from 1992 to 2011 were also quantified for the images depicted in Figure 6.6b and c. Threshold values of 800 and 10,000 s m^{-1} for r_s and conditional functions were applied to the image of the day/year 267/2009 for a vegetation classification. Following Teixeira (2012b), areas with r_s values lower than 800 s m^{-1} were considered irrigated crops; between this lower limit and 10,000 s m^{-1} they were assumed natural vegetation; and higher than this last value, they were considered features in the image that are not vegetation.

During these years the averaged fractions of R_n used as λE for the entire area increased from 8% to 32%, while in irrigated areas the corresponding fractions were from 50% to 80%. Taking into account only irrigation conditions the mean λE raised from 5.6 MJ m^{-2} day^{-1} in 1992 to 8.4 MJ m^{-2} day^{-1} in 2011, equivalent to ET rates from 2.3 to 3.4 mm day^{-1}. On an opposite situation, H/R_n values were 92% and 69% for the entire perimeter and 71% and 20% when taking into account only irrigated areas. In irrigation conditions, the H mean values ranged from 7.7 MJ m^{-2} day^{-1} in 1992 to 2.2 MJ m^{-2} day^{-1} in 2011.

Separating the natural vegetation from the entire irrigation perimeter area, it was observed that the energy partition is more steady, as in the absence of rains, the ET rates of the Caatinga species are very small, with $\lambda E/R_n$ and H/R_n around 0.06 and 0.93, respectively, during the driest conditions of the year. The main characteristic for the energy exchange from the natural vegetation in the Brazilian semiarid region is the conversion of large portions of the available energy into H flux.

6.6 Final Remarks

Considering the two algorithms for acquiring evapotranspiration analyzed in this chapter, the advantage of SAFER in relation to SEBAL is the absence of the need of neither crop classification nor hydrological extreme conditions, without the background knowledge in radiation physics. For rainy conditions, the assumption of zero latent heat flux by SEBAL is not realistic because of the absence of dry pixels. An additional advantage of SAFER is the possibility of using daily weather data from either conventional and/or automatic agrometeorological stations, making possible a historical evaluation of the energy balance components.

When comparing the SAFER and SEBAL algorithms with the traditional FAO methodology for water requirement quantification, it was concluded that under low soil cover, the first model showed more consistence with crop stages than the second one. After having more confidence in SAFER algorithm, coupling remote sensing parameters from MODIS and Landsat satellite images with weather data, the energy balance assessments were possible in a mixture of agroecosystems at the municipality and irrigation scheme scales, in the Brazilian semiarid region.

The largest evapotranspiration rates in the semiarid region of Brazil were in November for irrigated crops and in April for natural vegetation. The acquirements of the energy balance components outside the rainy period in the Nilo Coelho irrigation scheme allowed a better understanding of the energy fluxes on an irrigated perimeter level. These acquirements are important for appraising the impact of land use changes on the regional scale energy and water fluxes.

It could be concluded that net radiation values are very strongly dependent on global solar radiation independent of the type of vegetation. The averaged fractions of net radiation used as latent heat flux for the entire Nilo Coelho perimeter increased from 8% to 32% from 1992 to 2011, while in irrigated areas these fractions were from 50% to 80%, revealing that in the measure in which the irrigated area increases much more, the available energy is used in the evapotranspiration processes.

To make the SAFER model applicable together with weather data over a more diverse range of ecosystems, there is the need of adjusting the coefficients of the equation relating the ratio ET/ET_0 with the retrieving remote sensing parameters. After this calibration and with additional estimations on net radiation in conjunction with a network of agrometeorological stations, the energy exchanges in different agroecosystems can be well quantified and monitored on a large scale.

Acknowledgments

The research herein was supported by FACEPE and CNPq for the financial support to the actual project on Water Productivity in the São Francisco River Basin, Brazil.

References

Allen, R. G., L. S. Pereira, D. Raes, and M. Smith. 1998. *Crop Evapotranspiration: Guidelines for Computing Crop Water Requirements*. Food and Agriculture Organization, Rome.

Allen, R. G., M. Tasumi, A. Morse, R. Trezza, J. Wright, W. Bastiaanssen, W. Kramber, I. Lorite, and C. Robison. 2007. Satellite-based energy balance for mapping evapotranspiration with internalized calibration (METRIC)—Applications. *J. Irrig. Drain. Eng. ASCE* 133: 395–406.

Anderson, M. C., W. P. Kustas, J. G. Alfieri, F. Gao, C. Hain, J. H. Prueer, S. Evett, P. Colaizzi, T. Howell, and J. L. Chávez. 2012. Mapping daily evapotranspiration at Landsat spatial scales during BEAREX'08 field campaign. *Adv. Water Res.* 50: 162–177.

Bastiaanssen, W. G. M., R. A. L. Brito, M. G. Bos, R. A. Souza, E. B. Cavalcanti, and M. M. Bakker. 2001. Low cost satellite data for monthly irrigation performance monitoring: Benchmarks from Nilo Coelho, Brazil. *Irrig. Drain. Syst.* 15: 53–79.

Bastiaanssen, W. G. M., M. Menenti, R. A. Feddes, G. J. Roerink, and A. A. M. Holtslag. 1998. A remote sensing surface energy balance algorithm for land (SEBAL) 1. Formulation. *J. Hydrol.* 212–213: 198–212.

Bastiaanssen, W. G. M., H. Pelgrum, R. W. O. Soppe, R. G. Allen, B. P. Thoreson, and A. H. de C. Teixeira. 2008. Thermal infrared technology for local and regional scale irrigation analysis in horticultural systems. *Acta Hort.* 792: 33–46.

Bruin de, H. A. R., and J. N. M. Stricker. 2000. Evaporation of grass under nonrestricted soil moisture conditions. *Hydrol. Sci.* 45: 391–406.

Businger, J. A., J. C. Wyngaard, Y. Izumi, and E. F. Bradley. 1971. Flux-profile relationships in the atmospheric surface layer. *J. of Atm. Sci.* 28: 189–191.

Cleugh, H. A., R. Leuning, Q. Mu, and S. W. Running. 2007. Regional evaporation estimates from flux tower and MODIS satellite data. *Remote. Sens. Environ.* 106: 285–304.

Coll, C., and V. A. Caselles. 1997. Split-window algorithm for land surface temperature from advanced very high resolution radiometer data: validation and algorithm comparison. *J. Geophys. Res.* 102: 16,697–16,714.

Hernandez, F. B. T., A. H. de C. Teixeira, C. M. U. Neale, and S. Tahvaein (in press). Determining actual evapotranspiration and crop coefficient in large scale using weather station and remote sensing in the Northwest of the State of São Paulo, Brazil. *Acta Hort.*

Kalma, J. D., and D. L. B. Jupp. 1990. Estimating evaporation from pasture using infrared thermometry: Evaluation of a one-layer resistance model. *Agric. For. Meteorol.* 51: 223–246.

Kustas, W. P., M. C. Anderson, A. N. French, and D. Vickers. 2006. Using a remote sensing field experiment to investigate flux-footprint relations and flux sampling distributions for tower and aircraft-based observations. *Adv. Water Resour.* 29: 355–368.

Kustas, W. P., and J. M. Norman. 1999. Evaluation of soil and vegetation heat flux predictions using a simple two-source model with radiometric temperatures for partial canopy cover. *Agric. For. Meteorol.* 94: 13–29.

Liu, J., J. M. Chen, and J. Cihlar. 2003. Mapping evapotranspiration based on remote sensing: An application to Canada's landmass. *Water Resour. Res.* 39: 1189–1200.

Majumdar, T. J., R. Brattacharyya, and S. Chattejee. 2007. On the utilization of ENVISAT AATSR data for geological/hydrological applications. *Acta Astron.* 60: 899–905.

Menenti, M., and B. J. Choudhury. 1993. Parameterization of land surface evaporation by means of location dependent potential evaporation and surface temperature range. In *Exchange Processes at the Land Surface for a Range of Space and Time Scales* (H.-J. Bolle, R. A. Feddes, J. D. Kalma, Eds.), vol. 212. IAHS Press, Oxford, pp. 561–568.

Miralles, D. G., T. R. H. Holmes, R. A. M. De Jeu, J. H. Gash, A. G. C. A. Meesters, and A. J. Dolman. 2011. Global land-surface evaporation estimated from satellite-based observations. *Hydrol. Earth Syst. Sci.* 15: 453–469.

Norman, J. M., M. C. Anderson, W. P. Kustas, A. N. French, J. Mecikalski, R. Torn, G. R. Diak, T. J. Schmugge, and B. C. W. Tanner. 2003. Remote sensing of surface energy fluxes at 101-m pixel resolutions. *Water Resour. Res.* 39: 1221.

Norman, J. M., W. P. Kustas, J. H. Prueger, and G. R. Diak. 2000. Surface flux estimation using radiometric temperature: A dual-temperature-difference method to minimize measurement errors. *Water Resour. Res.* 36: 2263–2274.

Pôças, I., M. Cunha, L. S. Pereira, and R. G. Allen. 2013. Using remote sensing energy balance and evapotranspiration to characterize montane landscape vegetation with focus on grass and pasture lands, *Int. J. Appl. Earth Obs. Geoinf.* 21: 159–172.

Roerink, G. J., Z. Su, and M. Menenti. 2000. S-SEBI: A simple remote sensing algorithm to estimate the surface energy balance. *Phys. Chem. Earth* 25: 147–157.

Schmugge, T., S. I. Hook, and C. Coll. 1998. Recovering surface temperature and emissivity from thermal infrared multispectral scanner data. *Remote Sens. Environ.* 65: 121–131.

Schneider, K., and W. Mauser. 1996. Processing and accuracy of Landsat Thematic Mapper data for lake surface temperature measurement. *Int. J. Remote Sens.* 17: 2027–2041.

Smith, R. G. C., H. D. Barrs, and W. S. Meyer. 1989. Evaporation from irrigated wheat estimated using radiative surface temperature: An operational approach. *Agric. For. Meteorol.* 48: 331–344.

Su, Z. 2002. The Surface Energy Balance System (SEBS) for estimation of turbulent heat fluxes. *Hydrol. Earth Syst. Sci.* 6: 85–99.

Su, H., M. F. McCabe, E. F. Wood, Z. Su, and J. H. Prueguer. 2005. Modeling evapotranspiration during SMACEX02: Comparing two approaches for local and regional scale prediction. *J. Hydrometeorol.* 6: 910–922.

Tang, Q., E. A. Rosemberg, and D. P. Letenmaier. 2009. Use of satellite data to assess the impacts of irrigation withdrawals on Upper Klamath Lake, Oregon. *Hydrol. Earth Syst. Sci.* 13: 617–627.

Tasumi, M., and R. G. Allen. 2007. Satellite-based ET mapping to assess variation in ET with timing of crop development. *Agric. Water Manage.* 88: 54–62.

Teixeira, A. H. de C. 2009. *Water Productivity Assessments from Field to Large Scale: A Case Study in the Brazilian Semiarid Region*; LAP Lambert Academic Publishing, Saabrücken, Germany.

Teixeira, A. H. de C. 2010. Determining regional actual evapotranspiration of irrigated and natural vegetation in the São Francisco river basin (Brazil) using remote sensing and Penman-Monteith equation. *Remote Sens.* 2: 1287–1319.

Teixeira, A. H. de C. 2012a. Modelling evapotranspiration by remote sensing parameters and agro-meteorological stations. In *Remote Sensing and Hydrology* (C. M. U. Neale and M. H. Cosh, Eds.), vol. 352. IAHS Press, Oxford, pp. 154–157.

Teixeira, A. H. de C. 2012b. Determination of surface resistance to evapotranspiration by remote sensing parameters in the semi-arid region of Brazil for land-use change analyses. In *Remote Sensing and Hydrology* (C. M. U. Neale and M. H. Cosh, Eds.), vol. 352. IAHS Press, Oxford, pp. 167–170.

Teixeira, A. H. de C., W. G. M. Bastiaanssen, M. D. Ahmad, and M. G. Bos. 2008. Analysis of energy fluxes and vegetation-atmosphere parameters in irrigated and natural ecosystems of semi-arid Brazil. *J. Hydrol.* 362: 110–127.

Teixeira, A. H. de C., W. G. M. Bastiaanssen, M. D. Ahmad, and M. G. Bos. 2009a. Reviewing SEBAL input parameters for assessing evapotranspiration and water productivity for the Low-Middle São Francisco River basin, Brazil Part A: Calibration and validation. *Agric. For. Meteorol.* 149: 462–476.

Teixeira, A. H. de C., W. G. M. Bastiaanssen, M. D. Ahmad, and M. G. Bos. 2009b. Reviewing SEBAL input parameters for assessing evapotranspiration and water productivity for the Low-Middle São Francisco River basin, Brazil Part B: Application to the large scale. *Agric. For. Meteorol.* 149: 477–490.

Troufleau, D., J.-P. Lhomme, B. Monteny, and A. Vidal. 1997. Sensible heat flux and radiometric surface temperature over sparse sahelian vegetation: Is the kB^{-1} a relevant parameter? *J. Hydrol.* 189: 815–838.

Valiente, J. A., M. Nunez, E. Lopez-Baeza, and J. F. Moreno. 1995. Narrow-band to broad-band conversion for Meteosat visible channel and broad-band albedo using both AVHRR-1 and -2 channels. *Int. J. Remote Sens.* 16: 1147–1166.

7

Estimation of Land Surface Energy Fluxes from Remote Sensing Using One-Layer Modeling Approaches

Weiqiang Ma, Yaoming Ma, Hirohiko Ishikawa, and Zhongbo Su

CONTENTS

7.1 Introduction

The Earth's climate changes both at regional and global scales. Inevitably, this also causes a change of the Earth's water budget and its complex and dynamic energy balance with the Sun (Sorooshian et al. 2005). This is important for many regions around the globe, including the Asian Monsoon system over the Tibetan Plateau and northwest China. The Tibetan Plateau includes the world's highest elevation with an average elevation of about 4000 m. The area represents an extensive mass extending from subtropical to middle latitudes. Because of the role of the Tibetan Plateau terrain, the land surface absorbs a large amount of solar radiation energy and shows significant seasonal changes of land surface heat and water fluxes (Yanai et al. 1992; Ye and Wu 1998; Ma et al. 2006). The study on the energy exchanges between the land surface and atmosphere has also been of paramount importance for the CAMP/Tibet and WATER (Ma et al. 2006; Li et al. 2009). Some interesting detailed studies including *in situ* observation analysis and remote sensing in this area have already been conducted (Ma et al. 2002, 2004, 2006, 2008b, 2011a,b,c; Ma 2003a,b; Tanaka et al. 2001; Yang et al. 2002, 2003; Gao et al. 2004; Oku and Ishikawa 2004; Li et al. 2006; Ma and Ma 2006; Oku and Ishikawa 2010).

Combined with sparse field experimental stations, remote sensing provides a promising direction for obtaining the regional distribution of land surface heat fluxes over the Earth's land surface. Remote sensing allows obtaining consistent and frequent observations of spectral albedo and emittance of radiation at elements both in a patch landscape and on a global scale (Sellers et al. 1990), even on remote, otherwise inaccessible regions. The land surface parameters, such as land surface temperature T_{sfc}, surface hemispherical albedo r_0, normalized difference vegetation index (NDVI) (Rouse 1974), modified soil adjusted vegetation index (MSAVI) (Qi et al. 1994), vegetation coverage, leaf area index (LAI) (Jordan 1969), and surface thermal emissivity (ε) can be derived directly from satellite observations. The regional heat fluxes can be determined indirectly by means of these land surface parameters (Susskind et al. 1984; Chedin et al. 1985; Tucker and Hiernaux 1986; Becker and Li 1990; Schmugge and Andre 1991; Li and Becker 1993; Qi et al. 1994; Kustas and Norman 1997; Sobrino and Raissouni 2000; Su 2002; Ma et al. 2009, 2011a). Several approaches have been proposed in recent years for obtaining information about the regional distribution of surface heat fluxes. A review of these approaches was provided in Chapter 3 of the present volume. Generally, many of these methods require specification of the vertical temperature difference between the T_{sfc} and the air temperature (T_a) and an exchange resistance (Kustas and Norman 1997; Su 2002). Many remote sensing–based techniques have been developed to be implemented in homogeneous moist or semiarid regions, though with fewer studies evaluating the capability of those methods in deriving regional estimates of surface energy fluxes sparsely vegetated, heterogeneous landscapes of high altitudes such as in northwest China. Meanwhile, a large number of these modeling approaches have been implemented using satellite observations from low-resolution sensors such as the Moderate-Resolution Imaging Spectroradiometer (MODIS) and the Advanced Very High Resolution Radiometer (AVHRR) (Li and Becker 1993; Ma 2003a,b; Jia et al. 2011; Zhong 2010).

This chapter aims to perform an intercomparison of two widely used one-layer modeling approaches for deriving spatiotemporal estimates of energy fluxes from high-resolution imagery from the Advanced Spaceborne Thermal Emissions and Reflection radiometer (ASTER) sensor, acquired for the Tibetan Plateau and northwest China. To this end, the approaches used in our study included the Surface Energy Balance System (SEBS) and parameterization one-layer modeling techniques. To achieve our objectives, this chapter is structured as follows: First, a description of the study area and the datasets used in our study is provided. Subsequently, a detailed account on the workings of the two modeling approaches is furnished, providing detailed information on the parameterization of both approaches using in our case ASTER data; yet, the method can be implemented with any other type of remote sensing optical/thermal infrared data. Following this section, the results obtained from the intercomparison study and a critical reflection on the main findings observed are presented. Finally, we conclude by providing some suggestions for future work to be conducted toward obtaining more accurate estimates of energy fluxes by the approaches considered in our study.

7.2   Study Area and Dataset Description

7.2.1   Study Area

The intensive observation period (IOP) and long-term observation of the CAMP/Tibet have been done successfully over the past 12 years, resulting in a large amount of data being collected. The experimental region used in the present study consists of an area of approximately 150×250 km^2 that includes a variety of land surfaces such as a large areas with grassy marshland, some desertified grassland areas, many small rivers, and several lakes (Figure 7.1). A large number of stations located in the area were used in our study as they provided a large array of measurements suitable for meteorological experiments and related remote sensing work. BJ station is about 25 km south of Naqu City with an elevation of 4580 m above sea level. NPAM [NPAM is located north of Portable Automated Mesonet (PAM), so we called it NPAM) was chosen in a relatively larger open area about 20 × 30 km² with a flat plateau of grassy marshland surface (with grass height of about 10 cm). The elevation is 4765 m above sea level. Amdo station is about 6 km west of Amdo city, with an elevation of 4700 m above sea level. It was chosen in a relatively large open area about 15 × 50 km² with a flat Plateau grassy marshland surface (with grass height of about 5 cm) (Ma et al. 2008a).

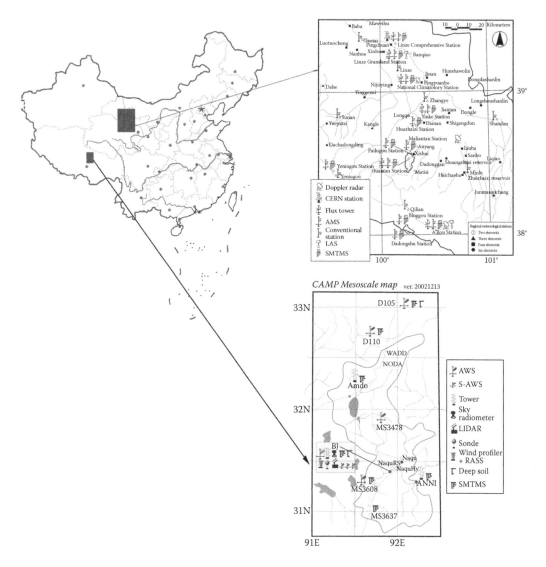

FIGURE 7.1
(See color insert.) Sketch map of study area and the sites during the CAMP/Tibet and WATER in China.

WATER (Li et al. 2009) is a multiscale land surface and hydrological experiment in a cold and arid region. In the present study, Yingke, Huazhaizi, Guantan, Maliantan, A'rou, Binggou, and Yakou are selected as weather stations to represent the middle and upper streams of the Heihe River Basin. Yingke station, located to the south of the city of Zhangye, is a typical irrigated farmland. The primary crops are corn and wheat. Huazhaizi station is located on desert steppe to the south of Zhangye oasis. Guantan is a forest station in the southern part of Zhangye oasis. Maliantan is a grassland station near Guantan station. A'rou station is located in the middle reaches of the Babao River Basin. There is a wide distribution of seasonally frozen soil. Binggou station is a high-mountain drainage system, and the mean depth of the seasonal snowpack is approximately 0.5 m, with a maximum of 0.8–1.0 m. Yakou station is located on a mountain pass. To ensure the consistency of the observation results at these stations, all instruments were demarcated before the experiments. Further details related to the experimental setting are provided by Li et al. (2009).

7.2.2 Datasets

Most of the *in situ* data used in the implementation of this study was collected at the CAMP/Tibet and WATER surface stations. These datasets were used in this study to assist in the parameterization method for land surface heat fluxes using from the ASTER remote sensing imagery acquired over our study region. They consisted of the surface radiation budget components, surface radiation temperature, surface albedo, vertical profiles of T_a, humidity, wind speed, and direction measured at the atmospheric boundary layer (ABL) towers, automatic weather stations (AWSs), radiosonde, turbulent fluxes measured by eddy covariance (EC) technique, soil heat flux (G_0), soil temperature profiles, soil moisture profiles, and the vegetation state (Ma et al. 2008a; Li et al. 2009).

Land surface data observed at CAMP/Tibet and WATER stations (BJ, Guantan, and Yingke) were also collected from an EC system, from which sensible heat flux (H) and latent heat flux (λE) were estimated. The turbulence data, observed with a sonic anemometer–thermometer and an infrared hygrometer, were processed using the eddy correlation methodology (Ma et al. 2011a). This final product formed our reference dataset. Before observation, calibration for all instruments was conducted in order to ensure consistency in our observation data.

ASTER acquires spectral data at 14 spectral bands ranging from the visible to the thermal infrared and at a high spatial, spectral, and radiometric resolution. The spatial resolution varies with wavelength: 15 m in the visible and near-infrared bands (VNIR, 0.52–0.86 μm), 30 m in the shortwave infrared bands (SWIR, 1.6–2.43 μm), and 90 m in the thermal infrared bands (TIR, 8.1–11.6 μm) (Yamaguchi et al. 1998). The level of ASTER data is level 1B (L1B); registered radiance at the sensor product contains radiometrically calibrated and geometrically coregistered data for each channel. In our work, five ASTER images over our study region were collected, including three Tibetan scenes bought from Earth Remote Sensing Data Analysis Center (ERSDAC) in Japan and two Heihe River scenes provided by Environmental and Ecological Science Data Center for West China, National Natural Science Foundation of China (http://westdc.westgis.ac.cn). For Tibetan scenes, the acquired date is July 24, 2001, November 29, 2001, and March 12, 2002. Meanwhile the Heihe River scenes are May 3 and June 4, 2008.

7.3 One-Layer Modeling Approaches

7.3.1 SEBS Algorithm

The SEBS is a one-layer modeling approach proposed by Su (2002) for estimating atmospheric turbulent fluxes and the evaporative fraction (EF) using remote sensing in combination with *in situ* observation.

The surface energy balance is commonly written as

$$R_n = G_0 + H + \lambda E \tag{7.1}$$

where R_n is the net radiation and G_0 is the soil surface heat flux. The unit of each of these terms is expressed in W m^{-2}.

SEBS consists of a set of input parameters for determining the land surface physical parameters, such as albedo, emissivity, temperature, and vegetation coverage, derived from spectral reflectance and radiance measurements. In addition, SEBS contains a model for determining the roughness length for heat transfer (Su 2002). The parameterization methodology applicable to this study implementation is made available below.

In our study, SEBS was applied to the ASTER images to evaluate the applicability of this approach to arid and cold regions of our study region. First, the ASTER data from the three ASTER instrument subsystems were reprojected to a common spatial resolution of 15 m. NDVI was derived from bands 2 and 3 of the ASTER data. The surface albedo for shortwave radiation was retrieved with the narrowband–broadband conversion in the work of Liang (2001). The land surface temperature T_{sfc} was derived using a method developed by Jiménez-Muñoz (Jiménez-Muñoz et al. 2006) from multispectral thermal infrared data using a split window method. Authors had evaluated a technique developed to extract emissivity information from multispectral thermal infrared data by adding vegetation information. A simplified method for the atmospheric correction (SMAC) (Rahman and Dedieu 1994) was used for atmospheric correction of the VNIR data. The input downward shortwave radiation was retrieved directly from observation data recorded by the WATER AWSs. R_n can be determined by combining the retrieved albedo, the surface emissivity, and T_{sfc} from the ASTER data with the downward short- and longwave thermal radiation from *in situ* measurements at the AWS stations. On the basis of the field observations, G_0 was estimated from R_n. H was estimated from T_{sfc}, T_a, and other parameters. The regional λE was derived using the EF, as described in Equations 7.8 through 7.10 (Su 2002).

R_n was estimated as

$$R_n(x,y) = k_\downarrow(x,y) - K_\uparrow(x,y) + L_\downarrow(x,y) - L_\uparrow(x,y)$$

$$= \left\{1 - r0(x,y)\right\} \cdot K_\downarrow(x,y) + \varepsilon_0(x,y)\left\{(L_\downarrow(x,y) - \sigma T_{sfc}^4(x,y)\right\} \tag{7.2}$$

where $\varepsilon_0(x, y)$ is the surface emissivity, which comes from NDVI values (Jiménez-Muñoz et al. 2006), $K_\downarrow$ (W m^{-2}) represents the downward shortwave components (0.3–3 μm), and $L_\downarrow$ (W m^{-2}) represents the downward longwave (3–100 μm) radiation components. The surface albedo $r_0(x, y)$ was derived using the narrowband–broadband conversion method in the work of Liang (2001). Liang found that the conversions are quite linear. The resultant linear equations are collated in the following:

$$r_0 = 0.484\alpha_1 + 0.335\alpha_3 - 0.324\alpha_5 + 0.551\alpha_6 + 0.305\alpha_8 - 0.367\alpha_9 - 0.0015 \tag{7.3}$$

The equation to calculate G_0 is given as follows (Su 2002):

$$G_0 = R_n \left[\Gamma_c + (1 - f_c) \cdot (\Gamma_s - \Gamma_c) \right] \tag{7.4}$$

where the constants $\Gamma_c = 0.05$ for a full vegetation canopy and $\Gamma_s = 0.315$ for bare soil (Su 2002). A linear interpolation was performed between these limiting cases using the fractional canopy coverage f_c. In this study, these constants for full canopy and bare soil conditions were evaluated for the Heihe River.

To derive H and λE, similarity theory was used. In the atmospheric surface layer, the similarity relationships for the profiles of the mean wind speed u and the mean potential temperature difference between the surface and the air, $\theta_0 - \theta_a$, are usually written in integral form as follows:

$$u = \frac{u_*}{k} \left[\ln\left(\frac{z - d_0}{z_{0m}} \right) - \Psi_m\left(\frac{z - d_0}{L} \right) + \Psi_m\left(\frac{z_{0m}}{L} \right) \right] \tag{7.5}$$

$$\theta_0 - \theta_a = \frac{H}{k u_* \rho C_p} \left[\ln\left(\frac{z - d_0}{z_{0h}} \right) - \Psi_h\left(\frac{z - d_0}{L} \right) + \Psi_h\left(\frac{z_{0h}}{L} \right) \right] \tag{7.6}$$

where z is the height above the surface, u_* is the friction velocity, C_p is the specific heat of air at constant pressure, ρ is the density of air, $k = 0.4$ is the von Kármán constant, d_0 is the zero plane displacement height, z_{0m} is the roughness height for momentum transfer, θ_0 is the potential temperature at the surface, θ_a is the potential T_a at height z, z_{0h} is the scalar roughness height for heat transfer, ψ_m and ψ_h are the stability correction functions for momentum and sensible heat transfer, and L is the Obukhov length. L is defined as follows:

$$L = \frac{\rho C_p u_*^3 \theta_v}{kgH} \tag{7.7}$$

where g is the acceleration due to gravity and θ_v is the virtual potential temperature near the surface. Other parameters can be found in the work of Su (2002). By combining Equations 7.5 through 7.7 using an iterative method, we can estimate H.

To estimate the EF, SEBS makes use of the energy balance at the limiting cases of the dry limit and the wet limit. So the relative evaporation (ratio of the actual evaporation to the evaporation at the wet limit) can be derived as

$$\text{EF}_r = 1 - \frac{H - H_{\text{wet}}}{H_{\text{dry}} - H_{\text{wet}}} \tag{7.8}$$

where H_{wet} is H at the wet limit and H_{dry} at the dry limit. Estimations of H_{wet} and H_{dry} are detailed in the work of Su (2002). In the dry limit cases, the latent heat becomes zero due to the limitation of soil moisture and H is at its maximum value. In the wet limit cases, where the evaporation takes place at potential rate, H takes its minimum value. The EF (ratio of λE to available energy) is estimated by the following equation:

$$\text{EF} = \frac{\lambda E}{R_n - G} = \frac{\text{EF}_r \cdot \lambda E_{\text{wet}}}{R_n - G} \tag{7.9}$$

where λE_{wet} is the λE at the wet limit (i.e., evaporation is only limited by the available energy under the given surface and atmospheric conditions). λE can then be calculated by

$$\lambda E = \mathrm{EF}(R_n - G_0) \tag{7.10}$$

7.3.2 Parameterization Method

In order to get the land surface heat fluxes from the ASTER images, another one-layer modeling approach is used. The method of calculating regional R_n is the same with that of SEBS method. However, the incoming shortwave radiation flux $K_\downarrow(x, y)$ in Equation 7.2 is derived from radiative transfer model Moderate Resolution Atmospheric Transmission (MODTRAN) (Kneizys 1988). The incoming longwave radiation flux $L_\downarrow(x, y)$ in Equation 7.2 is derived directly from MODTRAN. The model atmosphere used in MODTRAN is midlatitudes of summer with slant atmospheric paths. The model of execution is thermal radiance. The other, parameters such as temperature, pressure, water vapor, ozone, methane, nitrous oxide, carbon monoxide, other gases, and altitude profile, are set to midlatitudes summer. The aerosol model used is rural VIS = 23 km. Ground altitude above sea level is 4.5 km. Land surface temperature $T_{\mathrm{sfc}}(x, y)$ (K) in Equation 7.2 is derived from ASTER thermal infrared spectral radiance (Hook et al. 1992) (Figure 7.2).

The regional soil heat flux density $G_0(x, y)$ is determined through (Choudhury and Monteith 1988)

$$G_0(x, y) = \rho_s C_s [T_{\mathrm{sfc}}(x, y) - T_s(x, y)]/r_{\mathrm{sh}}(x, y) \tag{7.11}$$

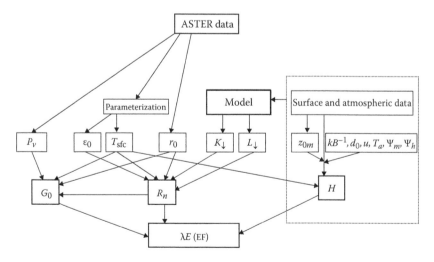

FIGURE 7.2
Diagram of parameterization procedure by combining ASTER data with field observations. Note that P_v is vegetation coverage; ε_0 is emissivity; T_{sfc} is land surface temperature; r_0 is land surface albedo; $K_\downarrow$ is downward shortwave radiation; $L_\downarrow$ is downward longwave radiation; z_{0m} is roughness height for momentum transfer; kB^{-1} is the excess resistance for heat transportation; d_0 is zero plane displacement height; u is mean wind speed; T_a is air temperature; c_m is stability correction function for momentum; and Ψ_h is stability correction function for sensible heat transfer.

where ρ_s (kg m^{-3}) is the soil dry bulk density, C_s (J kg^{-1} K^{-1}) is the soil specific heat, $T_s(x, y)$ (K) represents soil temperature of a determined depth, and $r_{sh}(x, y)$ (s m^{-1}) stands for resistance of soil heat transportation. The regional $G_0(x, y)$ cannot directly be mapped from satellite observations through Equation 7.11 because of the difficulty to derive the soil heat transportation resistance $r_{sh}(x, y)$ (s m^{-1}) and the soil temperature at a determined depth $T_s(x, y)$ (K) (Bastiaanssen et al. 1998). Many investigations have shown that the midday G_0/R_n ratio is reasonably predictable from spectral vegetation indices (Daughtry et al. 1990; Ma et al. 2008b, 2009, 2011a,b). The ratio is considered as a function F, which relates G_0/R_n to other variables (Ma 2003b). Some researchers have concluded that $G_0/R_n = F(\text{NDVI})$ (Clothier et al. 1986; Kustas and Norman 1997). A better ratio of $G_0/R_n = F(r_0, T_{sfc}, \text{NDVI})$ was also found (Bastiaanssen et al. 1998; Choudhury et al. 1987). The relationship between $G_0(x, y)$ and $R_n(x, y)$ found in the Tibetan Plateau area (Ma 2003b) will be used to determine regional soil heat flux over the CAMP/Tibet area here. It means that $G_0(x, y)$ is

$$G_0(x, y) = R_n(x, y) \cdot \frac{T_{sfc}(x, y)}{r_0(x, y)} \cdot \left(0.00029 + 0.00454\overline{r_0} + 0.00878\overline{r_0}^2\right) \cdot [1 - 0.964\text{MSAVI}(x, y)^4]$$

(7.12)

where $\overline{r_0}$ is the average albedo.

The regional distribution of H is calculated from

$$H(x, y) = \rho C_p \frac{T_{sfc}(x, y) - T_a(x, y)}{r_a(x, y)}$$

(7.13)

where the aerodynamic resistance is

$$r_a(x, y) = \frac{1}{ku_*(x, y)} \left[\ln\left(\frac{z - d_0(x, y)}{z_{0m}(x, y)}\right) + kB^{-1}(x, y) - \psi_h(x, y) \right]$$

(7.14)

where κ is the von Kármán constant, u_*(m s^{-1}) is the friction velocity, d_0 (m) is the zero-plane displacement height, z_{0m} (m) is the aerodynamic roughness, kB^{-1} is the excess resistance for heat transportation, and ψ_h is the stability correction function for heat. The friction velocity u_* is derived from

$$u_*(x, y) = ku(x, y) \left\{ \ln\left[\frac{z - d_0(x, y)}{z_{0m}(x, y)}\right] - \psi_m(x, y) \right\}^{-1}$$

(7.15)

where ψ_m is the stability correction function for momentum.

From Equations 7.13 through 7.15,

$$H(x, y) = \rho C_p \kappa^2 u(x, y) \frac{[T_{sfc}(x, y) - T_a(x, y)]}{\left[\ln\frac{z - d_0(x, y)}{z_{0m}(x, y)} + kB^{-1} - \psi_h(x, y) \right] \cdot \left[\ln\frac{z - d_0(x, y)}{z_{0m}(x, y)} - \psi_m(x, y) \right]}$$

(7.16)

where z is the reference height and u is the wind speed at the reference height. In the study, z and u are determined with the aid of the field measurements of the AWS system at the same height. $z_{0m}(x, y)$ in Equation 7.16 over the Northern Tibetan area is calculated using Jia's model (Li et al. 1999). d_0 is the zero-plane displacement, which is calculated using Stanhill's model (Stanhill 1969). $T_a(x, y)$ in Equation 7.16 is the regional distribution of T_a at the reference height; it is derived from a linear method:

$$T_a(x, y) = 0.7784T_{sfc}(x, y) + 60.1706 \qquad (7.17)$$

Values of the excess resistance to heat transfer kB^{-1} versus $u(T_{sfc} - T_a)$ over the northern Tibetan Plateau is calculated in the following equation (Ma et al. 2008b):

$$kB^{-1} = 0.062u(T_{sfc} - T_a) + 0.599 \qquad (7.18)$$

$\psi_h(x, y)$ and $\psi_m(x, y)$ in Equation 7.16 are the integrated stability functions. For an unstable condition, the integrated stability functions $\psi_h(x, y)$ and $\psi_m(x, y)$ are written as (Paulson 1970)

$$\begin{cases} \psi_m(x,y) = 2\ln\left(\dfrac{1+X}{2}\right) + \ln\left(\dfrac{1+X^2}{2}\right) - 2\arctan(X) + 0.5\pi \\ \psi_h(x,y) = 2\ln\left(\dfrac{1+X^2}{2}\right) \end{cases} \qquad (7.19)$$

where $X = \{1 - 16^* [z - d_0(x, y)]/L(x, y)\}^{0.25}$. For a stable condition, the integrated stability function $\psi_h(x, y)$ and $\psi_m(x, y)$ become (Webb 1970)

$$\psi_m(x,y) = \psi_h(x,y) = -5 \cdot \frac{z - d_0(x,y)}{L(x,y)} \qquad (7.20)$$

The stability function $[z - d_0(x, y)]/L(x, y)$ is calculated by Businger's method (Businger 1988):

$$\begin{cases} \dfrac{z - d_0(x,y)}{L(x,y)} = R_i(x,y) & \text{(unstable)} \\ \dfrac{z - d_0(x,y)}{L(x,y)} = R_i(x,y/[1 - 5.2R_i(x,y)] & \text{(stable)} \end{cases} \qquad (7.21)$$

where $R_i(x, y)$ is the Richardson number, and according to the definition of the Richardson number, the approximate analytical solutions of R_i found by Yang et al. (2001) will be used here.

The regional latent heat flux $\lambda E(x, y)$ is derived as the residual of the energy budget theorem (Ma et al. 2008b) for land surface:

$$\lambda E(x, y) = R_n(x, y) - H(x, y) - G_0(x, y) \qquad (7.22)$$

7.4   Results and Discussion

The regional distributions of land surface heat fluxes over the Tibetan Plateau area and northwest China were derived from the ASTER data and field observations using both the SEBS and parameterization one-layer modeling techniques. Both approaches are based on the land surface energy balance. SEBS involves many different variables to calculate the land surface physical parameters, including albedo, emissivity, T_{sfc}, and vegetation coverage. Furthermore, SEBS also includes a model for determining the roughness length for heat transfer. The parameterization method is calculating land surface heat flux based on land surface energy balance. In this chapter, the general idea is as follows: the first step is R_n using land surface radiation budgets. The second is to calculate G_0 with the relationship R_n. H is the most difficult one including a large number of parameters. The final parameter to be estimated is λE, calculated by land surface energy balance. However, SEBS is different from parameterization in several aspects. The most important one is that SEBS takes more careful physical processes into consideration. For example, SEBS calculates surface fluxes with the aid of energy balance at the dry limit and the wet limit.

As a case study to compare the two modeling approaches, two ASTER images over the WATER region were used. Figure 7.3 shows the spatial distribution of the surface heat fluxes for our study area derived from SEBS. Figure 7.4 shows the validation results for the derived R_n, H, and λE by comparing with ground measurements at the WATER *in situ* stations (i.e., Yingke, Huazhaizi, Guantan, Maliantan, A'rou, Binggou, and Yakou). The ground AWS measurements were averaged over a 10-min period, and the EC 30 min.

As it can be observed (Figure 7.4), the derived R_n using SEBS over the study area was in close agreement to the corresponding field measurements. The changes in R_n were comparable to the changes in T_{sfc}, the albedo, and the downward shortwave radiation. Here, T_{sfc}, the albedo, and the downward shortwave radiation agreed well with the measured data. R_n was also in reasonably good agreement with the *in situ* observations (Figure 7.4), with a root-mean-square error of 41.5 W m^{-2}. However, an error of R_n was evident at the Huazhaizi site. The Huazhaizi site is located nearby the Zhangye Oasis, so the advection between the oasis and Gobi desert could have affected the agreement. The boundary layer energy distribution was disarranged especially at noon. These factors would contribute to the R_n errors between remote sensing and land surface observation.

Overall, H flux derived from SEBS methods was not as good as the R_n, but still within a credible range. The parameterization scheme for H flux is the most complex in the entire land surface energy budget, especially for that in Guantan station. The station was located in the forest. Many uncertain factors may contribute to the deviations from credible values, for example, the land surface is forest in Guantan station, so land–atmosphere interaction shows complicated instead of pure land surface interactions. The resulting calculations can thus produce a large difference between the remote sensing results and observations at the station, because the calculation failed to take so many factors into consideration.

As regards the λE agreement, it can be seen (Figure 7.4) that the predicted λE is acceptable for the Yingke irrigation area, especially in June. June is the season of irrigation, and crop growth is substantial. The satellite retrieval results and the observations were similar on June 4. λE reached 350 W m^{-2} during this season. This value resulted from the fact that the Yingke irrigation area was largely occupied by irrigated corn. Because of field irrigation, the land surface was very moist. All of these factors led to a large surface λE. Evidently, the λE estimated from remote sensing were too large in comparison to the ground measurements at the Guantan site because of the large area homogeneity occupied mostly by

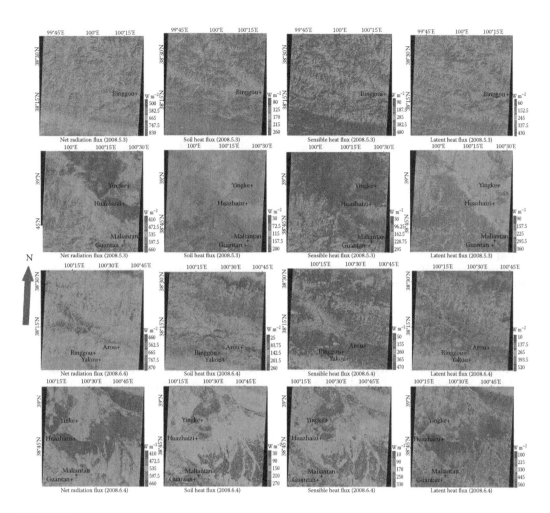

FIGURE 7.3
(See color insert.) Distribution maps of land surface heat fluxes using ASTER data with SEBS over the WATER area. (From Ma, W. et al., *Hydrology and Earth System Sciences Discussions*, 6, 4619–4635, 2011.)

the forest. On the contrary, the λE agreement was poor on May 3 at both the Yingke and Guantan sites, because of the large heterogeneity in the area. So the two reasons proposed for the poor results are, first, that it is the bad land surface observation for λE and, second, that the SEBS is not suitable for such a complicated condition. Overall, however, SEBS can be used to estimate the heterogeneous land surface λE distribution under limited conditions, because the method is used in only a few cases over the inhomogeneous regions.

Figure 7.5 shows the distribution of surface heat fluxes around the CAMP/Tibet area derived from the parameterization method, and Figure 7.6 shows their frequency distributions. The land surface heat fluxes (R_n, G_0, H, and λE) derived from satellite data were compared with the field measurements at BJ site. The measured surface G_0 was calculated from G_0 measured at –10 cm and the soil temperature measured at surface and –10 cm.

The derived land surface heat fluxes in 3 months over the study area were in good accordance with the land surface status. The experimental area includes a variety of land

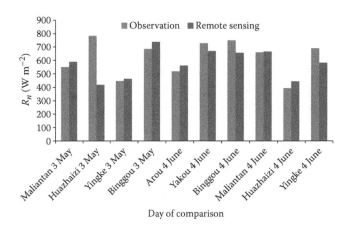

FIGURE 7.4
Validation of the derived net radiation from SEBS against ground measurements over the WATER stations, Yingke, Huazhaizi, Guantan, Maliantan, Arou, Binggou, and Yakou. Note that *x* axis is date and *y* axis is surface flux.

surfaces such as a large area of grassy marshland, some areas of desertified grassland, many small rivers, and several lakes; therefore these derived parameters show a wide range due to the complexity of surface features. R_n changed from 570 to 830 W m^{-2} in July and from 80 to 655 W m^{-2} in March and November. G_0 varied from 100 to 220 W m^{-2} in July and from 0 to 180 W m^{-2} in March and November. H was from 0 to 540 W m^{-2} in July and from 0 to 520 W m^{-2} in March and November, and λE varied from 0 to 736 W m^{-2} in July and 0 to 288 W m^{-2} in March and November (see Figure 7.5). The range of R_n was very similar as the same research area (Ma et al. 2002, 2006). Surface albedo, T_{sfc}, and H around the lake in the distribution maps were much higher in July, while R_n, G_0, and λE were lower in the area. The reason is that the predominant land cover around the lake is composed by desertified grass that was dry at the time of the measurement.

The derived pixel value (Figure 7.5) and average value (see Figure 7.6) of R_n, G_0, and λE in July were higher than that in March and November. It means that the evaporation in summer was much stronger than that in winter in the north Tibetan Plateau area. The heating density ($H + \lambda E = R_n - G_0$) in summer was also much higher than that in winter in the central Tibetan Plateau area. This is caused by the high solar radiation and lower albedo in the summer and conversely the lower solar radiation and higher albedo in the winter. This is because most of the land surface of the experimental area is covered by green grass in summer and by snow and ice during winter.

The derived R_n over the study area was very close to the field measurement with APD less than 3.1% as a result of the improvement on surface albedo and T_{sfc}. The regional soil heat flux derived from the relationship between G_0 and R_n is suitable for heterogeneous land surface of the CAMP/Tibet area, and the APD is less than 9.8% at the validation sites, because the relationship itself was derived from the same area.

The derived regional H and λE with APD less than 9.8% at the validation sites in the CAMP/Tibet area is in good agreement with field measurements. This is due to the fact that atmospheric boundary layer processes have been considered in more detail in our methodology and that the proposed parameterization for H and λE can be used over the north Tibetan Plateau area. ASTER's ranges and results are nearly same as AVHRR, MODIS, and landsat ETM+ in the Tibetan Plateau (Ma et al. 2002, 2011b; Ma 2003b).

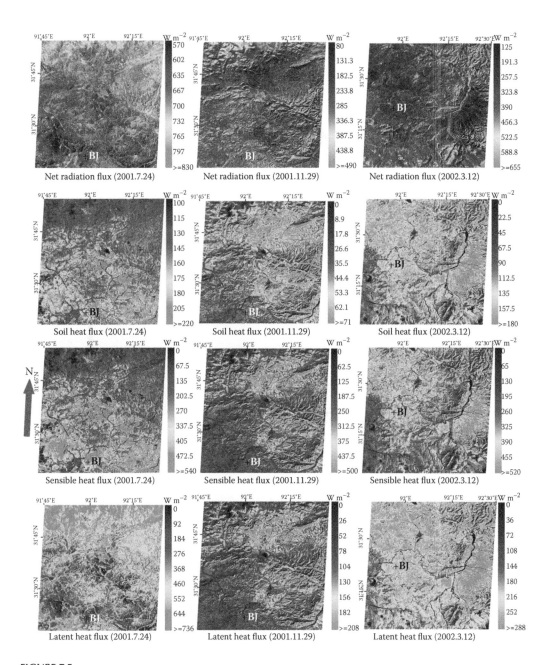

FIGURE 7.5
(See color insert.) Spatially distributed maps of land surface heat fluxes over the CAMP/Tibet area derived from the combination of ASTER data with the parameterization one-layer modeling technique. (From Ma, W. et al., *Hydrology and Earth System Sciences*, 13(1), 57, 2009.)

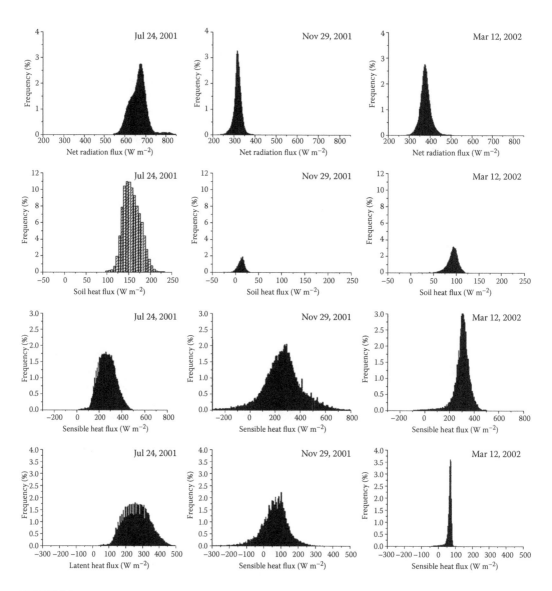

FIGURE 7.6
Frequency distribution of land surface heat fluxes for the CAMP/Tibet area using ASTER data with parameterization method at 1245 Beijing Time. (From Ma, W. et al., *Hydrology and Earth System Sciences*, 13(1), 57, 2009.)

7.5 Conclusions

In this chapter, the regional distributions of land surface heat fluxes (R_n, G_0, H, and λE) over the heterogeneous north Tibetan Plateau area and northwest China were derived using ASTER data and field observations by means of the widely used SEBS and parameterization one-layer modeling techniques. Both methods are based on the evaluation of the land surface energy balance. An intercomparison was conducted between the two methods, evaluating their ability to estimate key components of the energy balance closure in two regions in the

Tibetan Plateau, China. Reasonable results of land surface variables, vegetation variables, and land surface fluxes were obtained by both methods, in close agreement to *in situ* observations also available concurrently to the satellite data overpasses. Both methods can estimate land surface fluxes based on the land surface energy balance evaluation. Yet, there are many differences in the architectural development between the SEBS and parameterization methods. The most important one is that SEBS used more careful physical processes. For example, SEBS makes use of the energy balance of the dry limit and the wet limit to calculate surface fluxes.

The retrieval of regional land surface heat fluxes over heterogeneous landscape is not an easy task. The method presented in this study is, at the time of the writing this chapter, still in the development stage. For observation sites, it needs large financial and material resources to support the parameterization method. For analytical methods, some parameters are still based on observation or experience. All of these factors affect the results related to land surface heat flux. Only three ASTER images are used in the north Tibetan Plateau. To obtain more accurate regional land surface fluxes (daily to seasonal variations) over a larger area (the Tibetan Plateau or all the northwest China), more field observations (ABL tower and radiation measurement system, radiosonde system, turbulent fluxes measured by EC technique, soil moisture and soil temperature measurement system, etc.) and other satellite sensors such as MODIS and NOAA/AVHRR with more frequent temporal coverage have to be used.

Both methods implemented in our study are only applicable to clear-sky days. In order to extend its applicability to cloudy skies, the use of microwave remote sensing to derive surface temperature and other land surface variables should be considered. In future field works using the one-layer modeling approaches implemented herein, more attention should be paid to the vegetation variables, which can be measured in the coming experiments over the study area.

Acknowledgments

This work was carried out under the Chinese National Key Programme for Developing Basic Sciences (2010CB951701), Knowledge Innovation Project of Chinese Academy of Sciences (KZCX2-YW-QN309), and National Natural Science Foundation of China (41275010, 41275028, and 41175068).

References

Bastiaanssen, W., M. Menenti, R. Feddes, and A. Holtslag. 1998. A remote sensing surface energy balance algorithm for land (SEBAL). 1. Formulation. *Journal of Hydrology*, 212: 198–212.

Becker, F., and Z. Li. 1990. Towards a local split window method over land surfaces. *International Journal of Remote Sensing*, 11(3): 369–393.

Businger, J. 1988. A note on the Businger-Dyer profiles. *Boundary-Layer Meteorology*, 42(1): 145–151.

Chedin, A., N. Scott, C. Wahiche, and P. Moulinier. 1985. The improved initialization inversion method: A high resolution physical method for temperature retrievals from satellites of the TIROS-N series. *Journal of Climate and Applied Meteorology*, 24(2): 128–143.

Choudhury, B., S. Idso, and R. Reginato. 1987. Analysis of an empirical model for soil heat flux under a growing wheat crop for estimating evaporation by an infrared-temperature based energy balance equation. *Agricultural and Forest Meteorology*, 39(4): 283–297.

Choudhury, B., and J. Monteith. 1988. A four-layer model for the heat budget of homogeneous land surfaces. *Quarterly Journal of the Royal Meteorological Society*, 114(480): 373–398.

Clothier, B. et al. 1986. Estimation of soil heat flux from net radiation during the growth of alfalfa. *Agricultural and Forest Meteorology*, 37(4): 319–329.

Daughtry, C. et al. 1990. Spectral estimates of net radiation and soil heat flux. *Remote Sensing of Environment*, 32(2): 111–124.

Gao, Z. et al. 2004. Modeling of surface energy partitioning, surface temperature, and soil wetness in the Tibetan prairie using the Simple Biosphere Model 2 (SiB2). *Journal of Geophysical Research*, 109(D6): D06102.

Hook, S. J., A. Gabell, A. Green, and P. Kealy. 1992. A comparison of techniques for extracting emissivity information from thermal infrared data for geologic studies. *Remote Sensing of Environment*, 42(2): 123–135.

Jia, L., H. Shang, G. Hu, and M. Menenti. 2011. Phenological response of vegetation to upstream river flow in the Heihe River basin by time series analysis of MODIS data. *Hydrology and Earth System Sciences*, 15(3): 1047–1064.

Jiménez-Muñoz, J. C., J. A. Sobrino, A. Gillespie, D. Sabol, and W. T. Gustafson. 2006. Improved land surface emissivities over agricultural areas using ASTER NDVI. *Remote Sensing of Environment*, 103(4): 474–487.

Jordan, C. F. 1969. Derivation of leaf-area index from quality of light on the forest floor. *Ecology*, 663–666.

Kneizys, F. X. 1988. *Users Guide to LOWTRAN 7*, Air Force Geophysics Laboratory, Hanscom Air Force Base, MA.

Kustas, W. P., and J. M. Norman. 1997. A two-source approach for estimating turbulent fluxes using multiple angle thermal infrared observations. *Water Resources Research*, 33(6): 1495–1508.

Li, J., W. Jiemin, and M. Menenti. 1999. Estimation of area roughness length for momentum using remote sensing data and measurements in field. *Scientia Atmospherica Sinica*, 5: 013.

Li, M. et al. 2006. Analysis of turbulence characteristics over the northern Tibetan Plateau area. *Advances in Atmospheric Sciences*, 23(4): 579–585.

Li, X. et al. 2009. Watershed allied telemetry experimental research. *Journal of Geophysical Research*, 114(D22103): 11590.

Li, Z., and F. Becker. 1993. Feasibility of land surface temperature and emissivity determination from AVHRR data. *Remote Sensing of Environment*, 43(1): 67–85.

Liang, S. 2001. Narrowband to broadband conversions of land surface albedo I: Algorithms. *Remote Sensing of Environment*, 76(2): 213–238.

Ma, W., and Y. Ma. 2006. The annual variations on land surface energy in the northern Tibetan Plateau. *Environmental Geology*, 50(5): 645–650.

Ma, W. et al. 2011a. Estimating surface fluxes over middle and upper streams of the Heihe River Basin with ASTER imagery. *Hydrology and Earth System Sciences Discussions*, 6: 4619–4635.

Ma, W. et al. 2009. Estimating surface fluxes over the north Tibetan Plateau area with ASTER imagery. *Hydrology and Earth System Sciences*, 13(1): 57.

Ma, W., Y. Ma, and B. Su. 2011b. Feasibility of retrieving land surface heat fluxes from ASTER data using SEBS: A case study from the NamCo area of the Tibetan Plateau. Arctic, Antarctic, and Alpine. *Research*, 43(2): 239–245.

Ma, Y. 2003a. On measuring and remote sensing surface energy partitioning over the Tibetan Plateau—from GAME/Tibet to CAMP/Tibet. *Physics and Chemistry of the Earth, Parts A/B/C*, 28(1–3): 63–74.

Ma, Y. 2003b. Remote sensing parameterization of regional net radiation over heterogeneous land surface of Tibetan Plateau and arid area. *International Journal of Remote Sensing*, 24(15): 3137–3148.

Ma, Y. et al. 2008a. ROOF OF THE WORLD: Tibetan Observation and Research Platform. *Bulletin of the American Meteorological Society*, 89(10): 1487–1492.

Ma, Y., M. Menenti, R. Feddes, and J. Wang. 2008b. Analysis of the land surface heterogeneity and its impact on atmospheric variables and the aerodynamic and thermodynamic roughness lengths. *Journal of Geophysical Research*, 113(D8): D08113.

Ma, Y. et al. 2004. Remote sensing parameterization of regional land surface heat fluxes over arid area in Northwestern China. *Journal of Arid Environments*, 57(2): 257–273.

Ma, Y., Z. Su, Z. Li, T. Koike, and M. Menenti. 2002. Determination of regional net radiation and soil heat flux over a heterogeneous landscape of the Tibetan Plateau. *Hydrological Processes*, 16(15): 2963–2971.

Ma, Y. et al. 2011c. The characteristics of atmospheric turbulence and radiation energy transfer and the structure of atmospheric boundary layer over the northern slope area of Himalaya. *Journal of the Meteorological Society of Japan*, 89A: 345–353.

Ma, Y. et al. 2006. Determination of regional distributions and seasonal variations of land surface heat fluxes from Landsat-7 Enhanced Thematic Mapper data over the central Tibetan Plateau area. *Journal of Geophysical Research*, 111(D10): D10305.

Oku, Y., and H. Ishikawa. 2004. Estimation of land surface temperature over the Tibetan Plateau using GMS data. *Journal of Applied Meteorology*, 43(4): 548–561.

Oku, Y., and H. Ishikawa. 2010. Land surface energy budget over the Tibetan Plateau based on satellite remote sensing data. *Atmospheric Science*, 16: 147.

Paulson, C. A. 1970. The mathematical representation of wind speed and temperature profiles in the unstable atmospheric surface layer. *Journal of Applied Meteorology*, 9(6): 857–861.

Qi, J., A. Chehbouni, A. Huete, Y. Kerr, and S. Sorooshian. 1994. A modified soil adjusted vegetation index. *Remote Sensing of Environment*, 48(2): 119–126.

Rahman, H., and G. Dedieu. 1994. SMAC: A simplified method for the atmospheric correction of satellite measurements in the solar spectrum. *International Journal of Remote Sensing*, 15(1): 123–143.

Rouse, J. 1974. Monitoring vegetation systems in the Great Plains with ERTS. In *Proceedings of the Third ERTS Symposium*, vol. 1, pp. 309–371.

Schmugge, T., and J. C. Andre. 1991. *Land Surface Evaporation: Measurement and Parameterization*. Springer Verlag, New York.

Sellers, P., S. Rasool, and H. J. Bolle. 1990. A review of satellite data algorithms for studies of the land surface. *Bulletin of the American Meteorological Society*, 71: 1429–1447.

Sobrino, J., and N. Raissouni. 2000. Toward remote sensing methods for land cover dynamic monitoring: Application to Morocco. *International Journal of Remote Sensing*, 21(2): 353–366.

Sorooshian, S. et al. 2005. Water and energy cycles: Investigating the links. *World Meteorological Organization Bulletin*, 54(2): 58–64.

Stanhill, G. 1969. A simple instrument for the field measurement of turbulent diffusion flux. *Journal of Applied Meteorology*, 8: 509–513.

Su, Z. 2002. The Surface Energy Balance System (SEBS) for estimation of turbulent heat fluxes. *Hydrology and Earth System Sciences*, 6(1): 85–99.

Susskind, J., J. Rosenfield, D. Reuter, and M. Chahine. 1984. Remote sensing of weather and climate parameters from HIRS2/MSU on TIROS-N. *Journal of Geophysical Research*, 89(D3): 4677–4697.

Tanaka, K., H. Ishikawa, T. Hayashi, I. Tamagawa, and Y. Ma. 2001. Surface energy budget at Amdo on the Tibetan Plateau using GAME/Tibet IOP98 data. *Journal of the Meteorological Society of Japan*, 79(1B): 505–517.

Tucker, C. O., and P. H. Y. Hiernaux. 1986. Monitoring the grasslands of the Sahel using NOAA AVHRR data: Niger 1983. *International Journal of Remote Sensing*, 7(11): 1475–1497.

Webb, E. 1970. Profile relationships: the log-liner range and extension to strong stability. *Quarterly Journal of the Royal Meteorological Society*, 96(407): 67–90.

Yamaguchi, Y., A. B. Kahle, H. Tsu, T. Kawakami, and M. Pniel. 1998. Overview of advanced spaceborne thermal emission and reflection radiometer (ASTER). *IEEE Transactions on Geoscience and Remote Sensing*, 36(4): 1062–1071.

Yanai, M., C. Li, and Z. Song. 1992. Seasonal heating of the Tibetan Plateau and its effects on the evolution of the Asian summer monsoon. *Journal of the Meteorological Society of Japan*, 79(1B): 419–434.

Yang, K., T. Koike, H. Fujii, K. Tamagawa, and N. Hirose. 2002. Improvement of surface flux parameterizations with a turbulence-related length. *Quarterly Journal of the Royal Meteorological Society*, 128(584): 2073–2087.

Yang, K., T. Koike, and D. Yang. 2003. Surface flux parameterization in the Tibetan Plateau. *Boundary-Layer Meteorology*, 106(2): 245–262.

Yang, K., N. Tamai, and T. Koike. 2001. Analytical solution of surface layer similarity equations. *Journal of Applied Meteorology*, 40(9): 1647–1653.

Ye, D. Z., and G. X. Wu. 1998. The role of the heat source of the Tibetan Plateau in the general circulation. *Meteorology and Atmospheric Physics*, 67(1): 181–198.

Zhong, L., Y. Ma, Z. Su, and M. S. Salama. 2010. Estimation of land surface temperature over the Tibetan Plateau using AVHRR and MODIS data. *Advances in Atmospheric Sciences*, 27(5): 1110–1118.

8

Mapping Surface Fluxes and Moisture Conditions from Field to Global Scales Using ALEXI/DisALEXI

Martha C. Anderson, William P. Kustas, and Christopher R. Hain

CONTENTS

8.1 Introduction

Land surface temperature (LST) maps derived from thermal infrared (TIR) satellite data convey valuable information for detecting moisture stress conditions and for constraining diagnostic surface flux estimates based on remote sensing. Soil surface and vegetation canopy temperatures rise as available water in the surface layer and root zone is depleted, with thermal stress signals typically preceding significant change in vegetation structure or reduction in biomass (Moran 2003). Among surface moisture monitoring methods, thermal remote sensing is unique in its range of achievable spatial resolution—providing information at scales from individual farm fields to continental and global coverage. The spatial detail provided by TIR imaging provides a useful complement to coarser scale surface moisture information developed from passive microwave remote sensing.

Prognostic land surface models (LSMs) also have broad applications in drought monitoring and water resource management (Arsenault et al. 2003; Sheffield et al. 2004; Mo et al. 2011; Xia et al. 2012b). LSMs can provide time-continuous and interconsistent quantitative estimates for a full suite of hydrologic variables. However, they require extensive parameterization and critical physical processes may in some cases be inadequately represented, leading to regional and seasonal errors. Biases can occur because of inaccurate modeling assumptions, observational errors in the forcing data, and a reliance on surface parameter fields (e.g., soil texture and plant rooting depth) that may not be available with the required accuracy or spatial resolution (Betts et al. 1997; Schaake et al. 2004; Mo et al. 2012). Because they are principally constrained by the accuracy of the precipitation inputs, LSMs are typically limited in spatial resolution (several kilometers or coarser) and are only moderately portable to regions with sparse ground-based rain gauge networks required for accurate calibration. They are unable to capture surface flux response to changes in land cover conditions (e.g., deforestation), agricultural management practices (e.g., irrigation scheduling and crop rotation), or shallow water tables without significant *a priori* knowledge of when and where these processes are important.

In contrast, diagnostic indicators based on remote sensing can be generated at higher spatial resolution and with broad geographic coverage. Limited reliance on surface and subsurface information makes these techniques highly portable. Impacts of rooting depth, soil moisture storage capacity, surface disturbance, and groundwater contributions to the system evaporative flux may be effectively detected by the diagnostic LST observation with minimal need for *a priori* information. On the other hand, remote sensing methods generally have temporal sampling constraints, both in frequency and period of record. This is particularly true for TIR-based approaches, which require clear-sky conditions for a valid retrieval of LST. Data assimilation strategies have been developed to integrate *in situ* and remote sensing data into LSMs to reduce impacts of input biases and model parameterization errors in the prognostic model, while improving temporal characteristics of the diagnostic estimates (Hain et al. 2012; Houborg et al. 2012; Sheffield et al. 2012). This will likely be the optimal solution for future global hydrologic monitoring efforts. In preparation, intercomparisons between prognostic and diagnostic moisture indicators provide insight regarding relative regional and seasonal performance of different modeling systems.

In this chapter we describe a multiscale thermal modeling system for retrieving surface energy fluxes, evapotranspiration (ET), and available soil moisture, fusing information from geostationary and polar orbiting TIR imaging sensors to provide data at both high spatial and temporal resolution. Applications for model output in monitoring drought and crop condition are presented. The value of field-scale remotely sensed water use and availability information in data-sparse regions of the world is emphasized, particularly for improving climate resilience in vulnerable agroecosystems. These remote sensing data can both serve as an independent check that physical processes are adequately represented in prognostic hydrologic and atmospheric modeling systems, and as a means for monitoring land surface conditions in greater spatial detail than can be captured by mesoscale and global LSMs and forecasting systems.

8.2 Multiscale TIR Energy Balance System

8.2.1 TIR Imaging Sensors

Chapter 3 discusses satellite-based TIR imaging systems currently available for routine mapping applications. These systems span a broad range in spatial resolution and image collection frequency.

Coarse spatial resolution data (~3–10 km) are collected over much of the global land surface at 15 min to hourly temporal resolution by an international network of geostationary satellites, used primarily to support weather forecasting. Data from these systems are most useful between ±60° in latitude—at higher latitudes, obliqueness in the sensor viewing angle causes elongated ground footprints and excessive along-path atmospheric attenuation of the surface signal. Geostationary TIR data are uniquely useful because the high temporal frequency facilitates applications using time-differential LST data, reducing sensitivity to errors in absolute temperature retrieval. They also provide the diurnal net radiation information required to accurately estimate daily total evaporative fluxes. Hourly or subhourly insolation datasets generated from geostationary data are critical for high-resolution energy balance and land surface modeling.

Moderate resolution (~1 km) data are collected globally on approximately a daily basis with the Advanced Very High Resolution Radiometer (AVHRR) series, the Moderate Resolution Imaging Spectroradiometer (MODIS), and more recently the Visible Infrared Imaging Radiometer Suite (VIIRS). The Sea and Land Surface Temperature Radiometer (SLSTR) to be launched in 2013 on the Sentinel-3 platform will provide 1-km resolution TIR imagery at two viewing angles. At high latitudes, these systems provide multiple over-passes per day and have the potential for being used like limited geostationary systems.

High-resolution (defined here as ~100 m or finer) TIR data are collected routinely only by a limited number of satellites. The Landsat series have been acquiring TIR images at 120-m resolution or finer since Landsat 4 was launched in 1982. The Thermal Infrared Imaging System (TIRS) on the Landsat Data Continuity Mission (LDCM) launched in February 2013, provides improved quality TIR data in two spectral channels at 100-m resolution with a 16-day revisit and global coverage. Multispectral TIR data from the Advanced Spaceborne Thermal Emission and Reflection Radiometer (ASTER) have a pixel resolution of 90 m, but acquisition must be scheduled and these data are not routinely available at any given site. The proposed Hyperspectral Infrared Imager (HyspIRI) mission would provide 60-m-resolution multispectral TIR data at 5-day intervals, in addition to hyperspectral imaging every 19 days. This mission would significantly advance remote sensing studies of surface fluxes and vegetation stress, particularly in water-limited regions where the monitoring of water use and availability is critical for managing the competing demands from agriculture, human and industrial use, and the environment.

8.2.2 ALEXI/DisALEXI Modeling System

The Atmosphere-Land Exchange Inverse (ALEXI) model (Anderson et al. 1997, 2007b) and associated flux disaggregation scheme (DisALEXI; Norman et al. 2003; Anderson et al. 2004) is a multiscale modeling system (Figure 8.1) that can be applied to any of the classes of satellite-based TIR data streams described in Section 8.2.1, depending on the resolution required by a given application. This system was specifically designed to minimize the need for ancillary meteorological data while maintaining a physically realistic representation of land–atmosphere exchange over a wide range in vegetation cover conditions. It is one of the few diagnostic LSMs designed specifically to exploit the high temporal resolution of observations provided by geostationary satellites.

Here we provide a brief overview of this modeling framework, and in the following section, we introduce image sharpening and data fusion techniques that have been developed to improve spatiotemporal resolution in ET products by combining information from multiple satellites and wavebands.

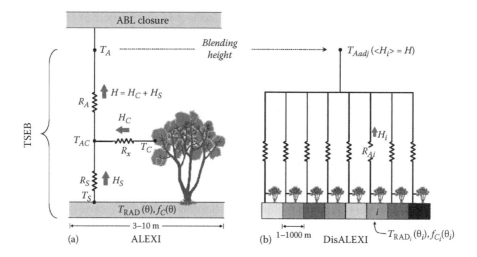

FIGURE 8.1
Schematic diagram representing the coupled (a) ALEXI and (b) DisALEXI modeling scheme, highlighting fluxes of sensible heat (H) from the soil and canopy (subscripts C and S) along gradients in temperature (T), and regulated by transport resistances R_A (aerodynamic), R_x (bulk leaf boundary layer), and R_S (soil surface boundary layer). DisALEXI uses the air temperature diagnosed by ALEXI near the blending height (T_A) to disaggregate 3- to 10-km ALEXI fluxes, given vegetation cover ($f_{C_i}(\theta_i)$) and directional surface radiometric temperature ($T_{RAD_i}(\theta_i)$) information derived from high-resolution remote sensing imagery at look angles θ_i. T_A is iteratively adjusted to ensure that DisALEXI H_i reaggregates on average to the ALEXI-derived H at the coarse ALEXI pixel scale.

8.2.2.1 Two-Source Energy Balance Model

Surface energy balance models estimate ET by partitioning the energy available at the land surface ($RN - G$, where RN is the net radiation and G is the soil heat conduction flux, in W m^{-2}) into turbulent fluxes of sensible and latent heating (H and λE, respectively, W m^{-2}):

$$RN - G = H + \lambda E \tag{8.1}$$

where λ is the latent heat of vaporization (J kg^{-1}) and E is the ET (kg s^{-1} m^{-2} or mm s^{-1}). Surface temperature plays a role in constraining RN, G, and H, while λE is generally computed either as a partial or total residual to Equation 8.1.

The land surface representation used in both ALEXI and DisALEXI is based on the series version of the Two-Source Energy Balance (TSEB) model of Norman et al. (1995), with subsequent revisions improving shortwave and longwave radiation exchange within the soil-canopy system and the soil-canopy energy exchange described by Kustas and Norman (1999a,b, 2000a,b). In the TSEB, the satellite-derived directional radiometric surface temperature, $T_{RAD}(\theta)$, is considered to be a composite of the soil and canopy temperatures, expressed as

$$T_{RAD}(\theta) \approx \left[f_C(\theta)T_C^4 + (1 - f_C(\theta))T_S^4 \right]^{1/4} \tag{8.2}$$

where T_C is the canopy temperature, T_S is the soil temperature, and $f_C(\theta)$ is the fractional vegetation cover observed at the radiometer view angle θ. For a canopy with a spherical leaf angle distribution and leaf area index (LAI) F,

$$f_C(\theta) = 1 - \exp\left(\frac{-0.5\Omega(\theta)F}{\cos\theta}\right) \tag{8.3}$$

where the factor $\Omega(\theta)$ indicates the degree to which vegetation is clumped from the view angle of the TIR sensor, as in row crops or sparsely vegetated shrubland canopies (Kustas and Norman 1999a, 2000b; Anderson et al. 2005). These component temperatures are used to compute the surface energy balance for the canopy and soil components of the combined land-surface system:

$$RN_S = H_S + \lambda E_S + G \tag{8.4}$$

$$RN_C = H_C + \lambda E_C \tag{8.5}$$

where RN_S is the net radiation at the soil surface and RN_C is the net radiation divergence in the vegetated canopy layer, H_C and H_S are the canopy and soil sensible heat flux, respectively, λE_C is the canopy transpiration rate, and λE_S is the soil evaporation.

Assuming the series resistance network schematically represented in Figure 8.1, H_C and H_S can be defined as a function of temperature differences, with

$$H_C = \rho C_P \frac{T_C - T_{AC}}{R_X} \tag{8.6}$$

and

$$H_S = \rho C_P \frac{T_S - T_{AC}}{R_S} \tag{8.7}$$

so that the total sensible heat flux $H = H_C + H_S$ is equal to

$$H = \rho C_P \frac{T_{AC} - T_A}{R_A} \tag{8.8}$$

where T_{AC} is an air temperature in the canopy air layer—closely related to the aerodynamic temperature, R_X is the total boundary layer resistance of the complete canopy of leaves, R_S is the resistance to sensible heat exchange from the soil surface, and R_A is the aerodynamic resistance. The original resistance formulations are described by Norman et al. (1995) with recent revisions described by Kustas and Norman (1999a,b, 2000a,b).

Given an estimate for canopy transpiration (described below), latent heat flux from the soil surface is solved as a residual in the soil energy balance equation:

$$\lambda E_S = RN_S - G - H_S \tag{8.9}$$

with G estimated as a fraction of the net radiation at the soil surface:

$$G = c_G RN_S \tag{8.10}$$

The value of the scaling factor c_G varies with time of day, due to the phase shift between G and RN_S over a diurnal cycle (Santanello and Friedl 2003).

In the original TSEB version (Norman et al. 1995), the Priestley–Taylor formula was used to initially estimate a potential rate for λE_C

$$\lambda E_C = \alpha_{PTC} f_G \frac{\Delta}{\Delta + \gamma} RN_C \tag{8.11}$$

where α_{PTC} is a variable quantity related to the so-called Priestley–Taylor coefficient (Priestley and Taylor 1972) but in this case defined exclusively for the canopy component (Tanner and Jury 1976). The variable α_{PTC} is normally set to an initial value ~1.3, f_G is the fraction of green vegetation, Δ is the slope of the saturation vapor pressure versus temperature curve, and γ is the psychrometric constant (~0.066 kPa C^{-1}). However, Equation 8.11 can result in nonphysical solutions, such as daytime condensation at the soil surface (i.e., $\lambda E_S < 0$) under conditions of moisture deficiency. In this case, the initial estimate of λE_C given by the Priestley–Taylor parameterization, which describes potential transpiration, overestimates actual transpiration. The higher λE_C leads to a cooler T_C and T_S must be accordingly larger to satisfy Equation 8.2. This drives H_S high, and the residual λE_S from Equation 8.9 becomes negative. If this condition is encountered by the TSEB scheme, α_{PTC} is iteratively reduced until $\lambda E_S \sim 0$ (expected for a dry soil surface). Validation experiments in a variety of natural and agro-ecosystems and a range in climate conditions have demonstrated that an initial value of $\alpha_{PTC} = 1.3$ provides reasonable estimates of system latent heat flux (Li et al. 2005; Anderson et al. 2005, 2007a; Cammalleri et al. 2012a), even under strongly advective conditions (Anderson et al. 2012b), although partitioning between canopy and soil evaporation may be less accurate in these extreme cases (Colaizzi et al. 2012).

Alternatively, a light-use efficiency (LUE) based version of the TSEB (Anderson et al. 2008) estimates coupled transpiration and carbon assimilation fluxes using an analytical expression for canopy conductance (Anderson et al. 2000), replacing the PT-based scheme for obtaining λE_C described above with a vapor pressure gradient-resistance equation. This has the benefit of both improving transpiration estimates and enabling carbon uptake mapping, but requires additional specification of nominal LUE by season and vegetation type. Work is underway to parameterize nominal LUE inputs to the TSEB using leaf chlorophyll retrievals obtained with a canopy reflectance inversion model (Houborg et al. 2011).

Kustas and Anderson (2009) discuss the advantages of the two-source approach used in the TSEB in comparison with simpler single-source models, which treat each pixel as a uniform surface. Generally, the TSEB outperforms single-source schemes under extreme conditions of very wet or dry surfaces, and very dense or sparse vegetation, accommodating this full range within a single modeling scheme. This is largely due to the nonunique relationship that exists between radiometric, $T_{RAD}(\theta)$, and aerodynamic temperature, T_{AC}, which most single-source modeling schemes are unable to accommodate unless an empirical adjustment based on vegetation density is included (Kustas et al. 2007). Nevertheless, even with these more sophisticated single-source schemes, they are found to have greater sensitivity to errors in key inputs, namely, $T_{RAD}(\theta)$ and leaf area F (Tang et al. 2011). Moreover, studies by Timmermans et al. (2007), Choi et al. (2009), and Tang et al. (2011) comparing TSEB and single-source schemes applied to the same thermal imagery determined that the largest discrepancies between the models—and the largest one-source model errors in comparison with flux observations—occurred in regions of partial vegetation cover. This is a condition where the radiometric–aerodynamic temperature relationship is particularly ill defined for the single-source models. Finally, the ability of two-source schemes like TSEB to partition ET into evaporation and transpiration components provides additional hydrologic information about the moisture status of the soil and canopy system, and about the vertical distribution of moisture in the soil profile (surface layer vs. root zone), which is not feasible with single-source approaches.

8.2.2.2 ALEXI

Regional-scale applications of the TSEB require specification of the near-surface air-temperature boundary condition across the domain (T_A in Figure 8.1), which is difficult to accomplish with adequate accuracy over large areas. Because near-surface atmospheric properties can be strongly coupled to local surface conditions, interpolation between synoptic weather network observations of T_A can result in large errors in the assumed surface-to-air temperature gradient ($T_{RAD} - T_A$) driving the modeled sensible heat flux. Therefore, for regional applications, the TSEB has been coupled with an atmospheric boundary layer (ABL) model to internally simulate land–atmosphere feedback on T_A. In the ALEXI model, the TSEB is applied at two times during the morning ABL growth phase (1 to 1.5 h after sunrise and before local noon), using radiometric temperature data obtained from a geostationary platform at spatial resolutions of ~3–10 km (Anderson et al. 1997). Energy closure over this interval is provided by a simple slab model of ABL development (McNaughton and Spriggs 1986), which relates the rise in air temperature in the mixed layer to the time-integrated influx of sensible heat from the land surface. As a result of this configuration, ALEXI uses only time-differential temperature signals, thereby minimizing flux errors due to absolute sensor calibration and atmospheric and spatial effects (Kustas et al. 2001). The primary radiometric signal is the morning surface temperature rise, while the ABL model component uses only the general slope (lapse rate) of the atmospheric temperature profile (Anderson et al. 1997), which is more reliably analyzed from synoptic radiosonde data than is the absolute temperature reference.

The ALEXI algorithm directly retrieves instantaneous fluxes at the two morning times bounding the ABL growth computation. To upscale to daily total fluxes, we employ a common approximation, assuming the evaporative fraction, $\lambda E/(RN - G)$, is relatively constant over the course of the day (Brutsaert and Sugita 1992). Hourly net radiation and soil heat flux are computed as described by Anderson et al. (2012b) and integrated to daily totals. Then daily latent heat flux is estimated from daily $RN - G$ using the evaporative fraction diagnosed by ALEXI at the second TSEB application time. Because TIR-based retrievals of LST are possible only under clear-sky conditions, a gap-filling algorithm is implemented to estimate ET during cloudy intervals, in this case conserving the ratio of actual-to-reference ET (f_{RET}). This ratio is computed for clear days using the Food and Agriculture Organization (FAO) Penman-Monteith formulation for reference ET (Allen et al. 1998). Time series of f_{RET} are interpolated to daily values using a Savitsky–Golay filter (Savitzky and Golay 1964) with a second-order smoothing polynomial. Finally, daily ET is estimated on cloudy days using gridded daily reference ET data and the smoothed f_{RET} values (Anderson et al. 2012b).

A complete ALEXI processing infrastructure has been developed to automatically ingest and pre-process all required input data, to execute the model, and to postprocess model output for visual display and use in other applications (Anderson et al. 2007b). The gap-filled model currently runs daily on a 10-km-resolution grid covering the continental

United States (CONUS) using data from the Geostationary Operational Environmental Satellites (GOES).

8.2.2.3 DisALEXI

ALEXI is constrained to operate at geostationary satellite resolutions of ~3–10 km, where time-differential LST observations are routinely available. Anderson et al. (2007a, 2012b) summarize ALEXI validation experiments employing a spatial flux disaggregation technique (DisALEXI; Norman et al. 2003; Anderson et al. 2004), which uses higher-resolution TIR imagery available from aircraft or polar orbiting systems such as Landsat, ASTER, or MODIS to downscale the geostationary satellite-based flux estimates to the scale of the flux measurement footprint (on the order of 30–1000 m). DisALEXI applies the TSEB in a spatially distributed mode over each ALEXI pixel, using an estimate of T_A computed at a nominal blending height by ALEXI along with LST and LAI information at the finer scale (Figure 8.1). T_A is iteratively adjusted until the average DisALEXI sensible heat flux within the coarse pixel matches the estimate from ALEXI, thereby ensuring reaggregation to a common flux basis established by the ALEXI analysis and, like ALEXI, avoiding the need for gridded air temperature observations as a principle boundary condition. Retrieved Landsat and MODIS fluxes can be upscaled to hourly and daily all-sky (clear and cloudy) flux time series using the gap-filling algorithms described above for ALEXI. Typical mean absolute differences with tower flux measurements of H and λE are about 15%–20% of the mean observed fluxes for 30-min averages, 10% at the daily time step, and ~5% for seasonal cumulative fluxes (Anderson et al. 2007a, 2012b; Cammalleri et al. 2012b).

8.2.2.4 Multiscale Mapping

Together, ALEXI/DisALEXI facilitates scalable flux and moisture stress mapping using TIR imagery from a combination of geostationary and polar orbiting satellites, zooming in from the global scale to sites of specific interest (Figure 8.2). There is utility in remotely sensed surface flux information at each of these spatial scales.

ET data at the global scale (from ALEXI) provide a diagnostic estimate of the evaporative component of the global water budget. This LST-driven energy balance estimate can serve as an independent check on other global ET products, determined using water balance or other approaches (e.g., Jimenez et al. 2011). At the continental scale, evaporative exchanges between the land and atmosphere are important in driving mesoscale weather patterns, so flux assessments at this scale may be useful for improving numerical weather forecasts. These regional maps also provide spatial context for finer-scale applications.

At the state or basin scale, water managers can benefit from ET maps generated with MODIS/VIIRS (DisALEXI) at 1-km resolution for hydrologic modeling and regional water use planning.

Fine resolution (<100 m) ET is critical for most agricultural applications, where decision makers need to know water use and crop condition on a field-by-field basis. The Landsat archive currently contains 30 years of TIR data, which can be mined to study changes in water use patterns at the scale of human intervention in the natural water cycle. With Landsat we can see how ET and water use has changed over the past three decades—as urban areas have expanded, and as land use has been converted from forest to agriculture and other uses (Anderson et al. 2012a).

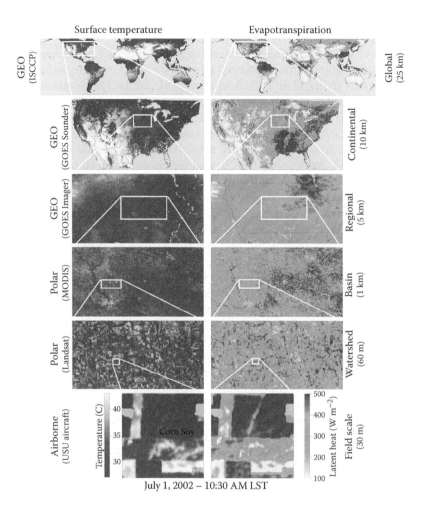

FIGURE 8.2
(See color insert.) Multiscale ET maps for July 1, 2002, produced with ALEXI/DisALEXI using surface temperature data from aircraft (30-m resolution), Landsat (60 m), MODIS (1 km), GOES Imager (5 km), and GOES Sounder (10 km), and an international suite of geostationary satellites (25 km), zooming into the Walnut Creek Watershed near Ames, Iowa, site of the SMEX02 Soil Moisture Experiment. The continental- and global-scale ET maps are 14-day composites of clear-sky model estimates.

8.3 Mapping ET at High Spatial and Temporal Resolution

To study time-continuous water use at the Landsat scale, methods must be employed to estimate daily ET between clear-sky Landsat overpasses. With a single Landsat, we receive a snapshot of surface conditions once every 8–16 days or longer depending on site location and cloud cover conditions. One possible approach to daily ET mapping is to fuse information from multiple platforms, merging in coarser spatial frequency structure developed from TIR data collected with higher temporal frequency systems. An example of such an ET data fusion process is described below, along with ancillary tools for sharpening the spatial resolution of TIR band data to that of higher resolution shortwave sensors on the same satellite platform.

8.3.1　ET Data Fusion

As described above, seasonal Landsat-scale estimates of daily ET are commonly obtained by temporally interpolating some conserved moisture-related quantity, such as the ratio of actual-to-reference ET (Allen et al. 2011; Anderson et al. 2012b). This method clearly suffers when discontinuities in ET signals arise owing to impulsive hydrological inputs (rainfall or irrigation) that occur between consecutive clear-sky Landsat TIR images.

Data fusion techniques can be used to integrate multi-sensor ET estimates in order to better capture changes in surface moisture status that occur at coarser scales (e.g., 1 km using daily MODIS ET retrievals). The Spatial Temporal Adaptive Reflectance Fusion Model (STARFM; Gao et al. 2006) compares one or more pairs of Landsat/MODIS-derived maps, collected on the same day, to predict maps at Landsat-scale on other MODIS observation dates, resulting in time-continuous daily Landsat-scale maps. The STARFM prediction for each Landsat pixel is based on a weighted function of the available Landsat and MODIS data reflecting spatial and spectral similarities with neighboring pixels. The flowchart in Figure 8.3 shows the processing steps involved in fusing daily (gap filled) ET at MODIS scale with periodic high-resolution retrievals from Landsat.

STARFM has been successfully applied over rainfed and irrigated agricultural areas in the United States to fuse ET maps generated with DisALEXI using TIR and LAI data from Landsat and MODIS. As discussed in Section 8.2.2.3, DisALEXI maps are constrained to reaggregate to a common basis flux distribution developed with ALEXI, thereby facilitating the fusion process which assumes reasonable consistency between products at different scales. The performance of STARFM has been evaluated in comparison with flux tower observations at subfield scales, and with benchmark cases obtained by temporally interpolating Landsat estimates using a spline function (Landsat-only case). At an irrigated site in the Texas Panhandle (United States) during the Bushland Evapotranspiration and Agricultural Remote Sensing Experiment of 2008 (BEAREX08; Evett et al. 2012), Landsat-MODIS yielded seasonal cumulative flux errors of 4.5% of the observed flux in comparison with 12% obtained with Landsat only. Improvement was more notable over rainfed sites, where moisture dynamics due to rainfall are better captured at the 1-km MODIS scale, than at irrigated sites where water applications are at substantially subpixel scale. In a fusion experiment conducted with

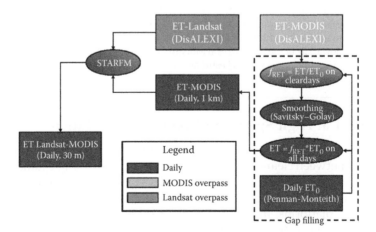

FIGURE 8.3
Flowchart describing the processing steps used to fuse daily (gap filled) ET evaluations at MODIS scale (1-km resolution) with periodic retrievals at 30 m generated with sharpened Landsat TIR imagery.

satellite imagery and flux observations collected over corn and soybean fields in central Iowa during the Soil Moisture Atmospheric Coupling Experiment of 2002 (SMACEX02; Kustas et al. 2005), seasonal errors were 4% for Landsat-MODIS and 8.5% for Landsat only (Cammalleri et al. in press). Figure 8.4 compares Landsat-only and fused Landsat-MODIS reconstructions of daily ET with flux observations in a soybean field. Incorporation of MODIS data at this site improved root-mean-square errors in daily ET retrievals from 1.1 to 0.6 mm day^{-1} and reduced errors in cumulative flux from 10% to 3% of the observed value. The impact of the MODIS data is most notable following a major rain event that occurred around day of year (DOY) 190, between Landsat overpasses on DOY 182 and 214. The Landsat-only solution has no information regarding the discontinuous increase in f_{RET} between these dates, but this signal is effectively captured by the daily MODIS TIR-based retrievals.

Figure 8.5 shows an example of fused ET maps obtained for the BEAREX08 irrigated agricultural site. The figure focuses on a time window when a rainfall event occurred (during the night between DOY 210 and 211) between Landsat overpasses. In this example, both MODIS (bottom row) and Landsat (middle row, first and last panels) estimates were obtained using the DisALEXI normalization procedure based on ALEXI daily ET maps at 10-km resolution over CONUS. Landsat-like maps between DOY 203 and 219 (middle row, except first and last panels) were obtained by applying STARFM. Images in the upper row zoom in on the region over the experimental site, demonstrating the capabilities of STARFM to combine spatial details captured by the Landsat data with the temporal dynamics observed in the MODIS datastream. Although STARFM was originally designed to fuse surface reflectance fields, it appears to have utility for high-resolution ET mapping as well.

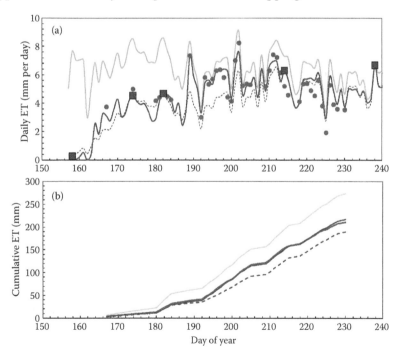

FIGURE 8.4

(See color insert.) Comparison of Landsat-only (red dashed lines) and fused Landsat-MODIS (solid red lines) reconstructions of (a) daily ET and (b) cumulative ET with flux observations (blue dots/line) in a soybean field (site 161) from the SMACEX02 field campaign in central Iowa (Cammalleri et al. in press). Also shown are reference ET (gray lines) and retrieved values on Landsat overpass dates (red boxes).

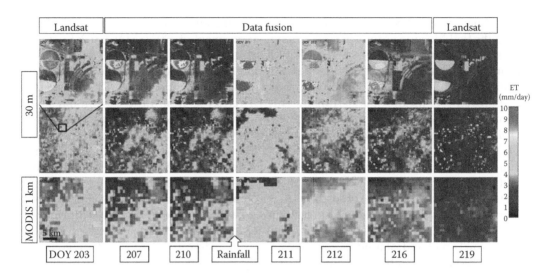

FIGURE 8.5
(See color insert.) Example of Landsat/MODIS ET data fusion, showing maps of daily ET from DisALEXI using MODIS at 1-km resolution (bottom row) and from the STARFM data fusion algorithm, fusing information from DisALEXI using Landsat TIR sharpened to 30-m resolution (middle row). Upper row shows in detail the maps obtained for BEAREX08 experimental fields.

8.3.2 Thermal Sharpening

Because of instrumental constraints, TIR imaging systems typically operate at coarser spatial resolution than that of shorter wave visible (vis) and near-infrared (NIR) band sensors carried on the same satellite platform. For example, MODIS collects vis/NIR data at 250 m and TIR at 1 km, while Landsat carries shortwave band instruments at 30-m resolution and a TIR imager at 60–120 m. The higher shortwave resolutions are more beneficial for many types of management applications, particularly over highly fragmented agricultural landscapes. In an attempt to improve the spatial resolution of TIR imagery, sharpening techniques have been developed exploiting the strong inverse relationship that typically exists between LST and vegetation indices (VIs) derived from vis/NIR data (e.g., Goward et al. 1985, 2002; Hope and McDowell 1992; Smith and Choudhury 1991; Prihodko and Goward 1997; Karnieli et al. 2010). This relationship reflects the fact that denser vegetation cover tends to be correlated with lower radiometric temperatures, due to cooling of the land surface by transpiration. Kustas et al. (2003) presented a simple generalized TIR image sharpening algorithm (TsHARP) using functional relationships between LST and the normalized difference vegetation index (NDVI) developed at the coarser TIR pixel resolution, and then applied at the finer shortwave resolution. While this simple technique recovers high spatial frequency structure in LST over many landscapes (Anderson et al. 2004; Agam et al. 2007b), success has been found to be limited in areas where NDVI and LST are not well correlated, for example, because of irrigation (Agam et al. 2007a, 2008) or the presence of senesced vegetation (Merlin et al. 2010).

Gao et al. (2012) improved on the TsHARP concept with a data mining sharpener (DMS) approach designed for application over widely varying landscapes. All shortwave band reflectances are considered in the DMS in order to better capture variations in LST that cannot be detected by simple VIs. The DMS approach builds a relationship between temperature (T) and reflectances (R) using a regression tree method, which can be applied to a broader range of landscapes and conditions and at different spatial resolutions. The spatial variation of the *T–R* relationship is described by a local model using a moving-window technique. The performance of the DMS and TsHARP approaches was evaluated at three sites including a rainfed corn and soybean production region in central Iowa (SMACEX02 site), an irrigated agricultural region growing primarily wheat, cotton, corn, and sorghum in the Texas High Plains (BEAREX08 site), and a complex heterogeneous naturally vegetated landscape in Alaska containing a variety of forests, open grass and shrublands, and open water surfaces. In each case, the DMS outperformed TsHARP, particularly in areas containing irrigated crops, water bodies, thin clouds, or rugged terrain (Figure 8.6).

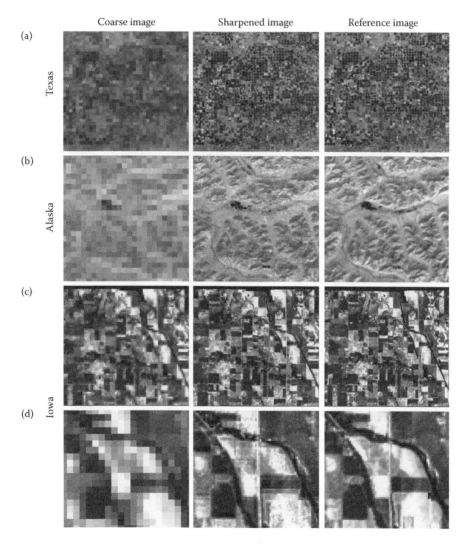

FIGURE 8.6

Examples of thermal image sharpening using the DMS tool applied to (a) Landsat data over the BEAREX08 irrigated agricultural site (Texas), aggregated to 960 m and sharpened to 60 m; (b) Landsat data over a site of high topographic relief (Alaska), aggregated to 960 m and sharpened to 60 m; (c and d) Aircraft imagery over the SMACEX02 rainfed agricultural site (Iowa), aggregated to 120 m and sharpened to 30 (Panel [d] shows detail). Left column shows coarse resolution image, middle column shows sharpened image, and right column shows observed temperature field at the sharpened resolution.

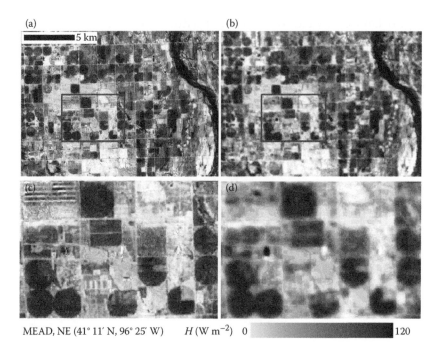

MEAD, NE (41° 11′ N, 96° 25′ W) H (W m^{-2})   0 ▭ 120

FIGURE 8.7
Sensible heat flux maps generated with DisALEXI over an irrigated and rainfed agricultural landscape near Mead, Nebraska, for September 12, 2003, at the time of the Landsat-5 overpass. Panels (b) and (d) show flux distributions computed using Landsat-5 TIR imagery at the native 120-m resolution, and panels (a) and (c) demonstrate the impact of sharpening to 30 m using the DMS. Panels (c) and (d) zoom in on subregion indicated by box in (a) and (b).

Figure 8.7 demonstrates the impact of DMS sharpening on retrieved sensible heat flux maps, generated using Landsat-5 imagery over a rainfed and irrigated agricultural landscape outside of Mead, Nebraska.

While thermal sharpening is a valuable tool for enhancing spatial information content in TIR imagery, it does not replace the need for TIR data collection at the subfield scale as currently provided by Landsat. Attempts to sharpen MODIS-scale TIR imagery at 1-km resolution down to Landsat scales can provide poor results, particularly over scenes with strong subpixel moisture variability that is uncorrelated with vegetation cover (Agam et al. 2007a, 2008; Anderson et al. 2011b; Gao et al. 2012).

8.4 Applications

Section 8.3 described techniques for generating daily remote sensing fields of ET at resolutions of 30 m to 10 km, covering areas from watershed to continental scales. Here we describe examples of how diagnostic ET information at multiple scales can be applied for purposes of monitoring soil moisture, drought, crop condition, and water use. In particular, we focus on synergistic use of diagnostic flux and moisture information derived from thermal remote sensing and prognostic evaluations from land surface modeling.

8.4.1 Monitoring Drought

Spatial and temporal variations in instantaneous ET at the continental scale are primarily due to variability in moisture availability (antecedent precipitation), radiative forcing (cloud cover, sun angle), vegetation amount, and local atmospheric conditions such as air temperature, wind speed, and vapor pressure deficit. Potential ET describes the evaporation rate expected when soil moisture is nonlimiting, ideally capturing response to all other forcing variables. To isolate effects due to spatially varying soil moisture availability, a simple Evaporative Stress Index (ESI) has been developed, describing the departure of model flux estimates of ET from a reference ET rate expected under nonmoisture limiting conditions (Anderson et al. 2007c, 2011a). The ESI is computed as standardized temporal anomalies in the f_{RET} ratio used in the daily interpolation algorithm described in Section 8.2.2.2 and shows good correspondence with standard drought metrics and with patterns of antecedent precipitation, but at significantly higher spatial resolution due to limited reliance on ground observations. This ratio has a value close to 1 when there is ample moisture/no stress, and a value of 0 when ET has been cut off because of stress-induced stomatal closure and/or complete drying of the soil surface. It therefore serves as a valuable proxy indicator for available soil moisture (Hain et al. 2009, 2012; Yilmaz et al. 2012; Anderson et al. 2012c). Where there is vegetation, the proxy reflects information over the full root zone, while it reflects surface moisture conditions (nominally the top 5 cm of soil profile) in areas of very sparse vegetation.

The ESI is particularly well suited for capturing rapid onset drought events, sometimes referred to as "flash droughts." Although drought is often thought of as a slowly developing climate phenomenon that can take several months or even years to reach its maximum intensity, drought onset can be very rapid if extreme atmospheric anomalies persist for several weeks. Vegetation health can deteriorate very quickly if moderate precipitation deficits are accompanied by an intense heat wave, strong winds, and sunny skies, as the enhanced ET quickly depletes root zone moisture (Mozny et al. 2012). The ALEXI energy balance scheme incorporates these important driving variables in its evaluation of ET. Furthermore, LST is a rapid response variable, which reflects increases in soil and canopy temperatures as soil moisture deficits and vegetation stress develop, and often prior to measurable reductions in shortwave vegetation indices (Moran 2003).

Diagnostic drought indicators like the ESI provide an independent check on prognostic precipitation-based indicators and in combination may improve confidence in signals of emerging drought—particularly during periods of rapid onset. Figure 8.8 compares the time evolution of drought classifications in the US Drought Monitor (USDM; Svoboda et al. 2002) with maps of ESI and anomalies in an ensemble averaged SM product (SM_{AV}) generated with the North American Land Data Assimilation System (NLDAS) LSM modeling suite (Xia et al. 2012a,c), sampled at monthly timesteps during the 2011 growing season. Also shown as independent verification are qualitative top-soil moisture observations (SM_{NASS_top}) collected by field representatives working with the National Agricultural Statistics Service (NASS). Given the importance of available soil moisture to crop yield, NASS has developed an extensive and long-term (2002–present) database of county-level observations of phenological development stage and condition of several major crop types, as well as an estimate of the soil wetness at two depths (top- and subsoil) during the growing season. These surveys are conducted weekly in crop growing countries from early April until late November and provide qualitative information that is grouped in five crop condition categories (very poor to excellent) and four soil moisture categories (very short to surplus).

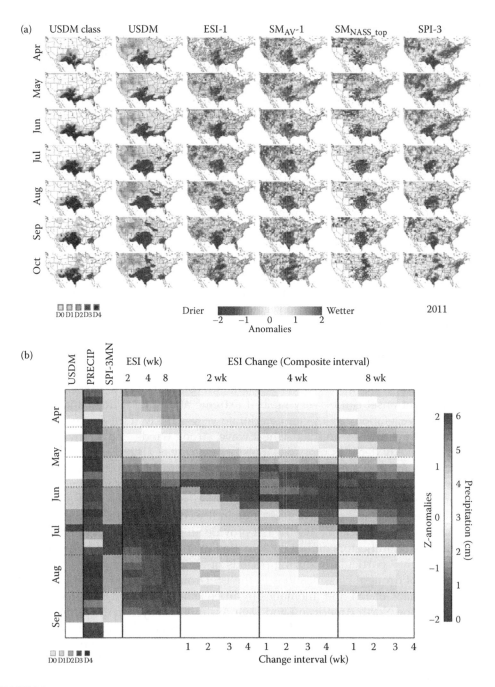

FIGURE 8.8
(See color insert.) (a) Monthly standardized anomalies in the USDM drought classes (USDM, 2nd column), the 1-month ESI composite (ESI-1), 1-month ensemble average SM anomalies (SMAV-1), anomalies in NASS county-level topsoil moisture observations (SMNASS_top), anomalies in 3-month standardized precipitation index (SPI), as well as USDM drought classes for the week closest to the end of each month (first column). (b) Drought evolution across eastern Oklahoma and western Arkansas during 2011. The USDM drought class is shown in the first column with the accumulated precipitation (cm) and 3-month SPI anomaly shown in columns 2 and 3. ESI for 2-, 4-, and 8-week composite periods are shown in columns 4–6. ΔESI for 1-, 2-, 3-, and 4-week differencing intervals are shown in columns 7–10, 11–14, and 15–18 for the ESI 2-, 4-, and 8-week composite periods, respectively.

The severe Texas drought of 2011 is captured with good spatial detail by both ESI and SM_{AV} (Figure 8.8a). In addition, a rapid onset or flash drought event occurred in 2011, beginning in June over Arkansas and spreading to Missouri and states to the north in July. While precipitation deficits were observed, they were not extreme. Rather, the spread of drought was fueled by strong winds and high temperatures that lead to higher ET demands and rapid depletion of soil moisture conditions. Advance warning of this expansion was indicated in products mapping ESI changes (ΔESI) from May to June and may have allowed for an earlier response. Figure 8.8b highlights ESI changes averaged over ALEXI grid points located in eastern Oklahoma and western Arkansas. Deteriorating conditions are first detected in 2-week changes near the end of May—a full month before the area was classified as D0 (abnormally dry). Routine generation of drought index change products, such as those shown in Figure 8.8b, could benefit state-of-the-art USDM classifications by allowing earlier identification of emerging areas of drought development.

8.4.2 Monitoring Crop Condition and Yield

Timely agricultural drought detection, monitoring, and prediction are important as limited water supplies in the root zone during the growing season trigger stress, which, in turn, leads to reduced yield production and food scarcity. Crop growth models simulate crop yield as a function of soil properties, agricultural practices, and weather conditions. Among these the amount of available heat and soil moisture has the greatest impact on the crop growth. ET is a direct indicator of the water used by the plants. Deviations from potential ET (PET) indicate that the plants are experiencing stress related to water deficiencies or other stressors. Timing (i.e., occurrence during the growing season) and duration of this deficit determine effects on crop health and consequently the yield, because crop susceptibility to drought varies over the growing season. Even short periods of intense water stress can lead to significant yield loss and reduced grain quality if they occur during sensitive stages in crop development such as emergence, pollination, and grain filling (e.g., Barnabás et al. 2008; Rotter and Geijn 1999; Saini and Westgate 1999; Li et al. 2009; Mishra and Cherkauer 2010; Pradhan et al. 2012). The simultaneous occurrence of depleted soil moisture and heat stress has an even larger detrimental effect on plant growth, reproduction, and yield than when either of these stresses occurs individually (e.g., Jiang and Huang 2001; Rizhsky et al. 2002; Mittler 2006; Prasad et al. 2011; Kebede et al. 2012).

Studies are underway to analyze correlations between spatiotemporal stress information conveyed by the ESI with monthly crop conditions and at-harvest yield datasets collected by NASS. Otkin et al. (in press) found good correlations between plumes of high ΔESI associated with rapid drought onset events, as depicted in Figure 8.8, and deterioration in crop condition in impacted areas. Monthly ESI for April through September shows reasonable correlation with state-level yield estimates for corn and soybean for 2002–2011. Predictive power increases through the growing season, with maximum correlation with annual yield observed in ESI maps from July to August.

Incorporating the thermal sharpening and data fusion techniques described in Section 8.3, daily time series required to compute ESI anomalies could be routinely generated over targeted areas at up to 30-m resolution. This field-scale information could significantly benefit drought and crop condition assessments because the response of different crop types to moisture stress can be resolved at these scales. Reliable precipitation data at these spatial scales are particularly difficult to obtain, underscoring the value of diagnostic TIR-based monitoring techniques.

8.4.3 Monitoring Water Use and Availability in Data-Sparse Regions

While capabilities of TIR remote sensing to provide water availability and drought information have been demonstrated over the United States and Europe (Anderson et al. 2011b), these techniques may be of even more value in data-sparse regions of the world where ground-based measurements of precipitation and surface fluxes are limited. ALEXI domains at 3- to 6-km resolution have been established over North Africa using land surface products generated by the Land Surface Analysis Satellite Applications Facility (http://landsaf.meteo.pt/) from imagery collected with the Meteosat Second Generation (MSG) geostationary satellite operated by EUMETSAT. The goal of this project is to combine prognostic hydrologic modeling with remotely sensed ET to provide improved information for water management and climate adaptation within the Nile Basin and the broader Middle East–North African (MENA) region. The remotely sensed ET from ALEXI/DisALEXI serves as an independent estimate of actual water use and will provide finer-scale information about water distribution and consumption than can be obtained with an LSM. It also provides a valuable independent check on prognostic estimates in regions where verification data are scarce or nonexistent.

Diagnostic models estimate energy balance fluxes without requiring ancillary datasets describing the surface and subsurface water conditions and management, which is particularly advantageous over areas where precipitation is not the only water source for ET. Figure 8.9 compares ALEXI ET with simulations from the prognostic Noah LSM (Ek et al. 2003) over the Nile River Basin in February 2009. Specifically, the LST inputs to ALEXI

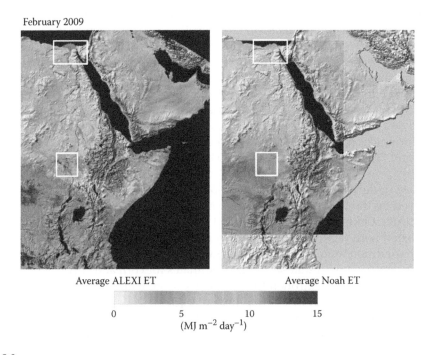

FIGURE 8.9
(See color insert.) Comparison of average ET over the Nile River basin over the month of February 2009 as estimated by the (left) diagnostic ALEXI model and the (right) prognostic Noah LSM. Boxes indicate the locations of the (top) irrigated Nile Delta and the (bottom) Sudd wetland.

capture enhanced ET over the Sudd wetlands in South Sudan and in intensively irrigated regions such as the Nile Delta. These features are missed in the Noah simulations, which had no *a priori* knowledge about irrigation or water table. For basin-scale hydrologic modeling, ALEXI can provide an objective estimate of water diverted within the basin in support of irrigated agriculture—information that is generally not publicly accessible through other means. Given limitations in availability of ground truth, analyses of temporal and spatial consistency between independently derived moisture products is an alternative to *in situ* verification. The high level of consistency between these two independent ET products over most of the modeling domain improves confidence in the results from both models.

The remote sensing data can also provide a level spatial detail that is not readily obtainable through prognostic modeling, particularly in data-sparse regions. TIR-based diagnostic models have the unique ability to solve for the surface fluxes at very fine scales (i.e., <100 m), while prognostic models generally use much coarser resolution atmospheric and precipitation forcing data (i.e., 5–20 km). Using Landsat TIR imagery sharpened to 30 m, we can study patterns of water use and availability at the field scale—a clear benefit for decision making regarding irrigation system management and land use planning.

These types of hydrologic mapping technologies have particular relevance for rural subsistence agricultural communities that are experiencing impacts of climate change. Such communities typically do not have adequate adaptive capacity survive prolonged periods of drought and receive insufficient advanced warning of impending flooding to take proper precautions. Better decision making will require better information about climate and land surface conditions—spatially distributed, and at scales of human activity (the field scale). Combined with land use/land cover, soils, and digital elevation data, hydrologic modeling and remote sensing ET tools can provide valuable data products for decision makers at local, regional, and national scales: spatially distributed estimates of high soil erosion potential and sediment load cycles, seasonal water use by various crops and agricultural practices, identification of areas of high and low drought susceptibility, estimates of water diverted for irrigated agriculture, and near-real-time monitors for use in drought and flood early warning (Zaitchik et al. 2012). Combined with climate model scenarios, downscaled to the county and the village scale, these tools can be used to investigate impacts of changing climate conditions and to test system capacity to weather expected shocks. Land use scenarios may allow identification of optimal farming locations given hillslope, water availability, local microclimate, and soil fertility conditions, as well as proximity to market infrastructure.

8.5 Future Work and Conclusions

This chapter describes a multisensor, multiscale approach to mapping ET and surface energy fluxes using thermal remote sensing data from both geostationary and polar orbiting satellite platforms. Global implementation of the ALEXI modeling system is an expansive endeavor, requiring collation and cross-calibration of imagery from multiple geostationary platforms operated by many different countries. Fortunately, archives of global geostationary data are now being constructed to support global monitoring applications. Two projects supporting this effort include the International Satellite Cloud Climatology Project (ISCCP) B1 data rescue project (Knapp 2008) instigated by NOAA and the Geoland2 project under the European Global Monitoring for Environment and Security (GMES) initiative (Lacaze et al. 2010). An ideal global geostationary dataset would include both LST and insolation at hourly timesteps and 5-km spatial resolution and would be updated in near real time for operational monitoring. Applications for global-scale diagnostic assessments of surface fluxes using ALEXI are discussed further in Chapter 19.

For field-scale monitoring, we are limited by the availability and temporal frequency of global high-resolution (100 m) TIR imaging systems. Starting in 2013, LDCM (Landsat 8) will provide high-quality TIR data globally with a 16-day revisit, supplementing datasets currently collected by Landsat 7 that have provided incomplete scene coverage due to a scan line corrector failure since 2003. However, the future of the Landsat program is currently in question. LDCM has a design life of 5 years, and currently there is no funding in place for construction and launch of a follow-up mission. We face a possible impending gap in global TIR coverage and water use estimates at this critical scale. The proposed HyspIRI sensor suite could help to alleviate this gap; however, this mission has not yet been approved for launch. Current uncertainty regarding long-term availability of high-resolution TIR data is significantly hindering adoption of satellite-based ET monitoring tools for operational management applications. Transitioning of the Landsat program from research to operational status, with a long-term platform acquisition and launch plan similar to that applied to geostationary weather satellites, would encourage the water resource management and agricultural sectors to actively integrate satellite-based ET retrievals into their operational management plans.

Future improvements to the ALEXI/DisALEXI modeling system include implementation of a two-source TIR-based energy balance scheme describing fluxes over snow-covered landscapes (Kongoli et al. 2013). This will extend application to year-round over much of the global land surface. At high latitudes, where geostationary view angles are too oblique to support accurate LST retrieval, alternative time-differential methodologies (e.g., Norman et al. 2000) are being explored using data from polar orbiting instruments like MODIS and AVHRR, which provide multiple overpasses per day near the poles. Finally, temporal gaps in LST-derived surface records due to cloud cover can be addressed in part by integrating microwave-based soil moisture information, which can be obtained under both clear and cloudy conditions, albeit at lower spatial resolutions. Chapter 14 describes efforts to combine periodic moisture information from TIR and microwave remote sensing retrievals with time-continuous assessments from prognostic LSMs through data assimilation. For example, Hain et al. (2012) showed that joint assimilation of TIR f_{RET} (from ALEXI) and microwave soil moisture into the Noah LSM provides better soil moisture estimates than does either retrieval method (TIR or microwave) in isolation. This multi-pronged approach to surface flux modeling could serve to maximize both spatial and temporal sampling of surface moisture conditions and would provide additional hydrologic information such as runoff, streamflow, and groundwater recharge.

Acknowledgments

This work was supported in part by funding from NOAA/CTB (Grant GC09-236) and NASA (NNH11ZDA001N-WATER).

References

Agam, N., W. P. Kustas, M. C. Anderson, F. Li, and P. D. Colaizzi (2007a). Utility of thermal sharpening over Texas High Plains irrigated agricultural fields. *J. Geophys. Res., 112*, D19110, doi:10.1029/2007JD008407.

Agam, N., W. P. Kustas, M. C. Anderson, F. Li, and P. D. Colaizzi (2008). Utility of thermal image sharpening for monitoring field-scale evapotranspiration over rainfed and irrigated agricultural regions. *J. Geophys. Res. Lett., 35*, doi:10.1029/2007GL032195.

Agam, N., W. P. Kustas, M. C. Anderson, F. Li, and C. M. U. Neale (2007b). A vegetation index based technique for spatial sharpening of thermal imagery. *Remote Sens. Environ., 107*, 545–558.

Allen, R. G., A. Irmak, R. Trezza, J. M. H. Hendrickx, W. G. M. Bastiaanssen, and J. H. Kjaersgaard (2011). Satellite-based ET estimation in agriculture using SEBAL and METRIC. *Hydrol. Process., 25*, 4011–4027.

Allen, R. G., L. S. Pereira, D. Raes, and M. Smith (1998). Crop evapotranspiration: Guidelines for computing crop water requirements, Irrigation and Drainage Paper 56. Food and Agriculture Organization, Rome, Italy, p. 300.

Anderson, M. C., R. G. Allen, A. Morse, and W. P. Kustas (2012a). Use of Landsat thermal imagery in monitoring evapotranspiration and managing water resources. *Remote Sens. Environ., 122*, 50–65.

Anderson, M. C., C. R. Hain, B. Wardlow, J. R. Mecikalski, and W. P. Kustas (2011a). Evaluation of a drought index based on thermal remote sensing of evapotranspiration over the continental U.S. *J. Clim., 24*, 2025–2044.

Anderson, M. C., W. P. Kustas, J. G. Alfieri, C. R. Hain, J. H. Prueger, S. R. Evett, P. D. Colaizzi, T. A. Howell, and J. L. Chavez (2012b). Mapping daily evapotranspiration at Landsat spatial scales during the BEAREX'08 field campaign. *Adv. Water Resour., 50*, 162–177.

Anderson, M. C., W. P. Kustas, and J. M. Norman (2007a). Upscaling flux observations from local to continental scales using thermal remote sensing. *Agron. J., 99*, 240–254.

Anderson, M. C., W. P. Kustas, J. M. Norman, C. R. Hain, J. R. Mecikalski, L. Schultz, M. P. Gonzalez-Dugo et al. (2011b). Mapping daily evapotranspiration at field to continental scales using geostationary and polar orbiting satellite imagery. *Hydrol. Earth Syst. Sci., 15*, 223–239.

Anderson, M. C., J. M. Norman, G. R. Diak, W. P. Kustas, and J. R. Mecikalski (1997). A two-source time-integrated model for estimating surface fluxes using thermal infrared remote sensing. *Remote Sens. Environ., 60*, 195–216.

Anderson, M. C., J. M. Norman, W. P. Kustas, R. Houborg, P. J. Starks, and N. Agam (2008). A thermal-based remote sensing technique for routine mapping of land-surface carbon, water and energy fluxes from field to regional scales. *Remote Sens. Environ., 112*, 4227–4241.

Anderson, M. C., J. M. Norman, W. P. Kustas, F. Li, J. H. Prueger, and J. M. Mecikalski (2005). Effects of vegetation clumping on two-source model estimates of surface energy fluxes from an agricultural landscape during SMACEX. *J. Hydrometeorol., 6*, 892–909.

Anderson, M. C., J. M. Norman, J. R. Mecikalski, J. A. Otkin, and W. P. Kustas (2007b). A climatological study of evapotranspiration and moisture stress across the continental U.S. based on thermal remote sensing: I. Model formulation. *J. Geophys. Res., 112*, D10117, doi:10110.11029/12006JD007506.

Anderson, M. C., J. M. Norman, J. R. Mecikalski, J. A. Otkin, and W. P. Kustas (2007c). A climatological study of evapotranspiration and moisture stress across the continental U.S. based on thermal remote sensing: II. Surface moisture climatology. *J. Geophys. Res., 112*, D11112, doi:11110.11029/12006JD007507.

Anderson, M. C., J. M. Norman, J. R. Mecikalski, R. D. Torn, W. P. Kustas, and J. B. Basara (2004). A multiscale remote sensing model for disaggregating regional fluxes to micrometeorological scales. *J. Hydrometeor., 5*, 343–363.

Anderson, M. C., J. M. Norman, T. P. Meyers, and G. R. Diak (2000). An analytical model for estimating canopy transpiration and carbon assimilation fluxes based on canopy light-use efficiency. *Agric. For. Meteorol., 101*, 265–289.

Anderson, W. B., B. F. Zaitchik, C. R. Hain, M. C. Anderson, M. T. Yilmaz, J. R. Mecikalski, and L. Schultz (2012c). Towards an integrated soil moisture drought monitor for East Africa. *Hydrol. Earth Syst. Sci., 16*, 2893–2913.

Arsenault, K. R., P. R. Houser, and D. A. Matthews (2003). Incorporating the Land Data Assimilation System into Water Resource Management and Decision Support Systems. *Earth Obs. Mag., 12*, 24–27.

Barnabás, B., K. Jäger, and A. Fehér (2008). The effect of drought and heat stress on reproductive processes in cereals. *Plant, Cell Environ., 31*, 11–38.

Betts, A. K., F. Chen, K. E. Mitchell, and Z. I. Janjic (1997). Assessment of the land surface and boundary layer models in two operational versions of the NCEP Eta model using FIFE data. *Mon. Weather Rev., 125*, 2896–2916.

Brutsaert, W., and M. Sugita (1992). Application of self-preservation in the diurnal evolution of the surface energy budget to determine daily evaporation. *J. Geophys. Res., 97*, 18,377–18,382.

Cammalleri, C., M. C. Anderson, G. Ciraolo, G. D'Urso, W. P. Kustas, G. La Loggia, and M. Minacapilli (2012). Applications of a remote sensing-based two-source energy balance algorithm for mapping surface fluxes without in situ air temperature observations. *Remote Sens. Environ., 124*, 502–515.

Cammalleri, C., M. C. Anderson, and F. Gao (in press). Fusion of Landsat-MODIS flux fields for mapping daily evapotranspiration at subfield scales. *Water Resources Res.*

Choi, M., W. P. Kustas, M. C. Anderson, R. G. Allen, F. Li, and J. H. Kjaersgaard (2009). An intercomparison of three remote sensing-based surface energy balance algorithms over a corn and soybean production region (Iowa, U.S.) during SMACEX. *Agric. For. Meteorol., 149*, 2082–2097.

Colaizzi, P. D., W. P. Kustas, M. C. Anderson, P. H. Gowda, S. A. O'Shaughnessy, T. A. Howell, and S. R. Evett (2012). Two-source model estimates of evapotranspiration using component and composite surface temperatures. *Adv. Water Resour., 50*, 134–151.

Ek, M. B., K. E. Mitchell, Y. Lin, E. Rogers, P. Grunmann, V. Koren, G. Gayno, and J. D. Tarpley (2003). Implementation of Noah land surface model advances in the National Centers for Environmental Prediction operational mesoscale Eta model. *J. Geophys. Res., 108*(D22), doi:10.1029/2002JD003296, 8851.

Evett, S. R., W. P. Kustas, P. H. Gowda, M. A. Anderson, J. H. Prueger, and T. A. Howell (2012). Overview of the Bushland Evapotranspiration and Agricultural Remote sensing EXperiment 2008 (BEAREX08): A field experiment evaluating methods quantifying ET at multiple scales, *Adv. Water Resour., 50*, 4–19.

Gao, F., W. P. Kustas, and M. C. Anderson (2012). A data mining approach for sharpening thermal satellite imagery over land. *Remote Sensing, 4*, 3287–3319.

Gao, F., J. Masek, M. Schwaller, and F. G. Hall (2006). On the blending of the Landsat and MODIS surface reflectance: Predicting daily Landsat surface reflectance. *IEE Trans. Geosci. Remote. Sens., 44*, 2207–2218.

Goward, S. N., G. D. Cruickshanks, and A. S. Hope (1985). Observed relation between thermal emission and reflected spectral radiance of a complex vegetated landscape. *Remote Sens. Environ., 18*, 137–146.

Goward, S. N., Y. Xue, and K. P. Czajkowski (2002). Evaluating land surface moisture conditions from the remotely sensed temperature/vegetation index measurements: An exploration with the simplified simple biosphere model. *Remote Sens. Environ., 79*, 225–242.

Hain, C. R., W. T. Crow, M. C. Anderson, and J. R. Mecikalski (2012). An EnKF dual assimilation of thermal-infrared and microwave satellite observations of soil moisture into the Noah land surface model. *Water Resour. Res., 48*, W11517, doi:10.1029/2011WR011268.

Hain, C. R., J. R. Mecikalski, and M. C. Anderson (2009). Retrieval of an available water-based soil moisture proxy from thermal infrared remote sensing. Part I: Methodology and validation. *J. Hydrometeorol., 10*, 665–683.

Hope, A. S., and T. P. McDowell (1992). The relationship between surface temperature and a spectral vegetation index of a tallgrass prairie: Effects of burning and other landscape controls. *Int. J. Remote Sens., 13*, 2849–2863.

Houborg, R., M. C. Anderson, C. S. T. Daughtry, W. P. Kustas, and M. Rodell (2011). Using leaf chlorophyll to parameterize light-use-efficiency within a thermal-based carbon, water and energy exchange model. *Remote Sens. Environ., 115*, 1694–1705.

Houborg, R., M. Rodell, B. Li, R. H. Reichle, and B. F. Zaitchik (2012). Drought indicators based on model-assimilated Gravity Recovery and Climate Experiment (GRACE) terrestrial water storage observation. *Water Resour. Res., 48*, W07525.

Jiang, Y., and B. Huang (2001). Drought and heat stress injury to two cool season turf grasses in relation to antioxidant metabolism and lipid peroxidation. *Crop Sci., 41*, 436–442.

Jimenez, C., C. Prigent, B. Mueller, S. I. Seneviratne, M. F. McCabe, E. T. Wood, W. B. Rossow et al. (2011). Global intercomparison of 12 land surface heat flux estimates. *J. Geophys. Res., 116*, D02102.

Karnieli, A., N. Agam, R. T. Pinker, M. C. Anderson, M. L. Imhoff, G. G. Gutman, N. Panov, and A. Goldberg (2010). Use of NDVI and land surface temperature for drought assessment: merits and limitations. *J. Climate, 23*, 618–633.

Kebede, H., D. K. Fisher, and L. D. Young (2012). Determination of moisture deficit and heat stress tolerance in corn using physiological measurements and a low-cost microcontroller-based monitoring system. *J. Agron. Crop Sci., 198*, 118–129.

Knapp, K. R. (2008). Scientific data stewardship of International Satellite Cloud Climatology Project B1 geostationary observations. *J. Appl. Remote Sens., 2*, 023548.

Kongoli, C., W. P. Kustas, M. C. Anderson, J. M. Norman, J. G. Alfieri, G. N. Flerchinger, and D. Marks (2013). Evaluation of a two-source snow-vegetation energy balance model for estimating surface energy fluxes in a rangeland ecosystem. *J. Hydrometeorol.*, in final review.

Kustas, W. P., and M. C. Anderson (2009). Advances in thermal infrared remote sensing for land surface modeling. *Agric. For. Meteorol., 149*, 2071–2081.

Kustas, W. P., M. C. Anderson, J. M. Norman, and F. Li (2007). Utility of radiometric-aerodynamic temperature relations for heat flux estimation. *Boundary Layer Meteorol., 122*, 167–187.

Kustas, W. P., G. R. Diak, and J. M. Norman (2001). Time difference methods for monitoring regional scale heat fluxes with remote sensing. *Land Surface Hydrol. Meteorol. Climate Obs. Model., 3*, 15–29.

Kustas, W. P., J. Hatfield, and J. H. Prueger (2005). The Soil Moisture Atmosphere Coupling Experiment (SMACEX): Background, hydrometeorological conditions and preliminary findings. *J. Hydrometeorol., 6*, 791–804.

Kustas, W. P., and J. M. Norman (1999a). Evaluation of soil and vegetation heat flux predictions using a simple two-source model with radiometric temperatures for partial canopy cover. *Agric. For. Meteorol., 94*, 13–29.

Kustas, W. P., and J. M. Norman (1999b). Reply to comments about the basic equations of dual-source vegetation-atmosphere transfer models. *Agric. For. Meteorol., 94*, 275–278.

Kustas, W. P., and J. M. Norman (2000a). Evaluating the effects of subpixel heterogeneity on pixel average fluxes. *Remote Sens. Environ., 74*, 327–342.

Kustas, W. P., and J. M. Norman (2000b). A two-source energy balance approach using directional radiometric temperature observations for sparse canopy covered surfaces. *Agron. J., 92*, 847–854.

Kustas, W. P., J. M. Norman, M. C. Anderson, and A. N. French (2003). Estimating subpixel surface temperatures and energy fluxes from the vegetation index-radiometric temperature relationship. *Remote Sens. Environ., 85*, 429–440.

Lacaze, R., G. Balsamo, F. Baret, A. Bradley, J.-C. Calvet, F. Camacho, R. D'Andrimont et al. (2010). Geoland2—Towards an operational GMES land monitoring core service; First results of the biogeophysical parameter core mapping service. In W. Wagner and B. Székely (Eds.), *ISPRS TC VII Symposium—100 Years ISPRS*. Vienna, Austria: ISPRS, pp. 354–359.

Li, F., W. P. Kustas, J. H. Prueger, C. M. U. Neale, and T. J. Jackson (2005). Utility of remote sensing based two-source energy balance model under low and high vegetation cover conditions. *J. Hydrometeorol., 6*, 878–891.

Li, Y. P., W. Ye, M. Wang, and X. D. Yan (2009). Climate change and drought: a risk assessment of crop-yield impacts. *Climate Res., 39*, 31–46.

McNaughton, K. G., and T. W. Spriggs (1986). A mixed-layer model for regional evaporation. *Boundary Layer Meteorol., 74*, 262–288.

Merlin, O., B. Duchemin, O. Hagolle, F. Jacob, B. Coudert, A. Chehbouni, G. Dedieu, J. Garatuza, and Y. Kerr (2010). Disaggregation of MODIS surface temperature over an agricultural area using a time series of Formosat-2 images. *Remote Sens. Environ., 114*, 2500–2512.

Mishra, V., and K. Cherkauer (2010). Retrospective droughts in the crop growing season: Implications to corn and soybean yield in the midwestern United States. *Agric. For. Meteorol., 150*, 1030–1045.

Mittler, R. (2006). Abiotic stress, the field environment and stress combination. *Trends Plant Sci., 11*, 15–19.

Mo, K. C., L. Chen, S. Shukla, T. J. Bohn, and D. P. Lettenmaier (2012). Uncertainties in North American Land Data Assimilation Systems over the Contiguous United States. *J. Hydrometeorol., 13*, 996–1009.

Mo, K. C., L. N. Long, Y. Xia, S. K. Yang, J. E. Schemm, and M. B. Ek (2011). Drought indices based on the Climate Forecast System Reanalysis and ensemble NLDAS. *J. Hydrometeorol., 12*, 181–205.

Moran, M. S. (2003). Thermal infrared measurement as an indicator of plant ecosystem health. In D. A. Quattrochi and J. Luvall (Eds.), *Thermal Remote Sensing in Land Surface Processes*. London: Taylor and Francis, pp. 257–282.

Mozny, M., M. Tnka, Z. Zalud, P. Hlavinka, J. Nekovar, V. Potop, and M. Virag (2012). Use of a soil moisture network for drought monitoring in the Czech Republic. *Theor. Appl. Climatol., 107*, 88–111.

Norman, J. M., M. C. Anderson, W. P. Kustas, A. N. French, J. R. Mecikalski, R. D. Torn, G. R. Diak, T. J. Schmugge, and B. C. W. Tanner (2003). Remote sensing of surface energy fluxes at 10^1-m pixel resolutions. *Water Resour. Res., 39*, 1221, doi:10.1029/2002WR001775.

Norman, J. M., W. P. Kustas, and K. S. Humes (1995). A two-source approach for estimating soil and vegetation energy fluxes from observations of directional radiometric surface temperature. *Agric. For. Meteorol., 77*, 263–293.

Norman, J. M., W. P. Kustas, J. H. Prueger, and G. R. Diak (2000). Surface flux estimation using radiometric temperature: A dual temperature difference method to minimize measurement error. *Water Resour. Res., 36*, 2263–2274.

Otkin, J. A., M. C. Anderson, C. R. Hain, I. E. Mladenova, J. B. Basara, and M. Svoboda (in press). Examining rapid onset drought development using the thermal infrared based Evaporative Stress Index. *J. Hydrometeorol.*

Pradhan, G. P., P. V. Prasad, A. K. Fritz, M. B. Kirkham, and B. S. Gill (2012). Response of Aegilops species to drought stress during reproductive stages of development. *Funct. Plant Biol., 39*, 51–59.

Prasad, P. V., S. R. Pisipati, I. Momcilivic, and Z. Ristic (2011). Independent and combined effects of high temperature and drought stress during grain filling on plant yield and chloroplast EF-Tu expression in spring wheat. *J. Agron. Crop Sci., 197*, 430–411.

Priestley, C. H. B., and R. J. Taylor (1972). On the assessment of surface heat flux and evaporation using large-scale parameters. *Mon. Weather Rev., 100*, 81–92.

Prihodko, L., and S. N. Goward (1997). Estimation of air temperature from remotely sensed surface observations. *Remote Sens. Environ., 60*, 335–346.

Rizhsky, L., H. Liang, and R. Mittler (2002). The combined effect of drought stress and heat shock on gene expression in tobacco. *Plant Physiol., 130*, 1143–1151.

Rotter, R., and S. C. van de Geijn (1999). Climate change effects on plant growth, crop yield, and livestock. *Climatic Change, 43*, 651–681.

Saini, H. S., and M. E. Westgate (1999). Reproductive development in grain crops during drought. *Adv. Agron., 68*, 59–96.

Santanello, J. A., and M. A. Friedl (2003). Diurnal variation in soil heat flux and net radiation. *J. Appl. Meteorol., 42*, 851–862.

Savitzky, A., and M. J. E. Golay (1964). Smoothing and differentiation of data by simplified least squares procedures. *Anal. Chem., 36*, 1627–1639.

Schaake, J. C., Q. Duan, V. Koren, K. E. Mitchell, P. R. Houser, E. F. Wood, A. Robock et al. (2004). An intercomparison of soil moisture fields in the North American Land Data Assimilation System (NLDAS). *J. Geophys. Res., 109*, doi:10.1029/2002JD00309.

Sheffield, J., G. Goteti, F. Wen, and E. F. Wood (2004). A simulated soil moisture based drought analysis for the United States. *J. Geophys. Res., 109*, D24108, doi:10.1029/2004JD005182.

Sheffield, J., Y. Xia, L. Luo, E. F. Wood, M. B. Ek, and K. E. Mitchell (2012). Drought monitoring with the North American Land Data Assimilation System (NLDAS): A framework for merging model and satellite data for improved drought monitoring. In B. Wardlow, M. Anderson, and J. Verdin (Eds.), *Remote Sensing of Drought: Innovative Monitoring Approaches*. London: Taylor and Francis, p. 422.

Smith, R. C. G., and B. J. Choudhury (1991). Analysis of normalized difference and surface temperature observations over southeastern Australia. *Int. J. Remote Sens., 12*, 2021–2044.

Svoboda, M., D. LeComte, M. Hayes, R. Heim, K. Gleason, J. Angel, B. Rippey et al. (2002). The drought monitor. *Bull. Amer. Meteorol. Soc., 83*, 1181–1190.

Tang, R., Z.-L. Li, Y. Jia, C. Li, X. Sun, W. P. Kustas, and M. C. Anderson (2011). An intercomparison of three remote sensing-based energy balance models using Large Aperature Scintillometer measurements over a wheat-corn production region. *Remote Sens. Environ., 115*, 3187–3202.

Tanner, C. B., and W. A. Jury (1976). Estimating evaporation and transpiration from a row crop during incomplete cover. *Agron. J., 68*, 239–242.

Timmermans, W. J., W. P. Kustas, M. C. Anderson, and A. N. French (2007). An intercomparison of the Surface Energy Balance Algorithm for Land (SEBAL) and the Two-Source Energy Balance (TSEB) modeling schemes. *Remote Sens. Environ., 108*, 369–384.

Xia, Y., M. B. Ek, H. Wei, and J. Meng (2012a). Comparative analysis of relationships between NLDAS-2 forcings and model outputs. *Hydro. Process., 26*, 467–474.

Xia, Y., K. E. Mitchell, M. B. Ek, B. A. Cosgrove, J. Sheffield, L. Luo, C. Alonge et al. (2012b). Continental-scale water and energy flux analysis and validation for North American Land Data Assimilation System project phase 2 (NLDAS-2): 2. Validation of model-simulated streamflow. *J. Geophys. Res., 117*, D03110.

Xia, Y., K. E. Mitchell, M. B. Ek, J. Sheffield, B. A. Cosgrove, E. F. Wood, L. Luo et al. (2012c). Continental-scale water and energy flux analysis and validation of the North American Land Data Assimilation System project phase 2 (NLDAS-2): 1. Intercomparison and application of model products. *J. Geophys. Res., 117*, D03109.

Yilmaz, M. T., W. T. Crow, M. C. Anderson, and C. R. Hain (2012). An objective methodology for merging satellite and model-based soil moisture products. *Water Resources Res., 48*, W11502.

Zaitchik, B. F., B. Simane, S. Habib, M. C. Anderson, M. Ozdogan, and J. D. Foltz (2012). Building climate resilience in the Blue Nile/Abay Highlands: A role for Earth System Sciences. *Int. J. Environ. Res. Public Health, 9*, 435–461.

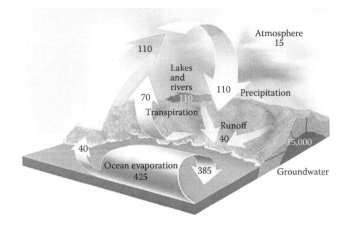

FIGURE 1.1
Global hydrological cycle in 1000 km^3 year^{-1}.

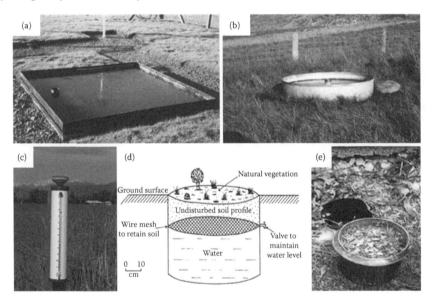

FIGURE 1.2
Different instrumentation used for the direct measurement of energy fluxes.

FIGURE 1.3
Example of instrumentation used in the eddy covariance system installed at Loobos site, located in the Netherlands.

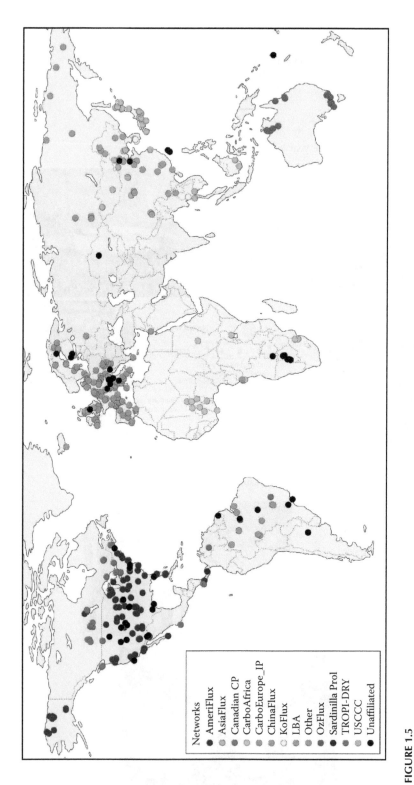

FIGURE 1.5

Global map of FLUXNET field sites. (Available from http://fluxnet.ornl.gov/introduction, Courtesy of Oak Ridge National Laboratory Distributed Active Archive Center, Oak Ridge, Tennessee.)

Networks

- AmeriFlux
- AsiaFlux
- Canadian CP
- CarboAfrica
- CarboEurope_IP
- ChinaFlux
- KoFlux
- LBA
- Other
- OzFlux
- Sardinilla Prol
- TROPI-DRY
- USCCC
- Unaffiliated

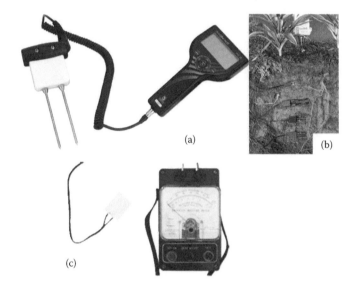

FIGURE 2.2
Examples of instruments used to measure soil moisture content: (a) a HydroSense II time domain reflectometer (Campell Scientific; http://www.campbellsci.com/hs2); (b) a sensor by Decagon Devices (http://www.decagon.com/), based on frequency domain technology; and (c) a gypsum block and sensor (http://vfd.ifas.ufl.edu/gainesville/irrigation/gypsum_block_probe.shtml).

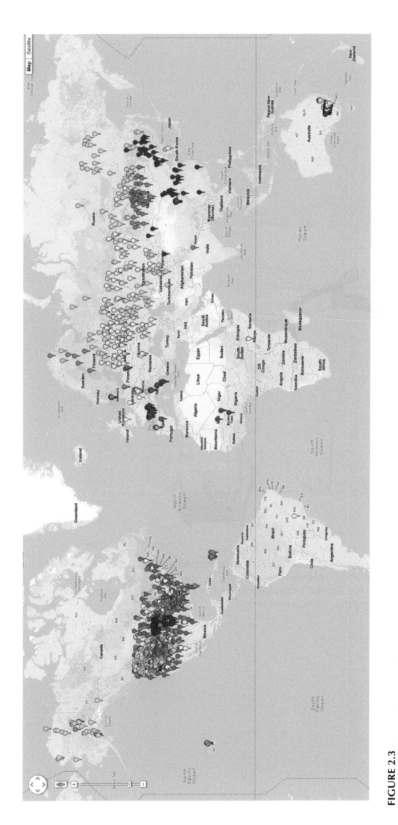

FIGURE 2.3

Geographical distribution of the ISMN sites. Place marks indicate the station coordinates of the different networks contained in the ISMN database. The different colors refer to the various networks that provide data.

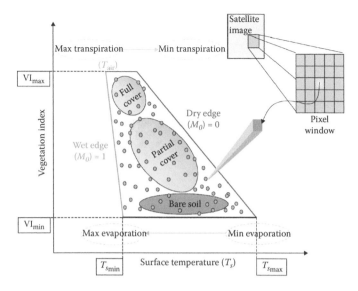

FIGURE 3.1
Key descriptors and physical interpretations of the T_s/VI feature space derived from remote sensing data.

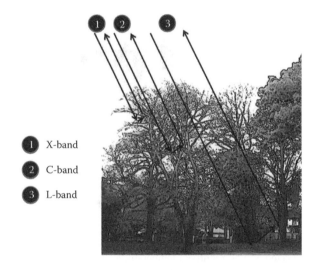

FIGURE 4.1
Vegetation influence on backscatter at three different wavelengths.

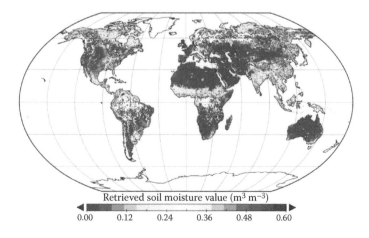

FIGURE 4.2
Global monthly average of the SMOS (ascending orbit) soil moisture values for September 2012.

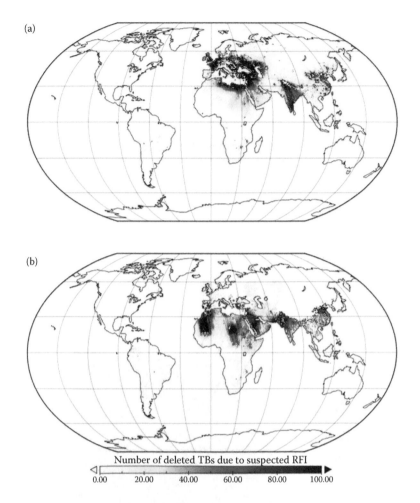

FIGURE 4.3
Global monthly count of deleted SMOS brightness temperatures (ascending orbit) due to suspected RFI for (a) September 2012 and (b) January 2010. Areas of highest RFI pollution are depicted in dark red. Data obtained from the Centre Aval de Traitement des Données SMOS (CATDS), operated for the Centre National d'Etudes Spatiales (CNES, France) by IFREMER (Brest, France).

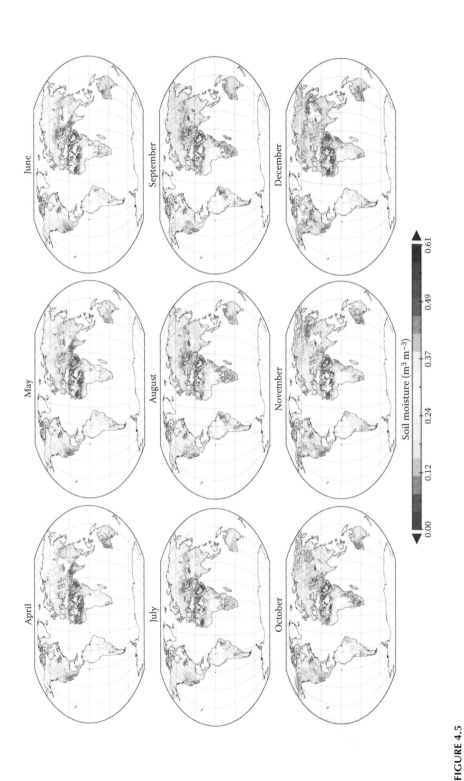

FIGURE 4.5
Global mean monthly volumetric soil moisture values for 2010 from the ECV product data set. (Data source: The ESA CCI SM project. Liu, Y. Y. et al., *Remote Sensing of Environment* 123:280–297, 2012; Wagner, W. et al., Fusion of active and passive microwave observations to create an Essential Climate Variable data record on soil moisture. Paper presented at XXII ISPRS Congress, Melbourne, Australia, 2012.)

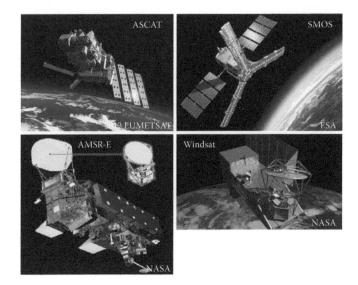

FIGURE 4.6
Examples of the satellite sensors providing operational estimates of soil moisture content.

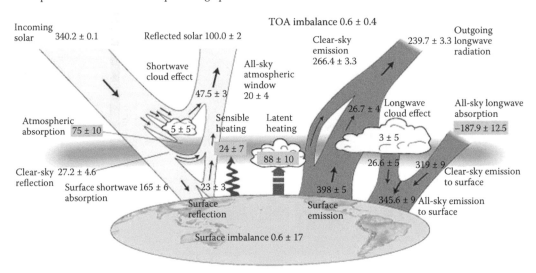

FIGURE 5.1
The global annual mean energy budget of Earth for the approximate period 2000–2010.

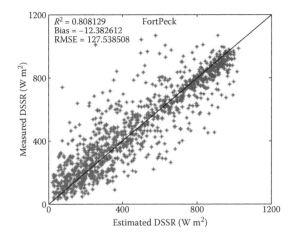

FIGURE 5.3
Validation results of the clear-sky and broadband cloudy-sky models at the SURFRAD station based on the MODIS products from Terra 2003–2005 (blue represents a clear sky; red represents a cloudy sky).

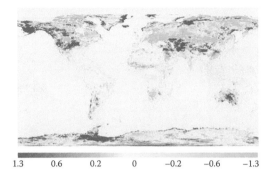

FIGURE 5.5
Global BHR variations from MODIS albedo product (2003–2008).

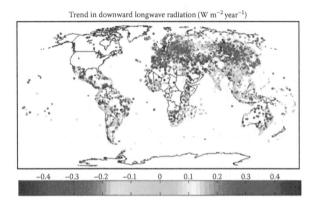

FIGURE 5.6
Linear trend of daily downward longwave radiation at 3200 global stations (1973–2008).

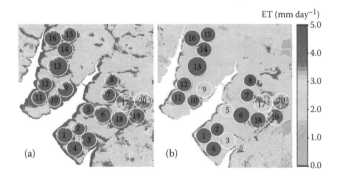

FIGURE 6.2
Daily evapotranspiration (ET) from (a) SEBAL and (b) SAFER in a Landsat image acquired in July 12, 2010, for northwestern São Paulo, in Brazil.

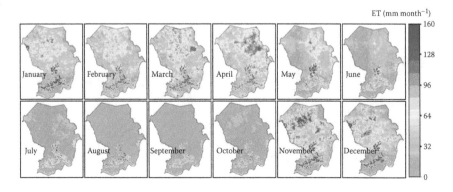

FIGURE 6.4
Spatial distribution of the monthly evapotranspiration from MODIS using SAFER model for 2011 in the Petrolina municipality, northeast Brazil.

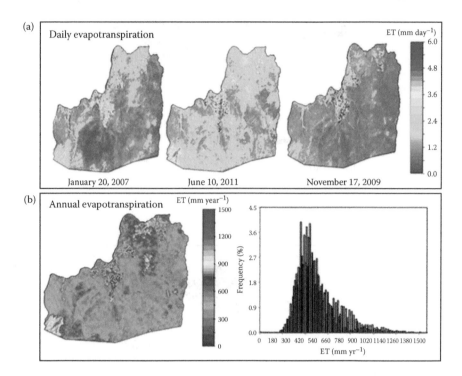

FIGURE 6.5
Spatial distribution of evapotranspiration from Landsat and SAFER in the Juazeiro municipality, northeast Brazil.

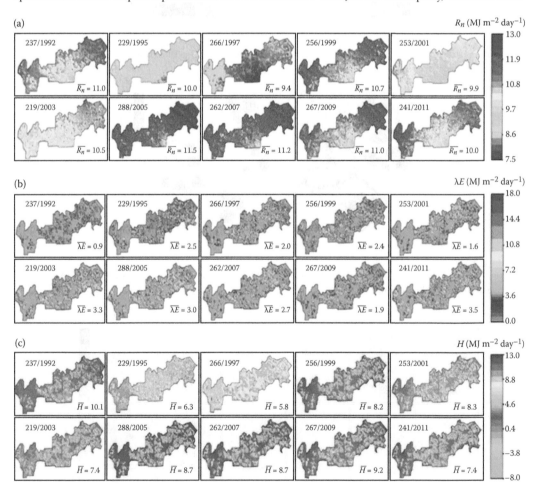

FIGURE 6.6
Spatial and temporal variation of the daily energy balance components: (a) R_n, (b) λE, and (c) H. from Landsat and SAFER and Slob equation, for the driest periods from 1992 to 2001, in the Nilo Coelho irrigation scheme, Brazil.

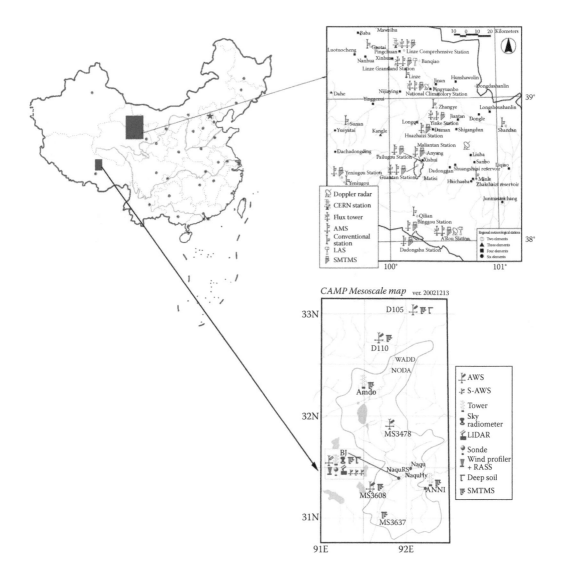

FIGURE 7.1
Sketch map of study area and the sites during the CAMP/Tibet and WATER in China.

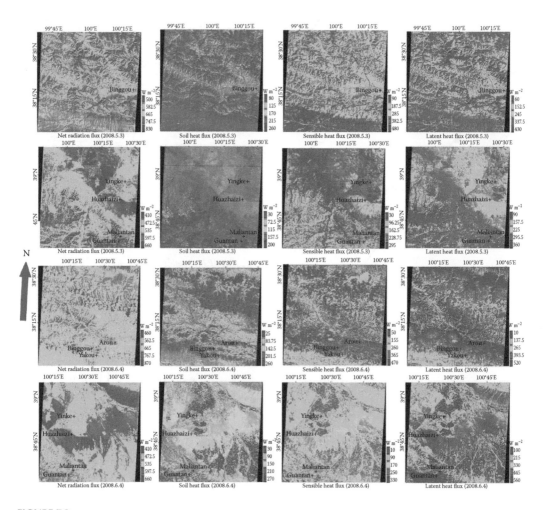

FIGURE 7.3
Distribution maps of land surface heat fluxes using ASTER data with SEBS over the WATER area. (From Ma, W. et al., *Hydrology and Earth System Sciences Discussions*, 6, 4619–4635, 2011.)

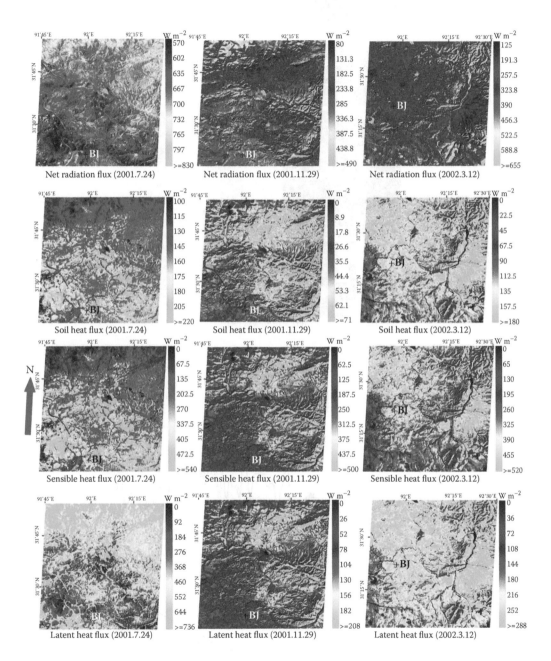

FIGURE 7.5
Spatially distributed maps of land surface heat fluxes over the CAMP/Tibet area derived from the combination of ASTER data with the parameterization one-layer modeling technique. (From Ma, W. et al., *Hydrology and Earth System Sciences*, 13(1), 57, 2009.)

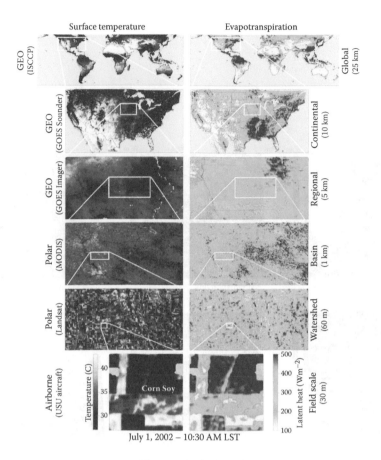

FIGURE 8.2
Multiscale ET maps for July 1, 2002, produced with ALEXI/DisALEXI using data from aircraft (30 m resolution), Landsat (60 m), MODIS (1 km), GOES Imager (5 km), and GOES Sounder (10 km), and an international suite of geostationary satellites (25 km), zooming into the Walnut Creek Watershed near Ames, Iowa site, USA.

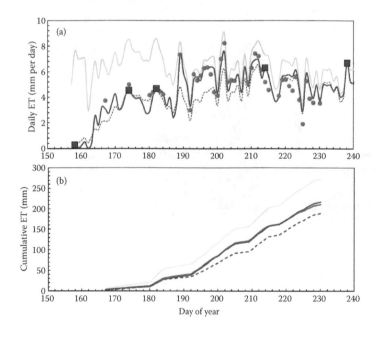

FIGURE 8.4
Landsat-only (red dashed lines) and fused Landsat-MODIS (solid red lines) reconstructions of (a) daily ET and (b) cumulative ET with flux observations (blue dots/line) in a soybean field from the SMACEX02 field campaign, Central Iowa, USA.

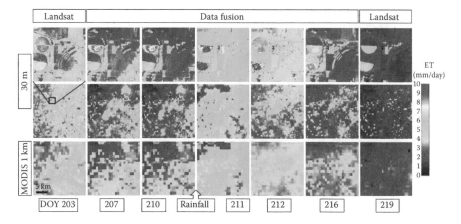

FIGURE 8.5

Example of Landsat/MODIS ET data fusion, showing daily ET from DisALEXI using MODIS at 1-km resolution (bottom row) and from the STARFM data fusion algorithm, fusing information from DisALEXI using Landsat TIR sharpened to 30-m resolution (middle row). Upper row shows maps obtained for BEAREX08 experimental fields.

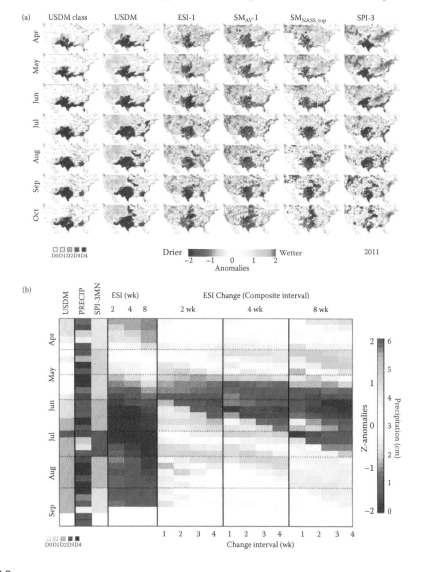

FIGURE 8.8

(a) Monthly standardized anomalies in the USDM drought classes (USDM), the 1-month ESI composite (ESI-1), 1-month ensemble average SM anomalies (SMAV-1), anomalies in NASS county-level topsoil moisture observations (SMNASS_top), anomalies in 3-month standardized precipitation index (SPI), as well as USDM drought classes for the week closest to the end of each month (first column). (b) Drought evolution across eastern Oklahoma and western Arkansas during 2011.

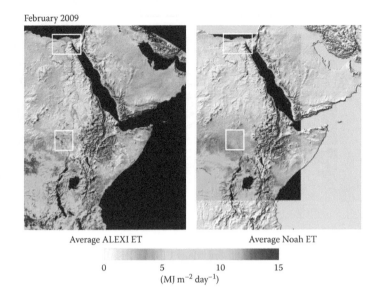

February 2009

Average ALEXI ET

Average Noah ET

0   5   10   15

(MJ m^{-2} day^{-1})

FIGURE 8.9

Comparison of average ET over the Nile River basin over February 2009 estimated by the (left) diagnostic ALEXI model and the (right) prognostic Noah LSM.

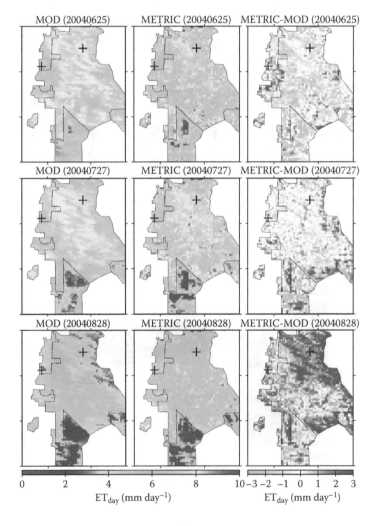

| MOD (20040625) | METRIC (20040625) | METRIC-MOD (20040625) |

| MOD (20040727) | METRIC (20040727) | METRIC-MOD (20040727) |

| MOD (20040828) | METRIC (20040828) | METRIC-MOD (20040828) |

0   2   4   6   8   10 -3 -2 -1 0 1 2 3

ET$_{day}$ (mm day^{-1})   ET$_{day}$ (mm day^{-1})

FIGURE 9.3

Spatial distribution of daily ET and difference between MODIS and METRIC-based approaches on June 25, July 27, and August 28, 2004. (Reproduced from Tang, Q. et al., *Journal of Geophysical Research* 114: D05114, 2009. With permission.)

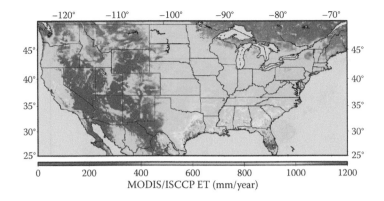

FIGURE 9.4
Mean annual ET in 2001–2006 over continental United States estimated from the MODIS-based T_s/VI approach.

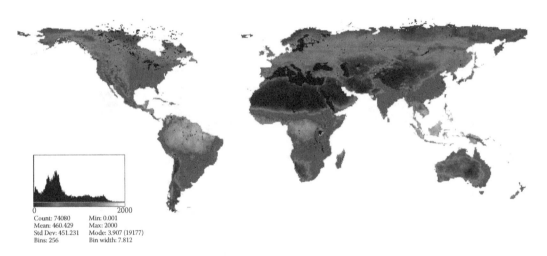

FIGURE 10.3
Total monthly evaporation (millimeters per month) from PT-JPL for 1994.

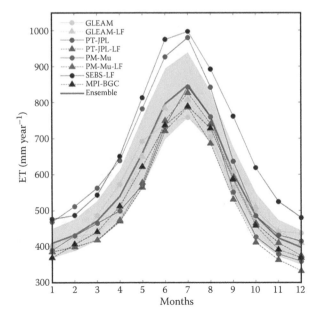

FIGURE 10.6
Annual cycles of global average evaporation for the year 2003.

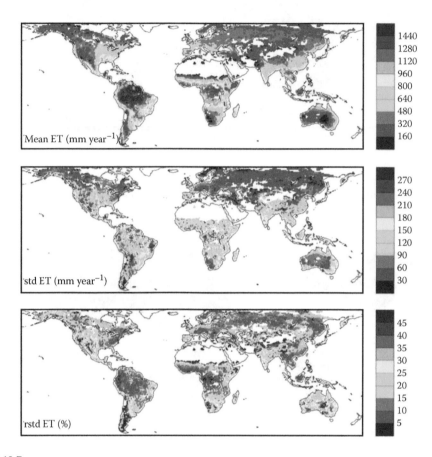

FIGURE 10.7
The 2003 all-product ensemble mean evaporation (top), standard deviation (std, middle), and relative standard deviation (rstd, bottom) expressed as a percentage of the pixel mean value.

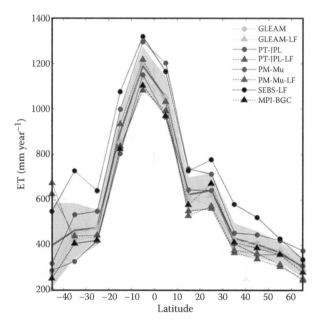

FIGURE 10.8
Zonal means of the 2003 evaporation calculated from the different model schemes and forcings.

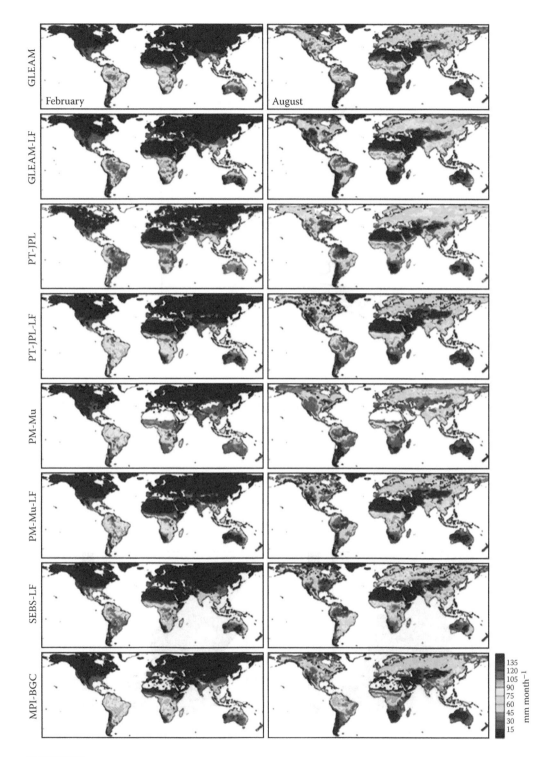

FIGURE 10.9
Monthly averaged evaporation for (left) February and (right) August 2003.

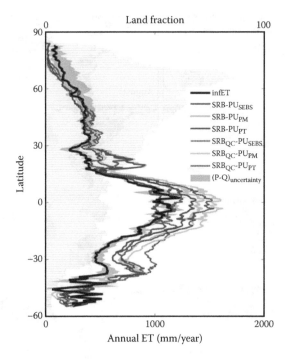

FIGURE 10.13
Zonal intercomparison of global E models against inferred E as determined from a selection of globally gridded precipitation data sets and the Global Runoff Data Centers Water Balance Method (GRDC-WBM) approach to calculate $P - Q$.

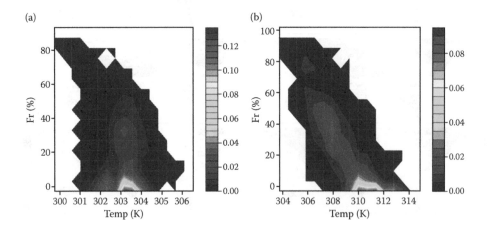

FIGURE 11.1
Joint probability density plots for the Landsat images for (a) June 3 and (b) August 6, 2007, over the CLASIC field site in central Oklahoma.

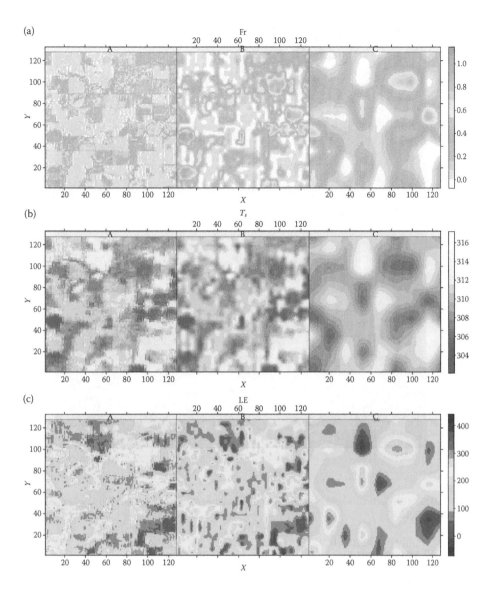

FIGURE 11.2
Wavelet decompositions of the August 6, 2007, Landsat imagery (a), fractional vegetation (%) (b), and surface temperature (K) (c) LE flux (W m^{-2}) for selected scales (A) 200 m (B) 800 m, and (C) 3200 m.

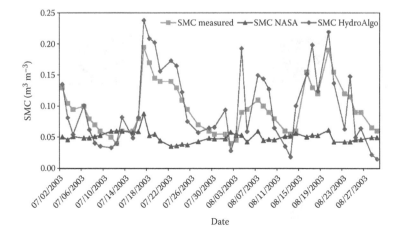

FIGURE 12.4
Comparison between SMC estimated by the HydroAlgo algorithm (magenta) and the standard NASA AMSR-E product (blue) with the SMC measured on ground (green), as a function of time, for two months of data from the Mongolian dataset.

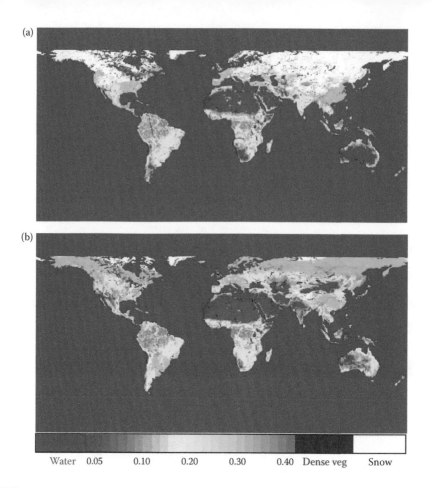

FIGURE 12.7
Global SMC maps of the whole world obtained in (a) December 2009 and (b) June 2010 by using HydroAlgo applied to AMSR-E data.

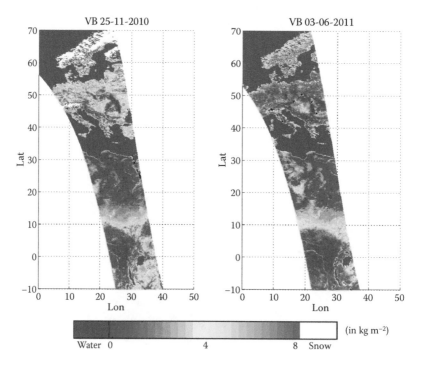

FIGURE 12.8
Vegetation maps (VB), expressed in PWC (in kg/m²), and generated from HydroAlgo on an AMSR-E orbit over Europe, in November 2010 and June 2011.

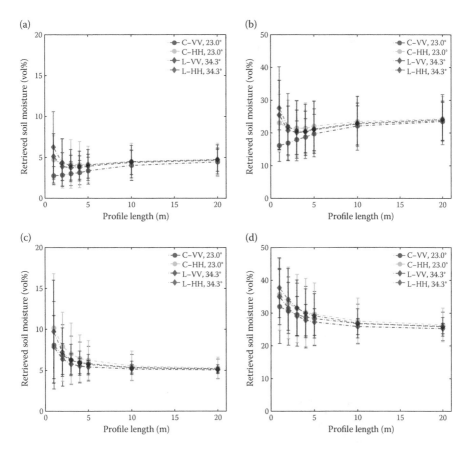

FIGURE 13.2
Mean and standard deviations of retrieved soil moisture values for different profile lengths for (*s*, *l*) equal to (a, b) (1 cm, 40 cm), (c, d) (2 cm, 5 cm), and initial moisture contents of (a, c) 5 vol% and (b, d) 25 vol%.

FIGURE 13.6
Illustration of the roughness classes (a) medium smooth and (b) stubbles.

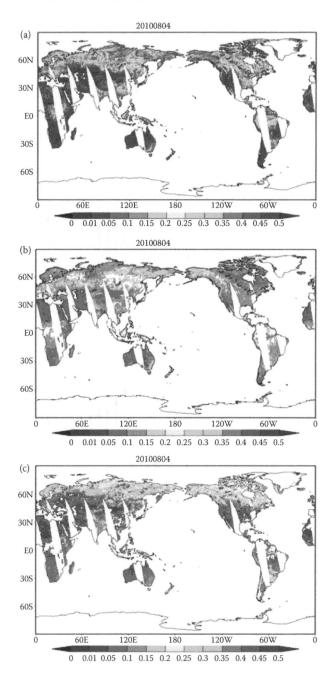

FIGURE 14.1
Soil moisture maps from AMSR-E observations using (a) SCA, (b) MCI, and (c) LPRM retrieval algorithms.

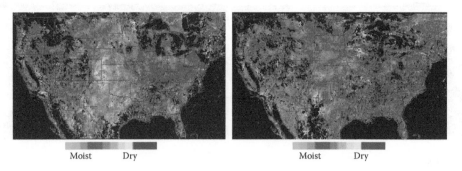

FIGURE 14.3
Example of the GOES-based dryness index estimated in (left) July 2011 and (right) July 2010.

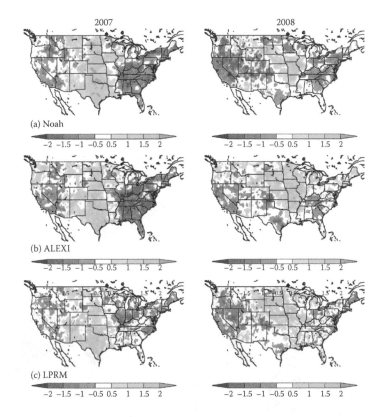

FIGURE 14.5
Yearly normalized soil moisture anomalies (unitless) for (a) Noah, (b) ALEXI, and (c) LPRM.

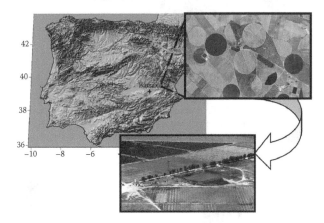

FIGURE 15.2
Barrax Cal/Val site.

FIGURE 15.3
(a) NDVI and (b) land surface temperature (in K) obtained from the AHS images during SEN3EXP field campaigning (for June 22, 2009).

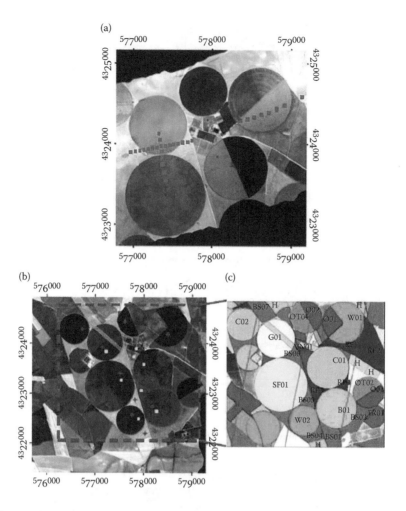

FIGURE 15.4

Main study area and location (UTM coordinates) of soil moisture *in situ* measurements (a) in the SEN2FLEX campaign (image at 2 × 2 m spatial resolution) and (b) in the SEN3EXP campaign (image at 6 × 6 m spatial resolution), and (c) land use map of the SEN3EXP study area, where the fields are pointed as C (corn), G (garlic), W (wheat), OT (oat), O (onion), BS (bare soil), SF (sunflower), B (barley), VN (vine), RF (reforestation), H (harvested), and FR (fruit). Squares point to the validation points and crosses to the calibration points. (Adapted from Sobrino, J. A. et al., *Remote Sensing of Environment*, 17, 415–428, 2012.)

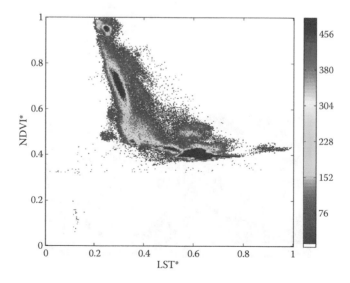

FIGURE 15.5

Normalized VI-T scatterplot derived from Figure 15.4. Scale bar represents the density of points.

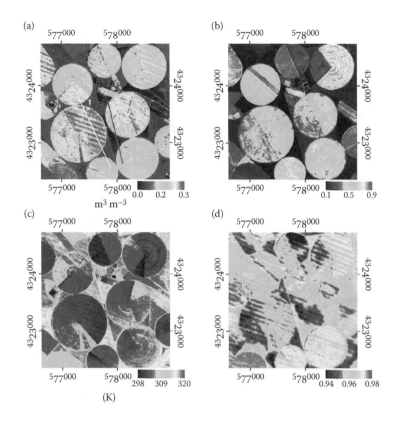

FIGURE 15.8

(a) Soil moisture (m³ m⁻³), (b) NDVI, (c) land surface temperature (K), and (d) emissivity (AHS band 73, 9.15 μm) images for June 20, 2009. (Adapted from Sobrino, J. A. et al., *Remote Sensing of Environment*, 17, 415–428, 2012.)

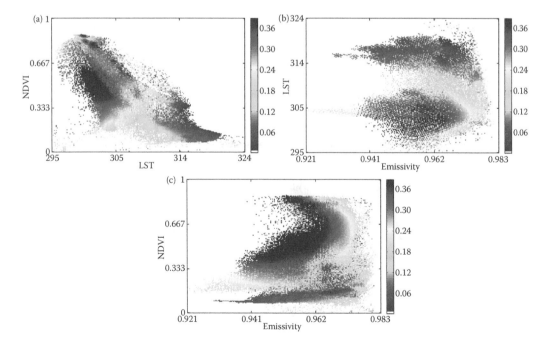

FIGURE 15.10

Soil moisture versus (a) surface temperature and NDVI, (b) surface temperature and emissivity, and (c) NDVI and emissivity. Plots were obtained from soil moisture values derived from Equation 15.11 and surface emissivity at AHS band 73 using the image acquired June 20, 2009, during the SEN3EXP campaign. (Adapted from Sobrino, J. A. et al., *Remote Sensing of Environment*, 17, 415–428, 2012.)

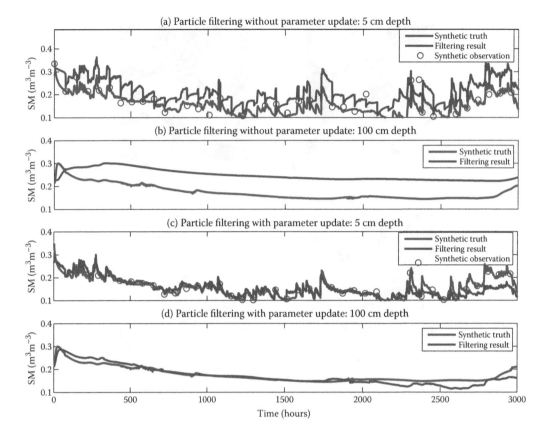

FIGURE 16.3
Results of a synthetic SSM DA study by Montzka et al. (2011). (a) and (b) show synthetic true soil moisture and state update only simulations as well as synthetic SMOS observations every 3 days. Incorrect parameterization leads to modeled soil moisture drifting away after caught back to the true state. This results in biased modeled soil moisture in deeper soil regions. (c) and (d) show synthetic true and dual state-parameter.

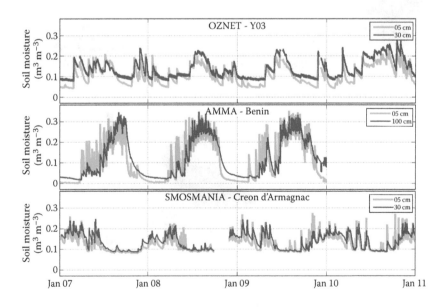

FIGURE 17.1
Time series of soil moisture *in situ* observations for the three sites used in this study.

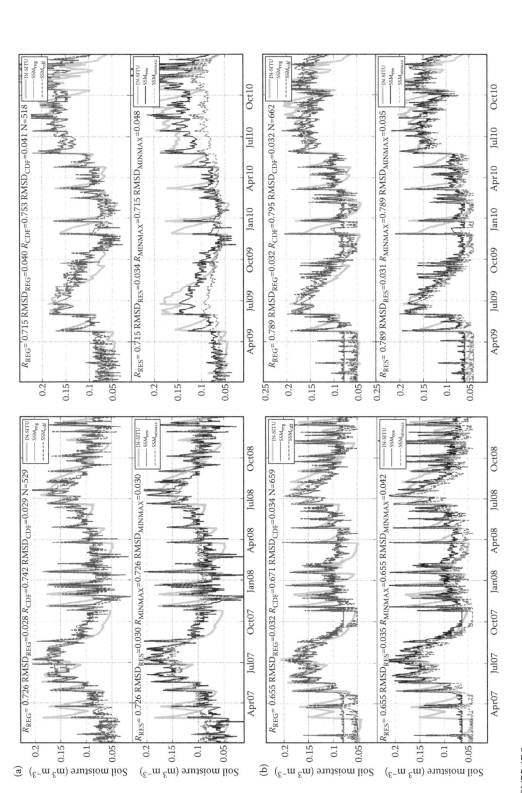

FIGURE 17.2

Comparison for the OZNET-Y03 site in Australia between *in situ* observations, IN-SITU, and scaled satellite data, SSM, with the different scaling methods and two satellite soil moisture products: (top) AMSR and (bottom) ASCAT. The calibration period is shown in the first column, while the validation is shown in the second one (REG, linear regression; CDF, CDF matching; RES, linear rescaling; MINMAX: Min/Max correction; R, correlation coefficient; RMSD, root-mean-square difference in m³ m⁻³).

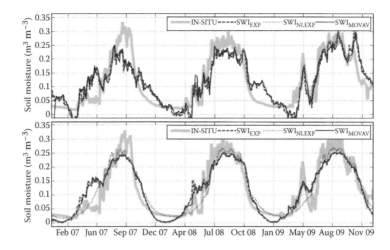

FIGURE 17.4
Comparison for the AMMA-Benin site, Africa, Africa between *in situ* and filtered satellite data, SWI, with the different filtering algorithms and two satellite soil moisture products: (top) AMSR and (bottom) ASCAT.

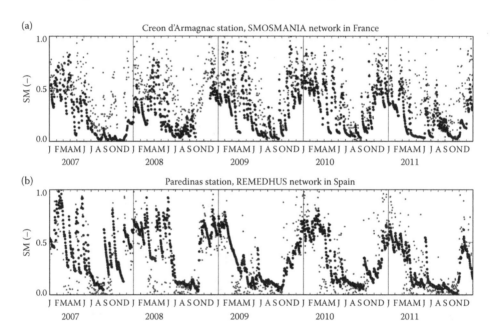

FIGURE 18.3
Temporal evolution of ASCAT scaled surface SM estimates (red dots) compared to (a) the *in situ* observations at 5 cm from the Creon d'Armagnac station (SMOSMANIA) and (b) the *in situ* at 5 cm from the Paredinas station (REMEDHUS) network over 2007–2011.

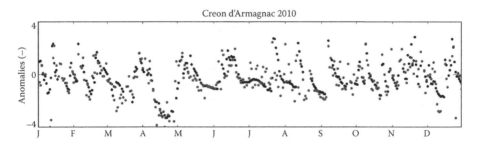

FIGURE 18.4
Temporal evolution of surface SM anomalies at the Creon d'Armagnac station (SMOSMANIA): ASCAT (red dots) and *in situ* observations at 5 cm (black dots) for the year 2010.

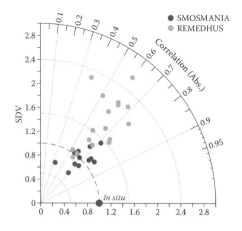

FIGURE 18.5
Taylor diagram illustrating the comparison between ASCAT and *in situ* surface SM for the stations of the SMOSMANIA network (red dots) and the REMEDHUS network (green dots) for 2007–2011.

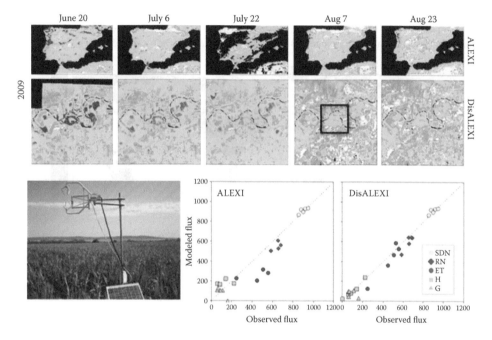

FIGURE 19.1
Example of ALEXI and DisALEXI retrievals over Spain, illustrating the importance of land cover classification and changing land use on flux estimation.

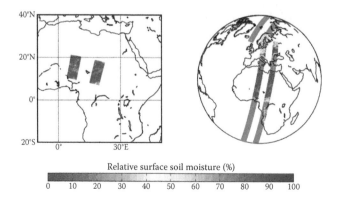

FIGURE 20.1
(Left) Example of an NRT ASCAT surface soil moisture data product, representing a 3-min data take. This data slice was acquired on April 1, 2012, 0845:01 UTC over Africa. (Right) Example of a full orbit ASCAT surface soil moisture data product from EUMETSAT's data archive. The orbit was acquired between April 1, 2012, 0833:00 UTC and 1014:59 UTC.

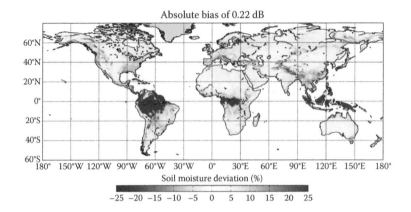

FIGURE 20.5
Simulated soil moisture errors assuming that the absolute calibration bias is constantly 0.22 dB all over the world. (Adapted from Hahn, S. et al., *IEEE Transactions on Geoscience and Remote Sensing*, 50, 2556–2565, 2012.)

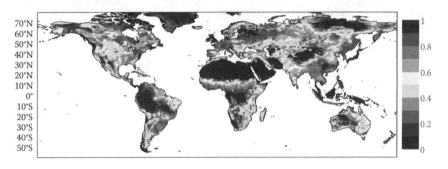

FIGURE 20.6
EUMETSAT H-SAF ASCAT root zone soil moisture in $m^3\ m^{-3}$ on November 25, 2012. ASCAT root zone soil moisture profile retrieval relies on ASCAT surface soil moisture data assimilation in the ECMWF land surface data assimilation system specifically adapted for the H-SAF retrieval of root zone soil moisture.

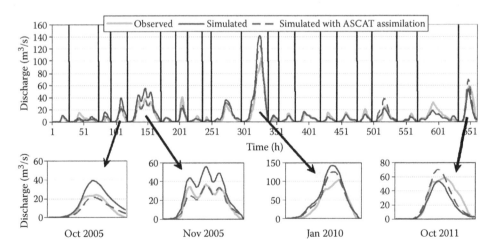

FIGURE 20.7
Comparison between observed and simulated discharge by assimilating the ASCAT-derived Soil Water Index for a sequence of 21 flood events that occurred in the Niccone basin (central Italy); the simulated discharge without assimilation is also shown as reference. The enlargement for four selected flood events is also shown in the lower panels where it highlighted the higher benefit due to the assimilation for the October and November 2005 events.

9

Spatiotemporal Estimation of Land Surface Evapotranspiration Rates from MODIS Data Based on a VI–T_s Approach

Qiuhong Tang, Lei Huang, Hui Liu, Pat Jen-Feng Yeh, and Jaeil Cho

CONTENTS

9.1 Introduction

According to a report by the United Nations (World Water Assessment Program 2009), the surging growth in global population has led to a rapid increase in the world's water demands. A large part of these water demands stem from food production requirements, irrigation being a prominent component in the requirement. Approximately 40% of crop production comes from the 16% of the agricultural land that is irrigated (Gleick 1993; Postel et al. 1996; Tilman et al. 2002). However, the agricultural land would not be sufficient for providing food for the world's population if not for water from rivers, lakes, reservoirs, and wells; in fact, about 60% of the world's freshwater withdrawal is used for irrigation purposes (Oki and Kanae 2006). Thus sustainable management of water resources is required to meet the growing global water demand.

Estimating the actual water consumption of crop fields, that is, evapotranspiration (ET), is essential for making informed decisions regarding efficient water use (Prajamwong et al. 1997; Cai et al. 2003; Lecina and Playan 2006), but adequate land surface ET data are currently lacking. One of the reasons for this lack of data is that it is technically difficult to obtain spatially distributed ET values using traditional surface measurements. Land surface ET depends on many heterogeneous factors such as meteorological conditions, vegetation characteristics, and soil moisture. Direct measurement of land surface ET is costly and generally only valid on a local scale (Yang et al. 2000; Brotzge and Crawford 2003). The actual ET over a large area is usually estimated using the reference ET and the potential rate. However, the actual ET can vary spatially, particularly in the case of large variations in vegetation, topography, and soil hydraulic properties. For these reasons, it is difficult to use traditional methods to estimate spatial and temporal variations in the actual ET based on ground observations of meteorological quantities (Tang et al. 2009a).

Although it is difficult to obtain a reliable estimate of spatially distributed ET using traditional field measurements, satellite remote sensing is a promising candidate for determining the pattern of ET over a large area (Bastiaanssen et al. 1998; Allen et al. 2007; Cleugh et al. 2007; Petropoulos et al. 2009; Vinukollu et al. 2011; Mu et al. 2011). However, a major drawback of most of these methodologies is that they require certain ground-based meteorological data, the availability of which is often limited in remote areas or in a near-real-time manner. Hence, although the obtained ET results are useful for planning studies in which retrospective estimates are sufficient, they have not yet been developed for use in near-real-time water management. Using data from the Moderate Resolution Imaging Spectroradiometer (MODIS), it is possible to estimate ET by exploiting the twice-daily temporal frequency and moderate spatial resolution of these data. In this chapter, we introduce a VI–T_s approach for estimating ET based entirely on satellite data in Section 9.2. Next, in Section 9.3, we present applications of the MODIS-based approach for near-real-time estimation of ET for the western United States and retrospective estimation over the continental United States. The MODIS-based ET estimates were assessed by comparison with other data obtained such as flux tower observations, the estimates from mapping ET at high resolution with internalized calibration (METRIC) using high-resolution Landsat data (Landsat 7 ETM+ SLC-off data), and the water balance estimates at the major river basins of the United States. The MODIS-based approach performed well in the assessments, showing a generally good agreement with the reference data. Thus it has the potential to be a useful tool in water management.

9.2 Methodology Overview

9.2.1 VI–T_s Approach

The main obstacle to using ET estimates in water management is the need for estimates on a daily to weekly timescale in a near-real-time manner. First, the input data and parameters used in the estimation must be routinely available for large regions, and thus, an approach using only satellite-based products could have global applicability. Second, the estimation needs to be insensitive to constraints imposed by the low temporal frequency of polar orbiting satellites and cloud blocks in the satellite observations. The estimation must take into account the diurnal variation of cloud effects on the ET estimation. Unlike other methods employed in the retrievals of energy fluxes, the VI–T_s method does not have the requirement of surface meteorological data and can be implemented to estimate the instantaneous ET at the satellite overpass time. Moreover, geostationary satellite products, which provide high temporal resolution observations of cloud and surface radiation fluxes, can also be used to extrapolate the instantaneous ET to the daily mean ET.

The VI–T_s method is based on a combination of the remotely sensed vegetation index (VI) and surface temperature (T_s). The surface water stress can be estimated from empirical

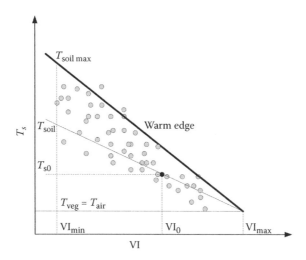

FIGURE 9.1
Schematic diagram of VI–T_s method. (Modified from Nishida, K. et al., *Journal of Geophysical Research* 108(D9): 4270, 2003. With permission.)

relations between satellite-based VI and T_s measurements (Jackson et al. 1977; Price 1982; Nagler et al. 2007; Glenn et al. 2007; Petropoulos et al. 2009). The classic diagram of the VI–T_s method is shown in Figure 9.1 (see also Sandholt et al. 2002; Nishida et al. 2003). A scatter plot of VI and T_s from a remote sensing window shows a linear or triangular form with a negative correlation between the VI and T_s values. The assumption is that land with dense vegetation (higher VI) has a lower surface temperature (lower T_s), whereas land with sparse vegetation (lower VI) has a higher surface temperature (higher T_s). VI–T_s relations have been widely used to estimate water stress from remote sensing data (Gillies et al. 1997; Jiang and Islam 2001, 2003). Further information can be found in Chapter 3 of the present volume.

In the VI–T_s diagram, the loci of the highest T_s at varying amounts of bare soil and vegetation (i.e., varying VI) delineate a sharply defined boundary, that is, the "warm edge," which is assumed to represent a lower limit of the surface soil water content (Carlson et al. 1995). The VI–T_s method may be used as well to estimate climatic parameters such as air temperature (Prince and Goward 1995). With increasing VI, the bare soil becomes increasingly masked by vegetation, resulting in a decrease in temperature. The surface temperature of a fully vegetated surface is consistently close to the air temperature because of the small aerodynamic resistance of the vegetation canopy (Carlson et al. 1995; Prince and Goward 1995). The maximum possible surface temperature of bare soil ($T_{\text{soil max}}$), where VI is at its minimum (VI_{min}), can be estimated from the upper left corner of the VI–T_s diagram (Nishida et al. 2003). Using the same logic for the derivation of $T_{\text{soil min}}$, the surface temperature of a full vegetation canopy (T_{veg}) can be extrapolated from VI_{max}. For example, at a point with a vegetation index of VI_0 and surface temperature of T_{s0}, the VI–T_s diagram is plotted using the data within a certain distance (21 km, following Nishida et al. 2003) from the point (see Figure 9.1). If the line representing the warm edge can be expressed as $T_s = a \, \text{VI} + b$, then $T_{\text{soil max}} = a \, \text{VI}_{\text{min}} + b$ and $T_{\text{veg}} = T_a = a \, \text{VI}_{\text{max}} + b$, where a is the slope of the line, b is the y intercept of the line, and T_a is the air temperature. The bare soil temperature (T_{soil}) of that point is estimated by extrapolating the line passing through VI_{max} and VI_0.

The evaporative fraction (EF) is defined as the ratio of ET (W m^{-2}) to available energy (Q, W m^{-2}) for ET (Shuttleworth et al. 1989). The EF for bare soil (EF$_{soil}$) can be estimated as (Nishida et al. 2003)

$$EF_{soil} = \frac{T_{soil\,max} - T_{soil}Q_{soil0}}{T_{soil\,max} - T_a Q_{soil}}$$

(9.1)

where Q_{soil0} is Q when T_{soil} is equal to T_a (W m^{-2}), Q_{soil} is Q over the bare soil surface (W m^{-2}), and the temperatures $T_{soil\,max}$, T_{soil}, and T_a (°C) are estimated using the VI–T_s diagram. $T_{soil\,max}$ is assumed to occur when ET is zero. Thus

$$Q_{soil} = H + ET = \rho C_P (T_{soil\,max} - T_a)/r_{a\,soil}$$

(9.2)

where ρ is the air density (kg m^{-3}), C_P is the specific heat of air under constant pressure (J kg^{-1} K^{-1}), and $r_{a\,soil}$ is the aerodynamic resistance of bare soil (s m^{-1}) (Allen et al. 1998). If Q_{soil} is known, $r_{a\,soil}$ can be inferred from Equation 9.2. An empirical equation can be used to estimate the wind speed, U_{1m}, at a height of 1 m using $r_{a\,soil}$ (Kondo 2000; Nishida et al. 2003).

$$U_{1m} = 667/r_{a\,soil}$$

(9.3)

The aerodynamic resistance (r_a, s m^{-1}) for cropland can be calculated using the following empirical formula (Kondo 2000):

$$1/r_a = 0.003U_{1m}$$

(9.4)

The canopy resistance (r_a, s m^{-1}) can be estimated following Jarvis (1976) and Nishida et al. (2003):

$$1/r_c = f_1(T_a)f_2(PAR)/r_{cMIN} + 1/r_{cuticle}$$

(9.5)

where PAR is the photosynthetic active radiation (μmol m^{-2} s^{-1}); r_{cMIN} is the minimum stomatal resistance (s m^{-1}) and is set to 33 for cropland, and $r_{cuticle}$ (s m^{-1}) is the canopy resistance related to diffusion through the cuticle layer of leaves and is set to 100,000 following the Biome-BGC model (White et al. 2000). Following Jarvis (1976), each component in Equation 9.5 is estimated as

$$f_1(T_a) = \left(\frac{T_a - T_n}{T_o - T_n}\right)\left(\frac{T_x - T_a}{T_x - T_o}\right)^{[(T_x - T_o)/T_o - T_n]}$$

(9.6)

$$f_2(PAR) = \frac{PAR}{PAR + A}$$

(9.7)

where T_n, T_o, and T_x (°C) are the minimum, optimal, and maximum temperatures for stomatal activity and are set to be 2.7, 31.1, and 45.3, respectively. The EF for vegetation (EF$_{veg}$) is calculated as

$$EF_{veg} = \frac{\alpha\Delta}{\Delta + \gamma\left(1 + r_c/2r_a\right)}$$

(9.8)

where α is the Priestley–Taylor parameter and is set to 1.26 (De Bruin 1983), Δ is the slope of the saturated vapor pressure in terms of temperature (Pa K^{-1}), and γ is the psychometric constant (Pa K^{-1}). ET for one remote sensing pixel is the sum of the ET values from the vegetated and bare soil parts of that pixel:

$$ET = f_{veg} \, Q_{veg} \, EF_{veg} + (1 - f_{veg}) \, Q_{soil} \, EF_{soil} \tag{9.9}$$

where f_{veg} is the proportion of vegetation in the remote sensing pixel and Q_{veg} is the available energy over the vegetation (W m^{-2}). The instantaneous EF_{soil} and EF_{veg} are assumed to be constant in the day (Crago 1996). With this constant EF hypothesis, the instantaneous EF estimate at MODIS daytime overpass time is used to estimate daily average ET, if daily Q can be estimated.

9.2.2 MODIS-Based Implementation of the VI–T_s Method

The MODIS-based ET estimation method described earlier (Section 9.2.1) does not require surface meteorological data. The satellite sensors/products used in the approach are MODIS land cover (MOD12Q1) (Strahler et al. 1999), surface reflectance (MOD09GQ) (Vermote and Vermeulen 1999), vegetation indices (MOD13Q1) (Huete et al. 1999), land surface temperature/emissivity (MOD11A1) (Wan 1999), and Albedo (MCD43A3) (Strahler and Muller 1999; Román et al. 2009). The spatial resolutions of these MODIS products range from 250 m to 1 km. All MODIS image products are resampled to 250-m grids for high-resolution implementations at small region. The corresponding MODIS products configured on a 0.05° latitude/longitude climate modeling grid (CMG) are used for coarse-resolution implementations at large region such as continental United States. The retrieval accuracy of each MODIS product varies and can be found from the above cited studies.

Surface radiative fluxes can be derived from satellite observations with reasonable accuracy (Pinker et al. 1992). The National Oceanic and Atmospheric Administration/National Environmental Satellite Data and Information Service (NOAA/NESDIS), University of Maryland, and NOAA/National Weather Service/National Centers for Environmental Prediction (NOAA/NWS/NCEP) have collaborated to estimate the surface radiation from geostationary satellite (GOES) observations (Pinker et al. 1992). Atmospheric data from the NCEP regional forecast model are delivered to the satellite data stream as inputs to a surface radiation budget (SRB) model, which is implemented at NOAA/NESDIS and provides surface radiation estimates in real time on an hourly basis for North America. The SRB data include instantaneous, hourly, and daily information on surface downward shortwave radiation, PAR, and cloud amount. The instantaneous radiation data are provided hourly at 15 min after the hour (i.e., at UTC 0115, 0215, etc.). The instantaneous radiation data can be used for instantaneous ET estimation. The hourly shortwave radiations, together with the longwave radiations estimated from temperatures (from the VI–T_s diagram), are used to estimate hourly available energy over vegetation and bare soil. The net shortwave radiation is estimated from the SRB downward shortwave radiation and MCD43A3 surface albedo. The effect of clouds on the ET estimate can be reasonably well accounted for by the SRB shortwave radiation. The MODIS surface temperature and air temperature derived from the VI–T_s diagram are interpolated to hourly values using a simple sine–cosine fit (Allen 1976). The upward (outgoing) longwave radiation is computed using the hourly temperature and the MOD11A1 surface emissivity. The downward longwave radiation is computed using an effective atmospheric temperature that is assumed to be 20 K lower than air temperature following Kondo (2000). The available energy is then calculated from net radiation (the sum of shortwave and longwave radiations) and soil heat flux. The soil heat flux is set to zero for instantaneous estimates over the vegetated part of the pixel and daily estimate of available energy. It is assumed to be 0.4 times the net radiation for instantaneous estimates over the bare soil part of the pixel (Idso et al. 1975). The available energy estimates, together with the constant *EF* hypothesis, allow this approach to estimate instantaneous and daily mean ET. As sensible heat flux (*H*) equals available energy minus latent heat flux (i.e., ET), this approach is capable to estimate *H* as well.

The NOAA/NESDIS SRB data have a latency of 2 days from the time of measurement. For near-real-time ET monitoring, the effective latency of the approach is determined by the MODIS products, which are typically available within 3 days to 1 week, although this could be reduced by using MODIS Rapid Response data. For retrospective estimation, radiation data from derived products archived by the International Satellite Cloud Climatology Project (ISCCP) were used to estimate land surface fluxes (Rossow and Schiffer 1999).

9.3 Case Studies

The MODIS ET approach detailed in the previous section was implemented at a spatial resolution of 250 m for a region of the Upper Klamath River Basin. In its upper reaches along the Oregon–California border, the Klamath River drains an arid interior basin (Tang et al. 2009a). The basin is a focal point for local and national discussions on water management and scarcity (Powers et al. 2005). Two Bowen ratio measurement systems were operated from April to October 2004 by Oregon State University at the KL03 and KL04 sites within the study domain (Tang et al. 2009a). The MODIS-based ET approach was also implemented at a coarse spatial resolution of 0.05° over the continental United States. The ET estimates at 15 river basins in the United States were compared with the observational estimates from the water balance method. The rivers include California (Sacramento and San Joaquin rivers), Colorado River, Columbia River, Great Basin, Missouri River, Ohio River, Arkansas-Red (Arkansas and Red rivers), Upper Mississippi river, Lower Mississippi river, Potomac River, Sabine-Neches (Sabine and Neches rivers), Mobile River, Altamaha River, and St. Johns River. According to the water conservation law, the ET of a river basin equals the precipitation minus the runoff and terrestrial water-storage change (Tang et al. 2009b). The water-storage change was estimated from the Gravity Recovery and Climate Experiment (GRACE) satellite data. The annual ET of a river basin could therefore be computed using observed precipitation from the PRISM Climate Group, Oregon State University (http://prism.ore gonstate.edu) and streamflow from the U.S. Geological Survey (http://waterdata.usgs.gov).

9.4 Results and Discussion

9.4.1 Validation with Ground Observations

The instantaneous ET observations at the Upper Klamath River Basin were compared with the MODIS-based estimates for the time when the MODIS satellite overpassed the pixels within which the measurement sites lay. The MODIS-based approach compared favorably

TABLE 9.1

Comparisons of Instantaneous ET and Daily Mean ET Rates from Flux Tower Observations at the Upper Klamath River Basin and the MODIS-Based Approach

Sites		Instantaneous ET (mm day⁻¹)		Mean Daily ET (mm day⁻¹)	
		All Days	Clear Days	All Days	Clear Days
KL03	Observation	7.93	8.89	3.03	3.42
	MODIS	7.12	8.71	2.64	3.14
	RMSE	2.75	2.22	1.02	0.88
KL04	Observation	11.67	13.61	4.55	5.11
	MODIS	10.90	13.26	4.02	4.76
	RMSE	3.14	2.68	1.06	0.92

Note: The comparison period is from April to October 2004.

with the ET observations over most of the evaluation period; Table 9.1 lists the mean values and root-mean-squared errors (RMSEs). These values indicate that the performance of the satellite-based method is generally similar for both clear-sky and cloudy conditions at both sites. The daily mean ET estimates of the MODIS-based approach were also compared with the flux tower observations. The RMSE of the MODIS-based estimates is generally less for the daily mean ET than for the instantaneous ET (Table 9.1). The error statistics for clear-sky and cloudy conditions are similar. This suggests that cloud effects on radiation partitioning are reasonably well-accounted for through the radiative products from geostationary satellites. The performance of the MODIS-based ET approach seems similar for both irrigated and nonirrigated sites. The biases between the observed and estimated values are small with relative biases generally less than 10%. However, the RMSEs are quite large with the RMSE fractions of the observation means ranging from 18% to 35%. The relatively large RMSEs are expected given the fact that the estimates derived from satellite products with moderate and coarse spatial resolutions are being compared with point observations. The small biases suggest that the agreement of the MODIS-based estimates and ground observations should improve with increasing averaging time.

The AmeriFlux network provides continuous observations of ET at flux tower sites in North America. The MODIS-based ET estimates were evaluated using the ET observations at the AmeriFlux sites when available. The estimated 8-day averaged ET values were directly compared with observations at the 0.05° pixel within which the AmeriFlux sites lie. Figure 9.2 shows the ET comparisons at AmeriFlux site US-NC1 from January 2005 to December 2006 and site US-Ne2 from April 2001 to July 2006. The US-NC1 (North Carolina Clearcut) site is located in a pine plantation among the mixed forests of the North Carolina lower coastal plain (Noormets et al. 2009), and the US-Ne2 (Mead Irrigated Rotation) site is irrigated with a center pivot system (Suyker and Verma 2008). At the US-NC1 site, the MODIS-based approach captures well the temporal variation of ET. However, at the US-Ne2 site, the MODIS-based approach underestimates the peak ET values in summer. Generally, the MODIS-based approach gives reasonable estimates at sites surrounded by a large area of relatively uniform vegetation (of the order of 5 km) (Table 9.2). Again, the biases between the observed and estimated values are small but the RMSEs are large. Furthermore, the RMSEs at the AmeriFlux sites are smaller than those at the sites in the Upper Klamath River Basin. The small biases and lower RMSEs confirm that the agreement of the MODIS-based estimates and ground observations will be better as the averaging time increases.

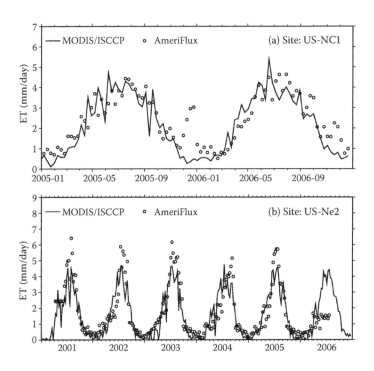

FIGURE 9.2
Comparison between MODIS-based ET estimates and AmeriFlux observations.

9.4.2 Intercomparisons with Other ET Estimates

The daily mean ET estimates at Upper Klamath River Basin from the MODIS-based approach were compared with the estimates from Mapping ET in High Resolution with Internalized Calibration (METRIC) using high-resolution Landsat data on 3 days (June 25, July 27, and August 28) in 2004 (Figure 9.3). METRIC is a satellite-based image-processing model for calculating ET as a residual of the surface energy balance (Allen et al. 2007). This model employed Landsat 7 ETM+ SLC-off images with a spatial resolution of 30 m. The Upper Klamath River Basin falls in Path4/Row30 of the Landsat images. METRIC does not use the VI–T_s relationship as does the MODIS-based approach. In the METRIC implementation, VI was used to estimate surface emissivity and T_s was used to estimate upward (outgoing) longwave radiation, soil heat flux, and momentum roughness length using semiempirical and theoretical relationships. The estimated ET patterns (with lower

TABLE 9.2

Comparisons of 8-day Averaged ET Rates at Selected Sites Belong to AmeriFlux Ground
Observations Network Observations and Estimates by the MODIS-Based Approach

	US-NC1			US-Ne2		
	Observation	MODIS	RMSE	Observation	MODIS	RMSE
8-day averaged ET (mm d^{-1})	2.42	2.17	0.75	1.75	1.66	0.88

Note: The comparison period is from January 2005 to December 2006 for US-NC1 site and from April 2001 to July 2006 for US-Ne2 site.

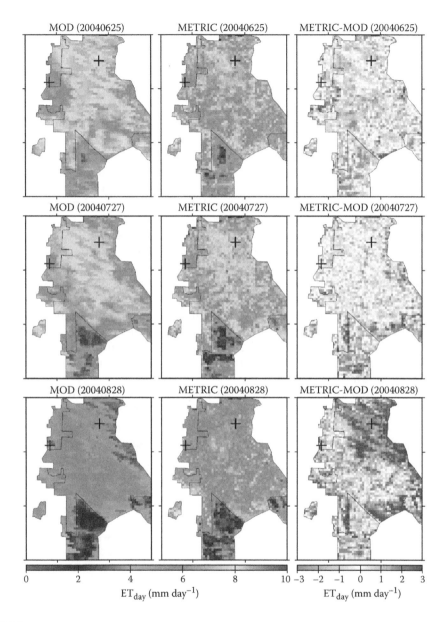

FIGURE 9.3
(See color insert.) Spatial distribution of daily ET and difference between MODIS and METRIC-based approaches on June 25, July 27, and August 28, 2004. (Reproduced from Tang, Q. et al., *Journal of Geophysical Research* 114: D05114, 2009. With permission.)

spatial resolution) from the MODIS-based approach were compared with the METRIC patterns (with higher spatial resolution). For the three comparison days, both MODIS-based and METRIC approaches showed that ET was lowest on August 28, 2004. Both approaches showed distinct ET values for the northern irrigated zone and southern non-irrigated zone. This indicates that the spatial and temporal variations of the estimated ET from the MODIS-based and METRIC approaches are broadly similar. There are many studies focusing on the effect of upscaling and downscaling remote sensing ET maps (Hong et al. 2009, 2011). It should be noted that certain spatial patterns may only show at specific scale (Seyfried and Wilcox 1995). It is possible that the decrease in spatial resolution results in a loss of information that may be valuable for particular applications (Carmel et al. 2001). The intercomparisons between MODIS-based and METRIC ET revealed that the two approaches showed similar spatial and temporal variations over the study area and period, although it cannot be expected that they provide the same information given their scale difference.

Figure 9.4 shows the mean annual ET in 2001–2006 estimated from the MODIS-based ET approach over the continental United States. The mean annual ET was smallest in the desert area around the Great Basin and largest in the southern United States. The ET pattern bears a resemblance to the pattern of mean annual precipitation, with a low trough in the central arid area and peaks over the western coast and eastern United States. However, the mean annual ET in the north is generally lower than that in the south. This feature is not seen in the mean annual precipitation pattern. From a surface energy balance perspective, the available energy is small and the energy constraints likely limit evaporation in the north. In the American Southwest where energy constraint is a less important factor, the ET pattern generally resembles the precipitation pattern. One exception is in southern California. Although southern California receives much less rainfall than northern California, the mean annual ET in southern California is comparable to that in northern California. The evaporative water in southern California seems to partly originate from other regions through the water transfer project (Israel and Lund 1995).

Figure 9.5 shows the annual basin-averaged ET time series from 2001 through 2006 at the river basins. The mean annual ET estimated from the MODIS-based approach matches very well with the water balance estimates for all the river basins. This suggests that the satellite-derived ET product, together with the other observable constituent variables (precipitation, streamflow, and terrestrial water storage change), can almost lead to the surface water balance at the basin scale. Among the studied basins, the mean annual ET is smallest

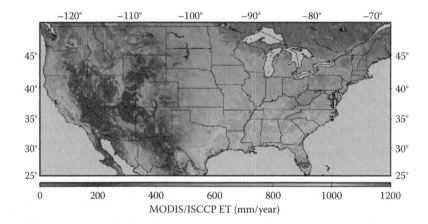

FIGURE 9.4
(See color insert.) Mean annual ET in 2001–2006 over the continental United States estimated from the MODIS-based approach. The MODIS-based approach was implemented daily at a spatial resolution of 0.05° over 2001–2006. The mean annual ET (mm year^{-1}) was computed from the daily ET estimates.

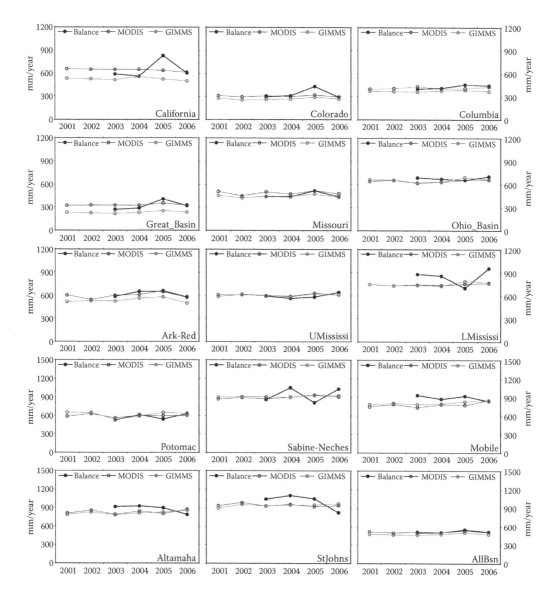

FIGURE 9.5
Annual basin-averaged ET time series from 2001 to 2006 at the river basins in the continental United States. The basin-averaged ET values were calculated from the MODIS-based ET estimates at a spatial resolution of 0.05°.

at the Great Basin and Colorado River Basin (~300 mm year^{-1}) and largest at the basins in the southern United States such as Sabine Neches, St. Johns, and Altamaha river basins (~900 mm year^{-1}). The MODIS-based ET estimates are close to the estimates derived from the Advanced Very High Resolution Radiometer (AVHRR) GIMMS NDVI and other satellite products (Zhang et al. 2010). The satellite-derived ET appears to have less interannual variation than the water balance ET. This is probably because of the limitations of the surface energy budget approach, including the VI–T_s method, which cannot completely account for the surface water budget (Ferguson et al. 2010). However, both satellite-derived and water balance products show very small inter-annual variation for the ET averaged over all basins. Previous attempts to close the surface water budget from remote sensing alone have generally been unsuccessful (Tang et al. 2009b). The general underestimation of combined ET and terrestrial water storage change from remote sensing is in part responsible for the unclosed water budget (Gao et al. 2010). If we can assume terrestrial water storage change is relatively small over a long time (Tang et al. 2010), our results indicate that the MODIS-based ET may close the surface water budget with observed precipitation and streamflow in the basins under variable climate conditions. It suggests the MODIS-based ET approach holds potential in land surface water budget studies via remote sensing.

9.5 Conclusions

Water management is a complex endeavor that can be potentially benefited from recent advances in remote sensing and hydrological modeling. The estimation of actual land surface ET, an important variable for water management, has become possible based entirely on satellite data. This chapter introduced a robust MODIS-based approach for the estimation of land surface ET. The approach uses only MODIS and geostationary satellite products and is appropriate for application at both local and regional scales. The ET estimates favorably agree with ground flux tower observations and estimates from the METRIC modeling approach using high-resolution Landsat data. The approach has good potential for operational applications given that both MODIS and geostationary satellite products are available in near real time. As compared with the METRIC approach, the MODIS-based approach is designed for ET estimation at a relatively large spatial scale with moderate spatial resolution and facilitates near-real-time ET monitoring from regional to continental scales. The MODIS-based approach has the potential, when combined with model simulations, to produce realistic estimates of the available water supply in real time and to more efficiently match water use with demand (Tang et al. 2009c).

The greatest advantage of satellite remote sensing is the ability to estimate land surface energy fluxes such as ET and H with moderate spatial resolution over large scale and without detailed knowledge of the specifics of land surface. However, the use of a single type of satellite/sensor for the fluxes estimation may be limited by the sensor's spatial resolution and temporal frequency. The MODIS-based approach takes advantage of the moderate spatial resolution of the MODIS sensor and the high temporal frequency of geostationary satellite to enable large-scale ET monitoring. This approach holds potential in future remote sensing–based modeling of ET when used in combination with useful information from multisensor or/and multisatellite observations.

Acknowledgments

This work was partially supported by the National Basic Research Program of China (no. 2012CB955403), the National Natural Science Foundation of China (no. 41171031), and the Hundred Talents Program of the Chinese Academy of Sciences.

References

Allen, J. C. 1976. A modified sine wave method for calculating degree days. *Environmental Entomology* 5: 388–396.

Allen, R. G., L. S. Pereira, D. Raes, and M. Smith. 1998. Crop evapotranspiration: Guidelines for computing crop water requirements. Irrigation and Drainage Paper 56. Rome: Food and Agricultural Organization.

Allen, R. G., M. Tasumi, and R. Trezza. 2007. Satellite-based energy balance for mapping evapotranspiration with internalized calibration (METRIC)—Model. *Journal of Irrigation and Drainage Engineering* 133(4): 380–394.

Bastiaanssen, W. G. M., M. Menenti, R. A. Feddes, and A. A. M. Holtslag. 1998. A remote sensing surface energy balance algorithm for land (SEBAL). 1. Formulation, *Journal of Hydrology* 212–213: 198–212.

Brotzge, J. A., and K. C. Crawford. 2003. Examination of the surface energy budget: A comparison of eddy correlation and Bowen ratio measurement systems. *Journal of Hydrometeorology* 4: 160–178.

Cai, X., D. C. McKinney, and L. S. Lasdon. 2003. Integrated hydrologic agronomic–economic model for river basin management. *Journal of Water Resources Planning and Management* 129(1): 4–17.

Carlson, T. N., R. R. Gillies, and T. J. Schmugge. 1995. An interpretation of methodologies for indirect measurements of soil water content. *Agricultural and Forest Meteorology* 77: 191–205.

Carmel, Y., D. J. Dean, and C. H. Flather. 2001. Combining location and classification error sources for estimating multi-temporal database accuracy. *Photogrammetric Engineering and Remote Sensing* 67: 865–872.

Cleugh, H. A., R. Leuning, Q. Mu, and S. W. Running. 2007. Regional evaporation estimates from flux tower and MODIS satellite data. *Remote Sensing of Environment* 106: 285–304.

Crago, R. D. 1996. Conservation and variability of the evaporative fraction during the daytime. *Journal of Hydrology* 180: 173–194.

De Bruin, H. 1983. A model for the Priestley-Taylor parameter α. *Journal of Applied Meteorology* 22: 572–578.

Ferguson, C. R., J. Sheffield, E. F. Wood, and H. Gao. 2010. Quantifying uncertainty in a remote sensing-based estimate of evapotranspiration over continental USA. *International Journal of Remote Sensing* 31(14): 3821–3865.

Gao, H., Q. Tang, C. R. Ferguson, E. F. Wood, and D. P. Lettenmaier. 2010. Estimating the water budget of major US river basins via remote sensing. *International Journal of Remote Sensing* 31(14): 3955–3978.

Gillies, R. R., W. P. Kustas, and K. S. Humes. 1997. A verification of the 'triangle' method for obtaining surface soil water content and energy fluxes from remote measurements of the Normalized Difference Vegetation Index (NDVI) and surface radiant temperature. *International Journal of Remote Sensing* 18(15): 3145–3166.

Gleick, P. 1993. Water and conflict: Fresh water resources and international security. *International Security* 18(1): 79–112.

Glenn, E. P., A. R. Huete, P. L. Nagler, K. K. Hirschboeck, and P. Brown. 2007. Integrating remote sensing and ground methods to estimate evapotranspiration. *Critical Reviews in Plant Sciences* 26: 139–168.

Hong, S.-H., J. M. H. Hendrickx, and B. Borchers. 2009. Up-scaling of SEBAL derived evapotranspiration maps from Landsat (30 m) to MODIS (250 m) scale. *Journal of Hydrology* 370: 122–138.

Hong, S.-H., J. M. H. Hendrickx, and B. Borchers. 2011. Down-scaling of SEBAL derived evapotranspiration maps from MODIS (250 m) to Landsat (30 m) scales. *International Journal of Remote Sensing* 32: 6457–6477.

Huete, A., C. Justice, and W. van Leeuwen. 1999. MODIS vegetation index (MOD 13), Algorithm theoretical basis document (ATBD), version 3, EOS Project Office, NASA Goddard Space Flight Center, Greenbelt, MD, pp. 1–120.

Idso, S. B., J. K. Aase, and R. D. Jackson. 1975. Net radiation—Soil heat flux relations as influenced by soil water content variations. *Boundary-Layer Meteorology* 9: 113–122.

Israel, M., and J. R. Lund. 1995. Recent California water transfers: Implications for water management. *Natural Resources Journal* 35: 1–32.

Jackson, R. D., J. L. Reginato, and S. B. Idso. 1977. Wheat canopy temperature: A practical tool for evaluating water requirements. *Water Resources Research* 13: 651–656.

Jarvis, P. G. 1976. The interpretation of the variations in leaf water potential and stomatal conductance found in canopies in the field. *Biological Science* 273: 593–610.

Jiang, L., and S. Islam. 2001. Estimation of surface evaporation map over southern Great Plains using remote sensing data. *Water Resources Research* 37(2): 329–340.

Jiang, L., and S. Islam. 2003. An intercomparison of regional latent heat flux estimation using remote sensing data. *International Journal of Remote Sensing* 24: 2221–2236.

Kondo, J. 2000. *Atmospheric Science Near the Ground Surface*. Tokyo: University of Tokyo Press.

Lecina, S., and E. Playan. 2006. Model for the simulation of water flows in irrigation districts. I: Description. *Journal of Irrigation and Drainage Engineering* 132(4): 310–321.

Mu, Q., M. Zhao, and S. W. Running. 2011. Improvements to a MODIS global terrestrial evapotranspiration algorithm. *Remote Sensing of Environment* 115: 1781–1800.

Nagler, P., E. Glenn, H. Kim, W. Emmerich, R. Scott, T. Huxman, and A. Huete. 2007. Seasonal and interannual variation of ET for a semiarid watershed estimated by moisture flux towers and MODIS vegetation indices. *Journal of Arid Environments* 70: 443–462.

Nishida, K., R. R. Nemani, S. W. Running, and J. M. Glassy. 2003. An operational remote sensing algorithm of land surface evaporation. *Journal of Geophysical Research* 108(D9): 4270.

Noormets, A., M. J. Gavazzi, S. G. McNulty, J.-C. Domec, G. Sun, J. S. King, and J. Chen. 2010. Response of carbon fluxes to drought in a coastal plain loblolly pine forest. *Global Change Biology* 16: 272–287.

Oki, T., and S. Kanae. 2006. Global hydrological cycles and world water resources. *Science* 313: 1068–1072.

Petropoulos, G., T. N. Carlson, M. J. Wooster, and S. Islam. 2009. A review of Ts/VI remote sensing based methods for the retrieval of land surface energy fluxes and soil surface moisture. *Progress in Physical Geography* 33: 224–250.

Pinker, R. T., and I. Laszlo. 1992. Modeling surface solar irradiance for satellite applications on a global scale. *Journal of Applied Meteorology* 31: 194–211.

Postel, S. L., G. C. Daily, and P. R. Ehrlich. 1996. Human appropriation of renewable fresh water. *Science* 271: 785–788.

Powers, K., P. Baldwin, E. Buck, and B. Cody. 2005. Klamath River Basin issues and activities: An overview. *CRS Report for Congress RL31098* 39.

Prajamwong, S., G. P. Merkley, and R. G. Allen. 1997. Decision support model for irrigation water management. *Journal of Irrigation and Drainage Engineering* 123(2): 106–113.

Price, J. 1982. Estimation of regional scale evapotranspiration through analysis of satellite thermal-infrared data. *IEEE Transactions on Geoscience and Remote Sensing* GE-20: 286–292.

Prince, S. D., and S. N. Goward. 1995. Global primary production: A remote sensing approach. *Journal of Biogeography* 22: 815–835.

Román, M. O. et al. 2009. The MODIS (Collection V005) BRDF/albedo product: Assessment of spatial representativeness over forested landscapes. *Remote Sensing of Environment* 113: 2476–2498.

Rossow, W. B., and R. A. Schiffer. 1999. Advances in understanding clouds from ISCCP. *Bulletin of the American Meteorological Society* 80(11): 2261–2287.

Sandholt, I., K. Rasmussen, and J. Andersen. 2002. A simple interpretation of the surface temperature/vegetation index space for assessment of surface moisture status. *Remote Sensing of Environment* 79: 213–224.

Seyfried, M. S., and B. P. Wilcox. 1995. Scale and the nature of spatial variability: Field examples having implications for hydrologic modeling. *Water Resources Research* 31(1):173–184.

Shuttleworth, W. J., R. J. Gurney, A. Y. Hsu, and J. P. Ormsby. 1989. The variation in energy partition at surface flux sites in remote sensing and large scale ground processes. In *Remote Sensing and Large-Scale Global Processes,* ed. A. Rango. IAHS Press, Wallingford, CT, pp. 67–74.

Strahler, A., D. Muchoney, J. Borak, F. Gao, M. Friedl, S. Gopal, J. Hodges et al. 1999. MODIS Land Cover Product, Algorithm Theoretical Basis Document (ATBD), version 5.0. Center for Remote Sensing, Department of Geography, Boston University, Boston, MA, pp. 1–72.

Strahler, A. H., and J. P. Muller. 1999. MODIS BRDF/Albedo Product: Algorithm Theoretical Basis Document, version 5.0. Boston University, Boston, MA, pp. 1–53.

Suyker, A. E., and S. B. Verma. 2008. Interannual water vapor and energy exchange in an irrigated maize-based agroecosystem. *Agricultural and Forest Meteorology* 148: 417–427.

Tang, Q., H. Gao, H. Lu, and D. P. Lettenmaier. 2009b. Remote sensing: Hydrology. *Progress in Physical Geography* 33: 490–509.

Tang, Q., H. Gao, P. Yeh, T. Oki, F. Su, and D. P. Lettenmaier. 2010. Dynamics of terrestrial water storage change from satellite and surface observations and modeling. *Journal of Hydrometeorology* 11:156–170.

Tang, Q., S. Peterson, R. H. Cuenca, Y. Hagimoto, and D. P. Lettenmaier. 2009a. Satellite-based near-real-time estimation of irrigated crop water consumption. *Journal of Geophysical Research* 114: D05114.

Tang, Q., E. A. Rosenberg, and D. P. Lettenmaier. 2009c. Use of satellite data to assess the impacts of irrigation withdrawals on Upper Klamath Lake, Oregon. *Hydrology and Earth System Sciences* 13: 617–627.

Tilman, D., K. G. Cassman, P. A. Matson, R. Naylor, and S. Polasky. 2002. Agricultural sustainability and intensive production practices. *Nature* 418: 671–677.

Vermote, E. F., and A. Vermeulen. 1999. Atmospheric correction algorithm: Spectral reflectances (MOD09), ATBD version 4.0. Department of Geography, University of Maryland, College Park, pp. 1–107.

Vinukollu, R. K., E. F. Wood, C. R. Ferguson, and J. B. Fisher. 2011. Global estimates of evapotranspiration for climate studies using multi-sensor remote sensing data: Evaluation of three process-based approaches. *Remote Sensing of Environment* 115: 801–823.

Wan, Z. 1999. MODIS Land-Surface Temperature Algorithm Theoretical Basis Document (LST ATBD), version 3.3. Institute for Computational Earth System Science, University of California, Santa Barbara, CA, pp. 1–75.

White, M. A., P. E. Thornton, S. W. Running, and R. R. Nemani. 2000. Parameterization and sensitivity analysis of the BIOME-BGC terrestrial ecosystem model: Net primary production controls. *Earth Interactions* 4(3): 1–85.

World Water Assessment Program. 2009. *The United Nations World Water Development Report 3: Water in a Changing World.* UNESCO, Paris.

Yang, J., B. Li, and S. Liu. 2000. A large weighing lysimeter for evapotranspiration and soil-water-groundwater exchange studies. *Hydrological Processes* 14: 1887–1897.

Zhang, K., J. S. Kimball, R. R. Nemani, and S. W. Running. 2010. A continuous satellite-derived global record of land surface evapotranspiration from 1983 to 2006. *Water Resources Research* 46: W09522.

10

Global-Scale Estimation of Land Surface Heat Fluxes from Space: Product Assessment and Intercomparison

Matthew McCabe, Diego Miralles, Carlos Jimenez, Ali Ershadi,
Joshua Fisher, Qiaozhen Mu, Miaoling Liang, Brigitte Mueller,
Justin Sheffield, Sonia Seneviratne, and Eric Wood

CONTENTS

10.1 Introduction

Improved representation of the spatial distribution and temporal development of surface latent (λE), sensible (H), and ground heat (G) fluxes is required not only to advance our understanding of important water and energy cycle processes but also to better describe the complex feedback mechanisms and interactions that occur between the land surface and the atmosphere. The latent heat flux provides the critical link between the water and energy cycles within the Earth system. Apart from enabling a description of water vapor exchanges for global water budget estimation, the accurate estimation of latent heat flux across a range of space- and timescales is needed in order to provide independent constraints on global climate models and other numerical approaches that seek to describe the hydrological response.

While precipitation and river runoff are reasonably quantifiable in the field, evaporation (*E*) is a phenomenon that is hard to measure directly. Since the beginning of the twentieth century, several techniques have been developed in an effort to measure the evaporative flux from both open water and terrestrial surfaces, with a focus on utilizing the often limited available meteorological measurements (e.g., Thornthwaite and Holzman 1939; Penman 1948; Swinbank 1951). Current *in situ* techniques are based on either measuring the net loss of water from the surface (local- and regional-scale water balances, evaporation pans, and lysimeters) or measuring the input of water vapor into the atmosphere (porometers, scintillometers, Bowen ratio methods, sap flow measuring devices, and eddy covariance instruments) (see Gash and Shuttleworth 2007). These measurement techniques either deal with one component of *E* or they do not differentiate between the different sources contributing to the total flux (i.e., transpiration, interception loss, soil evaporation, snow sublimation, and open-water evaporation).

Despite the multiple sources of error of these techniques (see Wang and Dickinson 2012 for a recent review), *in situ* measurements provide a reasonably accurate estimate of *E* at a given point in space and time. However, there is a necessity to routinely obtain *E* estimates over large areas for agricultural, hydrological, and climatological studies. Because of the reduced area sampled by ground observations, the application of satellite remote sensing to retrieve *E* has been explored over the past few decades. The general application of these techniques has developed from a need for regional-, continental-, and global-scale estimation of surface heat fluxes and particularly land surface evaporation.

As noted above, the natural trajectory of using remote sensing–based systems for flux retrieval is the move toward estimating variables at ever larger spatial scales. To this end, recent advances in the global estimation of land surface heat fluxes have seen the development of a number of methodologies that seek to characterize water vapor exchanges using a range of space-based data sources. These data provide a unique opportunity to allow for rapid and repeated assessment of flux behavior across vast spatial scales, while offering the capacity to develop an extended temporal record from existing data archives for longer-term climate assessment. However, there remain a number of issues that hinder the development of accurate long-term flux datasets, including (1) developing consistency within the forcing datasets required for any particular model approach; (2) resolving the intermodel and interproduct dependencies and sensitivities to enable reliable patterns and trends to emerge; and (3) building techniques for the robust assessment and evaluation of retrieved surface fluxes.

To address some of these issues, this chapter seeks to provide an overview of recent developments in the estimation of global terrestrial surface heat flux estimation from space and a review of community efforts to develop a harmonized and consistent heat flux record for long-term global assessment of land surface evaporation. This chapter does not seek to explicitly review developments in this technology for flux retrieval, as other chapters present studies relevant to recent advances in that field (see Section II). Extensive reviews of remote sensing–based methodologies can also be found in the works of Courault et al. (2005), Kalma et al. (2008), and Wang and Dickinson (2012). The contribution is divided according to these general themes, with a focus on specific elements relevant to developing global retrievals and the inherent challenges of these. For simplicity, the discussion is centered on the retrieval and development of evaporation (or latent heat flux), but the issues are equally pertinent to sensible heat flux estimation.

10.2 Global Surface Heat Flux Estimation

While space-based Earth observing systems have provided an unprecedented capacity for securing knowledge of terrestrial system processes and dynamics, neither E nor any of its components can be directly observed from space. Current methodologies for large-scale retrieval concentrate on the derivation of E by combining satellite physical variables that are linked to the evaporative process into existing physically based and empirical formulations. In all instances, there is an assumption that the models developed at the local scale are equally applicable at the field and larger scales: This is an occasionally heroic assumption, particularly where issues of strong land surface heterogeneity and paucity of meteorological forcing are prevalent. The assumed scale invariance remains a confounding problem, regardless of whether the model provides capacity for one-, dual-, or multiple-source descriptions of the land surface (see Kalma et al. 2008 for further details).

Following is an overview of how the state of science has evolved from early fundamental theoretical considerations to more recent approaches developed over the past few decades (see Chapter 3 for further details). A description of the most relevant of these methods as used in global product development is also provided, along with a summary of the data requirements and available sources of information used to drive these global retrievals.

10.2.1 Evaporation Measurement: A Historical Context

One of the earliest attempts to estimate E from atmospheric information dates from the beginning of the nineteenth century (Dalton 1802). Dalton conceptualized the flux of water vapor from the land to the atmosphere as a mass transfer proportional to the difference in humidity between land and the overlaying air. It is worth noting that his experiments precede by a number of decades the formulation of the first law of diffusion by Fick (1855). Implied in the proportionality constant used by Dalton was the consideration of the Earth–atmosphere interface as an electrical analog. The rate of exchange (latent heat flux) of a quantity (vapor pressure) between two media (the Earth and the atmosphere) is driven by a gradient of that quantity and controlled by a number of resistances. This theory of resistances has been extensively used over the last two centuries and is (implicitly or explicitly) formulated in the majority of existing methods that estimate turbulent heat fluxes (see Shuttleworth and Wallace 1985; Lhomme 1992).

Understanding that the evaporation process requires a source of energy, Penman (1948) combined the mass transfer equation from Dalton (1802) with the surface energy balance, that is, the net radiation at the surface (R_n) balanced by the sum of λE, H, and G. In addition, he introduced the concept of aerodynamic resistance (r_a): the resistance imposed by the air to the transfer of vapor. By linearizing the slope of the saturation vapor pressure versus temperature curve (Δ) and expressing this as a function of air temperature (T_a), Penman removed the requirement of measuring water vapor at two different vertical levels: a major advance for practical flux estimation. The Penman formulation was based on only T_a, relative humidity, R_n, and wind speed, which was used for the determination of r_a.

Penman's work served as a benchmark for further studies on evaporation and led to the development of specific formulations for the different components of the evaporative flux. In separate work, Monin and Obukhov (1945) provided new insight into the understanding of diffusive and turbulent transport of λE and H within the atmospheric boundary layer. Their equation for the transfer of vapor can be viewed as analogous to the Penman equation, but for two different levels in the atmosphere. It is worth noting that the work by Monin and Obukhov set the basis for modern *in situ* measurement from instruments such as the scintillometer and the eddy covariance technique (Foken 2006).

A few years later Monteith (1965) incorporated into Penman's equation the formulation of the resistances of vapor flow through stomata and from soil particles: jointly referred as r_s. In vegetated ecosystems, r_s depends mainly on the stomatal resistance. Under well-watered conditions, r_s will be dictated by atmospheric factors like the CO_2 concentration, air humidity, and photosynthetically active radiation or T_a, with the sensitivity to each factor responding differently for different vegetation types (Rana and Katerji 1998). As the soil dries out, r_s becomes more strongly controlled by the soil moisture deficit and less so by atmospheric factors (Baird 1999). However, if the calculation of r_a is problematic, the estimation and/or measurement of r_s represent a major challenge. Several solutions have been proposed for the estimation of r_s, the majority of which rely on empirical formulas that combine the environmental and physiological controls on r_s (soil moisture, humidity, CO_2 concentration, etc.) (e.g., Jarvis 1976; Ball 1987; Stewart 1988).

Today, the Penman-Monteith (PM) formula remains one of the most accurate and physically meaningful approaches to estimate E, provided the parameters are known (Allen et al. 1998). Unfortunately, this last condition is routinely difficult to satisfy, given the dependency of the method on a large variety of observations (at least wind speed, surface roughness, soil moisture induced evaporative stress, available energy [$R_n - G$], vapor pressure, and T_a)—a data challenge that is particularly pertinent when seeking to develop global retrievals.

Intending to address this data challenge, Priestley and Taylor (1972) developed an equation for the estimation of E in saturated grasslands that can be interpreted as a simplification of the PM equation. An empirically derived constant factor, referred to as the alpha (α_{PT}) coefficient, is defined to eliminate the requirement of information related to the properties and condition of the surface. Essentially, the estimation of E becomes feasible with radiation and temperature data only. In this approach, the surface is assumed to have limited control over the evaporative process. Not only is r_s neglected but so too are the potential effects of land surface in r_a (e.g., via roughness). The surface impact is restricted to indirect effects on the available energy and on T_a (e.g., via albedo or surface thermal properties).

The Priestley–Taylor (PT) equation still retains some of the important physics developed within the PM formula but seeks to limit the often onerous data requirements. One compromise in this approach relates to a simplification that ignores the fact that the rate of transpiration and that of soil evaporation are highly dependent on the availability of water in the soil (which rarely appears saturated). Over the past decades, there have been numerous attempts to extend the applicability of the PT equation by relaxing its major limitation: the nonconsideration of the effect of soil moisture deficit (which in the PM equation is incorporated via r_s). Some of the early modifications include the definition of moisture-dependent multiplicative factors, referred to as "stress factors" (Davies and Allen 1973; Jiang and Islam 1999). Norman et al. (1995) developed a unique application of PT by applying the formulation for the canopy elements only as an initial condition for unstressed canopy transpiration within a thermal-based two-source formulation. With surface temperature as the boundary condition, the model allows for stressed vegetation conditions by achieving radiative temperature and convective energy balance of the soil–vegetation canopy elements. It appears that this formulation is applicable to many landscapes and climates (Kustas and Anderson 2009; Agam et al. 2010).

The use of more complex model descriptions has been seen in the development of so called soil–vegetation–atmosphere transfer (SVAT) models. SVAT models have been used by both hydrologists interested in providing realistic estimates for well-instrumented catchments and atmospheric scientists concerned with the description of E for hydrometeorological and climate-related research. These models are generally based on some derivation of the PM equation. However, for agricultural practices and water management activities, the complexity of these models usually hampers their applicability. Partly because of the high input data requirements but also because of historical reasons, more empirical approaches (calibrated for local conditions) are often preferred in agricultural applications, perhaps as a response to their apparent local-scale reliability.

Despite recent progress toward the one-step application of the PM equation to estimate the needs for irrigation (Farahani et al. 2007), the conventional method in agriculture is to estimate E in a two-step approach. First, potential evaporation (or reference crop evaporation) is calculated, usually by applying the Penman or Priestley–Taylor equations in some form, and then crop-specific multiplicative factors are introduced (see Jensen 1968; Doorenbos and Pruitt 1975). Local calibration is essential as the daily and seasonal dynamics of E are heavily influenced by the practices themselves: Soil moisture is altered via irrigation, r_s by the choice of crop, and the partitioning of E on its different components as a consequence of harvesting.

10.2.2 Descriptions of Models Used for Global Retrievals

Meeting the needs of the agricultural, water resources management, and climate modeling communities remains a considerable challenge to those developing flux estimation approaches and one that has not been completely resolved as yet. The following is an overview of some of the major modeling approaches that have been used in recent years to develop global retrievals of surface heat fluxes. The list is not exhaustive but does cover the major variants and techniques that are in common use, particularly those that form part of community efforts such as the Global Energy and Water Exchanges (GEWEX) Data and Assessments Panel (GDAP) LandFlux project that is developing global land surface heat flux climatologies (see http://www.gewex.org/GDAP.html for more information). It is worth mentioning that more complex land surface model (LSM) and SVAT-based schemes are not included in this review, even though some of them may integrate satellite information into their model schemes (e.g., Rodell et al. 2004; Dirmeyer et al. 2006). The classification is hence exclusively dedicated to observation-based methodologies specifically created to derive E and with relatively low requirements for ancillary data. However, comparisons of primarily satellite derived estimates with LSM- and SVAT-based estimates are discussed briefly in Section 10.3.2.

10.2.2.1 Penman–Monteith–Based PM-Mu Approach

Monteith (1965) extended the Penman (1948) open-water evaporation formulation to vegetated surfaces by considering that the stomata vapor is saturated at the leaf temperature, the leaf surface is at the vapor pressure of the surrounding air, and there is a resistance that controls the transfer of vapor from the leaf to the surrounding air, that is, the leaf resistance is integrated up to the canopy resistance. The resulting PM equation can be expressed as

$$\lambda E = \frac{\Delta(R_n - G_0) + \rho_a c_p \text{VPD}/r_a}{\Delta + \gamma\left(1 + \dfrac{r_s}{r_a}\right)} \tag{10.1}$$

where λE is the latent heat flux, $R_n - G_0$ describes the available energy as the difference between the net radiation and ground heat flux, VPD is the vapor pressure deficit, Δ is the slope of the saturation vapor pressure–temperature curve (de/dT), ρ_a is the air density, c_p is the specific heat at constant pressure, γ is the psychrometric constant, and r_a and r_s are the bulk aerodynamic and surface resistances, respectively.

Various forms of the original PM model have been developed and used with differences in both the model structure and parameterization. The PM model of Monteith (1965) (the so-called big-leaf model) does not partition evapotranspiration between the contributing sources of soil/substrate evaporation, interception evaporation, and canopy transpiration. However, more detailed structural forms of the PM equation have been developed with consideration of E in different layers and/or sources, often targeted for patch to field-scale estimations of E (e.g., Shuttleworth and Wallace 1985; Farahani and Ahuja 1996; Brenner and Incoll 1997; Lhomme et al. 2012).

A recent development of the PM approach is the model of Mu et al. (2007). In Mu et al. (2007), the surface resistance scheme of Cleugh et al. (2007) was extended by considering the constraints of VPD and minimum temperature on stomatal conductance and using leaf area index (LAI) as a scalar for estimating canopy conductance as suggested by Jarvis (1976). This resistance parameterization was applied by Mu et al. (2007) using Moderate Resolution Imaging Spectroradiometer (MODIS) based vegetation and weather model data from the NASA Global Modeling and Assimilation Office (GMAO) to estimate global E at 5-km spatial resolution for 2001.

Ferguson et al. (2010) used the Mu et al. (2007) model in its original form to estimate E over the continental United States for 2003–2006. Later, Vinukollu et al. (2011b) used a modified form of Mu et al. (2007) model for global estimation of E at 1-km resolution with the aerodynamic resistance parameterization from SEBS (Su 2002) based on Monin–Obukhov similarity theory (Monin and Obukhov 1954) and the work of Massman (1987). In addition, Vinukollu et al. (2011b) added a canopy interception component to the Mu et al. (2007) model.

Mu et al. (2011) further developed the structure and resistance parameterization of their 2007 version model, with total evapotranspiration represented as the sum of evaporation from the wet canopy surface (λE_{wc}), transpiration from the dry canopy surface (λE_t), and evaporation from the soil surface (λE_s) as

$$\lambda E = \lambda E_{wc} + \lambda E_t + \lambda E_s \tag{10.2}$$

For the vegetated surface, evapotranspiration consists of the wet canopy surface and transpiration from plant tissue, with rates regulated by aerodynamic and surface resistances. Evaporation from the wet canopy (intercepted water) is modeled as

$$\lambda E_{wc} = f_{wet} \frac{\Delta A_c + f_c \rho c_p (e_s - e_a)/r_a^{wc}}{\Delta + \gamma \dfrac{r_s^{wc}}{r_a^{wc}}} \tag{10.3}$$

where f_c is the green canopy fraction, f_{wet} is the water cover fraction (based on the Fisher et al. [2008] derivation), A_c is the available energy in the canopy, and r_a^{wc} and r_s^{wc} are the aerodynamic and surface resistances against evaporation of the intercepted water.

r_a^{wc} and r_s^{wc} represent functions of T_a and LAI, with biome-specific values of leaf-scale conductance to heat and water vapor transfer (g_h and g_e). Canopy transpiration λE_t is calculated as

$$\lambda E_t = (1 - f_{\text{wet}}) \frac{\Delta A_c + f_c \rho c_p (e_s - e_a)/r_a^t}{\Delta + \gamma \dfrac{r_s^t}{r_a^t}} \tag{10.4}$$

where r_a^t and r_s^t are bulk aerodynamic and surface resistances against transpiration. r_a^t is determined similar to r_a^{wc}, whereas r_s^t is a function of stomatal conductance $\left(G_s^{st}\right)$, cuticular conductance $\left(G_s^{cu}\right)$, and canopy boundary layer conductance $\left(G_s^b\right)$. $\left(G_s^b\right)$ and G_s^{cu} are determined with biome-specific constant values, but G_s^{st} is a function of T_a, vapor pressure deficit, and LAI.

Evaporation from the soil surface (λE_s) is the sum of evaporation from wet soil (λE_{ws}) and evaporation from saturated soil (λE_{ss}) defined as $\lambda E_s = \lambda E_{ws} + \lambda E_{ss}$ with partitioning of soil surface to wet and saturated parts based on f_w. Hence evaporation from moist soil is

$$\lambda E_{ws} = f_{\text{wet}} \frac{\Delta(A_s - G_0) + (1 - f_c)\rho c_p \text{VPD}/r_a^s}{\Delta + \gamma \dfrac{r_s^s}{r_a^s}} \tag{10.5}$$

while evaporation from saturated soil is calculated as

$$\lambda E_{ss} = RH^{\text{VPD}/\beta}(1 - f_{\text{wet}}) \frac{\Delta A_s + (1 - f_c)\rho c_p \text{VPD}/r_a^s}{\Delta + \gamma \dfrac{r_s^s}{r_a^s}} \tag{10.6}$$

where r_a^s and r_s^s are the bulk aerodynamic and surface resistances against soil evaporation. $RH^{\text{VPD}/\beta}$ is a soil moisture constraint following Fisher et al. (2008) with β taken as a constant value. r_s^s is a function of biome-specific constant values of biological parameters and actual values of vapor pressure deficit.

To estimate global evapotranspiration using the above-mentioned Penman–Monteith algorithm, the required datasets include (1) vegetation characteristic data derived from Advanced Very High Resolution Radiometer (AVHRR) or MODIS products (land cover, vegetation fraction, LAI, and vegetation height); (2) meteorological inputs including T_a, humidity, and pressure; and (3) radiative data such as incoming shortwave and longwave radiation, surface emissivity, and albedo. Not all inputs are available from remote sensing systems, so reanalysis data have been used in some instances instead (see Mu et al. 2007, 2011).

A flow chart of the PM-Mu model with various inputs is provided in Figure 10.1 with further details on the meteorological forcing used in the scheme together with spatial and temporal resolution and extent of the recent global product provided in Section 10.2.3. Mu et al. (2007, 2011) and Vinukollu et al. (2011b) describe some efforts toward evaluating this particular scheme. The PM-Mu approach forms the basis of the official MODIS based

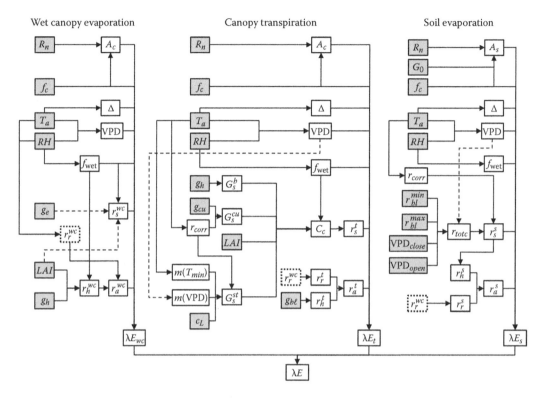

FIGURE 10.1
Model schematic of the Penman–Monteith–based Mu (2007) approach and the required data forcing (see also Table 10.2).

evapotranspiration product (see http://modis.gsfc.nasa.gov/data/dataprod/for further details).

10.2.2.2 Priestley–Taylor–Based PT-JPL Approach

The first global application of the Priestley–Taylor approach driven by satellite remote sensing observations is described in the PT Jet Propulsion Laboratory (JPL) algorithm by Fisher et al. (2008). PT-JPL (Fisher et al. 2008) is based on the formulation of the Priestley and Taylor (1972) equation, which eliminates the need to parameterize the stomatal and aerodynamic resistances and leaves only equilibrium evaporation multiplied by a constant value of 1.26 (the so-called alpha coefficient):

$$\lambda E_p = \alpha_{PT} \frac{\Delta}{\Delta + \gamma} R_n \tag{10.7}$$

To reduce λE_p to actual $E(\lambda E)$, Fisher et al. (2008) introduced a number of ecophysiological constraint functions (unitless multipliers referred to as f functions, with values between 0 and 1) based on atmospheric moisture (vapor pressure deficit VPD and relative humidity

RH) and vegetation indices (normalized difference and soil-adjusted vegetation indices, NDVI [Goward et al. 1991], and SAVI [Huete 1988], respectively). The driving equations in the model are

$$\lambda E = \lambda E_s + \lambda E_t + \lambda E_i \tag{10.8}$$

$$\lambda E_t = (1 - f_{\text{wet}}) f_g f_T f_M \alpha \frac{\Delta}{\Delta + \gamma} R_n^c \tag{10.9}$$

$$\lambda E_s = (f_{\text{wet}} + f_{SM}(1 - f_{\text{wet}})) \alpha \frac{\Delta}{\Delta + \gamma} \left(R_n^s - G_0 \right) \tag{10.10}$$

$$\lambda E_i = f_{\text{wet}} \alpha \frac{\Delta}{\Delta + \gamma} R_n^c \tag{10.11}$$

where λE_s, λE_t, and λE_i are evaporation from the soil, transpiration from canopy, and evaporation from intercepted water, respectively, each calculated explicitly and summing to total λE. f_{wet} is the relative surface wetness (RH[4]), f_g is the green canopy fraction ($f_{\text{APAR}}/f_{\text{IPAR}}$), and f_T is a plant temperature constraint.

$$f_T = \exp\left[-\left(\frac{T_a - T_{opt}}{T_{opt}} \right)^2 \right] \tag{10.12}$$

where f_M is a plant moisture constraint ($f_{\text{APAR}}/f_{\text{APARmax}}$) and f_{SM} is a soil moisture constraint (RH[VPD]). f_{APAR} is the fraction of absorbed photosynthetically active radiation (PAR), f_{IPAR} is the fraction of intercepted PAR, T_{max} is the maximum T_a, and T_{opt} is the optimum plant growth temperature estimated as the T_a at the time of peak canopy activity when the highest f_{APAR} and minimum VPD occur. No calibration or site-specific parameters are required of this approach. Figure 10.2 provides a schematic of the model structure and required input forcing.

Four general data inputs are required to drive the PT-JPL algorithm: (1) net radiation, (2) T_a, (3) actual vapor pressure, and (4) vegetation indices via NDVI or Enhanced Vegetation Index (EVI) (Jiang et al. 2008). These may be obtained from any number of sources, including *in situ* measurements (e.g., flux towers), merged ground stations (e.g., Climate Research Unit [CRU]), reanalyses (e.g., Modern Era Retrospective analysis for Research and Applications [MERRA]), and remote sensing or blended remote sensing products. The original Fisher et al. (2008) product was derived from surface radiation budget (SRB) R_n, CRU T_a and e_a, and surface reflectances from AVHRR. More recently, Vinukollu et al. (2011a,b) and Sahoo et al. (2011) drove PT-JPL entirely with remote sensing using R_n from Clouds and Earth's Radiant Energy System (CERES), meteorology from the Atmospheric Infrared Sounder (AIRS), and vegetation properties from MODIS. Badgley et al. (in press) ran PT-JPL with 19 different combinations of forcing data, including three sets of radiation products (SRB, CERES,

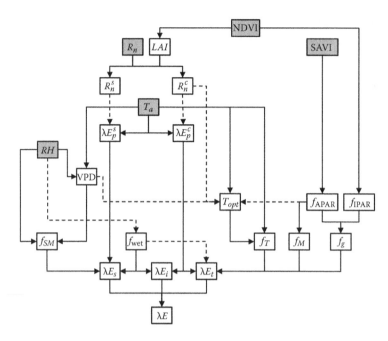

FIGURE 10.2
Schematic representation of the Jet Propulsion Laboratories modified Priestley–Taylor model.

ISCCP), three sets of meteorological datasets (CRU, AIRS, MERRA), and three vegetation index products (MODIS, GIMMS, FASIR), where ISCCP is the International Satellite Cloud Climatology Project, GIMMS is the Global Inventory Modeling and Mapping Studies, and FASIR is the Fourier-Adjusted Sensor and solar zenith angle corrected, interpolated and reconstructed.

PT-JPL has been run at both daily (Phillips et al. 2009) and monthly (Fisher et al. 2008; Sahoo et al. 2011; Vinukollu et al. 2011a,b; Badgley et al. in press) temporal resolutions and at 1° (Fisher et al. 2008), 0.5° (Fisher et al. 2009), 5 km (Sahoo et al. 2011; Vinukollu et al. 2011a,b), and 1 km spatial resolutions for up to 23 years (1984–2006) (Badgley et al. in press). The model code and output may be downloaded from josh.yosh.org.

PT-JPL has been tested against measured E from 36 FLUXNET (Baldocchi et al. 2001; Baldocchi 2008) sites worldwide, with a monthly average coefficient of determination (r^2) of 0.90 across all sites, RMSE of ~12 mm year^{-1}, and a slope/bias of 1.07 using *in situ* data (Fisher et al. 2008, 2009). Vinukollu et al. (2011a) recently conducted an independent inter-comparison of PT-JPL, PM-Mu (Mu et al. 2007), and SEBS (Su 2002) with common forcing data, showing PT-JPL to have the lowest bias and root-mean-squared deviation (RMSD), and highest Mann-Kendall (τ) across a range of comparison data, including watershed basins and flux towers. A Method of Moments (Hansen 1982; Rushdi and Kafrawy 1988; Madsen et al. 1997; Warnick and Chew 2004) analytical uncertainty assessment and resultant uncertainty map are provided by Fisher et al. (2008). Through a separate project, the JPL group is currently working on producing the PT-JPL product at 1 km^2 globally for the MODIS era and expanding the FLUXNET validation and error assessment to the full suite of sites in the La Thuile database. An example global map of E from PT-JPL is shown in Figure 10.3.

FIGURE 10.3
(See color insert.) Total monthly evaporation (millimeters per month) from PT-JPL for 1994.

10.2.2.3 Global Land Surface Evaporation: The Amsterdam Methodology (GLEAM)

GLEAM is a suite of algorithms that can be used to derive evaporation and its components (i.e., transpiration, interception loss, bare soil evaporation, sublimation, and open-water evaporation) at the global scale using traditional methods designed for local and regional scales. The method aims to combine the wide range of variables that are observable with satellite sensors and that contain information relevant to the evaporative flux in nature. Miralles et al. (2011b) describe the formulations in GLEAM and validate the outputs of the methodology against FLUXNET observations of E and Soil Climate Analysis Network (SCAN) measurements of soil moisture. Figure 10.4 details how the methodology is structured.

As a first step, a PT equation is applied to R_n and T_a observations to estimate potential evaporation. Separately, a multilayer running water balance is used to describe the infiltration of rainfall through the vertical soil profile and microwave surface soil moisture observations are assimilated into the top soil layer using a Kalman filter (Crow 2007). The purpose of this soil module is to convert observations of precipitation and surface soil moisture into estimates of soil water content at different depths across the root zone. Once these estimates have been derived, they are transformed into evaporative stress, parameterized in the form of a stress factor (ranging from 0 to 1).

The stress factor is multiplied with the PT estimates of potential evaporation to derive transpiration (for the vegetated regions) and soil evaporation (for bare soils). The conversion of the root zone soil moisture into estimates of evaporative stress is done through the use of empirical relationships and in combination with observations of vegetation water content (microwave vegetation optical depth) (see Liu et al. 2013). These empirical relations are specific for each cover type (i.e., tall and short vegetation and bare soil) and are based on the results of previous field experiments studying physiological response to drought.

Sublimation is calculated for regions covered by snow and ice (revealed by satellite observations of snow depth) based on a PT equation run with parameters designed for ice and super-cooled waters (Murphy and Koop 2005). Open-water evaporation is assumed

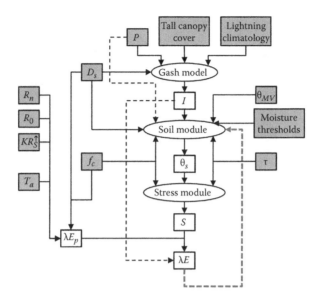

FIGURE 10.4
Remote sensing datasets used as input in GLEAM and a schematic of how they are combined within the four modules to derive the different components of evaporation.

to be PT potential evaporation and calculated using specific values of albedo and ground heat flux for open water (based on literature).

Interception loss is independently derived for tall vegetation using a Gash analytical model (Gash 1979; Valente et al. 1997). Following Gash's approach, a running water balance driven by rainfall observations is calculated for canopies and trunks, using parameters that describe the canopy cover, canopy and trunk storage, and average rainfall and evaporation rates during the time that canopies are saturated. In GLEAM, mean rainfall rates are derived using a remote sensing–based climatology of lightning frequency. Miralles et al. (2010) describe the derivation of the parameters, global implementation, and validation of this interception model. A set of interception loss measurements from past field studies over different forest ecosystems showed strong correlation and negligible bias against model results.

The methodology has recently been used to study the magnitude and global variability of land surface evaporation (Miralles et al. 2011a) and soil moisture–temperature feedbacks during heat waves (Miralles et al. 2012). The algorithms are currently being adapted to the regional scale by improving the realism of the soil module. Global fields of evaporation, evaporative stress, soil moisture, and evaporation components are currently available at a 0.25° resolution spanning the period from 1984 to 2007. The methodology has been run with different sets of input data, both at daily and three-hourly time steps.

10.2.2.4 SEBS Model

The SEBS model (Su 2002) was proposed for the estimation of atmospheric turbulent fluxes and evaporative fraction using satellite and *in situ* meteorological data based on the surface energy balance equation:

$$\lambda E = R_n - G_0 - H \tag{10.13}$$

The SEBS algorithm is structured around estimating the sensible heat flux based on the gradient of surface temperature and T_a, and the parameterization of the aerodynamic resistance. In order to derive the sensible heat flux, Monin–Obukhov similarity theory equations for the atmospheric surface layer are used:

$$u = \frac{u_*}{\kappa}\left[\ln\left(\frac{z-d_0}{z_{0m}}\right) - \Psi_m\left(\frac{z_{0m}}{L}\right)\right] \tag{10.14}$$

$$\theta_0 - \theta_a = \frac{H}{\kappa u_* \rho_a c_p}\left[\ln\left(\frac{z-d_0}{z_{0h}}\right) - \Psi_h\left(\frac{z-d_0}{L}\right) + \Psi_h\left(\frac{z_{0h}}{L}\right)\right] \tag{10.15}$$

where z is the height above the surface, u_* is the friction velocity, κ is the von Karman's constant (0.41), d_0 is the zero plane displacement height, z_{0m} and z_{0h} are the roughness heights for momentum and heat transfer, θ_0 is the potential land surface temperature, and θ_a is the potential T_a at height z. Ψ_m and Ψ_h are the stability correction functions for momentum and sensible heat transfer, respectively, and L is the Obukhov length defined as

$$L = -\frac{\rho_a c_p u_*^3 \theta_v}{\kappa g H} \tag{10.16}$$

where g is the acceleration due to gravity and θ_v is the potential virtual T_a at level z.

Two limiting cases are considered to derive evaporative fraction by constraining the sensible heat flux estimates between an upper and lower boundary. Under the dry limit, the latent heat becomes zero because of the extreme limitation of soil moisture and the sensible heat is at its maximum value as $H_{dry} = R_n - G_0$. Under the wet limit, evaporation takes place at a potential rate that is constrained only by the available energy under the given surface and meteorological conditions, and the sensible heat flux reaches its minimum value, H_{wet}, obtained via the PM equation:

$$H_{wet} = \frac{(R_n - G_0) - \dfrac{\rho_a c_p}{r_{ew}}\dfrac{VPD}{\gamma}}{1 + \dfrac{\Delta}{\gamma}} \tag{10.17}$$

where r_{ew} is the bulk external resistance at the wet limit and can be derived by

$$r_{ew} = \frac{1}{\kappa u_*}\left[\ln\left(\frac{z-d_0}{z_{0h}}\right) - \Psi_h\left(\frac{z-d_0}{L_w}\right) + \Psi_h\left(\frac{z_{0h}}{L_w}\right)\right] \tag{10.18}$$

and the wet-limit stability length is determined as

$$L_w = -\frac{\rho_a u_*^3}{0.61 \kappa g (R_n - G_0)/\lambda}$$ (10.19)

Thus the relative evaporation and evaporative fraction are given by

$$\Lambda_r = 1 - \frac{H - H_{\text{wet}}}{H_{\text{dry}} - H_{\text{wet}}}$$ (10.20)

$$\Lambda = \frac{\Lambda_r \lambda E_{\text{wet}}}{R_n - G_0}$$ (10.21)

The roughness parameters for heat and momentum are estimated using the kB^{-1} (inverse Stanton number) model as proposed by Massman (1999).

There have been numerous model evaluation and intercomparison exercises undertaken using the SEBS model, including the earlier works of Su et al. (2005) and McCabe and Wood (2006), as well as more recent contributions of McCabe et al. (2011) and Vinukollu et al. (2011b), who implemented a canopy interception component to SEBS in order to

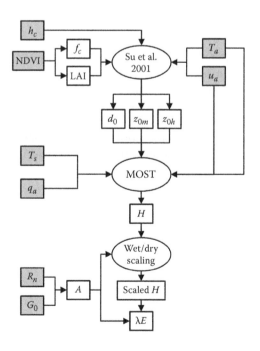

FIGURE 10.5
Schematic detailing processes and steps involved in the SEBS model, together with forcing data requirements (see Table 10.2 for further details).

estimate global evaporation using multisensor remote sensing data. Further details on the SEBS model can be found in the works of Su et al. (2001) and Su (2002). Figure 10.5 outlines a typical flowchart of the SEBS model.

10.2.3 Comparison of Model Descriptions and Forcing Data Requirements

Tables 10.1 and 10.2 provide a comparison of the major modeling schemes that form the basis of this chapter, focusing exclusively on those algorithms described in Section 10.2.2. Table 10.2 details the individual forcing data requirements of each of these models (used subsequently in Section 10.3), together with the source of data commonly used for each of the model variables. Table 10.2 is separated by forcing data type rather than model type, as not all models use the same data. It should be noted that the data source is application specific and generally refers to the product that has been provided to the LandFlux working group (by the product developers) as part of the evaluation and intercomparison exercise (see details in Section 10.3.2).

TABLE 10.1

Comparison of Modeling Schemes as Described in Section 10.2.2, with a Focus on Those Formulations that Form Part of the GEWEX LandFlux Project

Product	Method	Interception	Sublimation	Ground Flux	Resolution
PM-Mu Mu et al. (2011)	Penman–Monteith formulation with biome-specific aerodynamic and surface resistance	Modeled by aerodynamic resistance term	NEM	As function of daily temperature, R_n and vegetation fraction (f_c)	8 day 1 km × 1 km 2000–2011
GLEAM Miralles et al. (2011b)	Priestley and Taylor model with stress factor and a water budget module linking precipitation, soil moisture, and vegetation	Analytical model (Gash 1979; Valente et al. 1997) driven by global precipitation (Miralles et al. 2010)	PT equation with snow adapted constants from Murphy and Koop (2005)	As function of R_n with surface type specific variables	Daily 0.25° × 0.25° 1984–2007
PT-JPL Fisher et al. (2008)	Priestley and Taylor formulation with stress factors using biophysical metrics	Priestley and Taylor equation multiplied by fraction of time with wet surface	NEM	Not modeled; assumed 0 at monthly time steps	Monthly 0.5° × 0.5° 1983–2007
SEBS-LF Su (2002) *as run by Vinukollu et al. (2011b)	*E* as residual in energy balance after estimation of sensible flux from air-surface temperature gradient	*Mass-balance strategy based on the models of Valente et al. (1997)	*Penman equation as in work of Calder (1990)	As function of R_n with interpolation based on fractional canopy coverage	Daily 0.5° × 0.5° 1984–2007

Note: NEM, not explicitly modeled. SEBS-LF refers to the SEBS model run as part of the GEWEX LandFlux Version 0 product (described in Section 10.3.1).

TABLE 10.2

Summary of Model Forcing Data Requirements and Sources for Each of the Schemes Described in Table 10.1, Differentiated by Input Data Type

Product		Purpose and Description
Radiative Fluxes		
PM-Mu	MERRA GMAO	Daily downwelling shortwave (SW) fluxes, from 0.5° × 0.6° grid to 1-km MODIS pixels; upwelling SW using MODIS Col. 5 albedo, net longwave (LW) flux from Ta and estimates of surface and air emissivity
GLEAM	NASA/GEWEX SRB 3.0	Daily net SW and LW fluxes, from 1° to 0.25° grid
PT-JPL	NASA/GEWEX SRB	Monthly net SW and LW fluxes, from 1° to 0.5° grid
SEBS-LF	NASA/GEWEX SRB 3.0	3-hourly net SW and LW fluxes, from 1° to 0.5° grid
Surface Temperature		
SEBS-LF	VIC land-surface model (Sheffield and Wood 2007)	To estimate temperature gradient between the surface and overlying atmosphere (air temperature)
Surface Soil Moisture		
GLEAM	LPRM v04d (Owe et al. 2008)	Assimilated with the modeled water content of the first soil layer in water budget module; based on microwave observations
Precipitation		
GLEAM	CMORPH	Daily precipitation as input to water budget module, separated into rain and snow by snow-depth observations, from 0.07° to 0.25°; GPCP 4.0 to gap fill outside 60°S–60°N; Snow depth from National Snow and Ice Data Centre (NSIDC) Advanced Microwave Scanning Radiometer (AMSR-E)/Aqua daily L3 v001 (Kelly et al. 2003)
Air Temperature		
PM-Mu	MERRA GMAO	To estimate LW flux, slope of eSAT/T, VPD as function of T, and to parameterize aerodynamic and surface resistance, resolutions as described for the radiative fluxes.
GLEAM	ISCCP/AIRS	Use of AIRS near-surface air temperature for the period 2003–2007; ISCCP use for gap-filling and before 2003
PT-JPL	CRU TS 3.0	Slope of e_{sat}/T, plant temperature constraint, 0.5° × 0.5°
SEBS-LF	Princeton forcing (Sheffield et al. 2006)	Estimate surface-air temperature gradient; product is NCEP-NCAR temperature bias-corrected with CRU TS 2.0
Wind Speed		
SEBS-LF	Princeton forcing (Sheffield et al. 2006)	To estimate aerodynamic conductance, from ~2.0° to 0.5°
Water Vapor Pressure		
PM-Mu	MERRA-GMAO	To estimate VPD and relative humidity, and f_{wet} with resolutions as described for the radiative fluxes.
PT-JPL	CRU TS 3.0	To estimate VPD and relative humidity for soil moisture constraint and relative surface wetness, 0.5° × 0.5°
Vegetation Index		
PM-Mu	NASA MODIS Col5	FPAR for fractional veg cover, LAI to estimate resistance, 1 km; gap filled as in the work of Zhao et al. (2005)
GLEAM	LPRM v04d based on AMSR-E MW	Microwave vegetation optical depth as a proxy for vegetation water content at 0.25°; only applied to vegetated fractions of grid pixel

(*continued*)

TABLE 10.2 (Continued)

Summary of Model Forcing Data Requirements and Sources for Each of the Schemes Described in Table 10.1, Differentiated by Input Data Type

Product		Purpose and Description
PT-JPL	NASA MODIS Col5	NDVI, EVI to estimate fraction of PAR absorbed by green and total vegetation and derive green canopy fraction and plant moisture constraint
SEBS-LF	AVHRR Boston University (Tucker et al. 2005)	NDVI, LAI to derive the fractional vegetation cover and other parameters to be used in the determination of surface roughness height, from 8 km, 15 days
Land Cover/Soil Properties		
PM-Mu	MODIS Col. 4 (Friedl et al. 2010)	To determine land cover and fractional vegetation type, including 11 vegetation types
GLEAM	MOD44B Veg. Cont. Fields (Hansen et al. 2005) FAO's world soil database (Food and Agriculture Organization 2000)	To describe every pixel as a combination of tall canopy, herbaceous veg. and bare soil, at 1 km To define soil parameters (wilting point, critical soil moisture, and field capacity) for moisture module
PT-JPL	ISLSCP Initiative II common land/water mask	For land mask, no biome-specific parameters
SEBS-LF	MODIS-based land cover type (MOD12Q1) database	For land mask, no biome-specific parameters

Note: That a range of possible forcing data sources exist, but the information here refers specifically to the intercomparison study detailed in Section 10.3. SEBS-LF refers to the SEBS model run as part of the GEWEX LandFlux (LF) Version 0 product (described in Section 10.3.1).

10.3 Global Evaluation and Intercomparison Efforts

While the need for global heat flux estimates is apparent, the utility of such products is intrinsically linked to the accuracy and fidelity of the retrievals. Given that the capacity to undertake traditional point scale evaluation of competing approaches is limited by data availability, the representative scale of measurements, and the general lack of long-term evaluation datasets, alternative approaches to product assessment are required. Model intercomparison exercises provide an important step in assessing a product's capacity to reproduce expected responses, while also allowing insight into model dependencies, sensitivities, and limitations. Following is a brief description of recent efforts to address this task, focusing on the key methodologies described in Section 10.2.2 but also including other flux prediction schemes, such as land surface models and reanalysis datasets. These different estimation approaches allow a first-order assessment of uncertainty by providing an envelope of expected flux response.

10.3.1 Remote Sensing Model Intercomparison

The aim of the remote sensing–based intercomparison is to highlight the typical geographical and seasonal differences found when these products are compared across a given number of years. The period selected here is 2003–2006, as this date range provides improved capacity for evaluation against compilations of ground data records, relative to more recent periods. Most of these surface heat flux estimates are available as multidecade products, but a climatological evaluation of the long-term estimates is not intended in this section.

Products selected for analysis include published estimates from PT-JPL (but using MODIS for vegetation instead of AVHRR to cover 2003–2006 as reported by Badgley et al. [in press]), PM-Mu, SEBS, and GLEAM. Versions of PT-JPL, PM-Mu, and SEBS have been formulated by Sahoo et al. (2011) and Vinukollu et al. (2011b), with major differences related to a harmonization of those three methodologies (e.g., similar aerodynamic conductance or a common model for evaporation from canopy interception for the three algorithms; see Sahoo et al. [2011] and Vinukollu et al. [2011b] for the details). The Sahoo et al. (2011) and Vinukollu et al. (2011b) versions of PT-JPL, PM-Mu, and SEBS approaches have been rerun as part of the GEWEX LandFlux Project, along with a rerun of GLEAM using a consistent forcing dataset. The use of a common set of inputs allows the attribution of interproduct differences to the algorithms. However, the range of inputs required by each methodology is different, and therefore despite this effort, differences in driving data still exist. These new product simulations are also included in the comparison and are referred to by the same acronyms, but with a "-LF" suffix (i.e., GLEAM-LF, where LF refers to LandFlux) to distinguish them from the original products. A summary of these products and their associated forcing data can be found in Tables 10.1 and 10.2. Note that the forcings described for SEBS-LF also apply to -LF product runs.

The comparison is complemented by a global regression of eddy covariance measurements from FLUXNET by a machine-learning approach called model tree ensembles (Jung et al. 2009, 2011) and referred to here as the MPI-BCG model. The MPI-BCG model is driven by a long-term monthly fraction of absorbed photosynthetically active radiation (fAPAR) dataset, established by harmonizing AVHRR NDVI data (Vermote and Saleous 2006) with fAPAR from SeaWiFS (Gobron et al. 2006) and fAPAR from MERIS (Gobron et al. 2008), near-surface T_a from CRU, precipitation data from the Global Precipitation Climatology Center (GPCC) (Rudolf and Schneider 2005), and an estimate of the top of the atmosphere shortwave radiation.

To facilitate intercomparison, the different products have been aggregated to a common spatial and temporal resolution. First, the spatial resolution of the products has been downgraded to the 0.5° × 0.5° resolution by spatially averaging the original estimates. Next, the products were space-matched, that is, only pixels having estimates from all products were retained. Finally, the products were time matched, that is, only pixels having estimates for all months in the years are kept. This guarantees that differences in the statistics are not due to different spatial coverage or period of record. Note that Greenland and Antarctica are not included in the intercomparison.

After these operations, approximately 70% of the total land surface (not including Greenland or Antarctica) is retained, with most of the missing pixels over coastal regions and arid regions in Northern Africa and Central Asia. This implies that the reported globally averaged land surface fluxes will not be truly global (although for simplicity they will be referred to as global). Note also that not all products provide an estimate of the evaporation from snow, so this component is excluded from the products that explicitly estimate this component. This means that evaporation refers to here as the sum of evaporation for bare soil, transpiration from vegetation, and evaporation from canopy interception.

10.3.1.1 Global Seasonal Evaporative Response

The 2003 global evaporation annual cycles for the different products are plotted in Figure 10.6. Similar numbers are obtained for the other years, with small interannual

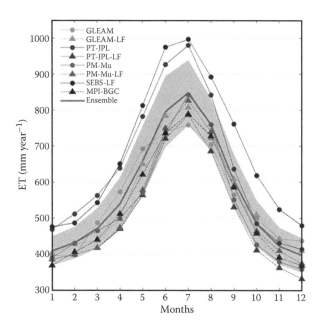

FIGURE 10.6
(**See color insert.**) Annual cycles of global average evaporation for 2003. The gray line and shadow display the ensemble mean and the standard deviation (±σ) of the individual product monthly means around the ensemble mean, respectively.

variation. All products have maximum global evaporation in July, illustrating the dominance of the northern hemisphere land areas. At the cycle maximum, there is a spread of around 250 mm year^{-1}, with an ensemble mean and standard deviation of approximately 840 and 100 mm year^{-1}, respectively. The evaporation ensemble mean and standard deviation of the annual means is around 590 and 70 mm year^{-1}, respectively. These figures are very similar to the average fluxes reported by Jiménez et al. (2011) (~45 and ~6 W/m^2, respectively) for a different period of time and using different products but for a similar percentage of land, suggesting some robustness in these estimates. Note that as the intercomparison here is displayed in units of millimeters, and it excludes a large fraction of arid and permanent snow-cover regions, the average global fluxes are anticipated to be somewhat lower than reported in Figure 10.6. Comparing the ensemble annual evaporation with the SRB dataset and Global Precipitation Climatology Project (GPCP) annual rainfall estimates give approximately 57% of the net radiation partitioned into latent heat fluxes and 65% of the precipitation is returned via the evaporative process over land.

The 2003 all-product ensemble evaporation annual average and the absolute and relative standard deviation (from the eight products annual means and normalized by the all-product ensemble average, respectively) are shown in Figure 10.7. In broad terms, the expected spatial structures related to the main climate regimes and topographical features are present in the ensemble average, suggesting that most products capture those structures. Nevertheless, the relatively large variability found in some regions suggests that the absolute values of the fluxes can be quite different from one product to the other. Examples of such regions include the Parana Basin and parts of the Brazilian eastern Atlantic region in South America and large regions in southern Asia.

Zonal means of evaporation for 2003 can be found in Figure 10.8. In broad terms, the shapes of the latitudinal distributions are generally consistent from one product to the

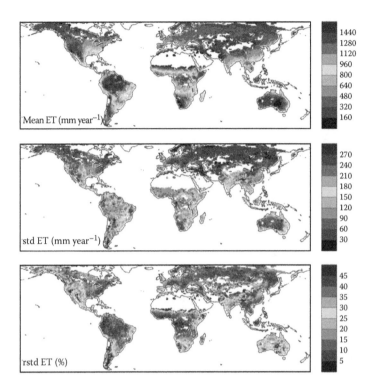

FIGURE 10.7
(See color insert.) The 2003 all-product ensemble mean evaporation (top), standard deviation (std, middle), and relative standard deviation (rstd, bottom) expressed as a percentage of the pixel mean value. Absence of data from some products precludes the computation of the averages over some regions (mainly arid regions).

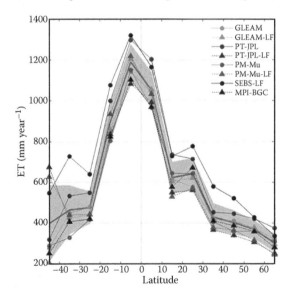

FIGURE 10.8
(See color insert.) Zonal means of the 2003 evaporation calculated from the different model schemes and forcings. The gray line and shadow display the ensemble mean and the standard deviation ($\pm\sigma$) of the individual product monthly means around the ensemble mean, respectively.

other, reflecting expected patterns in the annually averaged water and energy availability at different latitudes. Nevertheless, the averages at some latitudes can be quite different (i.e., large variability is observed below 30°S). All zonal means peak in the 10°S–0 latitude band, coinciding with the tropical region that supplies most of the global evaporation. However, relatively large differences can be seen within the zonal estimates, ranging from 1100 to 1500 mm year^{-1}. Similar zonal means (not plotted) are found for the remaining years.

10.3.1.2 Monthly Evaporation Patterns

Examples of monthly evaporation (February and August 2003) for the different products are provided in Figure 10.9. Differences between the February and August monthly mean reflect expected seasonal variations in radiation and precipitation patterns. As with the annual averages, the main geographical structures related to the major climatic regimes and geographical features are, in general, present across all products. Nevertheless, the differences in the absolute values can be large. Note that all products except PM-Mu and MPI-BGC were forced with similar net radiative fluxes (SRB 3.0), implying that the partitioning of the radiative fluxes into the sensible and latent heat flux components can be quite variable from one product to the other. For instance, the change of forcing and harmonization of the PT-JPL model in the work of Sahoo et al. (2011) and Vinukollu et al. (2011b) resulted in significant differences between the original PT-JPL and PT-JPL LandFlux forced estimate. Smaller differences are found between GLEAM and GLEAM-LF and PM-Mu and PM-Mu-LF. Compared with the other products, the extrapolation of FLUXNET estimates (MPI-BGC) presents some distinctive features in some regions (e.g., for the August estimates, the large fluxes in the transition region between tropical and northern arid regions in Africa, or the large estimates over extensive parts of India compared with the surrounding regions in southern Asia).

A comparison of 2003–2006 monthly mean estimates against tower-based eddy covariance measurements derived from the "open-use" FLUXNET data archive (http://www.fluxdata.org/SitePages/AboutFLUXNET.aspx) is summarized in Figure 10.10. The daily tower-based latent heat fluxes were averaged to monthly values and converted into water depth equivalents of mm month^{-1}. Twenty-eight stations were found to have fluxes in 2003–2006 within the regions considered in the intercomparison (approximately 500 total monthly mean matches). It should be noted that the global estimates spatially average over a very large area (~2500 km^2 for pixels close to the equator), compared with the typical footprint from the eddy covariance tower measurements (up to ~3 km, depending on tower and location). Note also that tower estimates were used as given, without any attempt to correct for the routinely reported underestimation of the turbulent fluxes as a response to closure related issues (e.g., Foken 2008).

Correlation (R), standard deviation of the differences (STD), and RMSD between the eddy covariance estimates and the individual products were calculated by way of a Taylor diagram, as shown in the work of Figure 10.10. Correlations range between 0.77 (PT-JPL-LF) and 0.84 (MPI-BGC), standard deviations between 17 (MPI-BGC) and 28 (SEBS-LF) mm month^{-1}, and RMSD between 31 (MPI-BGC) and 42 (SEBS-LF) mm month^{-1}. While MPI-BGC shows the best statistics, the response is not completely independent, as the model may be regressed (due to the particular methodology) by the same tower data against which it is being assessed. After MPI-BCG, GLEAM exhibits the next best response against tower measurements. Comparing the original products (PM-Mu, PT-JPL, and GLEAM) with the runs from the LandFlux data forced models (-LF), better agreement is found for

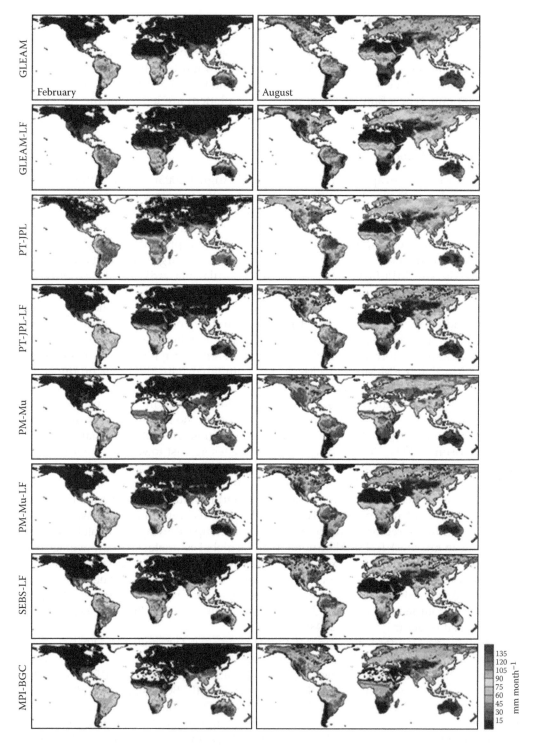

FIGURE 10.9
(**See color insert.**) Monthly averaged evaporation for (left) February and (right) August 2003.

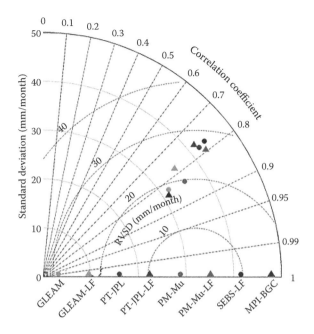

FIGURE 10.10

Taylor diagram summarizing the 2003–2006 comparison of eddy covariance measurement from FLUXNET. Individual evaporation products are described in Section 10.3.1.1.

the original products. It is worth noting that the likely underestimation of the eddy covariance measurements would penalize the products with higher fluxes (e.g., SEBS-LF and PT-JPL).

10.3.1.3 Basin-Averaged Evaporation

Monthly time series of the spatially averaged evaporation for a selection of four major river basins are presented by Figure 10.11. In South America, the Amazon Basin is representative of a tropical region with large rainfall and relatively small seasonal variability. Further south, the Parana River is more representative of wet–dry tropical/subtropical areas. The Volga Basin in Eurasia covers different climate zones that change significantly, from snowy winters and warm-humid summers in the north to little snow winters and hot-dry summers in the south. Finally, the Murray Basin in Australia represents a drier midlatitude region with large seasonal and interannual variability.

For the Amazon Basin, significant differences can be found in the way the different products model the seasonality, for example, see the contrasting seasonal patterns between GLEAM and PM-Mu. While PM-Mu seasonality in the Amazon follows the seasonality of solar radiation, in the case of GLEAM it is dictated by the strong seasonality of precipitation, given its effect via interception loss. Large differences in the Amazonian modeled latent heat flux annual cycle have also been reported elsewhere (e.g., Werth and Avissar 2004) where it was suggested that the differences develop via the way the vegetation controls the evapotranspiration in the models. Regarding ground measurements, time series of eddy covariance measurements in this basin (see the compilation reported by Fisher et

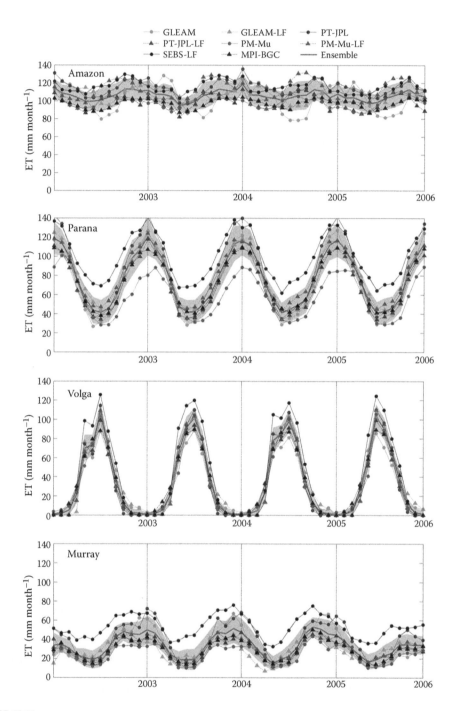

FIGURE 10.11
Spatially averaged evaporation for the years 2003–2006 for the Amazon, Parana, Volga, and Murray basins. The gray line and shadow display the ensemble mean and the standard deviation (±σ) of the individual product monthly means around the ensemble mean, respectively.

al. 2009) show that the seasonal changes depend on the location and water conditions at each specific site and year. Da Rocha et al. (2009) also reported large seasonal variations in some Amazonian sites at strongly water-limited regions with long dry seasons. It is therefore difficult to draw any firm conclusions regarding the expected seasonality for the basin-integrated estimates.

For the Parana Basin, the basin-integrated evaporation shows stronger seasonality. All products capture this seasonality, with maximum evaporation at the beginning of the year coinciding with the months of largest radiation. Nevertheless, large differences in the absolute values are found between the products with the smallest (PM-Mu) and the largest (SEBS-LF) estimates (e.g., differences of around 60 mm month^{-1} at the maximum of the basin-integrated evaporation). The Volga Basin shows very strong seasonal cycles, conditioned by the large annual variations in radiation and precipitation. This might imply that the estimates are more dependent on the forcing than on the model parameterizations (compared with the Amazonian Basin for example) and that the different estimates agree relatively well. For the Murray Basin, a strong seasonality is also displayed across all products, but with smaller annual variations. As for the Amazon Basin, the seasonal cycles from the different products do not always agree in their amplitude and phase (e.g., the end of 2006 with the maximum annual evaporation ranging from around 40 to 80 mm month^{-1} and peaking between October and January, depending on product).

10.3.2 Land Surface Model and Reanalysis Output

As noted previously, dedicated evaporation products derived from remote sensing require increasingly complex algorithms, because evaporation is not actually directly measured by satellites. On the one hand, this makes the comparison and evaluation of these satellite retrieved evaporation datasets, with (ideally) independent products, necessary in order to obtain an estimate of their validity (or uncertainty). On the other hand, remote sensing–based products such as those discussed in Section 10.3.1 should be seen as merged observation-model products. Global observation-based estimates of E can also be provided from other modeling-based environments using similar observational inputs. These model types include LSMs driven with observation-based forcing and reanalyses products. Parameterizations used in these model systems, which usually describe mass and energy budget considerations (as opposed to the single process based approaches described previously), are predominantly based on energy balance, aerodynamic approaches, and PM- or PT-type formulations. Typical input data to these schemes include precipitation, T_a, wind speed, specific humidity, radiation, and surface pressure. Observation-forced LSMs can be used to prepare reference products of high quality (such as those prepared within the Global Soil Wetness Project [GSWP] [Dirmeyer et al. 2006]).

Reanalysis products are based on coupled weather and climate models that assimilate (mostly atmospheric and satellite) observations. Even though land hydrological variables and precipitation are usually not assimilated in current reanalysis, the fact that the model state is updated with observations of near-surface temperature and humidity provides a strong constraint on the surface fluxes. Another family of coupled modeling schemes includes global climate models. These models can be driven by observations of temperature or carbon dioxide, if being run retrospectively. The atmospheric variables needed to force the land surface scheme are then, as compared to LSMs, a model product only and are likely to be less accurate than the observations-based forcing used for offline simulations. As a third model-data approach, one could also include "empirically derived" estimates, essentially resulting from the assimilation of observations in a statistical model, such as the MPI-BCG approach (Jung et al. 2010) detailed in Section 10.3.1 as part of the flux intercomparison.

The validation of E from observation-driven LSMs or reanalysis products can be performed against *in situ* observations, for instance, with eddy covariance data from the FLUXNET network as described by Figure 10.10. However, opportunities for comparisons with direct observations are generally limited, given the paucity of such data in both space and time. Therefore some recent initiatives have focused on the evaluation of evaporative fluxes based on intercomparison between a large number of observation-based datasets, LSMs and reanalysis. These gridded datasets are then used to evaluate E from global climate model simulations (e.g., IPCCAR4 simulations as in the work of Mueller et al. [2011]). The advantage of using these gridded estimates compared to *in situ* observations is a better match in scale, as well as the availability of observation-based information in regions not well covered by *in situ* measurements (which is much of the globe).

A significant challenge in evaporation evaluation is the nonexistence of a global benchmark dataset. This shortcoming has been addressed within the GEWEX supported LandFlux-EVAL initiative, which, following the intercomparison of existing global E datasets (Jiménez et al. 2011; Mueller et al. 2011), recently focused on the creation of multiproduct merged datasets to serve as benchmarks for model evaluation. There are two available datasets, spanning the periods 1989–1995 and 1989–2005, with each having 5 (or 7) diagnostic/dedicated datasets, 5 (or 28) LSMs and 4 reanalyses for the different years, respectively. This benchmarking effort aims at encompassing the whole range of available observation-based evaporation estimates.

Many of the currently available LSMs and reanalyses are listed in the work of Figure 10.12 together with their global land mean values of E. As a reference, we also include the LandFlux-EVAL (see the Web site listed in work of Figure 10.12) synthesis dataset (merged) that has been created from the most current diagnostic datasets, LSMs, and reanalyses. Some of the listed LSM simulations stem from projects within which different models were driven with the same forcing data, for example, the GSWP simulations (noted with GS and referenced in the work of Dirmeyer et al. [2006]) or the Water Model Intercomparison Project (WaterMIP) simulations (see, e.g., Haddeland et al. 2011). Apart from increasing the "plausibility" of the land surface estimates using a multitude of simulations, such projects can also assist in evaluating the uncertainty arising from the model structure through their use of identical forcing data.

The global mean evaporation from the remote sensing estimates from Section 10.3.1.1 amounts to 590 mm year^{-1} ($\pm$70 mm year^{-1}). The same estimate from the variety of LSMs and reanalyses in the work of Figure 10.12 ranges from 364 to 621 mm year^{-1}. This estimate is consistent with previous estimates, albeit of broader range. For example, Dirmeyer et al. (2006) estimated 488 mm year^{-1} (on average) for the GSWP simulations, while Haddeland et al. (2011) provided a range of 415 to 586 mm year^{-1} (or 60,000–85,000 km^3 year^{-1}) via the WaterMIP simulations. These uncertainties in absolute values of global E in current LSMs, and reanalyses are large and vary among different regions.

Some of these uncertainties underlying E from current observation-driven LSM estimates and reanalysis products are induced by the relatively large uncertainty of the forcing data. Precipitation, which is needed (and the dominant input globally) in every prognostic LSM for example, is measured routinely and with high accuracy in very few regions of the world. Further sources of uncertainty related to deficiencies or uncertainties in model parameterizations, for example, are discussed in Chapter 19.

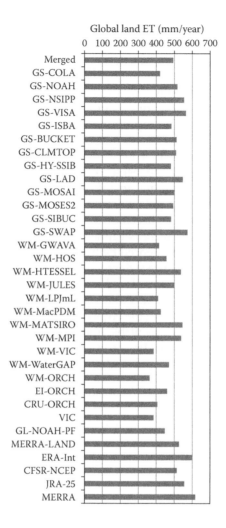

FIGURE 10.12
List of LSMs and reanalyses, including global land mean *E*. The first product in the list is a merged data product. A latitudinal weighting has been undertaken using the same formula as in the work of Haddeland et al. (2011; personal communication). Greenland is included in the data. Further details on the intercomparison can be found via the GEWEX LandFlux-Eval Web site at http://www.iac.ethz.ch/groups/seneviratne/research/LandFlux-EVAL.

10.3.3 Water Budget Assessment

As noted previously, one of the major issues in evaluation of modeled evaporation at large spatial scales is that there are no direct measurements of the flux either directly or independently from remote sensing (Kalma et al. 2008). To further complicate the evaluation task, point-scale FLUXNET observations cannot be simply extrapolated to the spatial scales of the remote sensing (or other model) resolution. While there have been some attempts to do this (e.g., MPI-BCG of Jung et al. [2010]), uncertainties in the approach are difficult to assess in regions with limited flux towers. One technique that has been proposed to assist in the evaluation exercise is the use of water budget estimates, or inferences, of evaporation (Vinukollu et al. 2011a,b). The water budget, as its name suggests, seeks to infer *E* as a residual component of the hydrological cycle via independent measurement of the remaining components.

The approach is designed to extend beyond point- or pixel-scale comparisons and to encompass the natural dynamics of catchment-scale hydrological processes. To infer evaporation using a water budget approach, one needs to ignore any storage changes within the hydrological system and then derive *E* by subtracting runoff (*Q*) from precipitation (*P*) as *E* = *P* − *Q*. As a result, the method is really only valid for long timescales (greater than or equal to annual) and also depends on the accuracy of the input fields (*P* and *Q*). At the catchment scale, *Q* can be evaluated using runoff records at the available stream flow gauging stations, but precipitation records are required from a network of rain gauges within the catchment or from gridded datasets. If the precipitation network is sufficiently dense and *P* and *Q* records are temporally consistent (with limited gaps), a good estimate of basin-scale *E* can be derived.

A common practice in case of gaps and insufficiency of precipitation data is to employ a spatiotemporal interpolation or extrapolation scheme. However, the development of such a scheme for runoff data is challenging because runoff stations are spatially sparse and temporally partial. In addition, runoff records cannot be directly interpolated or extrapolated in spatial extent because runoff is influenced by hydrological interactions between the saturated soil, vegetation interception, surface depressions, and storage pools. One method that employs spatial upscaling of runoff records is based on the water balance model (WBM) of Fekete et al. (2000). In this method, spatial scaling of the global runoff field is based on a simple balance model that extrapolates from the observed runoff of over 1000 worldwide gauged basins to ungauged regions, with data available from Global River Discharge Center (GRDC).

For the majority of important global basins (e.g., those used by Vinukollu et al. [2011b]), recent runoff data are not available and updating of the runoff records is challenging.

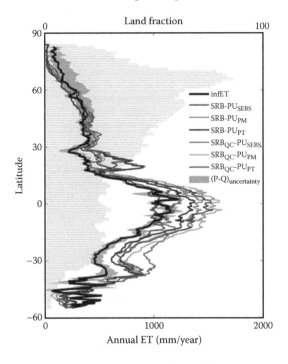

FIGURE 10.13
(See color insert.) A zonal intercomparison of global *E* models (see Section 10.2.2) against inferred *E* as determined from a selection of globally gridded precipitation datasets and the Global Runoff Data Centers Water Balance Method (GRDC-WBM) approach to calculate *P* − *Q*. Further details can be found in the work of Vinukollu et al. (2011).

For example, Vinukollu et al. (2011b) estimated climatological runoff (Q_{clim}) for 2003–2006 based on a simple runoff ratio concept. The authors calculated a mean annual precipitation using the GPCC precipitation data for 100 years (1901–2000) and assumed that the runoff ratio between 2003 and 2006 was related to that during 1901–2000, ignoring any water management changes. An estimated Q_{clim} for 2003–2006 was calculated by applying the runoff ratio of 1900–2000 to the climatological precipitation (P_{clim}) dataset. Inferred values of climatological evaporation λE_{clim} were then used to evaluate three remote sensing–based and seven operational analysis model–based estimates of E from basin to global scales. An example of this approach is shown by Figure 10.13, which is derived from the work of Vinukollu et al. (2011b).

10.4 Concluding Comments

With the number and variety of water and energy cycle products that are now (or in the near future) being developed, our ability to describe aspects of the Earth system dynamics has increased in an unprecedented manner. These developments have not come without their own problems though, foremost amongst a few being the issue of determining whether the increased information content of these products is actually informative: or more pertinently, which of the many competing model realizations is closest to the "truth." This itself is an obtuse target, as the numerous issues surrounding the validation and evaluation of model retrievals remain far from resolved.

For regional- and global-scale flux retrieval, it is unlikely that conventional evaluation methodologies will be appropriate, and thus the need to develop innovative techniques for model assessment and product evaluation (given the issues inherent in *in situ* measurements) to supplement (or supplant) traditional validation approaches is a necessarily high priority. The development of a common, standardized framework for the intercomparison and evaluation of global retrievals and competing model approaches is needed to facilitate such an effort.

Continuing efforts by the GEWEX LandFlux Initiative and other international groups and collaborators will ensure that the development of these global flux datasets will proceed at an exciting pace over the coming years. Such data will provide the Earth science community with an unprecedented capacity to monitor changes and variability in long-term flux behavior (at multiple spatial and temporal scales) and ultimately provide the information needed to characterize the drivers and mechanisms of these changes. Together with developments in global retrievals of other related water and energy cycle components, the community is primed to deliver insights and understanding of Earth system processes in ways that have not previously been possible.

References

Agam, N., W. P. Kustas, M. C. Anderson, J. M. Norman, P. D. Colaizzi, T. A. Howell, J. H. Prueger, T. P. Meyers, and T. B. Wilson (2010). Application of the Priestley-Taylor approach in a two-source surface energy balance model. *Journal of Hydrometeorology, 11*, 185–198.

Allen, R. G., L. S. Perista, D. Raes, and M. Smith (1998). Crop evapotranspiration—Guidelines for computing crop water requirements, Irrigation and Drainage Paper 56. Rome: Food and Agriculture Organization.

Badgley, G., J. B. Fisher, C. Jiménez, K. P. Tu, and R. K. Vinukollu (in press). On uncertainty in global evapotranspiration estimates from choice of input forcing datasets. *Geophysical Research Letters*.

Baird, A. (1999). *Eco-Hydrology*. London: Routledge.

Baldocchi, D. (2008). Turner review No. 15. 'Breathing' of the terrestrial biosphere: Lessons learned from a global network of carbon dioxide flux measurement systems. *Australian Journal of Botany, 56*, 1–26.

Baldocchi, D., E. Falge, L. Gu, R. Olson, D. Hollinger, S. Running, P. Anthoni et al. (2001). FLUXNET: A new tool to study the temporal and spatial variability of ecosystem–scale carbon dioxide, water vapor, and energy flux densities. *Bulletin of the American Meteorological Society, 82*, 2415–2434.

Ball, J. (1987). A model predicting stomatal conductance and its contribution to the control of photosynthesis under different environmental conditions. In *Progress Photosynthesis Reseach Proceedings of the International Congress 7th*, vol 4. Boston: Kluwer, pp. 221–224.

Brenner, A. J., and L. D. Incoll (1997). The effect of clumping and stomatal response on evaporation from sparsely vegetated shrublands. *Agricultural and Forest Meteorology, 84*, 187–205.

Calder, I. R. (1990). *Evaporation in the Uplands*. New York: John Wiley.

Cleugh, H. A., R. Leuning, Q. Mu, and S. W. Running (2007). Regional evaporation estimates from flux tower and MODIS satellite data. *Remote Sensing of Environment, 106*, 285–304.

Courault, D., B. Seguin, and A. Olioso (2005). Review on estimation of evapotranspiration from remote sensing data: From empirical to numerical modeling approaches. *Irrigation and Drainage Systems, 19*, 223–249.

Crow, W. T. (2007). A novel method for quantifying value in spaceborne soil moisture retrievals. *Journal of Hydrometeorology, 8*, 56–67.

Dalton, J. (1802). Experiments and observations to determine whether the quantity of rain and dew is equal to the quantity of water carried off by the rivers and raised by evaporation: With an enquiry into the origin of springs. Manchester: R & W Dean and Company, 30 pp.

Da Rocha, H. R., A. O. Manzi, O. M. Cabral, S. D. Miller, M. L. Goulden, S. R. Saleska, N. R. Coupe et al. (2009). Patterns of water and heat flux across a biome gradient from tropical forest to savanna in Brazil. *Journal of Geophysical Research, 114*, G00B12.

Davies, J., and C. Allen (1973). Equilibrium, potential and actual evaporation from cropped surfaces in southern Ontario. *Journal of Applied Meteorology, 12*, 649–657.

Dirmeyer, P. A., X. Gao, M. Zhao, Z. Guo, T. Oki, and N. Hanasaki (2006). GSWP-2: Multimodel analysis and implications for our perception of the land surface. *Bulletin of the American Meteorological Society, 87*, 1381–1397.

Doorenbos, J., and W. O. Pruitt (1975). Guidelines for predicting crop water requirements. *Irrigation and Drainage Paper 56*. Rome: Food and Agriculture Organization.

Food and Agriculture Organization (2000). Digital soil map of the world and derived soil properties, Rev. 1 (CD ROM). *FAO Land and Water Digital Media Series*. Rome.

Farahani, H. J., and L. R. Ahuja (1996). *Evapotranspiration Modeling of Partial Canopy/Residue-Covered Fields*. St. Joseph, MI: American Society of Agricultural Engineers.

Farahani, H., T. Howell, W. Shuttleworth, and W. Bausch (2007). Evapotranspiration: Progress in measurement and modeling in agriculture. *Transactions of the ASABE, 50*, 1627–1638.

Fekete, B., C. Vorosmarty, and W. Grabs (2000). Global, composite runoff fields based on observed river discharge and simulated water balances. Report. Federal Institute of Hydrology.

Ferguson, C. R., J. Sheffield, E. F. Wood, and H. Gao (2010). Quantifying uncertainty in a remote sensing-based estimate of evapotranspiration over continental USA. *International Journal of Remote Sensing, 31*, 3821–3865.

Fick, D. A. (1855). On liquid diffusion. *The London, Edinburgh, and Dublin Philosophical Magazine and Journal of Science, 10*, 30–39.

Fisher, J. B., Y. Malhi, D. Bonal, H. R. Da Rocha, A. C. De Araújo, M. Gamo, M. L. Goulden et al. (2009). The land–atmosphere water flux in the tropics. *Global Change Biology, 15*, 2694–2714.

Fisher, J. B., K. P. Tu, and D. D. Baldocchi (2008). Global estimates of the land-atmosphere water flux based on monthly AVHRR and ISLSCP-II data, validated at 16 FLUXNET sites. *Remote Sensing of Environment, 112,* 901–919.

Foken, T. (2006). 50 Years of the Monin–Obukhov Similarity Theory. *Boundary-Layer Meteorology, 119,* 431–447.

Foken, T. (2008). The energy balance closure problem: An overview. *Ecological Applications, 18,* 1351–1367.

Friedl, M. A., D. Sulla-Menashe, G. Tan, A. Schneider, N. Ramankutty, A. Sibley, and X. Huang (2010). MODIS Collection 5 global land cover: Algorithm refinements and characterization of new datasets. *Remote Sensing of Environment, 114,* 168–182.

Gash, J., and W. Shuttleworth (2007). Benchmark papers in hydrology: Evaporation. Wallingford: IAHS Press.

Gash, J. H. (1979). An analytical model of rainfall interception by forests quarterly. *Journal of Royal Meteorological Society, 105,* 43–45.

Gobron, N., B. Pinty, O. Aussedat, J. M. Chen, W. B. Cohen, R. Fensholt, V. Gond et al. (2006). Evaluation of fraction of absorbed photosynthetically active radiation products for different canopy radiation transfer regimes: Methodology and results using Joint Research Center products derived from SeaWiFS against ground-based estimations. *Journal of Geophysical Research, 111,* D13110.

Gobron, N., B. Pinty, O. Aussedat, M. Taberner, O. Faber, F. Mélin, T. Lavergne, M. Robustelli, and P. Snoeij (2008). Uncertainty estimates for the FAPAR operational products derived from MERIS—Impact of top-of-atmosphere radiance uncertainties and validation with field data. *Remote Sensing of Environment, 112,* 1871–1883.

Goward, S. N., B. Markham, D. G. Dye, W. Dulaney, and J. Yang (1991). Normalized difference vegetation index measurements from the Advanced Very High Resolution Radiometer. *Remote Sensing of Environment, 35,* 257–277.

Haddeland, I., D. B. Clark, W. Franssen, F. Ludwig, F. Voß, N. W. Arnell, N. Bertrand et al. (2011). Multimodel estimate of the global terrestrial water balance: Setup and first results. *Journal of Hydrometeorology, 12,* 869–884.

Hansen, L. P. (1982). Large sample properties of generalized method of moments estimators. *Econometrica: Journal of the Econometric Society, 50(4),* 1029–1054.

Hansen, M. C., J. R. G. Townshend, R. S. DeFries, and M. Carroll (2005). Estimation of tree cover using MODIS data at global, continental and regional/local scales. *International Journal of Remote Sensing, 26,* 4359–4380.

Huete, A. R. (1988). A soil-adjusted vegetation index (SAVI). *Remote Sensing of Environment, 25,* 295–309.

Jarvis, P. G. (1976). The interpretation of the variations in leaf water potential and stomatal conductance found in canopies in the field. *Philosophical Transactions of the Royal Society of London. Series B, Biological Sciences, 273,* 593–610.

Jensen, M. E. (1968). Water consumption by agricultural plants, Chapter 1, Water Deficits and Plants Growth, vol. 2. New York: Academic Press Inc.

Jiang, L., and S. Islam (1999). A methodology for estimation of surface evapotranspiration over large areas using remote sensing observations. *Geophysical Research Letters, 26,* 2773–2776.

Jiang, Z., A. R. Huete, K. Didan, and T. Miura (2008). Development of a two-band enhanced vegetation index without a blue band. *Remote Sensing of Environment, 112,* 3833–3845.

Jiménez, C., C. Prigent, B. Mueller, S. I. Seneviratne, M. F. McCabe, E. F. Wood, W. B. Rossow et al. (2011). Global intercomparison of 12 land surface heat flux estimates. *Journal of Geophysical Research, 116,* D02102.

Jung, M., M. Reichstein, and A. Bondeau. (2009). Towards global empirical upscaling of FLUXNET eddy covariance observations: Validation of a model tree ensemble approach using a biosphere model. *Biogeosciences, 6,* 2001–2013.

Jung, M., M. Reichstein, P. Ciais, S. I. Seneviratne, J. Sheffield, M. L. Goulden, G. Bonan et al. (2010). Recent decline in the global land evapotranspiration trend due to limited moisture supply. *Nature, 467,* 951–954.

Jung, M., M. Reichstein, H. A. Margolis, A. Cescatti, A. D. Richardson, M. A. Arain, A. Arneth et al. (2011). Global patterns of land-atmosphere fluxes of carbon dioxide, latent heat, and sensible heat derived from eddy covariance, satellite, and meteorological observations. *Journal of Geophysical Research*, 116, G00J07.

Kalma, J., T. McVicar, and M. McCabe (2008). Estimating land surface evaporation: A review of methods using remotely sensed surface temperature data. *Surveys in Geophysics*, 29, 421–469.

Kelly, R. E., A. T. Chang, L. Tsang, and J. L. Foster (2003). A prototype AMSR-E global snow area and snow depth algorithm. *IEEE Transactions on Geoscience and Remote Sensing*, 41, 230–242.

Kustas, W., and M. Anderson (2009). Advances in thermal infrared remote sensing for land surface modeling. *Agricultural and Forest Meteorology*, 149, 2071–2081.

Lhomme, J. P. (1992). Energy balance of heterogeneous terrain: Averaging the controlling parameters. *Agricultural and Forest Meteorology*, 61, 11–21.

Lhomme, J. P., C. Montes, F. Jacob, and L. Prévot (2012). Evaporation from heterogeneous and sparse canopies: On the formulations related to multi-source representations. *Boundary-Layer Meteorology*, 144, 243–262.

Liu, Y. Y., A. I. J. M. van Dijk, M. F. McCabe, J. P. Evans, and R. A. M. de Jeu (2013). Global vegetation biomass change (1988–2008) and attribution to environmental and human drivers. *Global Ecology and Biogeography*, 22(6), 692–770.

Madsen, H., P. F. Rasmussen, and D. Rosbjerg (1997). Comparison of annual maximum series and partial duration series methods for modeling extreme hydrologic events: 1. At-site modeling. *Water Resources Research*, 33, 747–757.

Massman, W. J. (1987). Heat transfer to and from vegetated surfaces: An analytical method for the bulk exchange coefficients. *Boundary-Layer Meteorology*, 40, 269–281.

Massman, W. J. (1999). A model study of kBH−1 for vegetated surfaces using 'localized near-field' Lagrangian theory. *Journal of Hydrology*, 223, 27–43.

McCabe, M., E. Wood, H. Su, R. Vinukollu, C. Ferguson, and Z. Su (2011). Multisensor global retrievals of evapotranspiration for climate studies using the surface energy budget system. In *Land Remote Sensing and Global Environmental Change*, B. Ramachandrin, C. Justice and M. Abrams (Eds.). New York: Springer, pp. 747–778.

McCabe, M. F., and E. F. Wood (2006). Scale influences on the remote estimation of evapotranspiration using multiple satellite sensors. *Remote Sensing of Environment*, 105, 271–285.

Miralles, D. G., R. A. M. De Jeu, J. H. Gash, T. R. H. Holmes, and A. J. Dolman (2011b). Magnitude and variability of land evaporation and its components at the global scale. *Hydrology and Earth System Sciences*, 15, 967–981.

Miralles, D. G., J. H. Gash, T. R. H. Holmes, R. A. M. de Jeu, and A. Dolman (2010). Global canopy interception from satellite observations. *Journal of Geophysical Research*, 115, D16122.

Miralles, D. G., T. R. H. Holmes, R. A. M. De Jeu, J. H. Gash, A. G. C. A. Meesters, and A. J. Dolman (2011b). Global land-surface evaporation estimated from satellite-based observations. *Hydrology and Earth System Sciences*, 15, 453–469.

Miralles, D. G., M. van den Berg, R. Teuling, and R. A. M. De Jeu (2012). Soil moisture-temperature coupling: A multiscale observational analysis. *Geophysical Research Letters*, 39, L21707.

Monin, A., and A. Obukhov (1954). Basic laws of turbulent mixing in the atmosphere near the ground. *Trudy Geofizicheskogo Instituta, Akademiya Nauk SSSR*, 24, 163–187.

Monin, A. S., and A. M. Obukhov (1945). Basic laws of turbulent mixing in the surface layer of the atmosphere. *Trudy Geofizicheskogo Instituta, Akademiya Nauk SSSR*, 24, 163–187.

Monteith, J. L. (1965). Evaporation and environment. *Symposia of the Society for Experimental Biology*, 19, 205–234.

Mu, Q., F. A. Heinsch, M. Zhao, and S. W. Running (2007). Development of a global evapotranspiration algorithm based on MODIS and global meteorology data. *Remote Sensing of Environment*, 111, 519–536.

Mu, Q., M. Zhao, and S. W. Running (2011). Improvements to a MODIS global terrestrial evapotranspiration algorithm. *Remote Sensing of Environment*, 115, 1781–1800.

Mueller, B., S. I. Seneviratne, C. Jimenez, T. Corti, M. Hirschi, G. Balsamo, P. Ciais et al. (2011). Evaluation of global observations-based evapotranspiration datasets and IPCC AR4 simulations. *Geophysical Research Letters, 38,* L06402.

Murphy, D. M., and T. Koop (2005). Review of the vapour pressures of ice and supercooled water for atmospheric applications. *Quarterly Journal of the Royal Meteorological Society, 131,* 1539–1565.

Norman, J. M., W. P. Kustas, and K. S. Humes (1995). Source approach for estimating soil and vegetation energy fluxes in observations of directional radiometric surface temperature. *Agricultural and Forest Meteorology, 77,* 263–293.

Owe, M., R. de Jeu, and T. Holmes (2008). Multisensor historical climatology of satellite-derived global land surface moisture. *Journal of Geophysical Research, 113,* F01002.

Penman, H. L. (1948). Natural evaporation from open water, bare soil and grass. *Proceedings of the Royal Society of London. Series A. Mathematical and Physical Sciences, 193,* 120–145.

Phillips, O. L., L. E. O. C. Aragão, S. L. Lewis, J. B. Fisher, J. Lloyd, G. López-González, Y. Malhi et al. (2009). Drought sensitivity of the Amazon rainforest. *Science, 323,* 1344–1347.

Priestley, C. H. B., and R. J. Taylor (1972). On the assessment of surface heat flux and evaporation using large-scale parameters. *Monthly Weather Review, 100,* 81–92.

Rana, G., and N. Katerji (1998). A measurement based sensitivity analysis of the Penman-Monteith Actual Evapotranspiration Model for crops of different height and in contrasting water status. *Theoretical and Applied Climatology, 60,* 141–149.

Rodell, M., J. Famiglietti, J. Chen, S. Seneviratne, P. Viterbo, S. Holl, and C. Wilson (2004). Basin scale estimates of evapotranspiration using GRACE and other observations. *Geophysical Research Letters,* 31, L20504.

Rudolf, B., and U. Schneider (2005). Calculation of gridded precipitation data for the global land-surface using in-situ gauge observations. In *Proceedings of the 2nd Workshop of the International Precipitation Working Group IPWG,* EUMETSAT, Monterey, October 2004, pp. 231–247.

Rushdi, A. M., and K. F. Kafrawy (1988). Uncertainty propagation in fault-tree analyses using an exact method of moments. *Microelectronics Reliability, 28,* 945–965.

Sahoo, A. K., M. Pan, T. J. Troy, R. K. Vinukollu, J. Sheffield, and E. F. Wood (2011). Reconciling the global terrestrial water budget using satellite remote sensing. *Remote Sensing of Environment, 115,* 1850–1865.

Sheffield, J., G. Goteti, and E. F. Wood (2006). Development of a 50-year high-resolution global dataset of meteorological forcings for land surface modeling. *Journal of Climate, 19,* 3088–3111.

Sheffield, J., and E. F. Wood (2007). Characteristics of global and regional drought, 1950–2000: Analysis of soil moisture data from off-line simulation of the terrestrial hydrologic cycle. *Journal of Geophysical Research, 112,* D17115.

Shuttleworth, W. J., and J. S. Wallace (1985). Evaporation from sparse crops-an energy combination theory. *Quarterly Journal of the Royal Meteorological Society, 111,* 839–855.

Stewart, J. (1988). Modelling surface conductance of pine forest. *Agricultural and Forest Meteorology, 43,* 19–35.

Su, H., M. F. McCabe, E. F. Wood, Z. Su, and J. H. Prueger (2005). Modeling evapotranspiration during SMACEX: Comparing two approaches for local- and regional-scale prediction. *Journal of Hydrometeorology, 6,* 910–922.

Su, Z. (2002). The Surface Energy Balance System (SEBS) for estimation of turbulent heat fluxes. *Hydrology and Earth System Sciences, 6,* 85–100.

Su, Z., T. Schmugge, W. P. Kustas, and W. J. Massman (2001). An evaluation of two models for estimation of the roughness height for heat transfer between the land surface and the atmosphere. *Journal of Applied Meteorology, 40,* 1933–1951.

Swinbank, W. (1951). The measurement of vertical transfer of heat and water vapor by eddies in the lower atmosphere. *Journal of Atmospheric Sciences, 8,* 135–145.

Thornthwaite, C. W., and B. Holzman (1939). The determination of evaporation from land and water surfaces. *Monthly Weather Review, 67,* 4–11.

Tucker, C. J., J. E. Pinzon, M. E. Brown, D. A. Slayback, E. W. Pak, R. Mahoney, E. F. Vermote, and N. El Saleous (2005). An extended AVHRR 8-km NDVI dataset compatible with MODIS and SPOT vegetation NDVI data. *International Journal of Remote Sensing, 26*, 4485–4498.

Valente, F., J. David, and J. Gash (1997). Modelling interception loss for two sparse eucalypt and pine forests in central Portugal using reformulated Rutter and Gash analytical models. *Journal of Hydrology, 190*, 141–162.

Vermote, E. F., and N. Z. Saleous (2006). Calibration of NOAA16 AVHRR over a desert site using MODIS data. *Remote Sensing of Environment, 105*, 214–220.

Vinukollu, R. K., J. Sheffield, E. F. Wood, M. G. Bosilovich, and D. Mocko (2011a). Multimodel analysis of energy and water fluxes: Intercomparisons between operational analyses, a land surface model, and remote sensing. *Journal of Hydrometeorology, 13*, 3–26.

Vinukollu, R. K., E. F. Wood, C. R. Ferguson, and J. B. Fisher (2011b). Global estimates of evapotranspiration for climate studies using multi-sensor remote sensing data: Evaluation of three process-based approaches. *Remote Sensing of Environment, 115*, 801–823.

Wang, K., and R. E. Dickinson (2012). A review of global terrestrial evapotranspiration: Observation, modeling, climatology, and climatic variability. *Reviews of Geophysics, 50*, RG2005.

Warnick, K. F., and W. C. Chew (2004). Error analysis of the moment method. *IEEE Antennas and Propagation Magazine, 46*, 38–53.

Werth, D., and R. Avissar (2004). The regional evapotranspiration of the Amazon. *Journal of Hydrometeorology, 5*, 100–109.

Zhao, M., F. A. Heinsch, R. R. Nemani, and S. W. Running (2005). Improvements of the MODIS terrestrial gross and net primary production global dataset. *Remote Sensing of Environment, 95*, 164–176.

11

The Role of Scale in Determining Surface Energy Fluxes from Remote Sensing

Nathaniel A. Brunsell and Leiqiu Hu

CONTENTS

11.1 Introduction

Humans have fundamentally altered the mass and energy fluxes between the land surface and the atmosphere. As the world's population continues to grow, meeting the basic food and water needs of the increased population will require accurate and near-real-time monitoring of water and diminishing resources. The only viable way to do this is through the use of satellite technologies (Wang and Dickinson 2012). However, these observations are collected at certain spatial resolutions that may (or may not) have anything to do with the spatial resolutions of the underlying processes controlling water and energy cycling between the surface and atmosphere. This is one particular aspect of the so-called scaling problem that must be addressed in order to accurately monitor and predict the cycling of water (Anderson et al. 2003; Brunsell and Gillies 2003b; McCabe and Wood 2006; Wu and Li 2009).

When discussing the issues related to scale and scaling, it is essential to have agreed upon definitions for the concepts. Here, we adopt the terminology of Bloschl (e.g., Bloschl 1999), which involves the so-called scaling triplet: extent, spacing, and support. The extent is the overall domain of the study (the image), while the spacing is the distance between consecutive observations (distance between pixels). Support is the area encompassed by an individual observation (resolution). Note that observations, model output, and processes each have their own scaling triplet, and ideally the observational and modeling scaling triplets are chosen to match those of the process triplet. Changing between certain aspects of the scaling triplet is quite easy, for example, decreasing the extent. However, changing the level of support (aggregation or downscaling) is often not trivial and is often the aspect of the scale problem people are most concerned with.

The scaling triplet of observations is often based on technological and financial considerations, such as what is the highest resolution (spatial and temporal) that can be achieved for a given investment. The modeling triplet can be a similar consideration, particularly for complex models such as general circulation models (GCMs). However, model formulations are often developed at a single scale and may be applied to a wide range of scales. When a model is run at a particular resolution, it is an assumption that all of the model physics are appropriately formulated for that spatial scale. This may induce errors in the model validation if the model is applied at a scale that is not valid. For example, consider the application of Darcy's law at the spatial resolution of 1° latitude. In the case of remote sensing, this may be a difficult error to assess, as there are few measurement systems (e.g., scintillometers) that are capable of measuring fluxes at the spatial and temporal resolution of satellites, particularly in the case of microwave soil moisture retrieval and other coarse resolution sensors.

It is important to consider that processes that operate across the land surface–atmosphere interface rarely occur at a single scale. In fact, the dominant scales of a process may, in fact, change over time due to the relative contributions of global and regional climate systems, vegetation dynamics, soil formation process, and anthropogenic modification of these processes. Thus disciplines that are focused on the same process but at different scales may develop different (and apparently contradictory) formulations for the same process (Jarvis and McNaughton 1986).

Therefore an observation at a particular scale is really a composite of different processes occurring over a range of time and space. These scales of transport range from microns (water flow through the soil matrix, atmospheric turbulent transport) to global-scale circulation patterns. These different scale processes also interact with one another in nonlinear ways. These nonlinear, multiscale interactions make predictability difficult. This is due to the fact that the insights gained in one location may not be directly transferrable to another location or possibly even the same location at a different time.

The relative importance of the different scales of interaction results in errors when model results are compared to observations that encompass the true multiscale nature of the underlying process. Therefore the goals of this chapter are to (1) highlight the importance of assessing the spatial scale of the inputs and the impact on the resultant surface energy fluxes and (2) discuss how changes in the spatial scale of controlling variables may alter the spatial scaling properties of the derived fluxes.

11.2 The Surface Energy Balance

The source of most of the energy for surface fluxes originates from the difference between incoming and outgoing solar and longwave radiation:

$$Rn = (1-\alpha)R_s + \varepsilon\left(L_u - \sigma T_s^4\right) \tag{11.1}$$

where Rn (W m^{-2}) is the net radiation available to do work at the surface, R_s is the incoming solar radiation (W m^{-2}), α is the surface albedo, L_u is the downcoming longwave radiation from the atmosphere (W m^{-2}), ε is the surface emissivity, σ is the Stefan–Boltzmann constant, and T_s is the surface radiometric temperature (K).

The net radiation is available to be partitioned into the surface turbulent and conductive fluxes:

$$Rn = H + LE + G \tag{11.2}$$

where H is the sensible heat transfer (W m^{-2}), LE is the latent heat (W m^{-2}), and G is the soil heat flux (W m^{-2}).

These fluxes are often assumed to be related to a vertical gradient in the associated scalar, for example,

$$H = -\frac{\rho c_p (T_0 - T_a)}{r_a} \tag{11.3a}$$

$$LE = -\frac{\rho (q_0 - q_a)}{r_q} \tag{11.3b}$$

where ρ is the air density (kg m^{-3}), T is the temperature (K), and q is the specific humidity (kg kg^{-1}) at the aerodynamic height (denoted by the 0 subscript) or in the lower atmosphere (denoted by the a subscript). Note that the heights where the aerodynamic and air temperature and humidity are measured are assumed to define the local gradient that determines the magnitude of the surface turbulent flux. There are conditions (e.g., within forest canopies) where countergradient fluxes develop and the above assumption fails.

A complication when using surface remote sensing products for assessing turbulent fluxes between the surface and the atmosphere is the fact that the effective source for the turbulent fluxes is not the actual surface but rather a slight distance above the surface called the aerodynamic roughness length. In addition, each flux has its own roughness length; thus there is a roughness length for heat (z_{0h}) and a separate one for humidity (z_{0q}). The practical result of this is that the temperature that drives the sensible heat flux (T_0) is not equal to the radiometric surface temperature (T_s). This is unfortunate in that the surface radiometric temperature is what can be determined from satellite observations.

There have been many attempts to relate the temperatures and the roughness lengths (e.g., Kustas et al. 2007). Generally, this relies on the determination of the kB^{-1} parameter, defined as the logarithm of the ratio of the roughness length for momentum to the roughness length of heat:

$$kB^{-1} = \ln \frac{z_0}{z_{0h}} \tag{11.4}$$

The roughness length for humidity is often assumed to be the same as that for heat.

Ideally, this parameter could be estimated as a function of vegetation canopy conditions (Lhomme et al. 1994) or as a function of Monin–Obukhov similarity theory (Sun and Mahrt 1995). In reality, however, the kB^{-1} parameter is highly variable in time and space and not easily related to controlling parameters of the surface that could be reasonably detected from satellite platforms (Kustas et al. 2007).

11.3 Surface Heterogeneity and Aggregation

The surface energy balance as outlined above applies to spatially homogeneous surfaces. Heterogeneity in surface properties such as vegetation cover and soil moisture leads to surface patches. When this variability occurs at the subpixel scale, this induces patchiness in the land surface that may be undetected by a satellite sensor. This is problematic from a flux retrieval perspective in that the true areal averaged flux must be aggregated from the individual component fluxes within the pixel and not the areal average of the controlling variables. Unfortunately, the linear average of the controlling variables (e.g., T_s) and parameters is exactly what a pixel value obtained from a satellite consists of.

Conservation of energy dictates that the areally averaged flux (F) is the linear average of the component fluxes. This necessitates the knowledge of all spatially distributed fluxes ($f(x)$) within the area of interest. Because of issues related to nonlinear averaging (e.g., Jensen's inequality), the areal averaged flux (F) is not the same as the average flux computed from the spatial average of the controlling parameters or variables ($F(X)$):

$$F = \overline{f(x)} \neq F(\overline{X}) \tag{11.5}$$

where the overbar denotes spatial averaging.

This is equivalent to stating that the spatially averaged value of the sensible and latent heat fluxes cannot be computed from the linear average of the resistance terms or the areal average of the temperature over heterogeneous terrain.

As mentioned above, the remote sensing of the surface energy fluxes is the fact that a satellite returns the linear average of the variables (e.g., surface temperature and reflectance) over the area of the pixel. This necessitates the application of a land surface model that uses the areally averaged value of the variable. Depending on the heterogeneity of the land surface, this may or may not cause significant issues with the flux retrieval (Giorgi and Avissar 1997). In general, the more heterogeneous the surface and the coarser the spatial resolution of the sensor, the more likely there are to be errors in the flux retrieval associated with the nature of the input data. This raises the following issue: how does one compute the heterogeneity of the surface and quantitatively determine if this is significant?

11.4 Multiple Scales of Interaction Across the Land Surface–Atmosphere Interface

Most applications of deriving fluxes from remote sensing use a one-dimensional single column model in which the surface interacts with the atmosphere directly above it. Some simple schemes (e.g., single source) assume everything within the pixel is spatially homogeneous. Most schemes assume at least two sources (a bare soil and a vegetated component). This assumes that each pixel interacts with the atmosphere above it independently of the upwind interactions.

The fact that most algorithms for determination of surface fluxes from satellites are one-dimensional (1D) implies that the surface properties as determined from the satellite are assumed to only interact with the atmosphere above the pixel. In fact, in many cases, this requires an assumption that the atmospheric properties (temperature, humidity, wind speed, etc.) over the spatial extent of the image are relatively homogeneous. Thus one of the largest issues associated with the spatial scale will be the extent to which the assumption of a 1D framework is valid.

As the spatial scale of the surface features becomes larger, the likelihood of a distinct microclimate developing increases. This has been observed in eddy covariance observations of the surface related to advection of dry air across an agricultural field over relatively small distances (Zermeno-Gonzalez and Hipps 1997). Through the application of a soil–vegetation–atmosphere transfer (SVAT) framework, this assumption of a relatively homogeneous atmosphere implies that each of the surface patches can be modeled independently (Koster and Suarez 1992). If there are several large (of the order of tens of kilometers) patches, then each may be associated with distinct atmospheric conditions that assist in determining the vertical gradient between the surface and the atmosphere, and this assumption may not be valid. A common scenario in which this assumption fails is the presence of mesoscale atmospheric circulations.

Because of the efficient nature of mixing in the atmospheric boundary layer (ABL), there is a height above which the individual patches of the surface patches no longer have a direct influence over the properties in the ABL, and the atmospheric properties are an areal average. This height (really a slow transition to the lack of surface impacts) is known as the blending height (Mason 1988).

The existence of a blending height raises the issue of how the lower atmosphere integrates the properties over a heterogeneous surface. To some extent, this is a function of the atmospheric turbulence in the surface layer [as estimated from the friction velocity, u_* (m s^{-1})], but it is also a function of the length scale of the surface heterogeneity and the magnitude of the surface heterogeneity fluctuations. Mahrt (2000) expressed this for conditions of microscale surface heterogeneity as

$$L_h \ll C_b \frac{U^2}{u_*} h \qquad (11.6)$$

where L_h is the length scale of the surface heterogeneity, C_b is a blending coefficient, U is the mean horizontal wind speed (m s^{-1}), and h is the height of the atmospheric boundary layer (m).

An important factor to consider here is that the spatial distribution of the surface fluxes is not solely related to the length scale of the surface heterogeneity when computing the land–atmosphere fluxes. It is the interaction between the surface features and the conditions of the lower atmosphere. In addition, the atmospheric conditions are not independent of the surface properties.

The use of large eddy simulation (LES) modeling has been very illustrative of the role of surface heterogeneity on coupled land–atmosphere fluxes. LES studies have shown that the length scale of surface patches will influence the development of the atmospheric boundary layer above and downwind of the patch (Albertson et al. 2001; Brunsell et al. 2011), which thus alters the scalar gradients and the resultant fluxes. The larger the patch is, the higher the influence will extend into the atmosphere. Many smaller patches may each exhibit their own internal boundary layer, above which the atmosphere may be well mixed. The height at which the atmosphere is well mixed is referred to as the blending height, and fluxes above that height are representative of the areal average (Mahrt 2000).

Recent results from LES studies have also shown how the length scales of surface heterogeneity can interact with entrainment and other processes impacting larger-scale circulations and interactions (Huang and Margulis 2009, 2010). This scale of interaction between the surface and the boundary layer may also help to explain the ubiquitous energy balance closure problem observed with eddy covariance measurements (Huang et al. 2007).

Attempts to characterize the degree of interaction between the surface and the atmosphere have focused on the concept of land–atmosphere coupling (Santanello et al. 2009; Findell and Eltahir 2003). These efforts have been applied to both observational and modeling frameworks to assess the relative roles of boundary layer interactions as a function of surface properties (e.g., wet and dry soil moisture state). While these have not specifically addressed spatial heterogeneity, assessing observational and modeling sensitivities through time and in response to initial surface conditions is essential to understanding the multiscale nature of land–atmosphere interactions.

11.5 Case Study: CLASIC

Determining the relative importance of different spatial and temporal scales to the resultant water and energy fluxes requires understanding the scalewise changes in the controlling variables. This is often quite difficult given a model is used and the data are specified at different scales. This increases the difficulty because it is inherently assuming that the relative variation between input and output does not vary with scale. To examine how different scales of surface vegetation, soil moisture, and radiometric temperature potentially impact the spatial scales of latent and sensible heat exchanges, we examined a case study from the Cloud Land Atmosphere Interaction Campaign (CLASIC) that took place in Oklahoma in the summer of 2007. CLASIC was a campaign that consisted of multiple surface and airborne measurements intended to investigate the role of land surface heterogeneity on boundary layer processes and cloud formation. Unfortunately, the weather during the campaign was generally less than ideal. In order to quantify the impact of scale on the resultant evaporative fluxes, here we focus on two Landsat scenes acquired on June 6 and August 8, 2006, over central Oklahoma. Even though these two Landsat scenes were actually acquired before and after the campaign, they can illustrate a number of issues associated with the determination of surface energy and mass fluxes from satellites.

Landsat data used for this analysis were obtained without terrain correction and were corrected for atmospheric effects using Modtran and radiosonde profiles. The Simsphere SVAT model (e.g., Brunsell and Gillies 2003a,b; Brunsell et al. 2008) was utilized for the mapping of the remotely sensed data into the surface energy fluxes using the methodology described next.

11.5.1 The "Triangle" Method

The "triangle" method is used to relate the remotely sensed vegetation index and land surface temperature to the resultant energy fluxes (Gillies et al. 1997; Carlson 2007; Petropoulos and Carson 2011) The vegetation index is the fractional vegetation (Fr), which is a scaled version of the normalized difference vegetation index (NDVI):

$$Fr = \left(\frac{NDVI_i - NDVI_{min}}{NDVI_{max} - NDVI_{min}} \right)^2 \tag{11.7}$$

where $NDVI_i$ is the pixel value of the NDVI and the max and min subscripts refer to the maximum and minimum values of NDVI in the scene.

When the Fr is plotted against the land surface temperature, a general triangular shape usually results. For example, Figure 11.1 illustrates the joint probability density plots between Landsat derived T_s and Fr for June 3 and August 6, 2007, over the CLASIC field site. This triangle is associated with several important features of the land surface, including a warm, nonvegetated area, progressing along a warm edge to a relatively warm fully vegetated region. The cool edge exhibits a relatively constant temperature that usually corresponds closely with the air temperature. For a given value of vegetation cover, there is a range of land surface temperatures observed in a given scene. These temperatures are assumed to be due to the variation in the near surface soil moisture (M_0).

As the near-surface soil moisture decreases, evapotranspiration begins to decrease because of water limitation. This, in turn, results in an increase in the radiometric temperature as more of the net radiation becomes partitioned into sensible heat. A SVAT scheme can be calibrated using additional information (e.g., a radiosonde profile) to match limiting cases of the water limitation (i.e., the corners of the triangle). The model is iterated over all possible combinations of initial soil moisture and vegetation fraction. The modeled radiometric temperature and surface energy fluxes are output at the time of satellite overpass, and a polynomial regression is constructed to relate the surface temperature and vegetation fraction to the fluxes:

$$F_x = \sum_{i=1}^{3} \sum_{j=1}^{3} a_{i,j} T^i Fr^j \tag{11.8}$$

where F_x is the flux of interest (latent heat LE, sensible heat H, etc.). This equation is then applied to the remotely sensed temperature and vegetation to determine the flux from the satellite imagery.

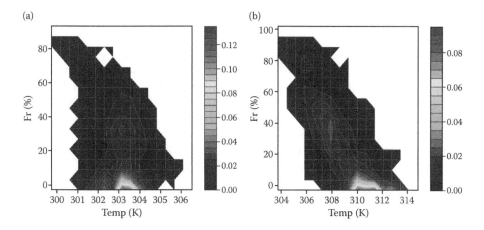

FIGURE 11.1
(See color insert.) Joint probability density plots for the Landsat images for (a) June 3 and (b) August 6, 2007, over the CLASIC field site in central Oklahoma.

11.5.2 Assessing Scale Issues

One possible way of understanding the relative importance of land surface heterogeneity is through the use of information theory metrics such as Shannon entropy, mutual information content, and relative entropy (Brunsell et al. 2008). These metrics are used to assess the amount of information within the observed signal, as well as the information redundancy and transfer between different data series.

The first step to examining these metrics was to examine the role of spatial scale on the remotely sensed data itself. For this purpose, we employed the technique of low-pass filters conducted with wavelet decomposition. This is similar to previous work conducted surrounding the Fort Peck Ameriflux site in Montana (Brunsell and Anderson 2011). This technique allows us to determine how the image would appear with coarser spatial resolutions by progressively removing the higher-frequency information. We utilized a dyadic decomposition through seven levels of decomposition, resulting in resolutions of 200 m to 12.8 km. Figure 11.2 shows selected scales of decomposition for the fractional vegetation, surface temperature, as well as the modeled latent heat fluxes for the August 6 image. For these illustrative purposes, we have only shown selected scales, corresponding to 200, 800, and 3200 m.

These images support the general idea of a reduction in variance with increasing spatial scale. While many features are collocated (e.g., more vegetated locations tend to have lower surface temperatures), these correlations are not perfect and there are some differences in the transition from fine to coarse resolution. However, it is difficult to quantify these relationships solely from visually inspecting the decomposed images.

In order to quantify the variability in the remotely sensed fields as well as the model output as a function of spatial scale, we calculated the wavelet spectra for each date. In addition, we can investigate the potential anisotropy in the images by calculating the wavelet spectra in the horizontal, diagonal, and vertical directions of the two-dimensional image. Figure 11.3 shows the three directional spectra for the M_0 and the derived LE flux for both the June and August images. The dominant length scale is defined as the scale at which the wavelet variance is highest. For most of these data fields this is of the order of 1600 m for both days. However, there is a clear difference in the anisotropy between the two dates. For example, consider the near-surface soil moisture M_0 on June 6 (Figure 11.3a), where each directional spectrum exhibits a different length scale ranging from 400 to 1600 m. While in the August case (Figure 11.3c), each direction is more or less the same.

The question then becomes, to what extent does this anisotropy in the controlling fields impact the spatial variability of the LE flux? This can also be seen in Figure 11.3b and d. The variability with spatial resolution on June 6 does not obviously match any of the controlling variables. Consider the vertical spectra, the LE flux peaks at 1600 m that most closely matches the pattern observed in the Fr field (not shown). While for the August case, the fields are mostly similar for the diagonal spectra, the resultant shape of the spectra most closely matches the T_s rather than the Fr. This could imply a shift in the physical dynamics controlling the evaporative process between these two dates.

While the wavelet spectra can illustrate the dominant length scale of the surface field and the associated anisotropy, it does not necessarily inform how much of the resultant signal (e.g., the LE flux) is due to the scalewise variability in the input fields. To ascertain this, we combined the wavelet decompositions with the information theory

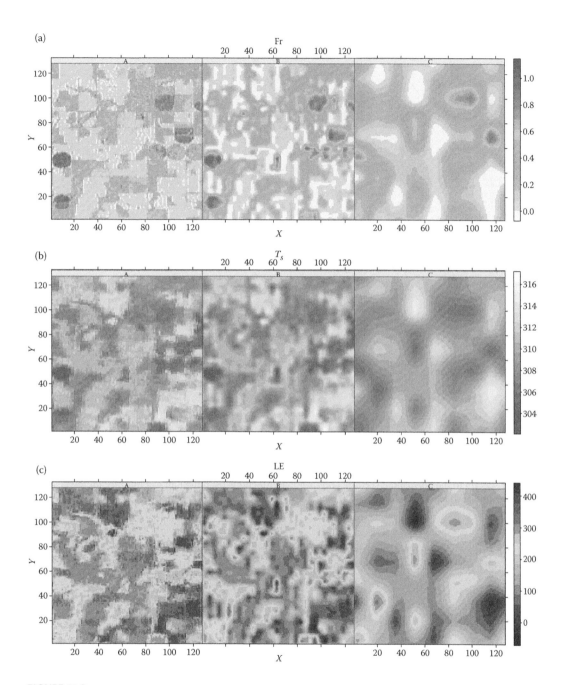

FIGURE 11.2
(See color insert.) Wavelet decompositions of the August 6, 2007, Landsat imagery (a), fractional vegetation (%) (b), and surface temperature (K) (c) latent heat flux (W m⁻²) for selected scales (A) 200 m, (B) 800 m, and (C) 3200 m.

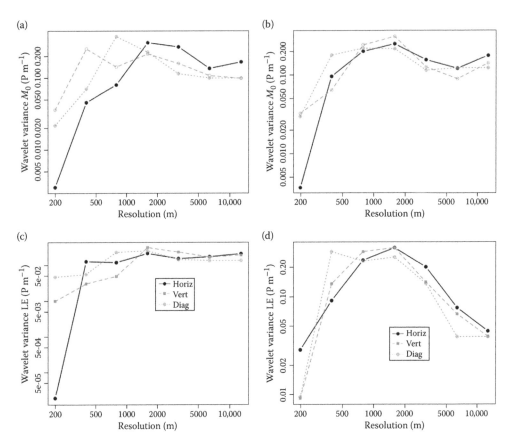

FIGURE 11.3
Wavelet spectra in the horizontal, vertical, and diagonal directions from the Landsat imagery for the (a, c) soil moisture and (b, d) latent heat flux for (a, b) June 3 and (c, d) August 6 images.

metrics entropy and relative entropy. This is done following the methodology developed previously (Brunsell et al. 2008; Brunsell 2010), where the Shannon entropy is

$$I = -\sum_{i=1}^{n} p(x_i) \log(p(x_i)) \tag{11.9}$$

where I is the entropy and p is the probability that the data value x falls within a histogram bin i. The entropy is a measure of the total information contained in the data series. The relative entropy ($Re(x,y)$) is a measure of the additional information needed to characterize x given the amount of information contained in y, based on their respective probability functions p and q:

$$Re(x,y) = \sum_{i=1}^{n} p_i \log\left(\frac{p_i}{q_i}\right) \tag{11.10}$$

These metrics provide insight into the information 'gained' from each scale, with the higher entropy values showing the most uniform distributions and having the highest information content. The spatial entropy spectra are shown in Figure 11.4a and b for the LE and the H fluxes, respectively. The LE shows a decrease in the entropy at almost all scales from June 3 to August 6. The H spectra, on the other hand, are generally the same up to approximately the 2000-m scale, at which point the spectra decline for both days with the June 3 case being lower. This illustrates that the scaling behavior of different fluxes cannot be assumed to be the same, even for the same scene.

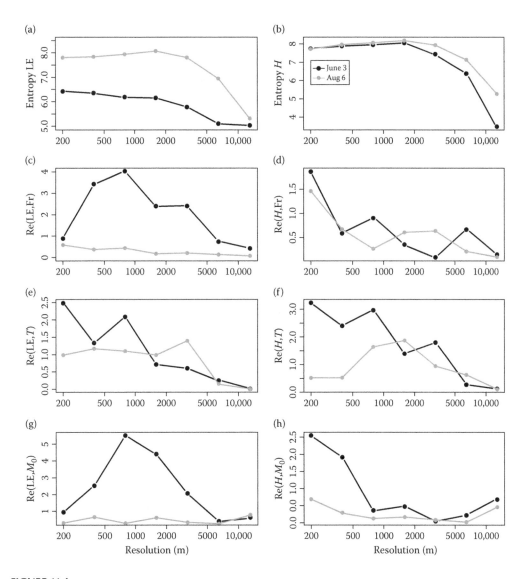

FIGURE 11.4
Information entropy for (a) latent heat flux and (b) sensible heat flux for June 3 and August 6 Landsat images over the CLASIC field campaign area. Panels (c) through (h) show the relative entropy between the controlling parameters of fractional vegetation, surface temperature, and soil moisture for each of the fluxes for each day.

In order to ascertain how the scalewise variability in the controlling fields of T_s, Fr, and M_0 relate to the changes in the fluxes across scale we utilized the relative entropy (Re) spectra. In this application, the relative entropy quantifies how much additional information is necessary to capture the output field given the input field. Thus the Re between the LE flux and the Fr would show to what extent we can predict the LE from the information contained in a particular scale of the Fr. Higher values indicate more information is needed, and therefore the ability to predict the total flux is more limited at that scale of the input data.

The Re for the LE fluxes are shown in the first column of Figure 11.4 (panels c, e, and g). For the June 3 date, the Fr exhibits higher Re in the intermediate spatial scales (e.g., 1000 m), with lower values at the extremely small and large scales. The Re(LE, T_s) shows a general decline in the Re from the smallest to the largest scales. The soil moisture influence is generally the same as the Fr. The August 6 image shows a different dynamic, where the Re(LE, Fr) and Re(LE, M_0) are relatively flat across all scales, indicating that each scale is relatively equally influenced by the Fr and M_0. The relative entropy with T_s shows a general decrease relative to the earlier image and is constant up to approximately 3200 m.

The sensible heat behaves differently with respect to the influence of the controlling variables across scale. The Re(H, Fr) is generally the same for the two dates and is highest at the smallest scale. The influence of T_s on H shows a generally decreasing Re similar to the LE variability for June 3, but the August 6 case shows a high Re in the intermediate scales (approximately 1600 m). The role of soil moisture changes depending upon the date, where the June 3 case shows a high value in the smallest scale and decreases to about 800 m, above which it is constant. While on August 6, the Re is basically constant with scale.

These results indicate that the dominant length scales are on the order of 1600 m for most cases, but the contribution to the fluxes by the dominant input variables changes depending on the date. To first order, the influence of the surface temperature is higher at the larger spatial scales (indicated by lower relative entropy), while the influence of the vegetation is the reverse. This is in agreement with prior results from the Southern Great Plains 1997 experiment (Brunsell and Gillies 2003a). However, this influence is not constant over time, which implies that the ability to scale fluxes based on a parameterization of the controlling variables is not likely (Brunsell and Anderson 2011).

11.6 Discussion and Recommendations

One of the significant findings of the above case study is that the role of surface heterogeneity varies depending on the time of data acquisition. More importantly, the relative contribution of the controlling variables such as soil moisture and vegetation varies in time and may be different for the different components of the surface energy balance. Note that this is not to be confused with simply saying that the seasonal pattern in time is important. This is illustrative that the role of the spatial length scales of the surface fields is changing in conjunction with the seasonal dynamics. Other research has also found that these patterns change in time and are also a function of the satellite used to quantify these patterns (Brunsell and Anderson 2011).

This is perhaps indicative of the nonlinear impacts between vegetation, soil, and boundary layer processes. Each of these factors is changing with their own timescales of variability and each have a relatively important role on the partitioning of surface energy fluxes.

When we consider the goal of quantifying the transfer of mass and energy between the surface and atmosphere, we must address how the processes of evaporation and sensible heat are a combination of processes acting at multiple spatial and temporal scales. It is well known that soil moisture and vegetation interact at a wide range of temporal scales to alter near-surface moisture and heat transfer (Katul et al. 2007; Seneviratne et al. 2010). At inter-annual timescales, woody encroachment and other changes in the vegetation can impact the surface pattern and thus directly alter the mass and energy fluxes (Huxman et al. 1986).

Changes over time with respect to vegetation phenology, seasonal patterns in precipitation and soil moisture, and longer-term soil formation processes all alter the timescales of surface processes. The fact that these processes are also varying spatially across the landscape implies that the spatial scaling characteristics are also temporally varying. Therefore any study that quantifies the spatial scaling characteristics at a single time may be of limited utility (Brunsell and Anderson 2011). There is additional evidence that changing the spatial resolution may result in more errors in modeling of vegetation processes (Kucharik et al. 2006).

Interactions between surface heterogeneity and processes above the boundary layer may also play an important role in the maintenance of preferred soil moisture–precipitation feedback regimes. Soil moisture–vegetation interactions have also been shown to be linked in particular regions to strong feedback with precipitation patterns both using remote sensing (Brunsell 2006) and modeling approaches (Koster et al. 2004; Jones and Brunsell 2009). The physical mechanisms underlying these feedback regions are linked to local evaporative fraction, energy balance partitioning, albedo, etc., which thus aid in cloud formation and precipitation (Huang and Margulis 2011).

The hydrological cycle is known to function at a wide range of scales from the molecular to the global. In operational settings, it is often necessary to choose specific resolutions while implicitly or explicitly ignoring the impacts of different scales. For example, in the case of monitoring evapotranspiration and other surface energy budget terms, it is often assumed that groundwater dynamics can be neglected. However, recent results have shown a direct impact of groundwater on surface fluxes (Kollet and Maxwell 2008). Thus longer-term geological processes may ultimately be important for truly understanding surface fluxes and thus the ability to quantify these interactions from satellite platforms even at short timescales. Ultimately, the only way to truly know if these interactions are important is to quantify the impact and then make the decision.

As we enter the anthropocene, it is essential that we also focus explicitly on the role of humans on altering the dynamics of land–atmosphere interactions. We have fundamentally altered the nature of hydrological cycling (Wohl et al. 2012) and have also altered many of the timescales associated with these cycles.

The human-induced nature of land cover change also directly impacts all of the processes outlined in this chapter (Foley et al. 2005; Feddema et al. 2005). One role for the use of remote sensing for quantifying these impacts is an explicit focus on the role of land cover and land cover change. For example, consider the case of irrigated agriculture (Twine et al. 2004). If we consider agricultural cropping selection patterns that are altered in response to energy needs (e.g., biofuel production), this will ultimately affect the water use requirements of the vegetation (VanLoocke et al. 2012), thus impacting irrigation, surface soil moisture, evapotranspiration, atmospheric humidity and precipitation, and all of the other impacts of the hydrological cycle. While the local impact of irrigation on the surface energy balance may appear quite easy to monitor from satellites in that it raises the evaporation to nonwater limited conditions, this has profound impacts on the meso- to regional-scale circulations in the region of irrigation (Adegoke et al. 2003). Irrigated sites can quickly create local oasis effects and profoundly alter the microclimates surrounding the area as well as the circulations. The microclimatic alterations to saturation deficit and air temperature may be quite difficult to ascertain from surface satellite observations, making the quantification of the surface fluxes problematic.

A secondary example of anthropogenic alteration of the landscape that may prove problematic for surface flux determination from satellite is the urban environment. As the world's landscape becomes more urbanized (Elvidge et al. 2004), understanding the role of the urban surface energy balance will become more vital. While previous efforts at addressing the nature of the surface energy balance in urban environments points out that a lot of progress has been achieved (Arnfield 2003), there is still a fundamental scale problem between the scale of urbanization and the ability to detect this from satellite platforms. Jin and Dickinson (1999) discussed the schemes to integrate urban effects in the land surface models using remote sensing techniques and concluded that satellite-observed urban information has highly potential simulated urban effects in climate models.

Even with high-resolution satellite data, there is still the issue of the interaction between the length scales of the surface and the atmosphere as outlined above. These interactions in urban environments also extend themselves well beyond the urban pixels through the urban heat island as well as alteration of the downwind precipitation patterns, etc.

The examples of irrigated agriculture and urban environments are particularly problematic for satellite monitoring due to the spatial scale of the land surface modification (i.e., the length scale of heterogeneity) but also the magnitude of the alteration on the local microclimate relative to the surrounding area. The spatial variances of the surface and lower atmosphere conditions increase with the intensity of the modification, often at incredibly fine spatial scales. Accurately quantifying the impacts on the surface energy balance requires understanding the length scale of the surface heterogeneity and the timescales of the surface alteration on surface properties like thermal inertia, heat capacity, etc., that all directly impact the partitioning of the net radiation into the component surface energy fluxes.

The surface alterations in these cases lead directly to profound impacts on the atmospheric conditions. Not only do the changes in the surface fluxes lead to changes in air temperature and saturation deficit, but they also induce circulations that will involve the advection of dry (moist) air over the irrigated (urban) area. This advection is induced from the changes in density over the area and the increase in buoyancy. This impact will be a result of the horizontal gradients in moisture and temperature that potentially can be estimated from the satellite observations of the surface.

From these examples, we can suggest that to increase the accuracy of remote sensing of the surface energy balance we should attempt to consider the role of spatial configuration of the surface patches on the pixel scale fluxes. Note that we are not referring to subpixel aggregation here but rather that airflow from adjacent pixels can alter the microclimate within a pixel. As discussed previously, LES simulations have shown that that magnitude and the configuration impacts the surface fluxes, and while the satellite can quantify the surface magnitude, little work has been done on explicitly incorporating remotely sensed land conditions into LES (Huang and Margulis 2010; Brunsell et al. 2011; Albertson et al. 2001).

Rather than conduct LES studies over all areas of the world using Landsat data, we recommend an explicit focus on the length scales of surface heterogeneity and the relationship between the surface variance and the induced variance in the atmospheric properties. Of course, the induced variability downwind will be a function of the spatial scale of the atmospheric grid, local conditions, etc. This would result in empirical scaling relationships such as those of Equation 11.6, which will vary in time and place. However, through enough studies, it may be possible to generalize these scaling relationships as a function of seasonal variation and dominant land cover class.

The above discussion often assumes that the local scale fluxes are known, and the difference between a ground-based observation and a remotely sensed estimate is due to various scale issues associated with the change in resolution. While this is true, the actual comparison of surface and satellite-based estimates is nontrivial. Surface observations such as eddy covariance towers and scintillometers have a resolution of their own. This resolution must be accounted for when comparing a surface observation to a satellite pixel. This usually consists of either (1) a direct comparison between the pixel in which the surface observation is located or (2) the use of a footprint model (e.g., Schmid 2002). This allows a weighting to be applied to the appropriate pixels and is most useful when the resolution of the satellite is finer than the area of the surface observation. Thus the surface observation encompasses more than one pixel. In the case of lower resolution satellites (e.g., MODIS), the entire footprint of the surface observation may be encompassed within a single pixel. Without additional information on the subgrid variability of that pixel, there is probably little gained by the application of a footprint model in these cases.

11.7 Conclusions

This chapter has highlighted some of the theoretical and practical considerations when attempting to address the scale issue for satellite monitoring of the surface energy balance. These issues range from the mismatch in theoretical scale of the equations to the practical scale of application at the satellite pixel, issues pertaining to the micrometeorological substitution of the radiometric surface temperature for the aerodynamic temperature, multiscale interactions between disparate processes governing vegetation, soil and atmospheric processes, as well as the interaction between spatial configuration and inducing mesoscale circulations. An example of a methodology for quantifying these variations was outlined using data from a field campaign that highlights how the dominant surface properties governing land–atmosphere exchanges can vary with date of satellite data acquisition. Ultimately, these results illustrate that the routine monitoring of water and energy fluxes from satellite data requires careful consideration of both surface and atmospheric interactions. We propose that future research should specifically focus on understanding the interaction as a function of spatial length scales and the variance of induced surface and atmospheric properties for determining scaling relationships that can hopefully allow enhanced monitoring of the water and energy fluxes using surface optical and thermal satellite data.

References

Adegoke, J. O., R. A. Pielke, J. Eastman, R. Mahmood, and K. G. Hubbard. 2003. Impact of irrigation on midsummer surface fluxes and temperature under dry synoptic conditions: A regional atmospheric model study of the U.S. high plains. *Monthly Weather Review*, 131, 556–564.

Albertson, J. D., W. P. Kustas, and T. M. Scanlon. 2001. Large-eddy simulation over heterogeneous terrain with remotely sensed land surface conditions. *Water Resources Research*, 37, 1939–1953.

Anderson, M. C., W. P. Kustas, and J. M. Norman. 2003. Upscaling and downscaling—A regional view of the soil-plant-atmosphere continuum. *Agronomy Journal*, 95, 1408–1423.

Arnfield, A. J. 2003. Two decades of urban climate research: A review of turbulence, exchanges of energy and water, and the urban heat island. *International Journal of Climatology*, 23(1), 1–26.

Bloschl, G. 1999. Scaling issues in snow hydrology. *Hydrological Processes*, 13, 2149–2175.

Brunsell, N., and R. Gillies. 2003a. Length scale analysis of surface energy fluxes derived from remote sensing. *Journal of Hydrometeorology*, 4(6), 1212–1219.

Brunsell, N., J. Ham, and C. Owensby. 2008. Assessing the multi-resolution information content of remotely sensed variables and elevation for evapotranspiration in a tall-grass prairie environment. *Remote Sensing of Environment*, 112(6), 2977–2987.

Brunsell, N. A. 2006. Characterization of land-surface precipitation feedback regimes with remote sensing. *Remote Sensing of Environment*, 100(2), 200–211.

Brunsell, N. A. 2010. A multiscale information theory approach to assess spatial–temporal variability of daily precipitation. *Journal of Hydrology*, 385(1–4), 165–172.

Brunsell, N. A., and M. C. Anderson. 2011. Characterizing the multi-scale spatial structure of remotely sensed evapotranspiration with information theory. *Biogeosciences*, 8(8), 2269–2280.

Brunsell, N. A., and R. R. Gillies. 2003b. Scale issues in land–atmosphere interactions: Implications for remote sensing of the surface energy balance. *Agricultural and Forest Meteorology*, 117(3–4), 203–221.

Brunsell, N. A., D. B. Mechem, and M. C. Anderson. 2011. Surface heterogeneity impacts on boundary layer dynamics via energy balance partitioning. *Atmospheric Chemistry and Physics*, 11, 3403–3416.

Carlson, T. 2007. An overview of the "Triangle Method" for estimating surface evapotranspiration and soil moisture from satellite imagery. *Sensors*, 7, 1612–1629.

Elvidge, C. D., C. Milesi, J. B. Dietz, B. T. Tuttle, P. C. Sutton, R. Nemani, and J. E. Vogelmann. 2004. U.S. constructed area approaches the size of Ohio. *Eos Transactions AGU*, 85(24), 233–234.

Feddema, J. J., K. W. Oleson, G. B. Bonan, L. O. Mearns, L. E. Buja, G. A. Meehl, and W. M. Washington. 2005. The importance of land-cover change in simulating future climates. *Science*, 310(5754), 1674–1678.

Findell, K. L., and E. A. B. Eltahir. 2003. Atmospheric controls on soil moisture-boundary layer interactions: Part I: Framework development. *Journal of Hydrometeorology*, 4, 552–569.

Foley, J., R. DeFries, G. Asner, C. Barford, G. Bonan, S. Carpenter, F. Chapin et al. 2005. Global consequences of land use. *Science*, 309, 570–574.

Gillies, R., T. Carlson, J. Cui, W. Kustas, and K. Humes. 1997. A verification of the "triangle" method for obtaining surface soil water content and energy fluxes from remote measurements of the Normalized Difference Vegetation Index (NDVI) and surface radiant temperature. *International Journal of Remote Sensing*, 18(15), 3145–3166.

Giorgi, F., and R. Avissar. 1997. Representation of heterogeneity effects in earth system modeling: Experience from land surface modeling. *Reviews of Geophysics*, 35(4), 413–437.

Huang, H.-Y., and S. A. Margulis. 2009. On the impact of surface heterogeneity on a realistic convective boundary layer. *Water Resources Research*, 45(4).

Huang, H.-Y. and S. A. Margulis. 2010. Evaluation of a fully coupled large-eddy simulation–land surface model and its diagnosis of land-atmosphere feedbacks. *Water Resources Research*, 46(6), W04425.

Huang, H.-Y., and S. A. Margulis. 2011. Investigating the impact of soil moisture and atmospheric stability on cloud development and distribution using a coupled large-eddy simulation and land surface model. *Journal of Hydrometeorology*, 12(5), 787–804.

Huang, J., X. Lee, and E. G. Patton. 2007. A modelling study of flux imbalance and the influence of entrainment in the convective boundary layer. *Boundary-Layer Meteorology*, 127(2), 273–292.

Huxman, T., B. Wilcox, D. Breshears, R. Scott, K. Snyder, E. Small, K. Hultine, W. Pockman, and R. Jackson. 2005. Ecohydrological implications of woody plant encroachment. *Ecology*, 86(2), 308–319.

Jarvis, P., and K. McNaughton. 1986. Stomatal control of transpiration: Scaling up from leaf to region. *Advances in Ecological Research*, 15(1), 1–49.

Jin, M., and R. Dickinson. 1999. Interpolation of surface radiative temperature measured from polar orbiting satellites to a diurnal cycle. 1. Without clouds. *Journal of Geophysical Research*, 104(D2), 2105–2116.

Jones, A. R., and N. A. Brunsell. 2009. Energy balance partitioning and net radiation controls on soil moisture–precipitation feedbacks. *Earth Interactions*, 13(2), 1–25.

Katul, G., A. Porporato, and R. Oren. 2007. Stochastic dynamics of plant-water interactions. *Annual Review of Ecology, Evolution, and Systematics*, 38(1), 767–791.

Kollet, S. J., and R. M. Maxwell. 2008. Capturing the influence of groundwater dynamics on land surface processes using an integrated, distributed watershed model. *Water Resources Research*, 44(2), W02402.

Koster, R., P. Dirmeyer, Z. Guo, G. Bonan, E. Chan, P. Cox, C. Gordon et al. 2004. Regions of strong coupling between soil moisture and precipitation. *Science*, 305(5687), 1138–1140.

Koster, R., and M. Suarez. 1992. Modeling the land surface boundary in climate models as a composite of independent vegetation stands. *Journal of Geophysical Research*, 97(D3), 2697–2715.

Kucharik, C., C. Barford, M. Maayar, S. Wofsy, R. Monson, and D. Baldocchi. 2006. A multiyear evaluation of a Dynamic Global Vegetation Model at three AmeriFlux forest sites: Vegetation structure, phenology, soil temperature, and CO_2 and H_2O vapor exchange. *Ecological Modelling*, 196(1–2), 1–31.

Kustas, W. P., M. C. Anderson, J. M. Norman, and F. Li. 2007. Utility of radiometric-aerodynamic temperature relations for heat flux estimation. *Boundary-Layer Meteorology*, 122(1), 167–187.

Lhomme, J. P., A. Chehbouni, and B. Monteny. 1994. Effective parameters of surface energy balance in heterogeneous landscape. *Boundary-Layer Meteorology*, 71, 297–309.

Mahrt, L. 2000. Surface heterogeneity and vertical structure of the boundary layer. *Boundary-Layer Meteorology*, 96(1), 33–62.

Mason, P. J. 1988. The formation of areally-averaged roughness lengths. *Quarterly Journal of the Royal Meteorological Society*, 114, 399–420.

McCabe, M., and E. Wood. 2006. Scale influences on the remote estimation of evapotranspiration using multiple satellite sensors. *Remote Sensing of Environment*, 105(4), 271–285.

Petropoulos, G., and T. N. Carlson. 2011. Retrievals of turbulent heat fluxes and soil moisture content by remote sensing. In *Advances in Environmental Remote Sensing: Sensors, Algorithms, and Applications*. Taylor and Francis, Oxford, pp. 667–502.

Santanello, J. A., Jr., C. D. Peters-Lidard, S. V. Kumar, C. Alonge, and W.-K. Tao. 2009. A modeling and observational framework for diagnosing local land–atmosphere coupling on diurnal time scales. *Journal of Hydrometeorology*, 10(3), 577–599.

Schmid, H. 2002. Footprint modeling for vegetation atmosphere exchange studies: A review and perspective. *Agricultural and Forest Meteorology*, 113, 159–183.

Seneviratne, S. I., T. Corti, E. L. Davin, M. Hirschi, E. B. Jaeger, I. Lehner, B. Orlowsky, and A. J. Teuling. 2010. Investigating soil moisture–climate interactions in a changing climate: A review. *Earth Science Reviews*, 99(3–4), 125–161.

Sun, J., and L. Mahrt. 1995. Relationship of surface heat flux to microscale temperature variations: Application to Boreas. *Boundary-Layer Meteorology*, 76, 291–301.

Twine, T., C. Kucharik, and J. Foley. 2004. Effects of land cover change on the energy and water balance of the Mississippi River basin. *Journal of Hydrometeorology*, 5(4), 640–655.

VanLoocke, A., T. E. Twine, M. Zeri, and C. J. Bernacchi. 2012. A regional comparison of water use efficiency for miscanthus, switchgrass and maize. *Agricultural and Forest Meteorology*, 164, 82–95.

Wang, K., and R. E. Dickinson. 2012. A review of global terrestrial evapotranspiration: Observation, modeling, climatology, and climatic variability. *Reviews of Geophysics*, 50(2), RG2005.

Wohl, E., A. Barros, N. Brunsell, N. A. Chappell, M. Coe, T. Giambelluca, S. Goldsmith et al. 2012. The hydrology of the humid tropics. *Nature Climate Change*, 2(9), 655–662.

Wu, H., and Z.-L. Li. 2009. Scale issues in remote sensing: A review on analysis, processing and modeling. *Sensors*, 9(3), 1768–1793.

Zermeno-Gonzalez, A., and L. E. Hipps. 1997. Downwind evolution of surface fluxes over a vegetated surface during local advection of heat and saturation deficit. *Journal of Hydrology*, 192, 189–210.

Section III

Remote Sensing of Soil Surface Moisture: Algorithms and Case Studies

12

Passive Microwave Remote Sensing Techniques for the Retrieval of Surface Soil Moisture from Space

Simonetta Paloscia and Emanuele Santi

CONTENTS

12.1 Introduction

Current climate changes, with weather-related natural disasters—such as floods, storms, cyclones, fires, and drought—have prompted scientific research to perform in-depth investigations of all components of the hydrological cycle and their subsequent interactions. Among parameters of the terrestrial water cycle, soil moisture content (SMC) emerges as a fundamental surface variable, controlling the water, energy, and carbon exchanges at the land–atmosphere interface. A precise evaluation of soil moisture on several spatial scales (from local to regional scales) can therefore be helpful in assessing the distribution of water between blue (surface water and ground water) and green (rainwater stored in the soil), the latter being essential for agricultural purposes (Liu et al. 2009; Liu and Yang 2010) or for identifying water wastes, due to human activities (Zang et al. 2012). The spatial distribution of soil moisture and its temporal evolution also play a significant role in the forecasting and management of flooding and landslides. However, a quantitative and accurate estimate of SMC on a global scale by using traditional techniques is intrinsically inadequate, since *in situ* measurement techniques are time consuming and the local SMC is highly variable in both space and time (Leese et al. 2001). Moreover, the use of hydrological models for estimating SMC and extending the forecast of moisture distribution over larger areas is not straightforward and depends heavily on the homogeneity of the selected areas and on the type and quality of information available (regarding their soil hydraulic characteristics and permeability, meteorological and climatological data, etc.).

For this reason, the estimate of SMC from satellite data became an extremely important topic from the moment the first satellite for Earth observations was launched into space (e.g., Eagleman and Lin 1976). Satellite observations are indeed very appealing because of their continuous monitoring of land surface over very large areas, with frequent revisiting time.

Microwave bands have proven particularly interesting for this purpose because the soil dielectric constant at these frequencies exhibits a noticeable dependence on the moisture content of the observed bodies, making a direct estimate of soil moisture possible (Ulaby et al. 1981, 1982, and 1986; Shutko 1982). At L band (1.4 GHz), for example, the real part of the dielectric constant ranges from 3 for dry soil to about 25 for saturated soil. Consequently, the related microwave brightness temperature can vary as much as several tens of K from wilting point to soil saturation conditions. Furthermore, microwave sensors can be operated day and night and are less affected by atmospheric conditions. On the basis of their energy source, microwave remote sensing techniques can be grouped into two categories: active (radar) sensors and passive (radiometer) sensors. However, this chapter will focus on passive microwave systems only. The estimate of SMC is a challenging task because its measurement requires low frequency observations and relatively high ground resolutions in order to be representative of the complex spatial variations. Microwave radiometers from satellite offer the advantage of overpassing the same surface almost every day; however, they are hampered by the very coarse spatial resolution, which is of the order of some tens of kilometers.

A significant amount of experimental and theoretical studies have been carried out since the late 1970s in several countries (e.g., Shutko 1982, 1986; Schmugge 1985; Schmugge et al. 1986; Jackson 1993) in order to examine in greater depth the mechanisms regulating the interactions between microwave emission and natural surfaces and to therefore identify the best methods for estimating soil moisture from microwave data. Research and theoretical activities carried out worldwide have demonstrated that sensors operating in the low-frequency portion of the microwave spectrum (from L, 1.4 GHz, to C, 6.8 GHz, bands) are the best for measuring the moisture of a soil layer, whose depth depends on the soil characteristics and moisture profile (e.g., Shutko 1982; Paloscia et al. 1993). However, because of the technological constraints limiting the use of the most suitable frequencies, a long series of satellites with onboard multipolarization microwave radiometers operating at frequencies higher than 10–19 GHz, were launched. Among them we can cite the Scanning Multichannel Microwave Radiometer (SMMR) on Nimbus 7, the Scanning Sensor Microwave Images (SSM/I) carried aboard the Defense Meteorological Satellite Program (DMSP) satellites, the Tropical Rainfall Measuring Mission's (TRMM) Microwave Imager (TMI), which includes an X-band (10 GHz) channel, but with limited latitude range. These sensors were essentially devoted to meteorological applications because of the presence of relatively high frequency sensors on board and the very coarse ground resolution. However, the possibility of obtaining surface feature information, especially for world areas where very few data are available (e.g., large boreal and equatorial forests, deserts, and poles), gave rise to both theoretical and experimental research.

The goal of estimating surface soil moisture became more attainable with the launch of sensors operating at C band (6.8 GHz), that is, the Advance Microwave Scanning Radiometer (AMSR and AMSR-E on board AQUA and AMSR2 on GCOM satellites), in 2003 and 2012.

More recently, the launch of the Soil Moisture Ocean Salinity (SMOS) satellite by European Space Agency in 2009, expressly designed to measure soil moisture and ocean salinity from space with a temporal resolution of 1–3 days, marked further significant improvement in this field because of its more suitable observation frequency (L band, i.e., 1.4 GHz) and improved ground resolution (40 km) (Kerr et al. 2010; Wigneron et al.

2000). One of the greatest challenges in using SMOS data (recorded by an interferometric L-band radiometer) in climatological and hydrological models involves downscaling from regional data to local scales, which requires further ground verification, calibration, and parameterization with soil and vegetation features (Balling et al. 2010; Wigneron et al. 1995). In spite of problems related to the radio frequency interferences (RFI, Skou et al. 2010), which can significantly degrade the signal, this sensor represented a solid step toward the generation of operational soil moisture maps.

In order to retrieve SMC, it is necessary to develop models that are capable of accounting for vegetation and surface roughness effects on the microwave signal. Models for simulating emission from a surface are usually based on the radiative transfer theory (RTT). One of the first radiative transfer models that described the physics of microwave radiation in the soil was developed by Wilheit (1975). Mo et al. (1982) developed a simplified form of this theory for vegetated surfaces, known as the "tau-omega" model. This important step in microwave research gave rise to consistent research activities of retrieving soil moisture. The additional step of inverting the model for estimating soil moisture parameters is accomplished in several ways, often using semiempirical or statistical inversion algorithms combined with ancillary data. In particular, a more recent inversion algorithm for the retrieval of SMC maps is the HydroAlgo, implemented at the Institute of Applied Physics of National Research Council in Italy (IFAC-CNR), which is based on RTT as well, and uses an artificial neural network (ANN) for the inversion. This algorithm, which is able to generate maps of SMC from AMSR-E/AMSR-2 data, was developed and implemented within the framework of the Japanese Aerospace Exploration Agency (JAXA) ADEOS-II/AMSR and GCOM/AMSR-2 programs, as well as within a project (PROSA) headed by the Italian Space Agency (ASI), which was devoted to civil protection from floods and landslides (Brogioni et al. 2010; Santi et al. 2010).

This chapter is organized as follows: Section 12.2, following a brief summary of the theory behind the SMC measurement from microwave sensors, summarizes some of the most suitable and recent methods for retrieving soil moisture, with emphasis on those that have proven most applicative in specific case studies. The investigation was restricted to the methods developed for the use of AMSR/AMSR-E/AMSR2 data. In Section 12.3, a relatively new algorithm, developed at IFAC-CNR (HydroAlgo), is described along with the test sites and datasets used for its validation. Section 12.4 contains some examples of SMC and vegetation biomass (VB) maps obtained on a global scale. Section 12.5 then provides a summary and a few concluding remarks with some considerations about the main pros/cons of the described methods. Some directions for future works conclude this chapter.

12.2 The SMC Estimate

One of the most influential spatial programs for generating several satellite-based soil moisture algorithms has been the AMSR-E on board the NASA satellite AQUA. AMSR-E is no longer operating after nearly a decade of use and has been replaced by the recently launched AMSR-2 instrument on the JAXA GCOM-W satellite. Over 10 years of research on the sensitivity of C-band emission (i.e., the lowest available frequency channel from AMSR-E) to moisture of low vegetated soils (e.g., Vinnikov et al. 1999; Jackson and Hsu 2001) demonstrated a significant capability in retrieving SMC from space-borne instruments. C band has, at least in Europe, the advantage of being less affected by the RFIs, which may severely limit the proper functioning of passive microwave systems (Skou et al. 2010; Njoku et al. 2005). RFI can be a serious problem, especially over densely populated areas, as it affects different frequencies depending on the country. For example, C-band data are significantly contaminated in the United States, Japan, and the Middle East, so that some algorithms for the retrieval of SMC employ higher-frequency data, despite the higher sensitivity to vegetation and surface roughness. In Europe, the problem is just the opposite because X-band data have been found to be those most affected by RFI (Njoku et al. 2005).

Both NASA and JAXA have supported the development of their own standard soil moisture algorithms (Njoku et al. 2003; Lu et al. 2009) and additionally supported common research by organizing experimental campaigns and stipulating research agreement for generating innovative algorithms devoted to the same task. Furthermore, JAXA continued to support two alternative research algorithms: the single channel algorithm (SCA, Jackson 1993; Jackson et al. 2010) and HydroAlgo, developed at IFAC-CNR by Paloscia et al. (2001, 2006), Paloscia and Santi (2009), and Santi et al. (2010, 2012).

Within the framework of these projects, a plethora of experiments for validating AMSR-E data over large areas characterized by different climatic conditions were carried out in equipped test areas located in Asia, the United States, and Australia, providing significant datasets for the test and validation of the inversion algorithms.

12.2.1 Theoretical Basis

As mentioned in Section 12.1, most algorithms capable of retrieving SMC are based on the RTT in the simplified form of the tau-omega model, which is capable to adequately describe the complex atmosphere–land–system interactions (Mo et al. 1982). These algorithms usually follow different approaches for inverting the RTT model by either using iterative minimizations of the root-mean-square error between model simulations and measurements assumptions or introducing some measured variables as ancillary information. Main differences usually lie in the methods used for correcting the effects of soil properties (surface roughness, texture), vegetation, and surface temperature.

The radiation from the land surface as observed from above the canopy may be expressed in terms of the radiative brightness temperature T_{bp} and is given as a radiative transfer equation in a rather simple formulation (Mo et al. 1982):

$$T_{bp} = T_s e \Gamma_p + (1 - \omega_p) T_v (1 - \Gamma_p) + (1 - e)(1 - \omega_p) T_v (1 - \Gamma_p) \Gamma_p \qquad (12.1)$$

where T_s and T_v are the thermodynamic temperatures of the soil and the canopy, respectively, ω_p is the single scattering albedo, and Γ_p is the vegetation transmissivity, ranging between 0 and 1. e is the surface emissivity. The subscript p denotes either horizontal (H) or vertical (V) polarization. The first term of the equation represents the radiation from the soil attenuated by overlying vegetation. The second term accounts for the upward radiation directly from the vegetation, while the third term defines the downward radiation from the vegetation, reflected upward by the soil and again attenuated by the canopy. At frequencies higher than 10 GHz the latter term can usually be disregarded. The transmissivity can be defined in terms of the optical depth τ and incidence angle θ such that

$$\Gamma p = \exp(-\tau/\cos\theta) \qquad (12.2)$$

The optical depth τ is a function of the VB, and it has been found that, for frequencies lower than 10 GHz, τ can be expressed as a linear function of vegetation water content

(Jackson et al. 1982), whereas for frequencies between 10 and 36 GHz the optical depth can be related to the VB by the following expression (Pampaloni and Paloscia 1986):

$$\tau = (k/\sqrt{\lambda}) \ln (1 + \text{PWC}) \tag{12.3}$$

where λ is the wavelength, k a crop factor (m½), experimentally found between 0.02 and 0.04 for agricultural vegetation, and PWC is the plant water content (in kg/m²). It was also shown that at C band, the signal becomes totally saturated at τ of about 1.5, although the sensitivity is already rather low above 0.75 (Owe et al. 2001).

ω describes the scattering of the soil emissivity by the vegetation and is a function of plant geometry. Computations of the scattering albedo gave a range of values from 0.04 to about 0.13 (Mo et al. 1982; Owe et al. 2001). ω has also been determined for corn and alfalfa in different physical conditions by using an iterative procedure, from which ω has been found to be very low and always lower than 0.1 at both 10 and 36 GHz (Pampaloni and Paloscia 1986).

From these investigations, several approaches to account for the major variables affecting microwave emission (effective temperature, surface roughness, and vegetation) were developed (Choudhury et al. 1979; Mo and Schmugge 1987; Jackson and Schmugge 1991; Pampaloni and Paloscia 1986; Paloscia and Pampaloni 1988).

12.2.2 Algorithm Review

The basis of the algorithms used was developed in the early 1970s, followed by a great deal of theoretical work to allow a better understanding of the interactions between microwave emission and soil and vegetation parameters. Local situations were investigated, taking into account not only different soil and vegetation characteristics (in terms of moisture and roughness conditions, texture, and different types of crops) but also different sensor system characteristics (i.e., frequency, polarization, and angular geometry), in order to improve microwave emission models from bare and vegetated soils and indicate the optimum system configuration for soil moisture monitoring. Furthermore, many experimental activities were developed in order to put together suitable datasets of microwave emission and ground data and to test and validate the theoretical models (e.g., Lee 1974; Wang et al. 1982; Paloscia et al. 1993; Shutko 1982; Jackson et al. 1982).

Among these algorithms, we have to cite the NASA standard algorithm (Njoku et al. 2003), the SCA (Jackson 1993; Jackson et al. 2010), the Land Parameter Retrieval Model (LPRM) (Owe et al. 2001; de Jeu and Owe 2003), the JAXA officially selected algorithm (Koike et al. 2004; Lu et al. 2009), and a more recent algorithm, which was implemented at IFAC-CNR (HydroAlgo; Santi et al. 2010, 2012).

These algorithms are all based on the same RTT formulation and utilize the tau-omega model by Mo et al. (1982) to represent the electromagnetic radiation from vegetated soils. They all use corrections for the surface temperature, vegetation, soil roughness, and atmospheric effects. A general assumption is that the thermodynamic temperatures of soil and vegetation are approximately the same ($T_v = T_s$). Soil reflectivity was related to the dielectric properties of the soil through the Fresnel equations. Finally, SMC was not retrieved over frozen soils, densely vegetated and snow-covered areas, which have been masked as well as open water. Moreover, RFI contaminated data have been disregarded.

The NASA Standard Algorithm was developed by Eni Njoku (Njoku and Li 1999; Njoku et al. 2003; and, more recently, Njoku and Chan 2006) following the launch of AMSR-E sensor, and it is based on the use of the X-band brightness temperature (T_b) data because of the very strong RFI affecting C-band data, especially in the United States. The algorithm uses all the low-frequency channels of the instrument (X, C, and Ka) in an iterative optimization scheme to simultaneously solve the RT equation for soil moisture, vegetation water content, and surface temperature. Correction for the effects of surface roughness is obtained by using an empirical correlation between the reflectivity of a rough soil surface and that of the equivalent smooth surface (Wang and Choudhury 1981).

The methodology used in the Land Surface Parameter Model (LPRM) (Owe et al. 2001, 2008; de Jeu and Owe 2003) is instead a nonlinear iterative procedure in a forward modeling method, which computes the canopy optical depth through an analytical approach, partitions the surface emission into the soil and the canopy emission, and then optimizes the soil dielectric constant through the Wang–Schmugge model (Wang and Schmugge 1980). The LPRM uses the model of Wang and Choudhury (1981) to correct emission for soil surface roughness effects. The emissivity is directly related to the dielectric constant of the soil, and this relation can be described with the Fresnel relations. The two techniques used to retrieve SMC from AMSR-E data that are described by Njoku et al. (2003) and Owe et al. (2008) were compared in the works of Wagner et al. (2007) and de Jeu et al (2008). The authors found that the NASA standard product (Njoku et al. 2003) provided a weaker performance than the LPRM and suggested that the NASA algorithm is not able to describe the effects of vegetation and/or surface temperature properly (see Table 12.1, where a summary of results obtained on some test areas located in France, Spain, and Australia is shown).

The first complete RTT-based algorithm that allowed direct T_b-soil moisture inversion is the SCA by Jackson (1993). The algorithm was initially developed to support the relatively simple instrument configurations that were available on aircraft platforms. This approach, as with the NASA standard algorithm, assumes that the single scattering albedo is negligible and that the atmospheric contributions are minimal, thus significantly simplifying the tau-omega model. The soil moisture is therefore estimated by sequentially performing temperature normalization, removing the attenuating effects of the overlaying canopy and atmosphere, and estimating the associated smooth (i.e., removed surface roughness effects) surface emissivity using ancillary data. The Fresnel equation is used to convert the emissivity to a dielectric constant, and then the resulting estimate of the dielectric constant is linked to soil moisture using a dielectric mixing model.

The JAXA officially selected algorithm, which was tested on the dry regions of Mongolia, also dealt with the low correlations between microwave emission and SMC in desert

TABLE 12.1

Comparisons of Performances Obtained in Some Test Areas, in Terms of R^2, RMSE, and Bias, from the Comparison between the LPRM Model (Owe et al. 2008) and the NASA Standard SMC Product (Njoku et al. 2003)

	LPRM			NASA		
Area	R^2	RMSE (m³/m³)	Bias (m³/m³)	R^2	RMSE (m³/m³)	Bias (m³/m³)
France 1	0.24	0.28	0.12	−0.01	0.37	0.16
France 2	0.61	0.19	0.08	0.01	0.36	0.11
Spain	0.69	0.06	–	0.00	0.12	–
Australia	0.62	0.03	–	0.29	0.05	–

Source: Data have been derived from R. A. M. de Jeu et al., *Surv. Geophys.* 29, 399–420, doi:10.1007/s10712-008-9044-0, 2008.

regions that are likely due to volume scattering or to the low dielectric dynamics within the soil. The volume scattering in dry soils causes a higher backscatter than the one in subsurface wet conditions. The volume scattering inside soil layers is calculated through the dense media radiative transfer theory (Wen et al. 1990; Tsang and Kong 2001), and the surface roughness effect is simulated by the advanced integration equation model (Chen et al. 2003). The optimal values of forward model parameters are estimated using *in situ* observation data and lower-frequency brightness temperature data. Afterward, a lookup table was generated that relates the variables of interest (i.e., SMC, soil physical temperature, vegetation water content, and atmosphere optical thickness) to microwave emission. Finally, SMC is estimated by linearly interpolating the brightness temperature or index into the inversed lookup table. A more detailed description of the algorithm can be found in the work of Lu et al. (2009).

12.3 HydroAlgo Description and Validation

Behind this algorithm lies the long-term experience of the Microwave Remote Sensing Group, at IFAC-CNR, which has investigated microwave emission and scattering from natural surfaces since the early 1980s. Beginning with the experiments carried out on agricultural surfaces with ground-based dual-frequency microwave radiometers (at X band, 10 GHz, and Ka band, 36 GHz) and combining measurements in thermal infrared bands, the first results were published (Paloscia and Pampaloni 1984, 1988, 1992), stating the relationships between microwave emission and soil and vegetation parameters. Later, the radiometric package was completed in an AMSR-E-like configuration, from C to Ka bands, with the addition of an L-band (1.4 GHz) radiometer, as well. Observations from airborne and ground-based platforms were performed on forest, snow cover, and various agricultural surfaces using this sensor package (e.g., Macelloni et al. 2001, 2003a, 2003b, 2005; Brogioni et al. 2009; Santi et al. 2009).

HydroAlgo is a newly developed algorithm for estimating SMC and VB, which was implemented and validated within the framework of the JAXA AMSR-E project and the PROSA (Products of Earth Observation for the Meteorological Alert) national project, funded by the ASI. HydroAlgo sprung out from this past experience and was already fully described by Santi et al. (2012). It uses a statistical approach based on ANNs for inverting the RTT modeling in order to estimate the SMC of bare or slightly vegetated surfaces. VB is also estimated as an auxiliary product. A flowchart of HydroAlgo is depicted in Figure 12.1. The algorithm has been optimized by using data from the AMSR-E sensor; however, its use can be extended to other sensors operating in similar frequency channels and, in particular, to AMSR-2 on board GCOM-W, which is the heir to the AMSR-E on AQUA platform.

Core of the algorithm are two ANNs, which have been defined and trained independently for ascending and descending orbits. ANN techniques offer the best compromise between retrieval accuracy and processing time for SMC estimate (Hornik 1989; Linden and Kinderman 1989). Training was performed by feeding the ANNs with a large set of brightness temperatures simulated by using the RTT model in the simplified form of tau-omega by Mo et al. (1982), increased with a number of experimental data acquired on three test areas in Mongolia and Australia, within the framework of the JAXA ADEOS-II/AMSR-E and GCOM-W/AMSR-2 programs.

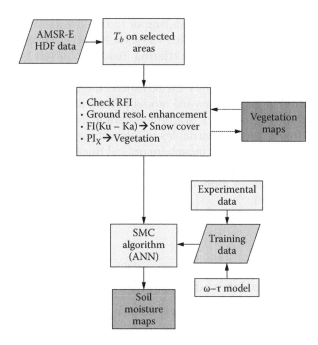

FIGURE 12.1
Flowchart of HydroAlgo.

Model simulations were iterated by randomly varying the input values of SMC and surface temperature T_s in a reasonable range of expected values (i.e., SMC from 0.05 to 0.5 m³/m³, and T_s from 275 to 320 K). The lower threshold of 275 K was selected in order to eliminate frozen soils. The effect of surface roughness was taken into account by including T_b data in the ANN training set, which corresponded to different surface roughness conditions. In the end, a dataset was generated that included 10,000 simulated values of T_b. The dielectric constant was derived from the input of SMC by means of the model from Dobson et al. (1985). The range of the other two inputs required by the model, that is, the optical depth (τ) and the equivalent single scattering albedo (ω), was also set in order to assure consistency between the model simulations and the experimental data.

The implemented ANNs are feed-forward multilayer perceptron, with two hidden layers of ten neurons between inputs and outputs, trained by using the back-propagation learning rule. This configuration was selected after completing several tests as the minimal architecture capable of providing an adequate fit of the data without overfitting, which is indeed the main drawback of the proposed approach (see Santi et al. 2012). When overfitting happens, owing to incorrect ANN sizing or bad training, the ANN returns wrong outputs when tested with input data outside the training range.

The architecture of the two trained ANNs is saved in a configuration file that is called by the algorithm and applied to the input T_b acquisitions during operational use.

First, ANN input is the brightness temperature at C band, in V polarization (T_{bVC}), owing to its sensitivity to the SMC of bare and slightly vegetated soils. The use of V polarization at the nominal incidence angle of AMSR-E (53°, close to Brewster angle) guarantees a relative independence of the soil surface roughness (e.g., Schwank et al. 2010), enhancing the sensitivity to the moisture content.

The second input is the polarization indexes at X band, namely, the difference between T_b in vertical (V) and horizontal (H) polarizations, normalized to their sum ($PI_X = 2 (T_{bVX} - T_{bHX})/(T_{bVX} + T_{bHX})$), which is able to provide information on vegetation optical depth, necessary in accounting for the effects of low vegetation in the retrieval process. PI_X ability to estimate the vegetation optical depth and to identify different levels of biomass was already established in past research carried out on agricultural fields (Choudhury 1989; Paloscia and Pampaloni 1992; Paloscia 1995; Wang and Choudhury 1981).

The polarization index at Ku band (PI_{Ku}), which is still sensitive to the VB, is considered to be a third input, which allows for a better estimation of the vegetation conditions, in order to enhance the retrieval accuracy.

Finally, the brightness temperature at Ka band, V pol (T_{bVKa}), is utilized to normalize for the daily and seasonal variations of the surface temperature owing to its strong relationship with the latter parameter for bare and scarcely vegetated soils (Owe and Van De Griend 2001; Paloscia et al. 2006).

Prior to being passed on to the ANNs, all microwave data necessary for the algorithm are checked for RFI contamination and disaggregated to the spatial resolution of Ka band (i.e., 10 km × 10 km), using the smoothing filter–based intensity modulation technique described in the work of Santi (2010).

Data preprocessing also includes a filtering of forests and densely vegetated areas, where the SMC retrieval is unrealistic. This operation is based on a threshold criterion applied to PI_X. Another threshold criterion masks out the snow-covered areas basing on the frequency index, defined as, basically, the difference between the brightness temperatures (T_b) at two frequencies (Ku and Ka): $FI = [(T_{bKuV} - T_{bKaV}) + (T_{bKuH} - T_{bKaH})]/2$, where V and H are the vertical and horizontal polarization, respectively. FI has been shown to be a good indicator of the presence of snow, owing to the different penetration power of the two wavelengths (Brogioni et al. 2009, 2010; Macelloni et al. 2003a; Santi et al. 2012).

The development, testing, and validation of the algorithm made use of large experimental datasets that were acquired on different test sites. An extensive and very useful dataset used for the development of the SMC algorithms was kindly provided by JAXA. This dataset consisted of 2 years of AMSR-E acquisitions (2003 and 2004) collected on two test sites in Mongolia and Australia. The Australian test area (latitude 35.10°S, longitude 147.70°E) was characterized by low to moderate vegetation conditions, with a marked seasonal vegetation cycle. Instead, the Mongolia site (latitude 46.25°N, longitude 106.75°E) was typified by semiarid conditions, with sparse vegetation and the presence of snow in winter. The AMSR-E acquisitions were co-located with direct measurements of volumetric SMC (Coordinated Enhanced Observing Period; http://www.ceop.net). Only the measurements collected simultaneously with the AMSR-E overpasses were considered in the dataset. The resulting dataset was composed of about 3000 measurements of brightness temperature (T_b) from C to Ka band and the corresponding SMC measurements in the range from 0.05 to ≈0.40 m³/m³, under different vegetation conditions (Santi et al. 2010, 2012). The experimental watersheds of the Agricultural Research Service (ARS) in the United States were selected for a further test of the algorithm. Ground SMC data to be compared with AMSR-E data were kindly provided by Tom Jackson of the USDA-ARS Hydrology and Remote Sensing Lab, in Beltsville, Maryland. These watersheds are well instrumented with multiple surface SMC and temperature sensors and have been the core sites for several AMSR-E validation campaigns. Overall, they represent a wide range of ground conditions and precipitation regimes. The four test areas (Little Washita, Little River, Walnut Gulch, and Reynolds Creek) were characterized by different meteorological, climatic, and consequently vegetation conditions and have been already described by

TABLE 12.2

Sketch of the Characteristics of ARS-US Test Area

Test Areas	No. of SM Stations	Type of Climate	Average Annual Rainfall (mm)	Crop Cover
Little Washita (OK)	20	Sub-humid	750	Rangeland and pasture plus some agricultural crops
Little River (GA)	29	Hot humid summers and short mild winters	1200	Heavily vegetated (forests, croplands, and pasture)
Walnut Gulch (AZ)	19	Semiarid climate	324	Brush and grass covered area
Reynolds Creek (ID)	19	Snow-dominated precipitation	Highly variable (250–1000)	Rangeland dominated area

Source: Data from T. J. Jackson et al., *IEEE Trans. Geosci. Remote Sens.* 48(12), 4256–4260, 2010,

Jackson et al. (2010) and more recently by Paloscia et al. (2012). A short overview of the test area characteristics is given in Table 12.2.

A preliminary analysis on the sensitivity of microwave emission to the soil and vegetation parameters was carried out. Figure 12.2 shows the direct correlation between T_b at C band, in V polarization and at incidence angle >50°, and SMC. Both JAXA and ARS datasets are represented, showing a noticeable sensitivity to SMC and that their regression lines are very close. The data spread indicates that the effect of other factors, and in particular vegetation, was important. Conversely, an evident correlation exists between PI_X and PWC, derived from NDVI measurements (Jackson et al. 2004; Santi et al. 2012) (Figure 12.3). This relationship, which is very similar to those found with ground-based sensors (Paloscia and Pampaloni 1988), points out that PI_X shows very high values on bare surfaces and a sharp decreasing trend as soon as the biomass increases. The diagram of Figure 12.3 confirms the hypothesis of correcting the vegetation effects by using PI_X.

The results obtained on the JAXA datasets by using HydroAlgo have been already published in the work of Santi et al. (2012) and can generally be summarized by the following

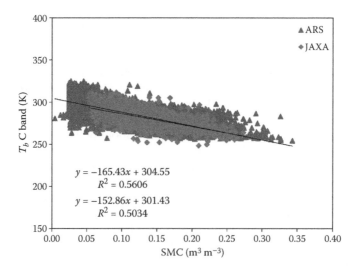

FIGURE 12.2

Brightness temperature (T_b), in V polarization, at C band, derived from AMSR-E sensor, as a function of SMC (m^3/m^3). The two clusters of points correspond to the ARS (triangles) and JAXA (diamonds) datasets, respectively.

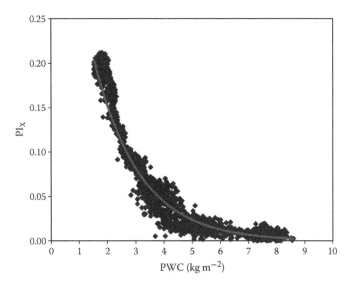

FIGURE 12.3
Polarization index at X band (PI_X), computed from AMSR-E data, as a function of the plant water content (PWC, in kg m^{-2}) derived from NDVI measurements. Solid line represents the regression equation ($PI_X = 0.5279e^{-0.62PWC}$, with $R^2 = 0.89$).

equation, which represents the comparison between the estimated and measured on ground SMC: $SMC_{est} = 0.76 \, SMC_{meas} + 4.98$, with an $R^2 = 0.8$, RMSE = 0.035 m^3 m^{-3}, and bias = 0.02 m^3 m^{-3}.

A comparison between HydroAlgo and NASA standard algorithm outputs was carried out by using 2 months of data extracted from the Mongolian dataset and not used for the training. The SMC generated by HydroAlgo is represented in Figure 12.4, where it is compared with ground data and the output of the standard NASA standard AMSR-E

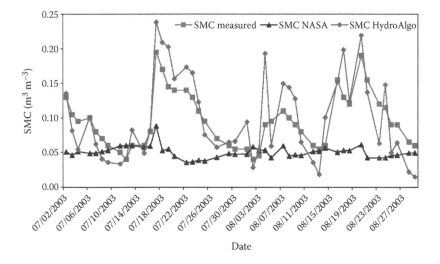

FIGURE 12.4
(**See color insert.**) Comparison between SMC (m^3 m^{-3}) estimated by the HydroAlgo algorithm (magenta) and the standard NASA AMSR-E product (blue) with the SMC measured on ground (green), as a function of time, by using two months of data extracted from the Mongolian dataset and not used for the training.

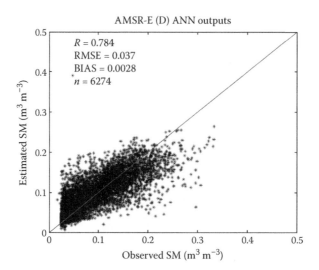

FIGURE 12.5
Comparison of SMC estimated by using the HydroAlgo algorithm on AMSR-E datasets from 2002 to 2009, with the SMC measured on ground, by using the ARS-US dataset, kindly provided by Dr. T. Jackson.

algorithm (Njoku et al. 2003). The HydroAlgo performances, by directly comparing estimated SMC with measured SMC, resulted in $R^2 = 0.72$, RMSE = 0.03 m^3 m^{-3}, and bias = 0.003 m^3 m^{-3}. However, the NASA standard algorithm results were $R^2 = 0.004$, RMSE = 0.06 m^3 m^{-3}, and bias = −0.043 m^3 m^{-3}. As already observed in other studies that made comparisons between different algorithms, NASA standard product usually tends to underestimate the actual SMC values and to smooth the oscillations between dry and moist conditions probably owing to the use of X band, which is undoubtedly less affected by RFI than C band but also physically less sensitive to SMC (de Jeu et al. 2008).

Figure 12.5 represents the result of the HydroAlgo obtained on the ARS-US test areas, where a scatterplot of estimated SMC is shown as a function of SMC measured on ground. The statistical parameters of the regression are the following: $R^2 = 0.61$, RMSE = 0.037 m^3 m^{-3}, and bias = 0.003 m^3 m^{-3} for descending orbits. In the case of ascending orbits, very similar results were obtained. An intercomparison between HydroAlgo and SCA was also carried out in this area and already published in the work of Paloscia et al. (2012). Average statistical values of this comparison are summarized in Table 12.3. R^2 is not very high for both algorithms and is in general between 0.3 and 0.4, whereas the RMSE is lower than 0.05 m^3 m^{-3}, with bias ≪ 0.01 m^3 m^{-3}.

TABLE 12.3

Statistical Values (R^2, RMSE, and BIAS) Obtained on ARS-US Test Areas from Both SCA and HydroAlgo Algorithms

Ascending Orbits	R^2		RMSE (m^3 m^{-3})		Bias (m^3 m^{-3})	
	SCA	HydroAlgo	SCA	HydroAlgo	SCA	HydroAlgo
Little Washita	0.46	0.37	0.047	0.055	−0.015	−0.019
Walnut Gulch	0.20	0.30	0.026	0.030	−0.016	0.001
Little River	0.46	0.28	0.038	0.039	0.018	0.013
River Creek	0.16	0.53	0.025	0.054	−0.014	0.012

Source: Data from S. Paloscia et al., *IEEE Trans. Geosci. Remote Sens.*, 2012.
Note: Similar results have been obtained for descending orbits as well.

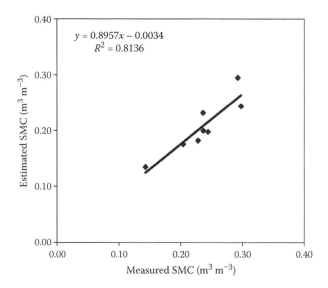

FIGURE 12.6
Direct comparison between SMC estimated by the HydroAlgo algorithm and the SMC measured on the ground in northern Italy.

An additional test was performed in northern Italy, in a flat agricultural area where a long-term experiment concerning the estimate of SMC was carried out using different sensors (Paloscia et al. 2008; Santi et al. 2012). In this area, AMSR-E images were gathered in different seasons from November 2003 to June 2009. In this case, ground measurements sampled over an area of 10×10 km^2 were compared with the output of the algorithm for a pixel centered on 45°N and 8.85°E. The obtained results are shown in Figure 12.6. The statistical parameters of the regression between estimated and measured SMC are $R^2 = 0.81$, RMSE = 0.035 m^3 m^{-3}, and bias = 0.09 m^3 m^{-3}.

12.4 Examples of Soil Moisture and VB Maps

An effort to test the validity of HydroAlgo on a larger scale was carried out. Although corresponding and adequate ground data are missing, this attempt can be useful for better understanding the algorithm's ability to reasonably estimate SMC and VB in other regions, compared to areas where it has been tested and can therefore verify its flexibility. This is particularly important also for evaluating the ANN's ability to generalize the training phase that was based on data derived from small areas.

Such an application of HydroAlgo was carried out over the entire terrestrial globe for SMC and on a portion of Europe for VB. In all these cases, an evaluation of the resulting maps was thus performed on the basis of climatic and meteorological characteristics of the regions investigated.

SMC maps (in m^3 m^{-3}) of the globe obtained on different dates (December 2009 and June 2010) are shown in Figure 12.7a and b. Snow cover and dense forests are masked in the images. At least four levels of SMC can easily be identified. Although no ancillary information is available, the results are in reasonable agreement with the climatic regions and the meteorological

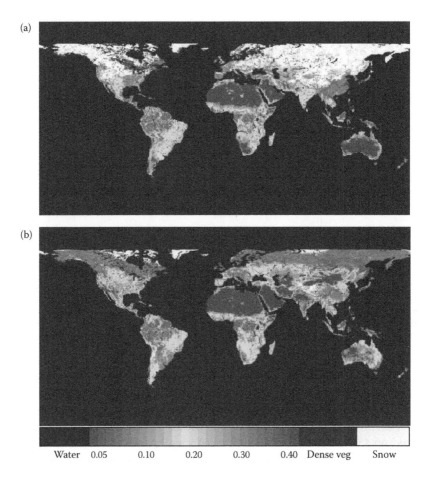

FIGURE 12.7
(See color insert.) Global SMC maps (in m³ m⁻³) of the whole world obtained in (a) December 2009 and (b) June 2010 by using HydroAlgo applied to AMSR-E data. Free water, dense vegetation, and snow areas have been masked.

conditions related to latitude and seasons. The slightly higher SMC values for the Arabian and Australian coasts correspond to the presence of sparse vegetation, as these regions are more humid than desert zones. The seasonal variation in SMC shows an opposite trend in the two hemispheres, for example, Australia is wetter in August than in February.

Maps of VB, expressed as PWC (kg m⁻²), are generated from PI_X and can represent an additional output of the algorithm. In Figure 12.8, an orbit of AMSR-E over Europe was selected in two different seasons: winter (November) and summer (June). Snow cover areas have been masked and represented as white. PWC ranges from 0 to 6 kg m⁻² and dense vegetation (PWC > 6 kg m⁻²) is represented as dark green. Deserts or in any case arid lands show very low values of PWC and are represented in orange/red, whereas the increasing vegetation cover changes from yellow to green. The contrast between winter and summer situations is evident: In November most of Europe is characterized by yellow/light-green colors, showing a high percentage of bare or slightly vegetated soils. Scandinavia is mostly covered by snow. In June, Europe is mostly green; Scandinavia is snow-free, except in the most northern area. Some differences can be observed in North Africa, too, although these areas, mainly covered

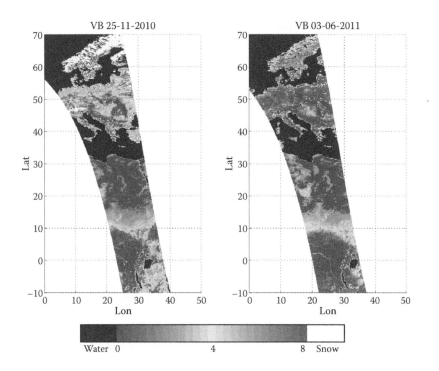

FIGURE 12.8
(See color insert.) Vegetation maps (VB), expressed in PWC (in kg m^{-2}), and generated from HydroAlgo on an AMSR-E orbit over Europe, in November 2010 and June 2011.

by deserts and dense forests, are most stable. Nonetheless, the area corresponding to Sahel shows some seasonal variations, with an increase in slightly vegetated areas in June.

According to these results, VB maps on a global scale can be reasonably generated by using PI$_X$ as a byproduct of the HydroAlgo, providing the advantage of using the same sensor for all applications.

12.5 Concluding Remarks

Since the early 1970s, the goal of estimating soil moisture at different spatial scales from space has slowly become a reality. Many suitable sensors have been launched in space since then, and a large work of experimental and theoretical research supported these spatial activities. At present, many algorithms are available that provide a good estimate of soil moisture and have therefore been adopted by the national space agencies for the creation of operational products to the users. As a general remark, it can be stated that the retrieval of SMC from satellite microwave radiometers has reached a good level of maturity in deriving almost accurate SMC maps though on a global scale. Nonetheless, research seeking for more precise and affordable algorithms continues through refining those available thus far and implementing new ones. One focal point for future research is the improvement of the methods for correcting surface roughness and vegetation effects, especially in case of well-developed canopy cover.

The main drawback of these products is the scarce ground resolution, which hampers the use of SMC maps obtained from spaceborne microwave radiometers in many applicative fields, such as agriculture management, precision farming, flood prevention, and so on, especially in countries characterized by heterogeneous landscapes, like Europe. However, the very frequent revisit time makes this type of sensors very useful, in spite of the poor ground resolution, for all countries with scarce information on their territory available.

This chapter first provided a brief review of the theoretical basis behind the retrieval of soil moisture from microwave sensors. Subsequently, it described some of the most suitable and recent methods for retrieving soil moisture, with a particular focus on those offering the greatest applicability. All these methods are based on the same theoretical basis, albeit using different approaches and approximations for the retrieval of SMC.

Particular attention was paid to the HydroAlgo algorithm, a new algorithm developed at IFAC-CNR, based on the RTT model using ANN for the inversion of the model and the retrieval of SMC. HydroAlgo has been implemented and validated in the framework of both JAXA and ASI preoperational programs. This recent algorithm, which also generates maps of vegetation cover as an auxiliary product, has been optimized to operate by using data from AMSR-E only and is able to produce daily maps of SMC at a spatial resolution comparable to the one of the 37-GHz frequency channel of this sensor (10 km). However, its operability can be extended to other sensors operating in similar frequency channels, such as AMSR-2. The use of ANN for the inversion of RTT model is its strength because of the robustness of this type of inversion methods and simultaneously its weakness because of the complexity of the training process. Very large sets of input data are, in fact, necessary for assuring training able to build a flexible algorithm.

On some areas, a comparison between LPRM and NASA, SCA and HydroAlgo, and also NASA and HydroAlgo algorithms was carried out. However, an actual comparison between the various algorithms is hampered by the scarce availability of shared datasets on common test sites. In general, we note that the NASA algorithm usually tends to underestimate the actual SMC values and to smooth the oscillations between dry and moist conditions. SCA and the JAXA official algorithms are more suitable for describing the SMC trends in arid or semiarid conditions, whereas LPRM and HydroAlgo seem to be able to reproduce fairly well the seasonal dynamic of SMC on a global scale. A major activity of intercomparison between the various inversion algorithms is highly desirable, with exchange of datasets and results among the different research groups, in order to obtain more unbiased results.

Some examples of SMC and VB maps on a global scale obtained by using HydroAlgo have been shown here. Although there are no ground data for a real validation of the maps, they are, nonetheless, in line with the climatic conditions of the different areas and the variations due to meteorological zones. It is pertinent to add that the generation of SMC and VB maps on a global scale is a very ambitious goal due to the very large variability in vegetation cover, climatic conditions, landscapes, and surface features of our globe.

Acknowledgments

This research was partially supported by the Italian Space Agency (ASI) through the PROSA project, the Japan Aerospace Exploration Agency (JAXA) through the ADEOS-II/AMSR and GCOM/AMSR2 programs, and finally, the CTOTUS project, which was cofunded by

Regione Toscana within the framework of the "Programma Operativo Regionale—obiettivo Competitività Regionale e Occupazione"—POR-CReOFESR2007-2013. The authors would also like to thank Dr. Thomas Jackson, who kindly provided data necessary for the comparison between HydroAlgo and SCA.

References

Balling, J., S. Søbjoerg, S. Kristensen, and N. Skou. 2010. RFI and SMOS: Preparatory campaigns and first observations from space. In *Proceedings of the 11th Specialist Meeting on Microwave Radiometry and Remote Sensing of the Environment (MicroRad)*, IEEE, New Brunswick, NJ, pp. 282–287.

Brogioni, M., G. Macelloni, E. Palchetti, S. Paloscia, P. Pampaloni, S. Pettinato, E. Santi, A. Cagnati, and A. Crepaz. 2009. Monitoring snow characteristics with ground-based multifrequency microwave radiometry. *IEEE Trans. Geosci. Remote Sens.* 47(11), 3643–3655, doi:10.1109/TGRS.2009.2030791.

Brogioni, M., G. Macelloni, S. Paloscia, P. Pampaloni, S. Pettinato, and E. Santi. 2010. Two operational algorithms for the retrieval of snow depth and soil moisture content from AMSR-E data. In *Proceedings of the 11th Specialist Meeting on Microwave Radiometry and Remote Sensing of the Environment (MicroRad)*. IEEE, New Brunswick, NJ, pp. 71–75, doi:10.1109/MICRORAD.2010.5559586.

Chen, K. S., T. D. Wu, L. Tsang, Q. Li, J. C. Shi, and A. K. Fung. 2003. Emission of rough surfaces calculated by the integral equation method with comparison to three-dimensional moment method simulations. *IEEE Trans. Geosci. Remote Sens.* 41, 90–101.

Choudhury, B. J., T. J. Schmugge, A. Chang, and R. W. Newton. 1979. Effect of surface roughness on the microwave emission from soils. *J. Geophys. Res.* 84, 5699–5706.

Choudhury, B. J. 1989. Monitoring global land surface using Nimbus-7 37 GHz data—theory and examples. *Int. J. Remote Sens.* 10(10), 1579–1605.

Dobson, M. C., F. T. Ulaby, M. T. Hallikainen, and M. A. El-Rayes. 1985. Microwave dielectric behavior of wet soil—Part II: Dielectric mixing models. *IEEE Trans. Geosci. Remote Sens.* 23(1), 35–46.

Eagleman, J. R., and W. C. Lin. 1976. Remote sensing of soil moisture by a 21-cm passive radiometer. *J. Geophys. Res.* 81(21), 3660–3666.

Hornik, K. 1989. Multilayer feed forward networks are universal approximators. *Neural Networks* 2(5), 359–366.

Jackson, T. J., and A. Y. Hsu. 2001. Soil moisture and TRMM microwave imager relationships in the Southern Great Plains 1999 (SGP99) Experiment. *IEEE Trans. Geosci. Remote Sens.* 39, 1632–1642.

Jackson, T., T. Schmugge, and J. Wang. 1982. Passive microwave remote sensing of soil moisture under vegetation canopies. *Water Resour. Res.* 18, 1137–1142.

Jackson, T. J. 1993. Measuring surface soil moisture using passive microwave remote sensing. *Hydrol. Process.* 7, 139–152.

Jackson, T. J., D. Chen, M. Cosh, Fuqin Li, M. Anderson, C. Walthall, P. Doriaswamy, and E. R. Hunt. 2004. Vegetation water content mapping using Landsat data derived normalized difference water index for corn and soybeans. *Remote Sens. Environ.* 92, 475–482.

Jackson, T. J., M. H. Cosh, R. Bindlish, P. J. Starks, D. D. Bosch, M. Seyfried, D. C. Goodrich, M. S. Moran, and Jinyang Du. 2010. Validation of Advanced Microwave Scanning Radiometer soil moisture products. *IEEE Trans. Geosci. Remote Sens.* 48(12), 4256–4260.

Jackson, T. J., and T. J. Schmugge. 1989. Passive microwave remote sensing system for soil moisture: Some supporting research. *IEEE Trans. Geosci. Remote Sens.* GE-27, 225–235.

Jackson, T. J., and T. J. Schmugge. 1991. Vegetation effects on the microwave emission of soils. *Remote Sens. Environ.* 36(3), 203–212.

de Jeu, R. A. M., and M. Owe. 2003. Further validation of a new methodology for surface moisture and vegetation optical depth retrieval. *Int. J. Remote Sens.* 24(22), 4559–4578.

de Jeu, R. A. M., W. Wagner, T. R. H. Holmes, A. J. Dolman, N. C. van de Giesen, and J. Friesen. 2008. Global soil moisture patterns observed by space borne microwave radiometers and scatterometers. *Surv. Geophys.* 29, 399–420, doi:10.1007/s10712-008-9044-0.

Kerr, Y. H., P. Waldteufel, J.-P. Wigneron, S. Delwart, F. Cabot, J. Boutin, M.-J. Escorihuela et al. 2010. The SMOS Mission: New tool for monitoring key elements of the global water cycle. *IEEE Proc.* 98(5), 666–687.

Koike, T., Y. Nakamura, I. Kaihotsu, G. Davva, N. Matsuura, K. Tamagawa, and H. Fujii. 2004. Development of an Advanced Microwave Scanning Radiometer (AMSR-E) algorithm of soil moisture and vegetation water content. *Ann. J. Hydraul. Eng.* 48(2), 217–222.

Lee, S. L. 1974. Dual frequency microwave radiometer measurements of soil moisture for bare and vegetated rough surfaces. Available at http://ntrs.nasa.gov/search.jsp?R=19740026660.

Leese, J., T. Jackson, A. Pitman, and A. Dirmeyer. 2001. Meeting summary, GEWEX/BAHC international workshop on soil moisture monitoring, analysis, and prediction for hydrometeorological and hydroclimatological applications. *Bull. Am. Meteorol. Soc.* 82, 1423–1430.

Linden, A., and J. Kinderman. 1989. Inversion of multilayer nets. *Neural Networks* 2, 425–430.

Liu, J., and H. Yang. 2010. Spatially explicit assessment of global consumptive water uses in cropland: green and blue water. *J. Hydrol.* 384, 187–197, doi:10.1016/j.jhydrol.2009.11.024.

Liu, J., A. J. B. Zehnder, and H. Yang. 2009. Global consumptive water use for crop production: The importance of green water and virtual water. *Water Resour. Res.* 45, W05428, doi:10.1029/2007WR006051.

Lu Hui, T. Koike, T. Ohta, D. N. Kuria, K. Y. H. Fujii, H. Tsutsui, and K. Tamagawa. 2009. Monitoring soil moisture from Spaceborne Passive Microwave Radiometers: Algorithm developments and applications to AMSR-E and SSM/I. In *Advances in Geoscience and Remote Sensing*, Gary Jedlovec (Ed.), Chapter 17. InTech, New York.

Macelloni, G., S. Paloscia, and P. Pampaloni. 2001. Airborne multifrequency L- to Ka- band radiometric measurements over forests. *IEEE Trans. Geosci. Remote Sci.* 39(11), 2507–2513.

Macelloni, G., S. Paloscia, P. Pampaloni, M. Brogioni, R. Ranzi, and A. Crepaz. 2005. Monitoring of melting refreezing cycles of snow with microwave radiometers: The Microwave Alpine Snow Melting Experiment (MASMEx 2002–2003). *IEEE Trans. Geosci. Remote Sci.* 43(11), 2431–2441.

Macelloni, G., S. Paloscia, P. Pampaloni, and E. Santi. 2003a. Global scale monitoring of soil and vegetation using active and passive sensors. *Int. J. Remote Sens.* 24(12), 2409–2425.

Macelloni, G., S. Paloscia, P. Pampaloni, E. Santi, and M. Tedesco. 2003b. Microwave radiometric measurements of soil moisture in Italy. *Hydrol. Earth Syst. Sci.* 7(6), 937–948.

Mo, T., B. J. Choudhury, T. J. Schmugge, J. R. Wang, and T. J. Jackson. 1982. A model for microwave emission from vegetation covered fields. *J. Geophys. Res.* 87, 11,229–11,237.

Mo, T., and T. J. Schmugge. 1987. A parameterization of the effect of surface roughness on microwave emission. *IEEE Trans. Geosci. Remote Sci.* 25(4), 481–486, doi:10.1109/TGRS.1987.289860.

Njoku, E. G., P. Aschroft, T. K. Chan, and L. Li. 2005. Global survey and statistics of radio-frequency interference in AMSR-E land observations. *IEEE Trans. Geosci. Remote Sci.* 43, 938–947.

Njoku, E. G., and S. K. Chan. 2006. Vegetation and surface roughness effects on AMSR-E land observations. *Remote Sens. Environ.* 100(2), 190–199.

Njoku, E. G., T. J. Jackson, V. Lakshmi, T. K. Chan, and S. V. Nghiem. 2003. Soil moisture retrieval from AMSR-E. *IEEE Trans. Geosci. Remote Sci.* 41, 215–229.

Njoku, E. G., and L. Li. 1999. Retrieval of land surface parameters using passive microwave measurements at 6–18 GHz. *IEEE Trans. Geosci. Remote Sci.* 37(1), 79–93.

Owe, M., R. de Jeu, and T. Holmes. 2008. Multisensor historical climatology of satellite-derived global land surface moisture. *J. Geophys. Res.* 113, F01002, doi:10.1029/2007JF000769.

Owe, M., R. de Jeu, and J. Walker. 2001. A methodology for surface soil moisture and vegetation optical depth retrieval using the microwave polarization difference index. *IEEE Trans. Geosci. Remote Sci.* 39, 1643–1654.

Owe, M., and A. A. Van De Griend. 2001. On the relationship between thermodynamic surface temperature and high-frequency (37 GHz) vertically polarized brightness temperature under semiarid conditions. *Int. J. Remote Sens.* 22(17), 3521–3532.

Paloscia, S. 1995. Microwave emission from vegetation. In *Passive Microwave Remote Sensing of Land-Atmosphere Interactions*, B. Choudhury, Y. Kerr, E. Njoku, and P. Pampaloni (Eds.). VSP BV, Utrecht, Netherlands, pp. 357–374.

Paloscia, S., G. Macelloni, and E. Santi. 2006. Soil moisture estimates from AMSR-E brightness temperatures by using a dual-frequency algorithm. *IEEE Trans. Geosci. Remote Sci.* 44, 3135–3144.

Paloscia, S., G. Macelloni, E. Santi, and T. Koike. 2001. A multifrequency algorithm for the retrieval of soil moisture on a large scale using microwave data from SMMR and SSM/I satellites. *IEEE Trans. Geosci. Remote Sci.* 39, 1655–1661.

Paloscia, S., and P. Pampaloni. 1984. Microwave remote sensing of plant water stress. *Remote Sens. Environ.* 16(3), 249–255.

Paloscia, S., and P. Pampaloni. 1988. Microwave polarization index for monitoring vegetation growth. *IEEE Trans. Geosci. Remote Sci.* 26(5), 617–621.

Paloscia, S., and P. Pampaloni. 1992. Microwave vegetation indexes for detecting biomass and water conditions of agricultural crops. *Remote Sens. Environ.* 40, 15–26.

Paloscia, S., P. Pampaloni, L. Chiarantini, P. Coppo, S. Gagliani, and G. Luzi. 1993. Multifrequency passive microwave remote sensing of soil moisture and roughness. *Int. J. Remote Sens.* 14, 467–484.

Paloscia, S., P. Pampaloni, S. Pettinato, and E. Santi. 2008. A comparison of algorithms for retrieving soil moisture from ENVISAT/ASAR images. *IEEE Trans. Geosci. Remote Sci.* 46(10), 3274–3284.

Paloscia, S., and E. Santi. 2009. Global Scale analysis of soil moisture and vegetation biomass by using AMSR-E data. *J. Remote Sens. Soc. Jpn.*, GLI-AMSR Special Issue, 29(1), 301–306.

Paloscia, S., E. Santi, S. Pettinato, I. Mladenova, T. Jackson, and R. Bindlish. 2012. A comparison between two algorithms for the retrieval of soil moisture on a regional scale by using AMSR-E data. Submitted to *IEEE Trans. Geosci. Remote Sci.*

Pampaloni, P., and S. Paloscia. 1986. Microwave emission and plant water content: A comparison between field measurement and theory. *IEEE Trans. Geosci. Remote Sci.* 24(6), 900–905.

Santi, E. 2010. An application of SFIM technique to enhance the spatial resolution of microwave radiometers. *Int. J. Remote Sens.* 31(9), 2419–2428.

Santi, E., S. Paloscia, P. Pampaloni, and S. Pettinato. 2009. Ground-based microwave investigations of forest plots in Italy. *IEEE Trans. Geosci. Remote Sci.* 47(9), 3016–3025.

Santi, E., S. Pettinato, M. Brogioni, G. Macelloni, F. Montomoli, S. Paloscia, and P. Pampaloni. 2010. A pre-operational algorithm for the retrieval of snow depth and soil moisture from AMSR-E data, In *Proceedings of the 2010 IEEE International Geoscience and Remote Sensing Symposium* (IGARSS 2010). IEEE, New Brunswick, NJ, pp. 3777–3780.

Santi, E., S. Pettinato, S. Paloscia, P. Pampaloni, G. Macelloni, and M. Brogioni. 2012. An algorithm for generating soil moisture and snow depth maps from microwave spaceborne radiometers: HydroAlgo. *Hydrol. Earth Syst. Sci.*, 16, 1–18, doi:10.5194/hess-16-1-2012.

Schmugge, T. J. 1985. Remote sensing of soil moisture In *Hydrological Forecasting*, M. Anderson and T. Burt (Eds.). John Wiley, New York, 101–124.

Schmugge, T. J., O. Neil, and P. E. Wang Jr. 1986. Passive microwave soil moisture research. *IEEE Trans. Geosci. Remote Sci.* 24, 12–22, doi:10.1109/TGRS.1986.289584.

Schwank, M., I. Völksch, J. P. Wigneron, Y. H. Kerr, A. Mialon, P. de Rosnay, and C. Mätzler. 2010. Comparison of two bare-soil reflectivity models and validation with L-band radiometer measurements. *IEEE Trans. Geosci. Remote Sci.* 48, 325–337.

Shutko, A. M. 1982. Microwave radiometry of lands under natural and artificial moistening. *IEEE Trans. Geosci. Remote Sci.* 20, 18–26.

Shutko, A. M. 1986. *Microwave Radiometry of Water and Terrain Surfaces*. Nauka, Moscow, Russia.

Skou, N., S. Misra, J. E. Balling, S. S. Kristensen, and S. S. Søbjoerg. 2010. L-band RFI as experienced during airborne campaigns in preparation for SMOS. *IEEE Trans. Geosci. Remote Sci.* 48, 1398–1407.

Tsang, L., and J. A. Kong. 2001. *Scattering of Electromagnetic Waves: Advanced Topics*. John Wiley, New York.

Ulaby, F. T., R. K. Moore, and A. K. Fung. 1981. *Microwave Remote Sensing: Active and Passive*, vol. I. *Microwave Remote Sensing Fundamentals and Radiometry*. Artech House, Norwood, MA.

Ulaby, F. T., R. K. Moore, and A. K. Fung. 1982. *Microwave Remote Sensing: Active and Passive*, vol. II. *Radar Remote Sensing and Surface Scattering and Emission Theory*. Artech House, Norwood, MA.

Ulaby, F. T., R. K. Moore, and A. K. Fung. 1986. *Microwave Remote Sensing: Active and Passive*, vol. III, *From Theory to Application*. Artech House, Norwood, MA.

Vinnikov, K., A. Robock, S. Qiu, J. Entin, M. Owe, B. Choudhury, S. Hollinger, and E. Njoku. 1999. Satellite remote sensing of soil moisture in Illinois, USA. *J. Geophys. Res.* 104, 4145–4168.

Wagner, W., V. Naeimi, K. Scipal, R. de Jeu, and J. Martínez-Fernández. 2007. Soil moisture from operational meteorological satellites. *Hydrogeol. J.* 15, 121–131.

Wang, J. R., and B. J. Choudhury. 1981. Remote sensing of soil moisture content over bare field at 1.4 GHz frequency. *J. Geophys. Res.* 86, 5277–5282.

Wang, J. R., J. E. McMurtrey III, E. T. Engman, T. J. Jackson, T. J. Schmugge, W. I. Gould, J. E. Fuchs, and W. S. Glazar. 1982. Radiometric measurements over bare and vegetated fields at 1.4-GHz and 5-GHz frequencies. *Remote Sens. Environ.* 12(4), 295–311.

Wang, J. R., and T. J. Schmugge. 1980. An empirical model for the complex dielectric permittivity of soils as a function of water content. *IEEE Trans. Geosci. Remote Sci.* 18(4), 288–295.

Wen, B., L. Tsang, D. P. Winebrenner, and A. Ishimura. 1990. Dense media radiative transfer theory: Comparison with experiment and application to microwave remote sensing and polarimetry. *IEEE Trans. Geosci. Remote Sci.* 28, 46–59.

Wigneron, J.-P., A. Chanzy, J.-C. Calvet, and N. Bruguier. 1995. A simple algorithm to retrieve soil moisture and vegetation biomass using passive microwave measurements over crop fields. *Remote Sens. Environ.* 51, 331–341.

Wigneron, J. P., P. Waldteufel, A. Chanzy, J. C. Calvet, and Y. Kerr. 2000. Two-dimensional microwave interferometer retrieval capabilities over land surfaces (SMOS mission). *Remote Sens. Environ.* 73, 270–282.

Wilheit, T. T. J. 1975. Radiative transfer in a plane stratified dielectric. Technical Report X-911-75-66, NASA Goddard Space Flight Center, Greenbelt, MD.

Zang, C. F., J. Liu, M. van der Velde, and F. Kraxner. 2012. Assessment of spatial and temporal patterns of green and blue water flows under natural conditions in inland river basins in Northwest China. *Hydrol. Earth Syst. Sci.* 16, 2859–2870.

13

Soil Moisture Retrieval from Synthetic Aperture Radar: Facing the Soil Roughness Parameterization Problem

Niko E. C. Verhoest and Hans Lievens

CONTENTS

13.1 Introduction

Although the total volume of soil moisture is very small compared to the total amount of water on Earth (i.e., only 0.0012%; Chow et al. 1988), it is a key variable within the hydrological cycle as it affects both water and heat fluxes. For instance, during precipitation events, the infiltration rate is controlled by soil moisture, and, consequently, it determines the amount of runoff produced. Also, the evapotranspiration rate is controlled by the wetness condition of the soil. Because of this, soil moisture is a crucial variable in the development of crops. Knowledge of the spatial and temporal variability of soil moisture is thus of high merit for watershed applications such as flood prediction, drought monitoring, and crop irrigation scheduling. Unfortunately, acquiring ground measurements of soil moisture is labor intensive. Hence measurements are often limited to a few locations in space and/or to some instances in time (e.g., field campaigns). This shortcoming is overcome with remote sensing, capable of providing spatial maps at regular instances in time.

During the past three decades, synthetic aperture radar (SAR) has demonstrated its ability to retrieve soil moisture at regional scales. A SAR sensor on board an aircraft or spacecraft emits microwaves that are scattered at or near the Earth's surface. The strength and direction of the scattering depend on the vegetation (fresh biomass, canopy structure, etc.), the dielectric constant of the soil and thus soil moisture (e.g., Hallikainen et al. 1985; Dobson et al. 1985; Mironov et al. 2004), the soil surface roughness, radar properties (frequency and polarization), and the incidence angle of the incoming microwave (Ulaby et al. 1996; Fung 1994). As such, the radar backscatter, which expresses the part of the signal that returns back to the sensor, provides information about the surface characteristics, including the soil moisture content.

In one of the first studies on soil moisture retrieval from radar signals, it was found that the soil moisture sensing is optimal when a copolarized (HH or VV) C-band radar is used which operates at a 7° to 15° incidence angle (Ulaby and Batlivala 1976), as for this configuration, the sensitivity of the backscattering coefficient to soil roughness is minimized. At higher incidence angles, a larger sensitivity to surface roughness is found (Ulaby et al. 1986; Baghdadi et al. 2002; Holah et al. 2005). From a hydrological point of view, soil moisture retrieval from sensors characterized by a shorter wavelength than C band is less interesting as the backscattering is influenced by the soil moisture content of the top few centimeters (Ulaby et al. 1996). Longer wavelengths ensure more soil profile information to be contained in the backscattered signal (Ulaby et al. 1996); however, a larger sensitivity of the backscattered signal to roughness is observed than for C band (Mattia et al. 1997). For very wet soils (moisture contents larger than 35 vol%), a lower sensitivity of the radar signal is observed with respect to soil moisture (Holah et al. 2005; Bruckler et al. 1988; Chanzy 1993), making it more difficult to accurately map higher soil moisture contents (Baghdadi et al. 2008).

Several techniques have been suggested for retrieving the soil moisture content from SAR, ranging from simple regression models (e.g., Oh et al. 1992; Taconet et al. 1996; Quesney et al. 2000; Zribi and Dechambre 2002; Le Hégerat-Mascle et al. 2002; Álvarez-Mozos et al. 2005; see also Chapter 4) over semiempirical modeling approaches (e.g., Oh et al. 1992, 2002; Oh 2004; Dubois et al. 1995a,b) to fully physically based backscatter models like the Integral Equation Model (IEM) (Fung et al. 1992; Fung 1994). Unfortunately, each technique is characterized by its inherent limitations. Where empirical models are restricted to the boundary conditions for which they were derived (Oh et al. 1992; Le Hégerat-Mascle et al. 2002; Moran et al. 2000), physically based models are hampered with the parameterization of surface roughness (Verhoest et al. 2008), which has a prevailing impact on the scattering process of the incoming microwaves on the surface. An accurate retrieval of soil moisture is thus only possible if *a priori* accurate surface roughness parameter values are available (Moran et al. 2004).

For natural soil surfaces, roughness is often anisotropic and can be approximated by the superposition of a single-scale process related to the tillage state with a multiscale random fractal process related to field topography (Davidson et al. 2000; Zhixiong et al. 2005) or a self-affine fractal process (Shepard and Campbell 1999; Dierking 1999; Shepard et al. 2001). However, in spite of a few attempts to incorporate a fractal surface description in backscattering models (e.g., Mattia and Le Toan 1999), most studies use a simplified surface description, as fractal approaches usually require a larger number of parameters causing difficulties for inverting the radar backscattering coefficient to soil moisture content (Zribi et al. 2000; Zribi and Dechambre 2002). As such, most backscatter models assume that surface roughness can be described as an isotropic zero-mean Gaussian random process that is characterized by the root-mean-square (RMS) height s, the correlation length l, and an autocorrelation function (ACF) of the surface height (e.g., Ulaby et al. 1996; Fung et al. 1992; amongst many others). These surface roughness parameters are often measured in the field, for example, by collecting surface profile samples. However, the characterization of soil roughness is not fully understood, and a large range of roughness parameter values can be obtained for the same surface depending on the method used (Verhoest et al. 2008; Lievens et al. 2009). Consequently, the parameterization of surface roughness can pose major problems for soil moisture retrieval (Mattia and Le Toan 1999; Davidson et al. 2000; Wagner et al. 2007). In order to circumvent these problems, different strategies have been suggested, consisting of multi-image or change detection analysis (e.g., Liu and Meyer 2002; Rahman et al. 2008; Lievens and Verhoest 2012), the use of SAR polarimetry (e.g., Mattia et al. 1997; Hajnsek et al. 2003; Marzahn and Ludwig 2009; Baghdadi et al. 2012), or the calibration of soil roughness parameters (e.g., Baghdadi et al. 2002, 2004, 2006; Lievens et al. 2011).

This chapter presents an overview of the difficulties encountered in the parameterization of surface roughness from field measurements and furthermore provides alternative ways for soil moisture retrieval from SAR circumventing roughness parameterization. The first section is devoted to roughness characterization and the impact of roughness measurement errors on soil moisture retrieval, based on a physically based backscatter model. In Section 13.2, alternative approaches to circumvent *in situ* soil roughness characterization are presented. In a case study, one of these methods is demonstrated. Finally, a conclusion is formulated.

13.2 Soil Roughness Parameterization

Generally, the surface roughness is characterized from the analysis of height variations along transects or profiles (Bryant et al. 2007) and defined by the RMS height, the correlation length, and an ACF. The RMS height identifies the standard deviation of surface heights along the profile with respect to a reference surface. For discrete one-dimensional profiles consisting of n points with surface height z_i, the RMS height s, is calculated as (Ulaby et al. 1982)

$$s = \sqrt{\frac{1}{n-1}\left[\left(\sum_{i=1}^{n} z_i^2\right) - n\bar{z}^2\right]}, \tag{13.1}$$

where

$$\bar{z} = \frac{1}{n}\sum_{i=1}^{n} z_i. \tag{13.2}$$

The correlation length *l* describes the horizontal distance over which the surface profile is correlated with a value larger than $1/e$ (Ulaby et al. 1982). The ACF of the profile is defined by

$$\rho(\xi) = \frac{\sum_{i=1}^{n-j} z_i z_{i+j}}{\sum_{i=1}^{n} z_i^2}, \tag{13.3}$$

with $\xi = j\Delta x$ and Δx the spatial resolution. A large number of propositions have been made for an analytical approximation of the ACF; however, most backscatter models offer only two standard types of ACF, that is, the exponential and the Gaussian ACF (Fung 1994). The exponential ACF is given by

$$\text{ACF}(\xi) = e^{-|\xi|/l}, \tag{13.4}$$

and the Gaussian ACF is defined as

$$\text{ACF}(\xi) = e^{-\xi^2/l^2}, \tag{13.5}$$

with *l* the correlation length. As the ACF is mostly assumed to be of the exponential or Gaussian type (Ogilvy 1991), the problem of the roughness characterization is restricted to the parameterization of only *s* and *l*. These surface roughness parameters are often measured in the field, for example, by collecting surface profile samples. Because surface roughness is a three-dimensional phenomenon, roughness measurements should optimally be derived from two-dimensional surface height measurements, for example, using terrestrial laser or photogrammetric instruments (Jester and Klik 2005). However, most radar remote sensing studies make use of one-dimensional surface height measurements for the parameterization of *s* and *l*, as such measurements are more easy to perform in the field. The standard procedure for such one-dimensional measurements consists of first defining a series of surface height points (the roughness profile) along a one-dimensional surface transect. Therefore, different instruments can be used, for example, meshboards, pin profilometers, or laser techniques (Mattia et al. 2003). Then, a linear trend is removed from this profile in order to compensate for the possibility that the instrument was not aligned perfectly parallel to a plane reference surface (Callens et al. 2006). Finally, the roughness parameters *s* and *l* are calculated from the series according to Equations 13.1 through 13.5.

In order to account for the large within-field variability in *s* and *l* values, generally several profiles are analyzed and average roughness parameters are used in retrieval studies (Davidson et al. 2000; Callens et al. 2006; Bryant et al. 2007). Although this method seems straightforward, different characteristics in the profile, such as total profile length, vertical measurement accuracies, and horizontal spacing between measurement points, lead to diverging roughness parameterizations (Ogilvy 1991; Bryant et al. 2007; Lievens et al. 2009). Furthermore, the linear trend removal as one of the analysis steps may not be adequate if longer profiles are measured in which topographic undulations are observed along the transect.

Lievens et al. (2009) performed an in-depth study of the error that can be expected in soil moisture retrieval based on the IEM due to different profile characteristics. In order to assess the impact of roughness parameterization techniques on the soil moisture retrieval, extremely long synthetic surface roughness series, characterized by a predefined RMS height and correlation length, were generated through a first-order autoregressive model assuming an exponential ACF (cfr. Lievens et al. 2009). From these series, surface roughness profiles were sampled with specific profile length, vertical accuracy, and horizontal spacing, depending on the profile characteristic under study.

13.2.1 Profile Length

On the basis of 500 repetitions per profile length, it was clearly found that roughness parameters increase with increasing profile length, as illustrated in Figure 13.1. Therefore short profiles can result in an underestimation of both s and l. These results confirm the findings of many other authors (e.g., Oh and Kay 1998; Mattia et al. 2003; Callens et al. 2006; among others). Furthermore, it is worth remarking that the increase of s and l with profile length is more pronounced for surfaces characterized by a larger correlation length. Finally, the figure also demonstrates that for increasing profile lengths, the variability in roughness parameters decreases.

As a consequence of this scaling behavior, different values for soil moisture may be obtained, depending on the profile length used for roughness parameterization. Figure 13.2 demonstrates the mean and standard deviation of the retrieved soil moisture contents for increasing profile lengths and different sensor configurations. It is found that the estimated soil moisture contents can be largely overestimated or underestimated if short profiles are used, especially if only one soil profile is used for soil roughness characterization. At high moisture contents, the errors that can be made are more pronounced. This can be attributed to the lower sensitivity of the backscatter signal to soil moisture if moist soils are illuminated. Because of this, small changes in roughness values can result in large differences in retrieved soil moisture (Lievens et al. 2009).

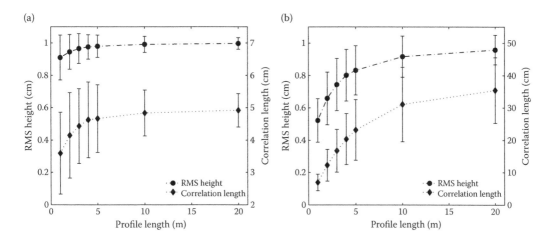

FIGURE 13.1
Mean and standard deviations of RMS height and correlation length for different profile lengths for (a) $(s, l) =$ (1 cm, 5 cm) and (b) $(s, l) = $ (1 cm, 40 cm).

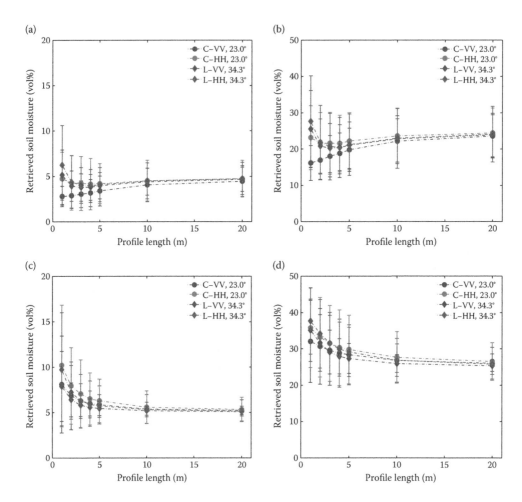

FIGURE 13.2
(See color insert.) Mean and standard deviations of retrieved soil moisture values for different profile lengths for (s, l) equal to (a, b) (1 cm, 40 cm), (c, d) (2 cm, 5 cm), and initial moisture contents of (a, c) 5 vol% and (b, d) 25 vol%.

Furthermore, the error made in the retrieval seems to depend on the roughness state of the surface: The parameterization of s and l on a rough soil causes that the soil moisture contents are overestimated, whereas smooth surfaces give rise to underestimated soil moisture contents. Unfortunately, there is no consensus on the spatial scale at which roughness parameters need to be defined in order to be most appropriate for describing scattering on rough surfaces. From the figures discussed, one can thus only conclude that major errors in the retrieval of soil moisture can be expected in function of different profile lengths for the same soil surface.

13.2.2 Number of Profile Measurements

Figures 13.1 and 13.2 show that, based on only one roughness profile, it is unlikely to obtain the expected average value of roughness parameters. By consequence, when these roughness values are used for the retrieval of soil moisture, large errors may be obtained. In order to reduce the variability in soil roughness parameter values, average values can be calculated from several profiles. Bryant et al. (2007) suggested that the surface RMS height should be averaged over at least 20 3-m profiles in order to be representative. Baghdadi et al. (2008) proposed averaging roughness values over 10 profiles (of 2 m). Such averaging results in RMS heights with a precision better than 5% and a precision for correlation lengths ranging between 5 and 15%.

To assess the impact of the number of profiles on the accuracy of roughness parameters and the retrieved soil moisture, extremely long roughness profiles were again simulated from which nonoverlapping profiles were extracted and roughness parameters were derived. Roughness parameters to be used for the inversion of the backscatter coefficient to soil moisture values are then determined by averaging n profiles of a predefined length (where n ranges from 1 to 20). The variability of the determined average roughness parameters was estimated through performing this procedure 1000 times for each profile length considered. Figure 13.3 demonstrates that the number of profiles needed to ascertain that the standard deviation on RMS height or correlation length is less than 10% of the mean decreases with increasing profile length. From Figure 13.3, it is also clear that surfaces exhibiting a larger correlation length need a larger number of profiles in order to reach the same precision. This can be attributed to the fact that for these surfaces, a higher variability in roughness parameterization is obtained (see Figure 13.1).

In order to assess the impact of the number of profiles used for estimating soil roughness on the soil moisture retrieval, we will only focus on 4-m profiles. As demonstrated in Figure 13.4, increasing the number of profiles for calculating averaged soil roughness parameters does not have a large impact on the soil moisture retrieval accuracy. More important are the sensor configuration, the wetness state of the soil, and the surface roughness condition. Note that the biases that are found in the soil moisture retrieval (see Figure 13.4) are attributed to the profile length used for estimating the roughness parameters (see section on profile length).

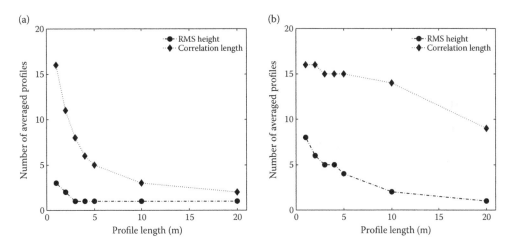

FIGURE 13.3
Number of profiles required to obtain a standard deviation of RMS height or correlation length less than 10% of the mean for different profile lengths for profiles with (s, l) equal to (a) (1 cm, 5 cm) and (b) (1 cm, 40 cm).

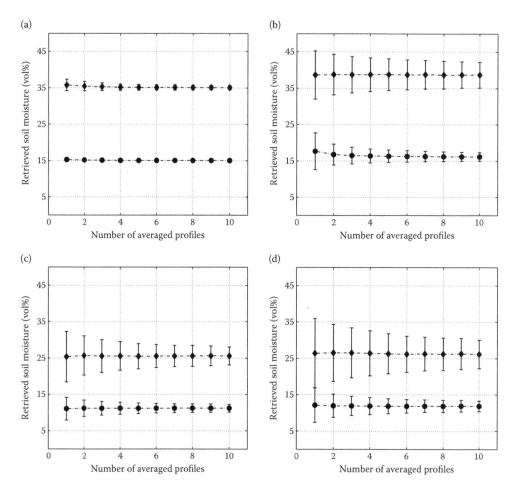

FIGURE 13.4
Mean and standard deviations of retrieved soil moisture values for different numbers of 4-m profiles used, for profiles with (s, l) equal to (a, b) (1 cm, 5 cm) and (c, d) (1 cm, 40 cm), and for (a, c) C-VV at 23° incidence angle, and (b, d) L-HH at 34.3° incidence angle. Considered initial moisture contents are 15 vol% (circles) and 35 vol% (diamonds).

13.2.3 Instrument Accuracy

The instruments that are used to measure surface roughness are characterized by a horizontal and a vertical resolution. Both resolutions have an impact on the roughness parameters as well as on the soil moisture retrieval.

13.2.3.1 Horizontal Resolution

On the basis of a theoretical experiment, Lievens et al. (2009) found that an increase in horizontal spacing causes a decrease in RMS height and an increase in correlation length, which are more pronounced for surfaces with small correlation lengths. Large spacings also cause a change in slope of the ACF around zero, as the high-frequency component (height deviations over very small horizontal distances) is not contained in the sampled surface profile. Using profiles from instruments with a coarser horizontal resolution may therefore cause that the ACF resembles a Gaussian function, whereas in reality the function has an exponential shape, as is often observed for agricultural soils (Callens et al. 2006). Such false description of the ACF may lead to large retrieval errors (Lievens et al. 2009). In general, instruments having a coarse horizontal resolution (spacing between measurement points >1 cm) give rise to large retrieval errors, which are more pronounced for high moisture contents and rough surfaces (Lievens et al. 2009).

13.2.3.2 Vertical Accuracy

The vertical accuracy of an instrument can be defined as the standard deviation of the height measurement if repetitive measurements would be made over the same point. Highly accurate instruments (e.g., a laser profiler) are characterized by accuracies up to 1 mm, whereas less accurate instruments (e.g., a meshboard or pin profilometer) reach up to 5 mm. It is obvious that larger uncertainties in the vertical measurement cause larger errors in the roughness parameters and by consequence large retrieval errors. When accurate instruments, characterized by vertical accuracies less than 2 mm are used, only marginal errors in roughness parameter estimation are obtained. Consequently, small variations in retrieved soil moisture can be expected. However, instruments with a noise level of 5 mm may result in soil moisture errors ranging from 0.5 vol% for dry and rough fields up to 8 vol% for wet and smooth fields (Lievens et al. 2009).

13.2.4 Detrending

As mentioned before in this chapter, detrending the measured profile is a standard processing step for roughness parameterization. Generally, this consists of removing a linear trend from the profile in order to compensate for the fact that the profile transect may have been slightly tilted with respect to the average surface plane. However, if a field is characterized by a slightly undulating surface, a very low frequency component will be introduced in the profile, which may highly affect the roughness parameters. According to Ulaby et al. (1982), only the high-frequency component should be maintained for the parameterization of roughness, as the low-frequency roughness component should be included directly in the backscatter model. Through comparing different detrending options (linear vs. nonlinear), Lievens et al. (2009) found that very large differences in roughness parameter values can be obtained. This consequently results in significantly different retrieved soil moisture values (>25 vol%), depending on the detrending technique used.

13.3 Circumventing Roughness Parameterization

Obviously, the characterization of soil roughness is at present not fully understood, and a large range of roughness parameter values can be obtained for the same surface depending on the method used (Verhoest et al. 2008; Lievens et al. 2009). Unfortunately, there is no protocol available that defines the required instrument characteristics (profile length, horizontal and vertical accuracy), the number of measurements needed, and the detrending techniques to be used in order to obtain reliable roughness parameters that lead to trustable soil moisture retrievals. Furthermore, one should be able to retrieve soil moisture from SAR imagery without the need of additional roughness measurements, as this violates the primary objective of remote sensing, that is, being able to monitor regions without performing additional field work (Verhoest et al. 2007). As such, a wide range of alternative approaches that circumvent roughness field measurements have emerged.

13.3.1 Multi-Image Methods

A promising way to eliminate the need for roughness measurements lies in combining information obtained from two or more SAR images. As was found by Oh et al. (1992), the difference in backscatter observed at 25° and 50° is much larger for a smooth surface (~9 dB) compared to a rough surface (~3 dB). On the basis of this observation, Srivastava et al. (2003) developed a model for soil moisture retrieval that incorporated the difference between high and low incidence backscatter, which enabled them to account for surface roughness. A second multi-image method was developed by Zribi and Deschambre (2002), who proposed a new parameter, namely, the Z index ($Zs = s^2/l$), which showed a strong correlation with the difference in backscatter between two acquisitions at different incidence angles. Rahman et al. (2007) demonstrated that the Z index could be used for deriving s and l separately when a SAR image acquired with dry soil conditions was available. This method allows for parameterizing the IEM to retrieve soil moisture without the need for ancillary data (Rahman et al. 2008). An obvious constraint of the latter approach is that it requires a SAR acquisition for dry soil conditions and two SAR acquisitions at different incidence angles. Unfortunately, there are currently no spaceborne C- or L-band SAR sensors capable of acquiring multi-incidence imagery. Hence subsequent acquisitions with different configurations need to be used. Furthermore, during the time span of these subsequent data takes, the soil moisture content needs to remain unchanged, which is a strong precondition given the long revisit time of the currently orbiting satellites. Fortunately, future missions such as the RADARSAT, SENTINEL, and SAOCOM constellations are expected to supply more sophisticated datasets at decreased revisit time, which may foster the use of multi-image based retrievals for hydrological applications (Kornelsen and Coulibaly 2012).

Another successful multi-image method that allows for circumventing roughness measurements is the use of change detection (Rignot and Van Zyl 1993). Change detection techniques are based on the reasoning that short-term changes in radar backscatter are related to changes in soil moisture only, because surface roughness and vegetation changes take place over longer timescales. Applications of this technique have particularly been successful using scatterometer observations at coarse scales (e.g., Wagner et al. 1999; Wagner and Scipal 2000), because the large swath widths of scatterometers enable frequent observations that capture the high temporal dynamics of soil moisture. Moreover, large scales allow for treating the surface roughness as temporally constant and simplify the seasonal description of vegetation. Nevertheless, change detection techniques have more recently also been applied using ENVISAT ASAR Global Monitoring (e.g., Pathe et al. 2009; Mladenova et al. 2010) and Wide Swath (e.g., Van Doninck et al. 2012) data. At high resolution, applications have mainly been restricted to airborne datasets, for example, from the 1999 Southern Great Plains (SGP99) experiment (Njoku et al. 2002), the 2002 Soil Moisture Experiment (SMEX02) (Narayan et al. 2006), and the AgriSAR 2006 campaign (Balenzano et al. 2011), because airborne campaigns mostly schedule frequent flights over short time periods without major changes in soil roughness. Because of a shortage of dense time stacks of acquisitions, only very few studies were able to use high-resolution spaceborne SAR data for evaluating soil moisture retrieval through change detection (e.g., Moran et al. 2000; Wickel et al. 2001; Yang et al. 2006; Lievens and Verhoest 2012). The application at a field scale level is further complicated when changes in roughness and vegetation growth may render the basic assumptions of the approach invalid. Future research is definitely needed to extend the applicability of change detection techniques at high resolution, for example, through combination with vegetation modeling and the incorporation of roughness changes.

13.3.2 SAR Polarimetry

An increasing interest has recently been drawn to the use of SAR polarimetry, which allows for retrieving surface parameters merely based on remote sensing observations. A polarimetric SAR device is designed to measure the complex scattering amplitudes of a target using four polarization combinations. On the basis of this information, a number of polarimetric features can be extracted that maximize the radar sensitivity to specific target properties, such as vegetation, soil roughness, or soil moisture content.

Mattia et al. (1997) demonstrated that the copolarized correlation coefficient or coherence, expressed in a circular polarization basis, strongly depends on surface roughness, while being independent of soil moisture. Schuler et al. (2002) provided that the real part of the circular coherence is actually more sensitive to soil surface roughness than the coherence itself. On the basis of the small perturbation model (SPM; Rice 1951), a new method for the inversion of soil moisture and surface roughness was introduced by Hajnsek et al. (2003) and evaluated using a series of airborne L-band polarimetric SAR images over test sites in Germany. The inversion of surface parameters is based on a set of three polarimetric parameters, that is, the scattering entropy, anisotropy, and alpha angle, which can be derived from the eigenvalue–eigenvector decomposition of Cloude and Pottier (1996). Marzahn and Ludwig (2009) compared the performance of several polarimetric features for the retrieval of the surface RMS height as measured by close-range photogrammetry and showed that the real part of the complex circular coherence outperforms the circular coherence and the anisotropy. Furthermore, they established a linear inversion scheme to retrieve RMS height from polarimetric SAR data. Although SAR polarimetry has meanwhile demonstrated its ability for L-band SAR data, the merit for C-band SAR still needs to be demonstrated. Because the approach of Hajnsek et al. (2003) is not applicable to C-band due to the restricted validity domain of the SPM ($ks < 0.3$, with k the wave number), Allain et al. (2004) introduced a polarimetric scattering model based on the IEM, with inclusion of the entropy, alpha angle, and a new parameter, that is, the eigenvalue relative difference. Baghdadi et al. (2012) investigated the use of the latter polarimeric features for surface parameter inversion based on C-band RADARSAT-2 data and concluded that its potential is very limited. They argue that, as opposed to L band, the low potential for C-band can be mainly related to a low sensitivity of polarimetric features for higher values of ks.

13.3.3 Roughness Calibration

Because the roughness parameters assessed from field measurements are generally inaccurate, parameters are often calibrated in order to improve soil moisture retrieval results. Baghdadi et al. (2002) proposed to calculate the correlation lengths from measured RMS heights, because the latter can be determined more accurately than the correlation lengths. They therefore derived an empirical relationship between RMS height and correlation length by fitting IEM results to radar observations:

$$l = e^{\alpha s + \beta} \tag{13.6}$$

where α and β are calibration constants, depending on incidence angle and polarization. In order to be applicable to a wider range of soil roughness conditions (RMS heights ranging between 0.25 and 5.5 cm) this relationship can be replaced by a power-type relationship (Baghdadi et al. 2004):

$$l = \alpha s^{\beta} \tag{13.7}$$

Su et al. (1997) circumvented the roughness parameterization problem by introducing the effective roughness approach. In this approach, backscatter observations and in situ soil moisture measurements are used for the estimation of roughness parameters. These calibrated "effective" roughness parameters are then used in the retrieval of soil moisture from subsequent acquisitions. However, Baghdadi et al. (2002, 2004, 2006) demonstrated that effective roughness parameters derived from subsequent acquisitions can show large differences. As such, the assumption of temporally constant parameters, as made by Su et al. (1997) may not be justified. Alternatively, Lievens et al. (2011) developed an empirical model that allows for calculating the effective roughness parameters of each new acquisition. In case of the IEM, two effective roughness parameters are needed: that is, an RMS height (s) and correlation length (l). Lievens et al. (2011) found that there are a large number of different (s, l)-combinations that yield a considerably small soil moisture retrieval error. In order to obtain one single effective (s, l)-combination, they therefore suggested to fix one roughness parameter and calibrate the other. As such, one can fix a value for s and determine the corresponding value of the correlation length that is referred to as l_{eff}. On the basis of a large set of 15°–30° incidence angle C-band backscatter observations over smooth to medium smooth bare soil fields and corresponding soil moisture measurements at several sites across Europe, Lievens et al. (2011) found that the behavior of the field-averaged effective correlation length varied linearly with the angular normalized backscatter coefficient σ_n^0. The normalization to a reference incidence angle can be performed for instance based on Lambert's law for optics (cf. Ulaby et al. 1982; Van Der Velde and Su 2009; Lievens et al. 2011; De Keyser et al. 2012). For a predefined value of s, for example, $s = 1$ cm, the effective correlation length can be modeled from a linear relationship:

$$l_{\text{mod}} = a\sigma_n^0 + b \tag{13.8}$$

where a and b correspond respectively to the slope and the intercept of the regression line. On the basis of an extensive cross validation, Lievens et al. (2011) demonstrated the robustness of this regression. They showed that through using this technique, soil moisture retrieval errors can be obtained, which are below 5 vol% for HH polarization and 6.5 vol% for VV polarization. In the following section, this method will be demonstrated for a case study over Flevoland, the Netherlands.

13.4 Case Study

The technique of effective roughness parameters is demonstrated based on RADARSAT-2 imagery acquired during the AgriSAR 2009 campaign supported by the European Space Agency. The campaign was organized in Flevoland, a flat agricultural polder, characterized by marine clay pedology, located in the center of the Netherlands (52°26′N, 5°33′E). The study uses three fully polarimetric descending (6 AM) RADARSAT-2 fine mode "Fine-Quad" (FQ) images that were acquired along with field measurements of soil moisture at the same day. The latter measurements have been carried out on August 12, September

TABLE 13.1

Overview of RADARSAT-2 Acquisition Characteristics

Nominal swath width	25 km
Number of looks	1
Projection	Slant range
Nominal slant range resolution	5.4 m
Nominal azimuth resolution	8.0 m
Descending/ascending pass	6 am/6 pm
NESZ	−36.5 ± 3 dB
Frequency	5.405 GHz
Polarization	HH, HV, VH, and VV
Beams	FQ2 to FQ21
Incidence angle range	20°–41°

5, and September 29, 2009. Table 13.1 gives an overview of the RADARSAT acquisition characteristics of the images used. The acquisition and processing of the RADARSAT-2 imagery have been performed by MacDonald, Dettwiler, and Associates Ltd. Geospatial Services. The RADARSAT-2 FQ mode data were acquired as Single-Look Complex data in slant range and were further processed through slant-to-ground range conversion, multilooking, and geocoding into geocoded terrain-corrected backscatter data (Caves et al. 2010a,b). Given the flat topography, it was not needed to calculate normalized backscattering coefficients for this study, as the incidence angle is nearly constant for all fields ($\theta = \pm 33°$).

In order to test the retrieval algorithm, soil moisture was measured at a large number (>65) of (nearly) bare soil fields within the Flevoland test site. The selected fields, displayed in Figure 13.5, have an average field size of about 10 ha. During each field campaign, soil

FIGURE 13.5
The selected study fields in Flevoland, the Netherlands.

FIGURE 13.6
(**See color insert.**) Illustration of the roughness classes (a) medium smooth and (b) stubbles.

moisture was measured using a 10-cm probe (Trime IT and MiniTrase) time domain reflec-tometry. In order to be able to measure as many fields as possible, soil moisture was only measured at one location in each field (close to the field boundary), with three repetitions to reduce uncertainty. It was found that the spatial variability in soil moisture between different fields was fairly low, that is, the standard deviation across the area was 5.73, 4.99, and 5.93 vol% for average soil moisture values of 32.93, 36.01, and 28.94 vol%, respectively, on August 12, September 5, and September 29.

Field average backscatter values, which were calculated by overlaying regions of interest with at least a one pixel buffer from field boundaries (Caves et al. 2011), were grouped in two different roughness classes (cf. Figure 13.6): medium (seedbed preparation, obtained by rotary tilling and smooth harrowing) and stubbles (corresponding to a smooth to medium smooth roughness onto which cereal harvest remains are found). For both roughness classes, the sensitivity of the field averaged backscattering coefficient toward soil moisture is demonstrated in Figure 13.7. This figure clearly shows that increasing soil moisture contents result in higher backscattering coefficients. For the medium surface roughness class,

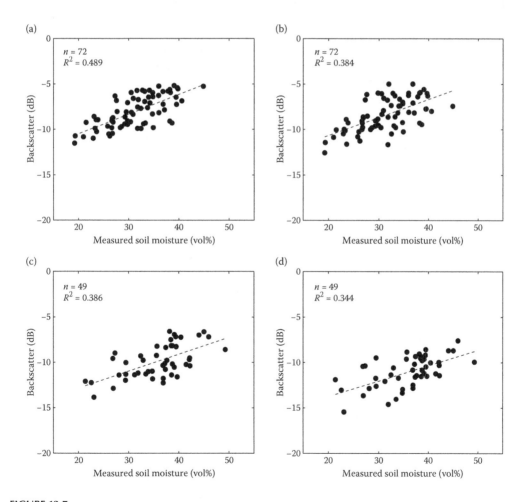

FIGURE 13.7
Sensitivity of RADARSAT-2 backscatter to soil moisture for medium roughness at (a) HH and (b) VV polariza-tion, and stubble fields at (c) HH and (d) VV polarization.

a strong correlation is found (R^2 approximately 0.5), whereas the stubble fields display a lower correlation. This decrease in correlation may be attributed to the presence of the stubbles that influence the scattering process.

To obtain effective roughness parameters, model parameters a and b from Equation 13.8 need to be fit. Therefore effective correlation lengths are first calculated for a predefined value of $s = 1.125$ cm for the medium rough class and $s = 0.75$ cm for the stubble fields, through inverting the IEM for correlation length, given field soil moisture, the predefined s value, and the field-averaged backscattering coefficient. These values for s were selected as they yield optimal soil moisture retrieval results for the roughness classes considered.

To assess the accuracy of the soil moisture retrieval, a leave-one-out cross validation method is performed for each of the roughness classes. In this technique, one data point is left out from the dataset as this will be used for validation, while the other data are used to fit the regression (Equation 13.8). The obtained regression is then used to model the effective roughness parameter (i.e., l_{mod}) based on the backscattering coefficient of the validation point and the soil moisture of the field is retrieved using the modeled effective roughness parameters to invert the IEM. The same procedure is repeated such that all data points of the dataset are used once as validation point. Subsequently, all retrieved soil moisture values are compared to the measured soil moisture values and retrieval statistics are calculated (see Table 13.2). As revealed by Table 13.2, model parameters differ between the polarization mode of the satellite in case of the stubble fields. This may indicate that stubbles have a different impact on the backscattering coefficient, depending on the polarization used.

Figure 13.8 plots the retrieved soil moisture versus the corresponding field measurements for both HH and VV polarization. For the medium smooth fields, the error on the soil moisture retrieval, expressed as RMSE equals 3.95 and 4.32 vol% for HH and VV polarization, respectively. As can be expected, soil moisture retrieval is slightly worse for stubble fields. This is especially the case for VV polarization, for which an RMSE of 6.26 vol% is found. HH polarization is less influenced by the stubbles, resulting in a retrieval error of 5.05 vol%.

These results confirm the large potential of the approach, as suggested by Lievens et al. (2011). Especially the fact that field measurements of surface roughness can be omitted if a single configuration SAR is used opens perspectives. Only general information about the roughness conditions is required as *a priori* information for soil moisture retrieval. Verhoest et al. (2007) suggests that such *a priori* information could for instance be derived from the tillage state of a field (e.g., through a crop calendar). However, it can be expected that such information may also be derived from polarimetric SAR data.

TABLE 13.2

95% Confidence Intervals for Regression Model Parameters a and b, Together with the F and p Values

Roughness	Polarization	n	a	b	F value	p Value
Medium	HH	72	[−4.749 to −4.193]	[−20.053 to −15.458]	1030.0	$<10^{-5}$
Medium	VV	72	[−4.510 to −3.715]	[−16.161 to −9.331]	425.4	$<10^{-5}$
Stubbles	HH	49	[−3.045 to −2.590]	[−18.738 to −14.111]	623.0	$<10^{-5}$
Stubbles	VV	49	[−5.569 to −4.988]	[−37.090 to −24.077]	296.7	$<10^{-5}$

Note: n represents the number of data points for each roughness class and polarization.

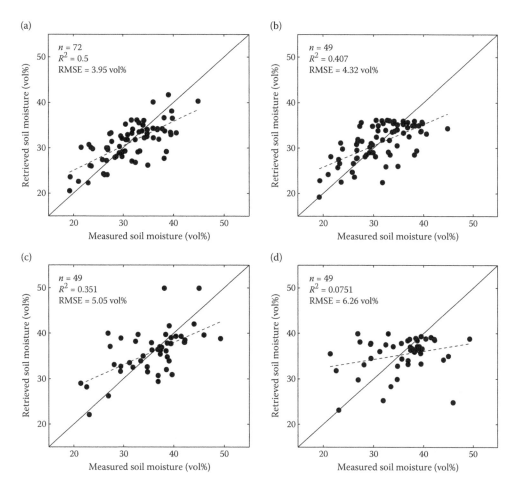

FIGURE 13.8
Measured versus retrieved soil moisture (vol%) for medium roughness at (a) HH and (b) VV polarization, and stubble fields at (c) HH and (d) VV polarization.

13.5 Conclusions

In this chapter, we have focused on the problems that may be encountered when using *in situ* measured surface roughness parameters in the retrieval of soil moisture from SAR data. It was shown that several characteristics of the roughness measurement device, such as profile length, instrument accuracy, or preprocessing of the measured roughness profiles, may have a significant impact on the characterization of the roughness parameters, and consequently on the soil moisture retrieval.

A number of alternative techniques to circumvent the problem of soil roughness characterization have been discussed. Multi-image methods have mainly demonstrated their skill for soil moisture retrieval using coarse resolution data, such as global monitoring and scatterometer observations. However, applications at the field scale are limited because of lack of dense and long-term time series of spaceborne SAR acquisitions and difficulties with changing surface roughness conditions and vegetation growth. Nevertheless, future missions such as SENTINEL and the RADARSAT constellation are expected to resolve for the high temporal coverage demands and may foster the use of multi-image approaches such as change detection for soil moisture retrieval. Recent years have shown significant advancements through the advent of polarimetric SAR, offering a number of features that are only proportional to surface roughness. Unfortunately, the majority of the operational SAR sensors are not capable of measuring in polarimetric mode. Furthermore, whereas the potential of L-band SAR data for surface parameter inversion has meanwhile become well established, the results based on C-band polarimetric parameters have so far been discouraging. Finally, the use of roughness calibration techniques has evidenced a large improvement over the use of field measurements; therefore calibration approaches have now become widespread both in active and passive microwave remote sensing of soil moisture.

A case study has been presented that focuses on a calibration method based on effective roughness parameters. Lievens et al. (2011) demonstrated that it is possible to model these effective roughness parameters from normalized backscattering coefficients. The technique is evaluated using RADARSAT-2 images over Flevoland, the Netherlands, yielding root-mean-square errors up to 5%, except when VV observations in stubble fields are used for soil moisture retrieval. In the latter case, errors of 6.3 vol% were found. This study demonstrates the large potential of this technique, which only relies on a few *a priori* assumptions about the general surface roughness state, while based on single configuration SAR.

References

Allain, S., L. Ferro-Famil, and E. Pottier. 2004. Two novel surface model based inversion algorithms using multi-frequency polSAR data, in *Proceedings of the IEEE International Geoscience and Remote Sensing Symposium*. IEEE, New Brunswick, NJ, pp. 823–826.

Álvarez-Mozos, J., J. Casalí, M. Gonzalez-Audícana, and N. E. C. Verhoest. 2005. Correlation between ground measured soil moisture and RADARSAT-1 derived backscatter coefficient over an agricultural catchment of Navarre (north of Spain), *Biosyst. Eng.*, 92, 119–133.

Baghdadi, N., O. Cerdan, M. Zribi, V. Auzet, F. Darboux, M. El Hajj, and R. Bou Kheir. 2008. Operational performance of current synthetic aperture radar sensors in mapping soil surface characteristics in agricultural environments: Application to hydrological and erosion modeling, *Hydrol. Proc.*, 22, 9–20.

Baghdadi, N., R. Cresson, E. Pottier, M. Aubert, M. Zribi, A. Jacome, and A. Benabdallah. 2012. A potential use for the C-band polarimetric SAR parameters to characterize the soil surface over bare agriculture fields, *IEEE Trans. Geosci. Remote Sens.*, 50(10), 3844–3858.

Baghdadi, N., I. Gherboudj, M. Zribi, M. Sahebi, C. King, and F. Bonn. 2004. Semi-empirical calibration of the IEM backscattering model using radar images and moisture and roughness field measurements, *Int. J. Remote Sens.*, 25, 3593–3623.

Baghdadi, N., N. Holah, and M. Zribi. 2006. Calibration of integral equation model for SAR data in C-band and HH and VV polarizations, *Int. J. Remote Sens.*, 27, 805–816.

Baghdadi, N., C. King, A. Bourguignon, and A. Remond. 2002. Potential of ERS and Radarsat data for surface roughness monitoring over bare agricultural fields: Application to catchments in northern France, *Int. J. Remote Sens.*, 23, 3427–3442.

Balenzano, A., F. Mattia, G. Satalino, and M. W. J. Davidson. 2010. Dense temporal series of C- and L-band SAR data for soil moisture retrieval over agricultural crops, *IEEE J. Select. Top. Appl. Earth Observ. Remote Sens.*, 3(4), 1–12.

Bruckler, L., H. Wittono, and P. Stengel. 1988. Near surface moisture estimation from microwave measurements, *Remote Sens. Environ.*, 26, 101–121.

Bryant, R., M. S. Moran, D. P. Thoma, C. D. Holifield Collins, S. Skirvin, M. Rahman, K. Slocum et al. 2007. Measuring surface roughness height to parameterize radar backscatter models for retrieval of surface soil moisture, *IEEE Trans. Geosci. Remote Sens. Lett.*, 4, 137–141.

Callens, M., N. E. C. Verhoest, and M. W. J. Davidson. 2006. Parameterization of tillage-induced single-scale soil roughness from 4-m profiles, *IEEE Trans. Geosci. Remote Sens.*, 44, 878–888.

Caves, R., G. Davidson, A. Ma, and G. Hui. 2010b. AGRISAR 2009 Data Simulation and Algorithm Report. Technical Report. MacDonald, Dettwiler and Associates Ltd., Eur. Space Agency, Contract 22689/09/NL/FF/ef.

Caves, R., G. Davidson, J. Padda, and A. Ma. 2011. AGRISAR 2009 Final Report, vol. 4, Data Analysis Field Level Database. Technical Report. MacDonald, Dettwiler and Associates Ltd., Eur. Space Agency, Contract 22689/09/NL/FF/ef.

Caves, R., A. Ma, and G. Hui. 2010a. AGRISAR 2009 Data Acquisition Report. Technical Report. Dettwiler and Associates Ltd., MacDonald. Eur. Space Agency, Contract 22689/09/NL/FF/ef.

Chanzy, A. 1993. Basic soil surface characteristics derived from active microwave remote sensing. *Remote Sens. Rev.*, 7, 303-319.

Chow, V. T., D. R. Maidment, and L. W. Mays. 1988. *Applied Hydrology*. McGraw-Hill, New York.

Cloude, S. R., and E. Pottier. 1996. A review of target decomposition theorems in radar polarimetry, *IEEE Trans. Geosci. Remote Sens.*, 34(2), 498–518.

Davidson, M. W. J., T. Le Toan, F. Mattia, G. Satalino, T. Manninen, and M. Borgeaud. 2000. On the characterization of agricultural soil roughness for radar remote sensing studies, *IEEE Trans. Geosci. Remote Sens.*, 38, 630–640.

De Keyser, E., H. Vernieuwe, H. Lievens, J. Álvarez-Mozos, B. De Baets, and N. E. C. Verhoest. Assessment of SAR-retrieved soil moisture uncertainty induced by uncertainty on modeled soil surface roughness, *Int. J. Appl. Earth Obs. Geoinf.*, 18, 176–182.

Dierking, W. 1999. Quantitative roughness characterization of geological surfaces and implications for radar signature analysis, *IEEE Trans. Geosci. Remote Sens.*, 37(5), 2397–2412.

Dobson, M. C., F. T. Ulaby, M. T. Hallikainen, and M. S. El-Rayes. 1985. Microwave dielectric behavior of wet soils: II. Dielectric mixing models, *IEEE Trans. Geosci. Remote Sens.*, 23, 35–46.

Dubois, P. C., J. van Zyl, and E. T. Engman. 1995a. Corrections to Measuring soil moisture with imaging radars, *IEEE Trans. Geosci. Remote Sens.*, 33, 1340.

Dubois, P. C., J. van Zyl, and E. T. Engman. 1995b. Measuring soil moisture with imaging radars, *IEEE Trans. Geosci. Remote Sens.*, 33, 915–926.

Fung, A. K. 1994. *Microwave Scattering and Emission Models and Their Applications*. Artech House, Boston, MA.

Fung, A. K., Z. Li, and K. S. Chen. 1992. Backscattering from a randomly rough dielectric surface, *IEEE Trans. Geosci. Remote Sens.*, 30, 356–369.

Hajnsek, I., E. Pottier, and S. Cloude. 2003. Inversion of surface parameters from polarimetric SAR, *IEEE Trans. Geosci. Remote Sens.*, 41(4), 727–744.

Hallikainen, M. T., F. T. Ulaby, M. C. Dobson, M. A. El-Rayes, and L. K. Wu. 1985. Microwave dielectric behavior of wet soils, Part I: Empirical models and experimental observations, *IEEE Trans. Geosci. Remote Sens.*, 23, 25–34.

Holah, N., N. Baghdadi, M. Zribi, A. Bruand, and C. King. 2005. Potential of ASAR/ENVISAT for the characterization of soil surface parameters over bare agricultural fields, *Remote Sens. Environ.*, 96, 78–86.

Jester, W., and A. Klik. 2005. Soil surface roughness measurement-methods, applicability, and surface representation, *Catena*, 64, 174–192.

Kornelsen, K. C., and P. Coulibaly. 2012. Advances in soil moisture retrieval from synthetic aperture radar and hydrological applications, *J. Hydrol.*, 476, 460–489.

Le Hégerat-Mascle, S., M. Zribi, F. Alem, and A. Weisse. 2002. Soil moisture estimation from ERS/SAR data: Toward an operational methodology, *IEEE Trans. Geosci. Remote Sens.*, 40, 2647–2658.

Lievens, H., and N. E. C. Verhoest. 2012. Spatial and temporal soil moisture estimation from Radarsat-2 imagery over Flevoland, the Netherlands, *J. Hydrol.*, 456–457, 44–56.

Lievens H., N. E. C. Verhoest, E. De Keyser, H. Vernieuwe, P. Matgen, J. Álvarez-Mozos and B. De Baets. 2011. Effective roughness modelling as a tool for soil moisture retrieval from C- and L-band SAR, *Hydrol. Earth Syst. Sci.*, 15(1), 151–162.

Lievens H., H. Vernieuwe, J. Alvarez-Mozos, B. De Baets, and N. E. C. Verhoest. 2009. Error in SAR-derived soil moisture due to roughness parameterization: An analysis based on synthetical surface profiles, *Sensors*, 9(2), 1067–1093.

Liu, Z., and D. J. Meyer. 2002. Study of high SAR backscattering caused by an increase of soil moisture over a sparsely vegetated area: Implications for characteristics of backscatter, *Int. J. Remote Sens.* 23, 1063–1074.

Marzahn, P., and R. Ludwig. 2009. On the derivation of soil surface roughness from multi parametric polsar data and its potential for hydrological modelling, *Hydrol. Earth Syst. Sci.*, 13(3), 381–394.

Mattia, F., M. W. J. Davidson, T. Le Toan, C. M. F. D'Haese, N. E. C. Verhoest, A. M. Gatti, and M. A. Borgeaud. 2003. A comparison between soil roughness statistics used in surface scattering models derived from mechanical and laser profilers, *IEEE Trans. Geosci. Remote Sens.*, 41, 1659–1671.

Mattia, F., and T. Le Toan. 1999. Backscattering properties of multi-scale rough surfaces, *J. Electromagn. Waves Appl.*, 13(4), 491–526.

Mattia, F., T. Le Toan, J. C. Souyris, G. De Carolis, N. Floury, F. Posa, and G. Pasquariello. 1997. The effect of surface roughness on multifrequency polarimetric SAR data, *IEEE Trans. Geosci. Remote Sens.*, 33, 915–926.

Mironov, V., M. Dobson, V. Kaupp, S. Komarov, and V. Kleshchenko. 2004. Generalized refractive mixing dielectric model for moist soils, *Soil Sci. Soc. Am. J.*, 42(4), 773–785.

Mladenova, I., V. Lakshmi, J. P. Walker, R. Panciera, W. Wagner, and M. Doubkova. 2010. Validation of the ASAR global monitoring mode soil moisture product using the NAFE'05 dataset, *IEEE Trans. Geosci. Remote Sens.*, 48(6), 2498–2508.

Moran, M. S., D. C. Hymer, J. Qi, and E. E. Sano. 2000. Soil moisture evaluation using multitemporal synthetic aperture radar (SAR) in semiarid rangeland, *Agric. Forest Meteorol.*, 105, 69-80.

Moran, M. S., C. D. Peters-Lidard, J. M. Watts, and S. McElroy. 2004. Estimating soil moisture at the watershed scale with satellite-based radar and land surface models, *Can. J. Remote Sens.*, 30, 805—826.

Narayan, U., V. Lakshmi, and T. J. Jackson. 2006. High-resolution change estimation of soil moisture using L-band radiometer and radar observations made during the SMEX02 experiments, *IEEE Trans. Geosci. Remote Sens.*, 44(6), 1545–1554.

Njoku, E. G., W. J. Wilson, S. H. Yueh, S. J. Dinardo, F. K. Li, T. J. Jackson, V. Lakshmi, and J. Bolten. 2002. Observations of soil moisture using a passive and active low-frequency microwave airborne sensor during SGP99, *IEEE Trans. Geosci. Remote Sens.*, 40(2), 2659–2673.

Ogilvy, J. A. 1991. *Theory of Wave Scattering from Random Rough Surfaces*. IOP Publishing, Redcliffe Way, Bristol, UK.

Oh, Y. 2004. Quantitative retrieval of soil moisture content and surface roughness from mulitipolarized radar observations of bare soil surfaces, *IEEE Trans. Geosci. Remote Sens.*, 42, 596–601.

Oh, Y., and Y. C. Kay. 1998. Condition for precise measurement of soil surface roughness, *IEEE Trans. Geosci. Remote Sens.*, 36, 691–695.

Oh, Y., K. Sarabandi, and F. T. Ulaby. 1992. An empirical model and an inversion technique for radar scattering from bare soil surfaces, *IEEE Trans. Geosci. Remote Sens.*, 30, 370–381.

Oh, Y., K. Sarabandi, and F. T. Ulaby. 2002. Semi-empirical model of the ensemble-averaged differential Mueller matrix for microwave backscattering from bare soil surfaces, *IEEE Trans. Geosci. Remote Sens.*, 40, 1348–1355.

Pathe, C., W. Wagner, D. Sabel, M. Doubkova, and J. B. Basara. 2009. Using ENVISAT ASAR global monitoring mode data for surface soil moisture retrieval over Oklahoma, USA, *IEEE Trans. Geosci. Remote Sens.*, 47(2), 468–480.

Quesney, A., S. Le Hégerat-Mascle, O. Taconet, D. Vidal-Madjar, J. P. Wigneron, C. Loumagne, and M. Normand. 2000. Estimation of watershed soil moisture index from ERS/SAR data, *Remote Sens. Environ.*, 72, 290–303.

Rahman, M. M., M. S. Moran, D. P. Thoma, R. Bryant, E. E. Sano, C. D. Holifield Collins, S. Skirvin, C. Kershner, and B. J. Orr. 2007. A derivation of roughness correlation length for parameterizing radar backscatter models, *Int. J. Remote Sens.*, 28, 3994–4012.

Rahman, M. M., M. S. Moran, D. P. Thoma, R. Bryant, C. D. Holifield Collins, T. Jackson, B. J. Orr, and M. Tischler. 2008. Mapping surface roughness and soil moisture using multi-angle radar imagery without ancillary data, *Remote Sens. Environ.*, 112, 391–402.

Rice, S. O. 1951. Reflection of electromagnetic waves from slightly rough surfaces, *Commun. Pure Appl. Math.*, 4, 361–378.

Rignot, E. J. M., and J. J. Van Zyl. 1993. Change detection techniques for ERS-1 SAR data, *IEEE Trans. Geosci. Remote Sens.*, 31(4), 896–906.

Shepard M. K., and B. A. Campbell. 1999. Radar scattering from a self-affine fractal surface: Near-nadir regime, *Icarus*, 141(1), 156–171.

Shepard, M. K., B. A. Campbell, M. H. Bulmer, T. G. Farr, L. R. Gaddis, and J. J. Plaut. 2001. The roughness of natural terrains: A planetary and remote sensing perspective, *J. Geophys. Res.*, 106(E12), 32,777–32,795.

Schuler, D. L., J. S. Lee, D. Kasilingam, and G. Nesti. 2002. Surface roughness and slope measurements using polarimetric SAR data, *IEEE Trans. Geosci. Remote Sens.*, 40(3), 687–698.

Srivastava, H. S., P. Patel, M. L. Manchanda, and S. Adiga. 2003. Use of multi-incidence angle RADARSAT-1 SAR data to incorporate the effects of surface roughness in soil moisture estimation, *IEEE Trans. Geosci. Remote Sens.*, 41, 1638–1640.

Su, Z., P. A. Troch, and F. P. De Troch. 1997. Remote sensing of soil moisture using EMAC/ESAR data, *Int. J. Remote Sens.*, 18(10), 2105–2124.

Taconet, O., D. Vidal-Madjar, C. Emblanch, and M. Normand. 1996. Taking into account vegetation effects to estimate soil moisture from C-band radar measurements, *Remote Sens. Environ.*, 56, 52–56.

Ulaby, F. T., and P. P. Batlivala. 1976. Optimum radar parameters for mapping soil moisture, *IEEE Trans. Geosci. Remote Sens.*, 14, 81–93.

Ulaby, F. T., P. C. Dubois, and J. van Zyl. 1996. Radar mapping of surface soil moisture, *J. Hydrol.*, 184, 57–84.

Ulaby, F. T., R. K. Moore, and A. K. Fung. 1982. *Microwave Remote Sensing: Active and Passive*, vol. II. Artech House, Boston, MA.

Ulaby, F. T., R. K. Moore, and A. K. Fung. 1986. *Microwave Remote Sensing: Active and Passive*, vol III. Artech House, Boston, MA.

Van Der Velde, R., and Z. Su. 2009. Dynamics in land-surface conditions on the Tibetan Plateau observed by Advanced Synthetic Aperture Radar (ASAR), *Hydrol. Sci. J.*, 54, 1079–1093.

Van Doninck, J., J. Peters, H. Lievens, B. De Baets, and N. E. C. Verhoest. 2012. Accounting for seasonality in a soil moisture change detection algorithm for ASAR Wide Swath time series, *Hydrol. Earth Syst. Sci.*, 16, 773–786.

Verhoest, N. E. C., B. De Baets, F. Mattia, G. Satalino, C. Lucau, and P. Defourny. 2007. A possibilistic approach to soil moisture retrieval from ERS synthetic aperture radar backscattering under soil roughness uncertainty, *Water Resour. Res.*, 43, W07435.

Verhoest, N. E. C., H. Lievens, W. Wagner, J. Álvarez-Mozos, M. S. Moran, and F. Mattia. 2008. On the soil roughness parameterization problem in soil moisture retrieval of bare surfaces from synthetic aperture radar, *Sensors*, 8(7), 4213–4248.

Wagner, W., G. Blöschl, P. Pampaloni, J.-C. Calvet, B. Bizzarri, J.-P. Wigneron, and Y. Kerr. 2007. Operational readiness of microwave remote sensing of soil moisture for hydrologic applications, *Nordic Hydrol.*, 38, 1–20.

Wagner, W., G. Lemoine, M. Borgeaud, and H. Rott. 1999. A study of vegetation cover effects on ERS scatterometer data, *IEEE Trans. Geosci. Remote Sens.*, 37(2), 938–948.

Wagner, W., and K. Scipal. 2000. Large-scale soil moisture mapping in Western Africa using the ERS scatterometer, *IEEE Trans. Geosci. Remote Sens.*, 38(4), 1777–1782.

Wickel, A. J., T. J. Jackson, and E. F. Wood. 2001. Multitemporal monitoring of soil moisture with RADARSAT SAR during the 1997 Southern Great Plains hydrology experiment, *Int. J. Remote Sens.*, 22, 1571–1583.

Yang, H., J. Shi, Z. Li, and H. Guo. 2006. Temporal and spatial soil moisture change pattern detection in an agricultural area using multi-temporal Radarsat ScanSAR data, *Int. J. Remote Sens.*, 27(19), 4199–4212.

Zhixiong, L., C. Nan, U. D. Perdok, and W. B. Hoogmoed. 2005. Characterisation of soil profile roughness, *Biosyst. Eng.*, 91, 369–377.

Zribi, M., V. Ciarletti, O. Taconet, P. Boissard, P. Chapron, and B. Rabin. 2000. Backscattering on soil structure described by plane facets, *Int. J. Remote Sens.*, 21, 137–153.

Zribi, M., and M. Dechambre. 2002. A new empirical model to retrieve soil moisture and roughness from radar data, *Remote Sens. Environ.*, 84, 42–52.

14

On the Synergistic Use of Microwave and Infrared Satellite Observations to Monitor Soil Moisture and Flooding

Marouane Temimi, Christopher R. Hain, Xiwu Zhan, Robert Rabin,
Martha C. Anderson, Claudia Notarnicola, Jan Stepinski, and Amelise Bonhomme

CONTENTS

14.1 Introduction

Extreme hydrological processes are often very dynamic and destructive. In recent years, we have observed major flooding events such as in 2008 in Iowa and Hurricane Irene in 2011, which impacted the eastern coast of the United States. Several other events in the world were very destructive as well, like the major flood in Pakistan in 2010. It is important to promptly monitor these extreme processes and forecast their behavior to mitigate their impact and consequences. A better understanding of these processes requires an accurate mapping of key variables that control them. In this regard, soil moisture is perhaps the most important parameter that impacts the magnitude of flooding events, as it controls the partitioning of rainfall into runoff and infiltration. It is therefore crucial to monitor the spatial and temporal variability of soil moisture conditions. In this context, remote sensing–based techniques have shown a great efficacy.

Several sensors have been used to monitor hydrological processes from space, typically falling into one of two categories.

1. Microwave (MW) sensors, active and passive, have been widely used because of their high sensitivity to liquid water, which is the consequence of a significant gap between the dielectric constant of water and soil. Passive MW sensors measure the soil radiances that are useful to infer information on soil moisture, quantitatively through radiative transfer modeling or qualitatively through wetness indices (Njoku et al. 2003; Sippel et al. 1994; Basist et al. 1998). Nonetheless, the spatial resolution of passive MW observations is usually coarse, on the order of 25 to 50 km, which makes them appropriate for regional and global studies but not suitable for monitoring local scale processes. Active MW sensors, commonly known as radar systems, are equally sensitive to liquid water but are also influenced by vegetation and roughness and provide observations at higher spatial resolution. Future missions like the NASA Soil Moisture Active Passive (SMAP) mission (Entekhabi et al. 2010) intend to exploit the capabilities of active MW sensors along with the higher sensitivity of passive MW to water to retrieve a blended soil moisture product that provides a compromise between spatial resolution and accuracy.

2. Sensors used to infer information on soil moisture and inundation include instruments on geostationary and polar orbiting satellites like SEVIRI, GOES, AVHRR, and MODIS, which operate in the visible and infrared (IR) wavelengths. They have been used to delineate inundated zones and infer information about soil moisture (Verstraeten et al. 2006; Scheidt et al. 2010) despite the negative impact of cloud coverage on these images. In the case of frequent revisit cycle, particularly in the case of geostationary instruments like SEVIRI, image compositing and gap-filling techniques may mitigate the impact of clouds. The direct sensitivity to liquid water of IR and optical sensors is relatively low in comparison to MW sensors. However, IR observations can be obtained at significantly higher spatial resolutions. Hence it is expected that multisensor approaches that combine observations from optical and IR sensors, on one hand, with observations from MW passive and active sensors, on the other, should improve our capabilities to retrieve soil moisture and monitor extreme hydrological processes by overcoming the individual limitations of each sensor.

This chapter addresses the use of satellite imagery to monitor soil moisture and inundation from space with a particular emphasis on the potential of the synergistic use of observations in the MW and IR domains. First, MW-based techniques for soil moisture and inundation monitoring are discussed. Then, examples of soil moisture related products that are based on the use of IR observations are cited and discussed. Furthermore, the agreement between IR- and MW-based products is analyzed. Finally, we present examples of blended products and demonstrate the added value of the use of observations from multiple sensors.

14.2 Use of MW Observation to Monitor Soil Moisture and Flooding

14.2.1 MW Observations for Soil Moisture Retrieval

MW soil moisture remote sensing has been explored for several decades (Mo et al. 1982; Njoku and Entekhabi 1996; Njoku et al. 2003; Njoku and Kong 1977; Njoku and Li 1999; Wang and Choudhury 1981). The fundamental principle of MW soil moisture remote sensing is the large contrast in the dielectric properties of water and soil; water has a relative complex dielectric constant of about 80 for the real part as compared to about 3.5 for dry soil. Thus the real part of relative dielectric constant for wet soil can be between 3.5 and 40 for a soil of 50% porosity. This large dielectric constant difference between wet and dry soil correspondingly impacts the soil emissivity (passive MW) and soil reflectivity (active MW), and hence the thermal MW emission [brightness temperature (TB)] and backscattering cross section from the land surface, respectively.

To retrieve soil moisture from TB observations of MW satellite sensors, two main categories of algorithms have been employed: the single-channel retrieval (SCR) algorithms and the multiple-channel inversion (MCI) algorithms. The SCR algorithm was proposed by Jackson (1993) and has been used for generating soil moisture data product in NOAA's Soil Moisture Operational Product System (SMOPS) (Zhan et al. 2012). This algorithm was also selected to be the baseline algorithm for the soil moisture data product of NASA's SMAP mission. The MCI algorithm was introduced by Njoku and Entekhabi (1996) and used for the soil moisture data product from the AMSR-E (Njoku et al. 2003). Under the MCI category of algorithms, the LPRM algorithm was introduced by Owe et al. (2001) and De Jeu et al. (2003) for generating soil moisture data products from SMMR, TMI, and AMSR-E. Although all of these algorithms are based on the MW radiation transfer model (frequently called tau-omega model), their soil moisture retrievals are significantly different. Figure 14.1 shows the results of three different algorithms applied to the same AMSR-E observations on August 4, 2010.

Satellite sensors for soil moisture observations are also different in terms of wavelength, footprint size, and observation time, which results in significant differences among the different satellites. For convenience of NOAA numerical weather prediction users and other potential customers of soil moisture data developers, NOAA's SMOPS are generating a global soil moisture data product that merges soil moisture retrievals from WindSat, ASCAT, and SMOS. A soil moisture data product is being generated from AMSR2 at NOAA and will be included in the merged global product in SMOPS. The merged soil moisture data product has larger geographic coverage.

14.2.2 Active/Passive MW Retrieval of Soil Moisture

It has been shown that a passive MW radiometer with less than 1 K radio brightness noise sensitivity can measure near-surface soil moisture with a root-mean-square error of 1%–2% vol/vol for bare soil (Njoku and Entekhabi 1996). However, this 1 K radio brightness noise sensitivity can only be obtained for spaceborne radiometer footprints larger than 40 km with current satellite antenna technology (Njoku et al. 2003; Njoku and Li 1999), greatly limiting their applicability.

Experiments using active MW (radar) have indicated that 3.5% vol vol^{-1} soil moisture retrieval accuracy for spatial resolutions down to 1 km may be achievable for soil surfaces with vegetation cover shorter than 15 cm (Dobson and Ulaby 1986; Shi et al. 1997; Wang

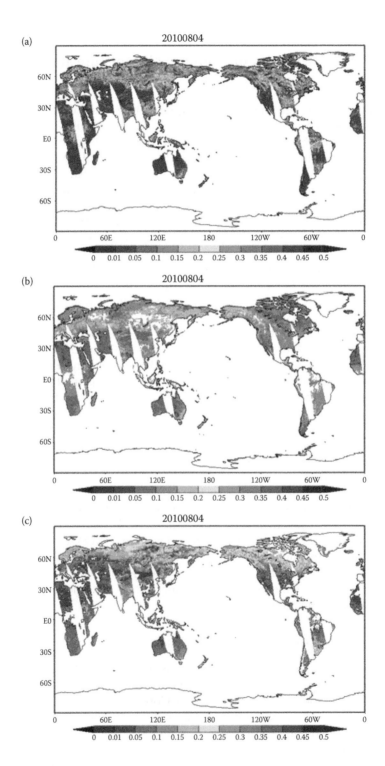

FIGURE 14.1
(See color insert.) Soil moisture (m^3/m^{-3}) maps from AMSR-E observations using different retrieval algorithms: (a) SCA, (b) MCI, and (c) LPRM.

et al. 1986). However, because of the greater impact of vegetation, surface roughness, and topography on radar signals than on radiometer observations, typical root-mean-square error values of such fine spatial resolution soil moisture retrievals from radar are significantly higher (Dobson and Ulaby 1986).

To overcome the individual limitations of the two approaches (passive radiometric remote sensing has higher accuracy but lower spatial resolution, and active radar remote sensing has higher spatial resolution but lower accuracy), the NASA SMAP mission is combining these two technologies. SMAP will use both an approximately 40-km L-band MW radiometer and 3-km radar to provide global-scale spaceborne land surface soil moisture observations at 3-day repeat intervals (Entekhabi et al. 2004). SMAP will provide coarse, fine, and medium resolution soil moisture products. The coarse resolution (40 km) product will be derived from the radiometer land surface brightness temperature observations using both single-channel methods with ancillary data to correct for the impacts of soil temperature, vegetation moisture content and surface roughness, and multichannel methods with reduced reliance on ancillary data (Crow et al. 2005). The fine resolution (3 km) product will be derived from the radar observations (Entekhabi et al. 2004; Das et al. 2011), while the medium resolution (10 km) product will be derived from combining the coarse resolution radiometer and fine resolution radar observations (Entekhabi et al. 2004). Zhan et al. (2006) described a Bayesian approach to merging these radiometer and radar observations as an Observing System Simulation Experiment (OSSE) of the SMAP mission (previously called the Hydros mission). The preliminary background field used in the study (Zhan et al. 2006) was a single-channel passive MW radiometer inversion. On the basis of the Hydros OSSE datasets with low and high noise added to the simulated observations or model parameters, the Bayesian method performed better than direct inversion of either the brightness temperature or radar backscatter observations alone. The root-mean-square errors of 9-km soil moisture retrievals from the Bayesian merging method were reduced by 0.5% and 1.4% vol/vol from the errors of direct radar inversions for the entire OSSE domain of all 34 consecutive days for the low and high noise datasets, respectively.

14.3 Inferring Soil Moisture Using IR Observations

An alternative methodology for monitoring soil moisture conditions exploits observations of land surface temperature (LST) in the thermal IR atmospheric window (10–12 μm). In theory, evolution of LST in the morning hours is strongly dependent on soil moisture states; for example, wet soil or well-watered (i.e., nonstressed) vegetation will heat up more slowly, and dry soil or stressed vegetation will heat up more rapidly. This has led to the development of several strategies that have been developed to potentially use TIR LST as a tool for monitoring soil moisture conditions (Anderson et al. 2007; Rabin and Schmit 2006; Scheidt et al. 2010; Verstraeten et al. 2006).

Trends in surface temperature are related to the diurnal amplitude, known also as the diurnal temperature range (DTR). It has been demonstrated in a study by Dai et al. (1997) that DTR has been decreasing worldwide owing to an increase in nighttime minimum temperature that surpasses the increase in daytime maximum temperature. These changes can be partially attributed to changes in soil water content, which affects the thermal inertia and therefore the daily temperature range. This assumption has been investigated using *in situ* observations from the NOAA-CREST soil moisture observation

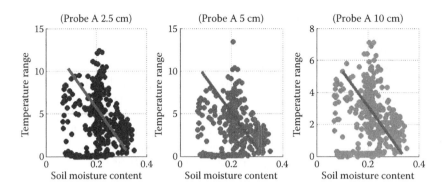

FIGURE 14.2
Comparison of *in situ* soil moisture values (m^3/m^{-3}) at 2.5, 5, and 10 cm to DTR in Kelvin measured in Millbrook, New York.

network that is deployed in the region of Millbrook, New York. The network comprises six sites equipped with six probes each and measuring soil moisture and temperature at three different depths, namely, 2.5, 5, and 10 cm. The observations used here were taken from June 2011 to June 2012. All sites are located in grass-covered open fields. Figure 14.2 shows an existing inverse relationship between *in situ* observed soil moisture and DTR at the three observation depths. The DTR decreases as the soil moisture increases. The correlation between both variables reached −0.57 and was in agreement with the correlation level that was obtained using satellite product, which implies that changes in soil DTR, or similar proxies, can be used to infer soil moisture variability at the watershed scale.

14.3.1 Using the Atmosphere Land Exchange Inverse Model to Estimate Soil Moisture

In this particular application, a TIR-based soil moisture monitoring methodology exploits surface flux observations retrieved from the Atmosphere Land Exchange Inverse (ALEXI) surface energy balance algorithm (Anderson et al. 1997, 2007; Hain et al. 2009, 2011). ALEXI was formulated as an extension to the Two-Source Energy Balance (TSEB) model of Norman et al. (1995), which, in turn, was developed to address many of the documented disadvantages of monitoring surface energy fluxes from remote sensing platforms. Both TSEB and ALEXI are described in greater detail in Chapter 8 of this book and are reviewed only briefly here.

The two-source approximation treats the radiometric temperature (T_{RAD}) of a vegetated surface as the weighted average of the individual temperatures of soil (T_s) and vegetation (T_c) subcomponents, partitioned by the fractional vegetation cover (fc(φ)) apparent from the sensor view angle (φ), which can be expressed in a linearized form as

$$T_{RAD} \approx \{f(\varphi)\ T_c + [1 - f(\varphi)]\ T_s\} \tag{14.1}$$

where fc(φ) is derived from the observed leaf area index (LAI) using Beer's law:

$$f(\varphi) = 1 - \exp\left(\frac{-0.5\text{LAI}}{\cos\varphi}\right) \tag{14.2}$$

The TSEB separately balances energy budgets for the soil and vegetation components of the system and also solves for total system fluxes of net radiation (RN), latent heat (LE), sensible heat (*H*), and soil heat conduction (*G*). In ALEXI, regional application is achieved by coupling the TSEB with an atmospheric boundary layer (ABL) model to internally simulate land–atmosphere feedback on near-surface air temperature (T_a) (Anderson et al. 1997; 2007). In this coupled mode, the ALEXI model simulates air temperature at the blending height internally within the ABL model, ensuring that T_a is consistent with the modeled surface fluxes from the TSEB. The TSEB is applied at two times during the morning hours, at approximately 1.5 h after local sunrise (t_1) and 1.5 h before local noon (t_2). The ABL component of ALEXI relates the rise in Ta within the mixed layer over the time interval (t_1 to t_2) to the time-integrated influx of *H* from the surface, thus providing a means for surface energy closure (Anderson et al. 1997; McNaughton and Spriggs 1986). A complete description of ALEXI can be found in the works of Anderson et al. (1997, 2007).

The TSEB component of ALEXI partitions the total system LE flux into soil evaporation (LE_s) and canopy transpiration (LE_c) subcomponents. These fluxes, in turn, are largely controlled by SM in the surface layer and the root zone layer, respectively. In general, wet soil moisture conditions lead to increased LE (decreased *H*) and a depressed morning surface temperature amplitude, while dry soil moisture conditions lead to decreased LE (increased *H*) and an increased diurnal surface temperature amplitude. Unlike an LSM, ALEXI is a purely diagnostic model lacking any prognostic state calculations. Therefore, in a data assimilation context, it can be mathematically treated as a retrieval algorithm that converts higher-level satellite products into a geophysical variable suitable for assimilation. Anderson et al. (2007) and Hain et al. (2009, 2011) outline a technique for simulating the effects of soil moisture on LE estimates from ALEXI through a soil moisture stress function based on a derived estimate of the fraction (f_{PET}) between actual ET and potential ET. In many LSMs, a semiempirical linear or nonlinear relationship is defined for the relationship between f_{PET} and soil moisture to account for the effects of soil moisture depletion on surface evaporative fluxes. Following Hain et al. (2011), we assume this relationship to be of the form

$$\theta_{ALEXI} = (\theta_{fc} - \theta_{wp})^* f_{PET} + \theta_{wp} \tag{14.3}$$

where θ_{ALEXI} is the retrieved soil moisture value (in volumetric m^3/m^3 units) based on ALEXI estimates of f_{PET} and θ_{fc} and θ_{wp} are the volumetric soil moisture contents at field capacity and permanent wilting point, respectively. Derived values of θ_{ALEXI} from diagnosed evaporative fluxes (the inverse of the prognostic approach) should provide a reasonable estimate of true soil moisture when soil moisture is between θ_{fc} and θ_{wp}. However, ALEXI can lose sensitivity when soil moisture is above θ_{fc} and below θ_{wp}. This upper region of low soil moisture sensitivity, in particular, is likely to cause reduced retrieval accuracy outside of the growing season. Furthermore, it is assumed that the retrieval of soil moisture from ALEXI provides an instantaneous estimate of current SM conditions. While Equation 14.3 requires specification of soil texture-specific values of θ_{fc} and θ_{wp}, these local constants (usually obtained from a soil texture map) cancel out in the computation of grid cell–based standardized anomalies describing deviations from mean conditions at each pixel and for each day of the study period. Standardized anomalies of θ_{ALEXI} can therefore be computed directly from f_{PET} without soil texture information. Sensitivity to θ_{fc} and θ_{wp} can also be circumvented via the common land data assimilation preprocessing step of scaling remotely sensed soil moisture retrievals to match the climatology of modeled soil moisture prior to the assimilation of ALEXI soil moisture retrievals [see e.g., Hain et al. (2012) and Reichle and Koster (2004, 2005)].

Values of θ_{ALEXI} have a distinctly different vertical support than surface soil moisture retrievals that can be acquired from MW sensors. For this methodology it is assumed that the contributions to θ_{ALEXI} from the surface and root zone soil layers are related to the observed fraction of green (e.g., actively transpiring) vegetation derived from remotely sensed estimates of LAI. For example, over bare soil, the ALEXI signal is dominated by direct soil evaporation and reflects soil moisture conditions in only the first several centimeters of the soil profile, similar to an effective sensing depth of current MW sensors. However, over dense to full vegetation cover ($f_c > \sim 60\%$) under well-watered conditions, the ALEXI signal is predominantly partitioned to canopy transpiration, and soil evaporation becomes negligible in comparison. In this case, the signal is governed by soil moisture conditions in the root zone. Between these two extremes, θ_{ALEXI} should be interpreted as a composite of both surface and root zone soil moisture.

14.3.2 A GOES-Based Dryness Index as Proxy for Soil Moisture

Remote thermal IR measurements of the land surface (radiative temperatures or their temporal change) have been largely used to estimate the surface energy budget and soil moisture (Brutsaert et al. 1993; Hall et al. 1992). The relationship between the observed daytime rise in surface radiative temperature, derived from the Geostationary Operational Environmental Satellites (GOES) sounder clear-sky data, and modeled soil moisture was explored over the continental United States (Rabin and Schmit 2006). The motivation was to provide an IR satellite-based index for soil moisture, which has a higher resolution than possible with the MW satellite data. The daytime temperature rise was negatively correlated with soil moisture in most areas. Anomalies in soil moisture and daytime temperature rise were also negatively correlated on monthly timescales. However, a number of exceptions to this correlation existed, particularly in the western United States. In addition to soil moisture, the capacity of vegetation to generate evapotranspiration influences the amount of daytime temperature rise as sensed by the satellite. In general, regions of fair to poor vegetation health correspond to the relatively high temperature rise from the satellite. Regions of favorable vegetation matched locations of lower than average temperature rise.

A dryness index was developed to provide information on the relative amount of moisture in the biomass (vegetation canopy) or the soil surface in the case of bare ground (or partially covered soil), with the goal to support fire weather forecasting. This index is based on daytime heating rates of the land surface as observed from clear-sky GOES imagery IR measurements. Higher (lower) heating rates are associated with drier (wetter) surfaces and higher (lower) ratios of sensible to latent heat flux (Bowen ratio). In areas of moderate to high vegetation cover, the dryness index should be negatively correlated with normalized difference vegetation index (NDVI). Both NDVI and the dryness index can be used together to help assess dryness. The dryness index may provide independent information on dryness where NDVI is relatively small (sparse vegetation cover).

Specifically, the dryness index is based on the surface temperature (T_s) rise between 1000 and 1300 local solar time. This value is normalized (divided) by the incoming solar radiation at the surface (S). Both T_s and S are obtained from hourly GOES Surface and Insolation Products (GSIP) produced at the NESDIS Center for Satellite Applications and Research (STAR). The horizontal resolution of the GOES dryness index is dictated by that of the GSIP

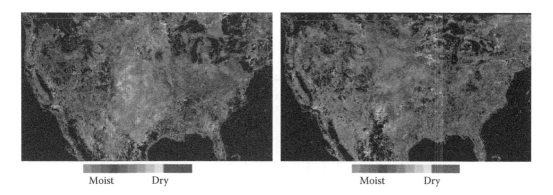

FIGURE 14.3
(See color insert.) Example of the GOES-based dryness index (unitless) estimated in (left) July 2011 and (right) July 2010 for northern United States.

products (1/8th degree). The dryness index is produced on a daily basis. Owing to limited coverage due to cloud cover (and to facilitate comparison with NDVI), 7- and 14-day averages are computed on a daily basis. The dryness index is only evaluated where sky conditions are deemed to be clear. Figure 14.3 shows examples of dryness index maps developed using GOES observations acquired in July 2011 and July 2010. Despite persistent gaps because of cloud coverage, the dryness index maps allow to determine the spatial distribution of the dryness across the United States. Central regions were dryer in 2011 than 2010.

14.4 On the Agreement between MW- and IR-Based Estimates of Soil Moisture

14.4.1 Evaluation of the GOES-Based Dryness Index

The agreement between soil moisture products derived from MW observations with those obtained using IR observations is assessed in this study. The selected MW-based product was the NASA soil moisture product that is developed using observation from the AMSR-E. Soil moisture estimates from the NASA AMSR-E product were compared to values from the GOES-based dryness product introduced in Section 14.3.2. The AMSR-E product is in the EASE grid projection at a spatial resolution of 25 km, which is coarser than the GOES dryness product. To allow for a quantitative comparison of both products, dryness index maps were reprojected and resampled to match the AMSR-E product projection and spatial resolution. The analysis of 2010 and 2011 values from both products demonstrated the negative relationship between dryness index and AMSR-E soil moisture values. The correlation between both products reached −0.60 (Figure 14.4). Noteworthy, the agreement between IR and passive MW products showed sensitivity to land cover and land use. High correlation values were obtained where vegetation density is low. The agreement between both products varied throughout the seasons as land surface conditions change. The analysis of longer time series is needed to investigate the seasonal variability of the agreement between soil moisture estimation from MW and IR data.

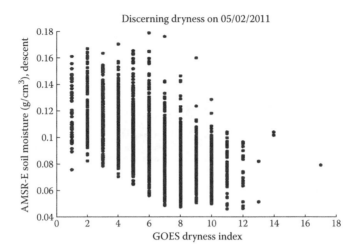

FIGURE 14.4
Example of negative correlation between GOES-based dryness index (unitless) and soil moisture estimates (g/cm³) from the NASA AMSR-E product.

14.4.2 Assessment of ALEXI Soil Moisture Estimates

Hain et al. (2009) evaluated ALEXI fraction of available water retrievals in comparison with SM observations over a multiyear period (2002–2004) from the Oklahoma Mesonet and found reasonable temporal and spatial agreement (RMSE values around 20% of mean observed soil moisture). A follow-on study by Hain et al. (2011) conducted a multiyear intercomparison study among LSM predictions from the Noah LSM, soil moisture retrievals from ALEXI, and soil moisture retrievals from the Land Parameter Retrieval Model (LPRM; Owe et al. 2001, 2008), an algorithm that uses brightness temperature observations from AMSR-E. A series of analyses were used to study relationships between the three products: (1) An analysis of spatial anomaly correlations on a seasonal timescale showed that seasonal composites were spatially consistent between the three datasets, except for the degradation in AMSR-E anomaly detection over moderate to dense vegetation, mainly occurring over the eastern half of the CONUS. The ALEXI SM maps showed better spatial correspondence with Noah SM than did the MW SM retrievals over the period studied (2007–2008; see Figure 14.5). The average spatial anomaly correlation for Noah and ALEXI was $r = 0.70$, while the average for Noah and AMSR-E was $r = 0.59$; and (2) An analysis was conducted to identify relationships between the three soil moisture products as a function of time at a given grid point (e.g., time series anomaly correlation). The temporal correlation coefficients computed over CONUS between weekly products (April to September, 2003–2008) for Noah and ALEXI and for Noah and AMSR-E show, in general, that ALEXI and Noah exhibit good temporal agreement over the southeastern and most of the central and western CONUS, while AMSR-E is more strongly correlated in time with Noah over the central and western CONUS with poor performance over heavily vegetated sections of the eastern CONUS. ALEXI and AMSR-E both show statistically significant r with Noah over a large majority of the CONUS: ALEXI exhibits statistically significant r at 93.8% of all pixels, while AMSR-E shows statistically significant r at 80.3% of pixels.

In general, the intercomparison of TIR-based and MW-based soil moisture showed that the two datasets provided complementary information about the current soil moisture

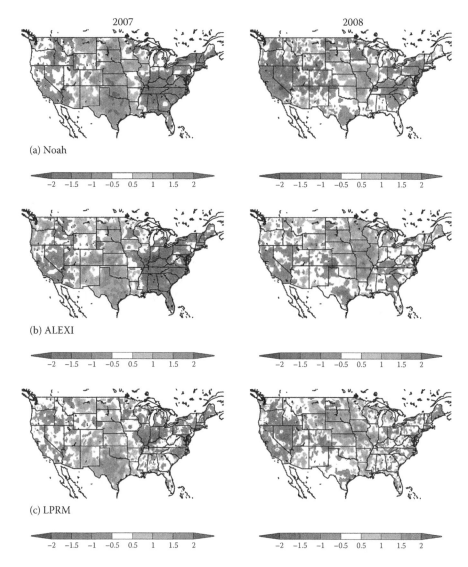

FIGURE 14.5
(See color insert.) Yearly normalized soil moisture anomalies (unitless) for (a) Noah, (b) ALEXI, and (c) LPRM.

state. TIR methods can provide soil moisture information over dense vegetation, a large gap in current MW methods, while serving as an additional independent source of information over low to moderate vegetation. The complementary nature of the two retrieval methods leads to the potential of integration within an advanced data assimilation system, provided that accurate representation of relative errors are available for each method. As shown here, these error specifications will optimally vary spatially and temporally tied to remotely sensed information about cover fraction and type and to average sensor repeat cycle.

14.5 Multisensor-Based Approaches for Improved Soil Moisture Retrieval and Flood Monitoring

The advantage of IR observations with respect to those from MW sensors such as ASCAT or AMSR-E is their finer spatial resolution, 1–4 km. Their main limitation, however, is that they can operate only under cloud-free conditions. Many of the IR-based approaches used to retrieve soil moisture assume an inverse relationship between soil moisture and diurnal soil temperature gradient or other proxies like thermal inertia. This implies that it is important to determine the daily minimum and maximum of surface temperature to accurately determine the DTR. However, most sensors' overpass times do not coincide with the occurrence of the lowest and highest daily temperatures that may introduce an error in inferring soil moisture from the daily temperature gradient. Moreover, the combination of multisatellite information can help to disentangle the various contributions to the total radiation arising from the vegetation and the soil, thus deriving more reliable and robust soil moisture estimates. It is expected that the use of ancillary observation from MW sensors along with IR data may enhance the quality of these products and mitigate the impact of their limitations. For instance, improving the spatial resolution of the MW-based products using information on soil moisture from IR observation clearly demonstrated the added value of opting toward a synergistic use of these different products (Merlin et al. 2008, 2010). This section introduces examples of blended TIR and MW soil moisture products and their use in a number of applications.

14.5.1 Using IR and MW Observation to Infer Soil Moisture

An approach was recently introduced to infer soil moisture variability using a combination of MW and IR observation (Notarnicola et al. 2012). The approach identifies different levels of soil moisture from MODIS images using the Apparent Thermal Inertia (ATI) concept along with passive MW data. The innovative aspect of the proposed technique resides in the synergistic use of MODIS and AMSR-E data. Specifically, the ATI time series is filtered by using AMSR-E derived soil moisture estimates. Because of the coarse resolution of AMSR-E soil moisture estimates (25 km) with respect to ATI estimates (1 km), it is expected that the former reflects the overall trend of soil moisture values. Hence it can be used as a low pass filter in order to reduce the irregularity due to different acquisition time, outliers, and other effects that may affect the IR soil moisture estimates. The method was applied over two test sites located in Italy and France where long time series of *in situ* soil moisture are available.

The use of MW observation along with IR data in the retrieval process has improved the accuracy of the derived product with respect to IR estimates. In the Italian test site, the determination coefficients R^2 increase from 0.58 to 0.72 when NDVI < 0.4 and from 0.45 to 0.56 when NDVI > 0.4. For the French test site, R^2 values increase from 0.61 to 0.70 for NDVI < 0.4 and from 0.23 to 0.43 for NDVI > 0.4. These data state on one side the improvement due to the filtering technique, while on the other side it clearly indicates the influence of vegetation on the ATI values. The presence of the vegetation reduces the sensitivity of ATI to soil moisture and the possibility of distinguishing between different soil moisture clusters. In the French test sites where NDVI is higher, only three soil moisture classes were distinguishable instead of the four found in the less vegetated areas in the Italy site.

It is important to add that the NASA AMSR-E sensor failed in late 2011 and stopped delivering observations. However, follow-on missions like AMSR 2, SMOS, ASCAT, and other recent MW sensors like the NOAA NPP/JPSS ATMS sensor will continue to provide MW observation suitable for soil moisture retrieval and therefore foster the development of blended IR- and MW-based approaches. Future mission like SMAP will also support the use of MW observations in conjunction with observations in other channels like the visible and IR. Finally, the use of geostationary data, for instance, from SEVIRI, or the future NOAA GOES-R, may allow for a more accurate determination of the diurnal cycle of skin temperature and therefore higher accuracy of ATI-based products.

14.5.2 A Multisensor Approach to Monitor Inundation and Soil Moisture Using MW and IR/Optical Observations

MW and IR observations from space exhibit different sensitivities to liquid water at or near the surface of the soil. MW observations, active and passive, are sensitive to the total amount of water that includes soil moisture with a penetration depth that depends on the frequency and the soil texture and also the open water, which comprises flooded areas and permanent water bodies. However, single observations in IR/visible domain can only sense open water and inundated areas, although diurnal change in the IR signal can lead to the assessment of soil moisture as demonstrated previously. So, it is expected that information on soil moisture may lay in the difference between IR/visible (sensitive to open-water bodies and inundated areas) estimates on one side and MW estimates (sensitive to open-water bodies, inundated areas, and soil moisture) on the other.

This difference has led to the definition of a new basin wetness index (BWI), which can be determined according to the following relationship (Temimi et al. 2007):

$$BWI = \frac{WSF(AMSR) - WSF(MODIS)}{1 - WSF(MODIS)} \tag{14.4}$$

where WSF(AMSR-E) is the water surface fraction derived from AMSR-E data and WSF(MODIS) is the water surface fraction derived from MODIS data. The WSF(MODIS) can be written as a function of discharge observations using a rating curve relationship (Temimi 2005). The proposed BWI was tested over the Mackenzie River Basin and was sensitive to changes in observed soil moisture and precipitation (Temimi et al. 2007, 2010). The index estimated the difference between fractions of open water from MODIS and AMSR-E. It is noticed that AMSR-E values of fraction of open water were systematically higher than those derived from MODIS because the MW signal is sensitive to an additional amount of water (i.e., soil moisture) that is not present in the values derived from MODIS estimates.

14.5.3 A Blended Coastal Flood Monitoring System Using IR and Active MW Observations

Coastal areas and wetlands play a major role in both hydrologic processes and climate trends; therefore it is crucial to understand the hydrodynamics of these systems and the interplay between the ocean, estuaries, rivers, and surrounding wetlands. Satellite images (optical and MW) have also shown an interesting potential to monitor changes that occur over long and short timescales. Active microware sensors measure backscatters and return signal amplitude, phase, and polarization that are independent of sun angles and weather conditions. SAR interferometry (InSAR) measures the corresponding phase difference resulting from the difference in distances to the same target in two SAR images. It is possible based on this concept to determine changes in water levels in coastal region, lakes, and rivers as well. Combining interferometric radar observations with a digital elevation model of the river region allows for the determination of water extent. In a recent study (Chaouch et al. 2011), Radarsat 1, Landsat imagery, and aerial photography for the Apalachicola region in Florida have shown to be useful for inundation mapping and have a great potential for evaluating wetting/drying algorithms of inland and coastal hydrodynamic models. A change detection approach was implemented to emphasize differences in high- and low-tide inundation patterns. To alleviate the effect of inherent speckle in the SAR images ancillary optical data were used. The flood-prone area for the site was delineated *a priori* through the determination of lower and higher water contour lines with Landsat images combined with a high-resolution digital elevation model. This masking technique improved the performance of the proposed algorithm with respect to detection techniques using the entire Radarsat scene. The resulting inundation maps agreed well with historical aerial photography as the probability of detection reached 83%. The combination of SAR data and optical images, when coupled with a high-resolution digital elevation model, has clearly demonstrated the advantage of blended mapping approaches.

14.5.4 Synergy of Thermal IR and MW Soil Moisture Retrievals in a Land Data Assimilation Scheme

Most studies in the literature have mainly focused on the assimilation of active and passive MW retrievals of soil moisture (Crow and Wood 2003; De lannoy et al. 2007; Draper et al. 2009, 2011; Kumar et al. 2009; Margulis et al. 2002; Reichle and Koster 2005; Reichle et al. 2007; Ryu et al. 2009; Walker and Houser 2001; Merlin et al. 2006). However, as shown by Hain et al. (2011), retrievals from thermal IR method have the ability to provide unique soil moisture information, especially over moderate to dense vegetation, a known deficiency in MW-based methods. Hain et al. (2012) examined the assimilation using an ensemble Kalman filter (EnKF) to assimilate a thermal IR-based product based on ALEXI evaporative flux estimates (see Section 14.3.1) and a MW-based soil moisture retrieval from the LPRM (Owe et al. 2008, 2001) into the Noah land surface model. The methodology employed a data denial framework to quantify the ability of the two different soil moisture retrieval datasets to correct for errors in precipitation forcing (using a satellite-based precipitation product) in an open-loop simulation as compared to a control simulation forced with a higher-quality, gauge-based precipitation dataset. Therefore the evaluation is meant to quantify the ability of remotely sensed soil moisture retrievals to compensate for the impact of errors in satellite-based precipitation products. The study found that the joint assimilation of TIR and MW soil moisture into the Noah LSM provides better soil moisture estimates than does either retrieval method (TIR or MW) in isolation. The added value of TIR assimilation over MW alone is most significant in areas of moderate to dense vegetation cover (>60%), where MW retrievals have little sensitivity to soil moisture at any depth. These conditions characterize much of the central and eastern United States during the warm season (May–August). Joint assimilation of both TIR and MW soil moisture into a prognostic land surface model serves to maximize both spatial and temporal sampling of soil moisture conditions and provide important applications to numerical weather prediction, drought monitoring, and hydrological modeling.

14.6 Conclusion

This chapter addressed the use of MW and IR satellite observation to estimate soil moisture and map inundated areas. It particularly focused on the synergistic use of both sources of data and highlighted the added value of blended MW and IR products for soil moisture and inundation mapping. First, it was demonstrated that the synergistic use of multisatellite data is fostered by an existing agreement between MW- and IR-based products. It was shown that both products are sensitive to changes in soil moisture. It is worth noting that vegetation cover may impact the sensitivity of these signals to soil moisture in different ways. In the case of MW signal, the sensitivity to soil moisture under a vegetated cover depends on the frequency and the opacity of the canopy. Lower MW frequencies like the L band are less impacted by vegetation and penetrate to deeper soil layers. Scattering is more significant in higher frequencies where the vegetation water content significantly affects the MW signal. The IR signal, on the other, seems to show better sensitivity under vegetation cover. IR-based products like ALEXI can sense water in root zone when the soil is vegetated. In the case of bare soils, ALEXI is sensitive, like MW products, to moisture in the top layer of soil. The merging of MW and IR observation should improve our capability to monitor soil moisture particularly in areas with dense vegetation cover, like in the eastern part of the United States.

Second, it was demonstrated in this chapter that the potential of merging multisatellite data in the MW and IR domain may also improve the performance of inundation mapping techniques. The synergistic use of IR and MW exploits the higher spatial resolution of IR observation to improve the delineation of inundated areas. Prompt assessment of water lateral extent during flooding events is very important. The all-sky capabilities of MW sensors can ensure a continued access to impacted areas. One of the challenges that face the integration of observations from multiple sensors in a synergistic framework is data latency. This issue causes a delay in obtaining data from one sensor with respect to another and reduces the value of merging tall data together. The development of more elaborated ground infrastructure may contribute to reducing the latency. In addition, the difference in overpass times of different satellites may affect the sensitivity of their sensors to surface conditions (Dorigo et al. 2012). An accurate intercomparison of products from MW and IR requires that both observations be collected at the same day. The availability of data from satellites equipped with both types of sensors like AQUA that has MODIS and AMSR-E or the recent NPP satellite that carries ATMS and VIIRS helps to achieve this goal.

Acknowledgments

This study was supported by the National Oceanic and Atmospheric Administration (NOAA) under grant number NA06OAR4810162. Support was also provided by the NSF REU Program under NSF Grant no. ATM-0755686. The statements contained in this article are not the opinions of the funding agencies or government but reflect the views of the authors.

References

Anderson, M. C., J. M. Norman, G. R. Diak, W. P. Kustas, and J. R. Mecikalski. 1997. A two-source time-integrated model for estimating surface fluxes using thermal infrared remote sensing. *Remote Sensing of Environment* 60(2):195–216.

Anderson, M. C., J. M. Norman, J. R. Mecikalski, J. A. Otkin, and W. P. Kustas. 2007. A climatological study of evapotranspiration and moisture stress across the continental United States based on thermal remote sensing: 2. Surface moisture climatology. *Journal of Geophysical Research-Atmospheres* 112(D11), doi:10.1029/2006jd007507.

Basist, A., N. C. Grody, T. C. Peterson, and C. N. Williams. 1998. Using the special sensor microwave/imager to monitor land surface temperatures, wetness, and snow cover. *Journal of Applied Meteorology* 37(9):888–911.

Brutsaert, W., A. Y. Hsu, and T. J. Schmugge. 1993. Parameterization of surface heat fluxes above forest with satellite thermal sensing and boundary-layer soundings. *Journal of Applied Meteorology* 32(5):909–917.

Chaouch, N., M. Temimi, S. Hagen, J. Weishampel, S. Medeiros, and R. Khanbilvardi. 2011. A synergetic use of satellite imagery from SAR and optical sensors to improve coastal flood mapping in the Gulf of Mexico. *Hydrological Processes* 26(11):1617–1628.

Crow, W. T., S. T. K. Chan, D. Entekhabi, P. R. Houser, A. Y. Hsu, T. J. Jackson, E. G. Njoku, P. E. O'Neill, J. C. Shi, and X. W. Zhan. 2005. An observing system simulation experiment for Hydros radiometer-only soil moisture products. *IEEE Transactions on Geoscience and Remote Sensing* 43(6):1289–1303.

Crow, W. T., and E. F. Wood. 2003. The assimilation of remotely sensed soil brightness temperature imagery into a land surface model using Ensemble Kalman filtering: A case study based on ESTAR measurements during SGP97. *Advances in Water Resources* 26(2):137–149.

Dai, A., A. D. DelGenio, and I. Y. Fung. 1997. Clouds, precipitation and temperature range. *Nature* 386(6626):665–666.

Das, N. N., D. Entekhabi, and E. G. Njoku. 2011. An algorithm for merging SMAP radiometer and radar data for high-resolution soil-moisture retrieval. *IEEE Transactions on Geoscience and Remote Sensing* 49(5):1504–1512.

De Jeu, R. A. M., and M. Owe. 2003. Further validation of a new methodology for surface moisture and vegetation optical depth retrieval. *International Journal of Remote Sensing* 24(22):4559–4578.

De Lannoy, G. J. M., R. H. Reichle, P. R. Houser, V. R. N. Pauwels, and N. E. C. Verhoest. 2007. Correcting for forecast bias in soil moisture assimilation with the ensemble Kalman filter. *Water Resources Research* 43(9), doi:10.1029/2006wr005449.

Dobson, M. C., and F. T. Ulaby. 1986. Active microwave soil-moisture research. *IEEE Transactions on Geoscience and Remote Sensing* 24(1):23–36.

Dorigo, W., R. de Jeu, D. Chung, R. Parinussa, Y. Liu, W. Wagner, and D. Fernandez-Prieto. 2012. Evaluating global trends (1988–2010) in harmonized multi-satellite surface soil moisture. *Geophysical Research Letters* 39, doi:10.1029/2012gl052988.

Draper, C., J. F. Mahfouf, J. C. Calvet, E. Martin, and W. Wagner. 2011. Assimilation of ASCAT near-surface soil moisture into the SIM hydrological model over France. *Hydrology and Earth System Sciences* 15(12):3829–3841.

Draper, C. S., J. F. Mahfouf, and J. P. Walker. 2009. An EKF assimilation of AMSR-E soil moisture into the ISBA land surface scheme. *Journal of Geophysical Research* 114, D20104.

Entekhabi, D., E. G. Njoku, P. Houser, M. Spencer, T. Doiron, Y. J. Kim, J. Smith et al. 2004. The hydrosphere state (Hydros) satellite mission: An Earth system pathfinder for global mapping of soil moisture and land freeze/thaw. *IEEE Transactions on Geoscience and Remote Sensing* 42(10):2184–2195.

Entekhabi, D., E. G. Njoku, P. E. O'Neill, K. H. Kellogg, W. T. Crow, W. N. Edelstein, J. K. Entin et al. 2010. The Soil Moisture Active Passive (SMAP) mission. *Proceedings of the IEEE* 98(5):704–716.

Hain, C. R., W. T. Crow, M. C. Anderson, and J. R. Mecikalski. 2012. An EnKF dual assimilation of thermal-infrared and microwave satellite observations of soil moisture into the Noah Land Surface Model. *Water Resources Research* 48, W11517.

Hain, C. R., W. T. Crow, J. R. Mecikalski, M. C. Anderson, and T. Holmes. 2011. An intercomparison of available soil moisture estimates from thermal infrared and passive microwave remote sensing and land surface modeling. *Journal of Geophysical Research-Atmospheres* 116, doi:10.1029/2011jd015633.

Hain, C. R., J. R. Mecikalski, and M. C. Anderson. 2009. Retrieval of an available water-based soil moisture proxy from thermal infrared remote sensing. Part I: Methodology and validation. *Journal of Hydrometeorology* 10(3):665–683.

Hall, F. G., K. F. Huemmrich, S. J. Goetz, P. J. Sellers, and J. E. Nickeson. 1992. Satellite remote sensing of surface-energy balance: Success, failures, and unresolved issues in FIFE. *Journal of Geophysical Research* 97(D17):19,061–19,089.

Jackson, T. J. 1993. Measuring surface soil moisture using passive microwave remote sensing. *Hydrological Processes* 7(2):139–152.

Kumar, S. V., R. H. Reichle, Randal D. Koster, Wade T. Crow, and Christa D. Peters-Lidard. 2009. Role of subsurface physics in the assimilation of surface soil moisture observations. *Journal of Hydrometeorology* 10(6):1534–1547.

Margulis, S. A., D. McLaughlin, D. Entekhabi, and S. Dunne. 2002. Land data assimilation and estimation of soil moisture using measurements from the Southern Great Plains 1997 Field Experiment. *Water Resources Research* 38(12), doi:10.1029/2001wr001114.

McNaughton, K. G., and T. W. Spriggs. 1986. A mixed-layer model for regional evaporation. *Boundary-Layer Meteorology* 34(3):243–262.

Merlin, O., A. Al Bitar, J. P. Walker, and Y. Kerr. 2010. An improved algorithm for disaggregating microwave-derived soil moisture based on red, near-infrared and thermal-infrared data. *Remote Sensing of Environment* 114(10):2305–2316.

Merlin, O., A. Chehbouni, G. Boulet, and Y. Kerr. 2006. Assimilation of disaggregated microwave soil moisture into a hydrologic model using coarse-scale meteorological data. *Journal of Hydrometeorology* 7(6):1308–1322.

Merlin, O., A. Chehbouni, J. P. Walker, R. Panciera, and Y. H. Kerr. 2008. A simple method to disaggregate passive microwave-based soil moisture. *IEEE Transactions on Geoscience and Remote Sensing* 46(3):786–796.

Mo, T., B. J. Choudhury, T. J. Schmugge, J. R. Wang, and T. J. Jackson. 1982. A model for microwave emission from vegetation-covered fields. *Journal of Geophysical Research* 87(C13):1229–1237.

Njoku, E. G., and D. Entekhabi. 1996. Passive microwave remote sensing of soil moisture. *Journal of Hydrology* 184(1–2):101–129.

Njoku, E. G., T. J. Jackson, V. Lakshmi, T. K. Chan, and S. V. Nghiem. 2003. Soil moisture retrieval from AMSR-E. *IEEE Transactions on Geoscience and Remote Sensing* 41(2):215–229.

Njoku, E. G., and J. A. Kong. 1977. Theory for passive microwave remote-sensing of near-surface soil-moisture. *Eos Transactions, American Geophysical Union* 58(6):554–554.

Njoku, E. G., and L. Li. 1999. Retrieval of land surface parameters using passive microwave measurements at 6-18 GHz. *IEEE Transactions on Geoscience and Remote Sensing* 37(1):79–93.

Norman, J. M., W. P. Kustas, and K. S. Humes. 1995. A two-source approach for estimating soil and vegetation energy fluxes from observations of directional radiometric surface temperatures. *Agriculture and Forest Meteorology* 77:263–293.

Notarnicola, C., L. Caporaso, F. Di Giuseppe, M. Temimi, B. Ventura, and M. Zebisch. 2012. Inferring soil moisture variability in the Mediterranean Sea area using infrared and passive microwave observations. *Canadian Journal of Remote Sensing* 38(1):46–59.

Owe, M., R. de Jeu, and T. Holmes. 2008. Multisensor historical climatology of satellite-derived global land surface moisture. *Journal of Geophysical Research* 113(F1), doi:10.1029/2007jf000769.

Owe, M., R. de Jeu, and J. Walker. 2001. A methodology for surface soil moisture and vegetation optical depth retrieval using the microwave polarization difference index. *IEEE Transactions on Geoscience and Remote Sensing* 39(8):1643–1654.

Rabin, R. M., and T. J. Schmit. 2006. Estimating soil wetness from the GOES sounder. *Journal of Atmospheric and Oceanic Technology* 23(7):991–1003.

Reichle, R. H., and R. D. Koster. 2005. Global assimilation of satellite surface soil moisture retrievals into the NASA Catchment land surface model. *Geophysical Research Letters* 32(2), doi:10.1029/2004gl021700.

Reichle, R. H., R. D. Koster, P. Liu, S. P. P. Mahanama, E. G. Njoku, and M. Owe. 2007. Comparison and assimilation of global soil moisture retrievals from the Advanced Microwave Scanning Radiometer for the Earth Observing System (AMSR-E) and the Scanning Multichannel Microwave Radiometer (SMMR). *Journal of Geophysical Research* 112(D9), doi:10.1029/2006jd008033.

Reichle, R. H., and R. D. Koster. 2004. Bias reduction in short records of satellite soil moisture. *Geophysical Research Letters* 31(19), doi:10.1029/2004gl020938.

Ryu, D., W. T. Crow, X. Zhan, and T. J. Jackson. 2009. Correcting unintended perturbation biases in hydrologic data assimilation. *Journal of Hydrometeorology* 10(3):734–750.

Scheidt, S., M. Ramsey, and N. Lancaster. 2010. Determining soil moisture and sediment availability at White Sands Dune Field, New Mexico, from apparent thermal inertia data. *Journal of Geophysical Research* 115, doi:10.1029/2009jf001378.

Shi, J. C., J. Wang, A. Y. Hsu, P. E. Oneill, and E. T. Engman. 1997. Estimation of bare surface soil moisture and surface roughness parameter using L-band SAR image data. *IEEE Transactions on Geoscience and Remote Sensing* 35(5):1254–1266.

Sippel, S. J., S. K. Hamilton, J. M. Melack, and B. J. Choudhury. 1994. Determination of inundation area in the Amazon river floodplain using the SMMR 37 GHz polarization difference. *Remote Sensing of Environment* 48(1):70–76.

Temimi, M., R. Leconte, F. Brissette, and N. Chaouch. 2007. Flood and soil wetness monitoring over the Mackenzie River Basin using AMSR-E 37 GHz brightness temperature. *Journal of Hydrology* 333(2–4):317–328.

Temimi, M., R. Leconte, N. Chaouch, P. Sukumal, R. Khanbilvardi, and F. Brissette. 2010. A combination of remote sensing data and topographic attributes for the spatial and temporal monitoring of soil wetness. *Journal of Hydrology* 388(1–2):28–40.

Verstraeten, W. W., F. Veroustraete, C. J. van der Sande, I. Grootaers, and J. Feyen. 2006. Soil moisture retrieval using thermal inertia, determined with visible and thermal spaceborne data, validated for European forests. *Remote Sensing of Environment* 101(3):299–314.

Walker, J. P., and P. R. Houser. 2001. A methodology for initializing soil moisture in a global climate model: Assimilation of near-surface soil moisture observations. *Journal of Geophysical Research* 106(D11):11,761–11,774.

Wang, J. R., and B. J. Choudhury. 1981. Remote-sensing of soil-moisture content over bare field at 1.4 GHZ frequency. *Journal of Geophysical Research* 86(C6):5277–5282.

Wang, J. R., E. T. Engman, J. C. Shiue, M. Rusek, and C. Steinmeier. 1986. The SIR-B observations of microwave backscatter dependence on soil-moisture, surface-roughness, and vegetation covers. *IEEE Transactions on Geoscience and Remote Sensing* 24(4):510–516.

Zhan, X., J. Liu, X. Wang, L. Zhao, K. Jensen, F. Weng, and M. Ek. 2012. NESDIS Soil Moisture Data Products, Their Validation and Applications. Paper presented at AMS 18th Conference on Satellite Meteorology, Oceanography and Climatology/First Joint AMS-Asia Satellite Meteorology Conference. New Orleans, LA.

Zhan, X. W., P. R. Houser, J. P. Walker, and W. T. Crow. 2006. A method for retrieving high-resolution surface soil moisture from hydros L-band radiometer and radar observations. *IEEE Transactions on Geoscience and Remote Sensing* 44(6):1534–1544.

15

On the Synergy between Optical and TIR Observations for the Retrieval of Soil Moisture Content: Exploring Different Approaches

José Sobrino, Cristian Mattar, Juan Carlos Jiménez-Muñoz,
Belen Franch, and Chiara Corbari

CONTENTS

15.1 Introduction

During the past century, scientists have developed several approaches on remote sensing strategies for estimating surface soil moisture. Soil moisture is one of the most relevant surface parameters that influences the Earth energy budget and hydrological process (Kerr 2007). Moreover, surface soil moisture is a key variable in describing the water and energy exchanges at the land surface–atmosphere interface (Wigneron et al. 2003).

Recently, the Soil Moisture and Ocean Salinity (SMOS) mission was successfully launched to retrieve surface soil moisture in the 0–5 cm with an error precision better than 4% $m^3 m^{-3}$ (Kerr et al. 2001; Kerr 2007). The coarse spatial resolution of this sensor was improved recently from 40 to 1 km (Merlin et al. 2008; Piles et al. 2011). However, in the near future, the Soil Moisture Active Passive (SMAP) mission presents a novel approach in soil moisture estimation. This sensor will include the radars capable of high spatial resolution and passive L-band radiometers with high sensitivity to soil moisture even in the presence of moderate vegetation (Entekhabi et al. 2010). Despite these advantages using passive or active microwave radiometric information, it is also necessary to analyze the land surface cover, such as surface temperature or vegetation cover, which is a basic parameter and mandatory for soil moisture estimation.

The intrinsic relationships between soil moisture and land surface parameters like surface temperature and vegetation cover (usually estimated as an index or a fraction) can be physically correlated to obtain enhanced soil moisture retrievals at combining different remote sensing sources, such as thermal, optical, and passive or active microwave. The relation Vegetation Index–Temperature (VI-T) widely known as the "triangle method" is one of the most used techniques to relate land cover parameters and soil moisture (Carlson et al. 1994). Besides, one of the most important applications to the VI-T interactions is to describe the interaction between soil moisture and land surface flux such as evapotranspiration (ET).

The scope of this chapter is to provide to the reader a current status of the VI-T method and its recent improvements using remote sensing techniques. Case studies developed over an agricultural test site and further applications to soil moisture and fluxes are also presented.

15.2 Vegetation–Temperature as a Key Interaction for Soil Moisture Retrievals

The main keyword in VI-T interaction is "synergy". The concept of synergy comes from the Greek word *synergia*, which means joint work and cooperative action. Thus synergy is created when things work in concert together to create an outcome that is in some way of more value than the total of what the individual inputs are. Synergy has been applied in several areas such as business, economic and administrative sciences, and remote sensing. In particular, the synergy concept in remote sensing is defined when the use of two or more spectral information, from different ranges of the electromagnetic spectra, generates better results than only one range at estimating one or more biophysical variables. For instance, soil moisture or water status could be measured by using passive microwave in the L band. However, when this data source is properly combined with VI or thermal parameters (spectral information from the optical and thermal range), the soil moisture retrievals have to be more accurate than the same retrieval but using just one spectral source (visible, thermal, or L band). In this context, the surface temperature–VI envelope is an optical-thermal synergistic approach widely used in remote sensing to analyze water status, flux energy exchange, and land cover classification.

The potential of obtaining information about the energy and water status of a surface for classification of land cover through the relation of VI-T has been initially applied by Watson (1971) to help locate mineral deposits. Then, Carlson et al. (1981) were among the first to underline a potentially discernible relationship among LE fluxes, soil moisture content, and fractional vegetation cover (Fr). Later, in the 1990s, several works were developed to estimate surface turbulent energy fluxes and surface soil water content. Many other papers, following along on similar lines, emerged during the 1990s, keeping the basic idea behind all these techniques that surface radiant temperature (Tir)—and by association the surface turbulent energy fluxes—are sensitively dependent on the surface soil water content (Nemani and Running 1989; Kustas 1990; Stewart et al. 1994; Kustas and Norman 1996; Bastiaanssen et al. 1998; Mecikalski et al. 1999). However, the concept of the "triangle," as a useful tool to relate energy fluxes and soil moisture based on the VI-T relation, was initially introduced by Price (1990) and later elaborated on by Carlson et al. (1994,

1995), Gillies and Carlson (1995), Lambin and Ehrlich (1996), Gillies et al. (1997), Owen et al. (1998), and Jiang and Islam (2003).

The concept of the triangle method was further implemented for applications about crop stress indexes (Moran et al. 1994; Sandholt et al. 2002), thermal inertia (Stisen et al. 2008), energy fluxes, and modeling (Gillies and Temesgen 2004; Margulis et al. 2005), among others. A completed review of the triangle (trapezoid) method based on VI-T appears very well detailed in the work of Carlson (2007) and Petropoulos et al. (2009a).

The next section shows the mechanism of the VI-T envelope a new empirical algorithm recently developed to relate VI-T with soil moisture and further biophysical variables, such as surface emissivity, albedo, or passive microwave brightness temperature.

15.2.1 Background in Soil Moisture Estimation of Synergic Methods

During the 1990s a new approach to mapping both the land surface moisture and the surface turbulent energy fluxes was developed (Carlson et al. 1994) and verified by Gillies et al. (1997). This method, referred to as the universal triangle (or trapezoid), allows the pixel distribution from the image to fix the boundary conditions for the model, thereby largely bypassing the need for ancillary atmospheric and surface data. The triangle method is based on an interpretation of the image (pixel) distribution in surface temperature/vegetation fraction space. If a sufficiently large number of pixels are present (and representative of heterogeneous areas) and when cloud and surface water and outliers are removed, the shape of the pixel envelope resembles a triangle. Simulations with a soil vegetation atmosphere transfer (SVAT) model have also demonstrated that if an image contains a full range of soil water content and fractional vegetation cover, the shape of the surface radiant temperature and vegetation fraction values will also form a triangular (or slightly truncated trapezoidal) shape. Basically, a triangle emerges because the range of surface radiant temperature decreases as the vegetation cover increases, its narrow vertex attesting to the narrow range of surface radiant temperature over dense vegetation.

In principle, the triangle method could work with any reasonable SVAT model. By linking such one-dimensional SVAT model to the spatialized information provided by airborne or satellite Earth observation (EO) data, a potentially more powerful synergistic avenue has been developed to take advantage of the benefits of both the modeling and EO-based approaches (Olioso 1992). Overviews of some of the numerical and data assimilation approaches linking remote sensing to hydrological models, including SVAT models, can be found in the works of McLaughlin (1995), Van Loon and Troch (2001), and recently in the work of Moradkhani (2008), whereas Olioso et al. (1999) provide a comprehensive discussion on the variety of methods used for incorporating remote sensing specially into SVAT models. One of the widely used SVAT models is the SimSphere detailed by Petropoulos et al. (2009b).

Generally speaking, SVAT models serve merely to create the relationship between measured surface temperature, vegetation fraction, evaporative fraction, and soil moisture values. The main approach is to use the model once to create a matrix of normalized surface temperature and evaporative fraction (Shuttleworth et al. 1989) values for an entire range of input vegetation cover and soil moisture values, given a fixed set of ancillary input variables such as surface albedo and stomatal resistance. From this matrix, a set of polynomials are generated, which are then used to calculate soil moisture and evaporative fraction for all pixel values.

$$\text{SM} = \sum_{i=0}^{N} \sum_{j=0}^{N} a_{ij} T^{*i} \, \text{Veg}^{*j} \tag{15.1}$$

where T^* is the normalized surface temperature and Veg is a surface vegetation indicator such as the fractional vegetation cover (FVC), the normalized difference vegetation index (NDVI), or others. However, this method has suffered several modifications in order to improve the soil moisture estimation based on the temperature–vegetation relationship. One of them was developed by Sandholt et al. (2002), who introduced a new concept, the temperature vegetation dryness index (TVDI), using surface temperature and NDVI (Figure 15.1). The empirical estimation of TVDI is based on assumptions that (1) soil moisture is the main source of variation for surface temperature (Ts) and (2) TVDI is related to surface soil moisture due to changes in thermal inertia and evaporative control (evaporation and transpiration) on net radiation partitioning (energy balance).

For the operational estimation of TVDI from satellite data, a number of error sources exist: (1) no account of view angle effects on Ts and NDVI, which affects the fraction of bare soil and vegetation visible to the sensor; (2) the triangle may not be determined correctly from the EO data if the area of interest does not include a full range of variability in land surface conditions (e.g., dry bare soil, saturated bare soil, water stressed vegetation, and well-watered vegetation); (3) no account of errors in estimation of Ts (unknown and varying land surface emissivity and atmospheric effects); (4) no account of clouds, shadows, and associated variation in net radiation; (5) decoupling of the top surface soil layer from lower layers (Capehart and Carlson 1997); and (6) dependence of Ts and NDVI on surface type due to differences in aerodynamic resistance (Friedl and Davis 1994; Lambin and Ehrlich 1995).

A novel improvement that strengthened the relation of VI-T was proposed by Chauhan et al. (2003), who introduced a new concept into the universal triangle scheme. The surface broadband albedo is added to strengthen the relationship between the high end of soil moisture and measurable land parameters. Thus, as a practical application, the equation proposed by Chauhan et al. (2003) is detailed as follows:

$$\text{SM} = \sum_{i=0}^{N} \sum_{j=0}^{N} \sum_{k=0}^{N} a_{ij} T^{*i}\, \text{NDVI}^{*j}\, \alpha^{*k} \tag{15.2}$$

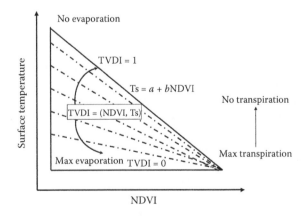

FIGURE 15.1
TVDI. (Adapted from Sandholt, I. et al., *Remote Sensing of Environment*, 79, 213–224, 2002.)

where α is the surface albedo. However, in order to compute the surface albedo, a high number of angular measurements have to be carried out and these data are not always available. Moreover, several constraints have been kept at using surface albedo. The surface albedo is defined as the ratio between the upwelling and downwelling incident irradiance upon a surface. While reflectance is defined as this same fraction for a single incident angle, albedo is the directional integration of reflectance over all sun view geometries (Pinty and Verstraete 1992).

The bidirectional reflectance distribution function (BRDF) describes with directional dependence of the land reflectance as function of the sun-target-sensor geometry. Therefore the surface albedo is derived integrating the BRDF model over all view and illumination angles. Consequently, a high number of angular measurements are required to calculate the surface albedo.

Focusing to strengthen the relationship detailed in Equation 15.2 and also to avoid the demand of the high number of angular measurements, Sobrino et al. (2012) replaced the albedo by the surface emissivity (ε) described as follows:

$$\text{SM} = \sum_{i=0}^{N}\sum_{j=0}^{N}\sum_{k=0}^{N} a_{ij} T^{*i}\, \text{NDVI}^{*j}\, \varepsilon^{*k} \tag{15.3}$$

as ε^* is the normalized emissivity by the maximum and the minimum values. Thermal emissivity could characterize the land surface based on the intrinsic kinetic properties of each other. Surface emissivity is one of the most important key parameters to estimate land surface radiance and therefore the land leaving flux energy. Surface emissivity has been correlated with surface cover properties such as vegetation index (NDVI and SAVI). The mechanism of how emissivity, NDVI, and temperature can be related with soil moisture will be described in the next section.

However, Equations 15.1 through 15.3 have only presented two spectral ranges as predicting variables. Those are the visible and thermal ranges. However, the first use of three different spectral ranges for predicting soil moisture using the universal triangle was developed by Piles et al. (2011), which introduces the bipolarized passive microwave brightness temperature (TB) measured by SMOS into Equation 15.3 resulting in

$$\text{SM} = \sum_{i=0}^{N}\sum_{j=0}^{N}\sum_{k=0}^{N} a_{ij} T^{*i}\, \text{NDVI}^{*j}\, \text{TB}^{*k} \tag{15.4}$$

The use of TB in Equation 15.4 is needed to capture soil moisture variability at a high resolution and to reproduce changes in soil moisture due to rain events being detected by the passive microwave range but not by visible or thermal range.

In this context, the combination of SMOS data with higher-resolution data coming from other sensors offers a potential solution to decompose or disaggregate global soil moisture estimates to the higher resolution required. Visible/infrared sensors are commonly used to provide an indirect measurement of soil moisture but not to retrieve it. As an alternative to these empirically based approaches, a physically based algorithm that includes a complex surface process model and high-resolution multispectral data and surface variables involved in a land surface–atmosphere model is presented by Merlin et al. (2005). This method is simplified using an energy balance model in the work of Merlin et al. (2008) to downscale low-resolution soil moisture data from passive microwave. However, the applicability of these algorithms to the upcoming spaceborne observations is limited to the availability of the soil and vegetation parameters they need at a global scale.

15.2.2 Role of Emissivity on Retrieving Soil Moisture Content

One of the main problems estimating surface soil moisture is the high frequency of rainfall events that cannot be easy to analyze when using vegetation fraction. Theoretically, emissivity increases when soil moisture increases (Hulley et al. 2010; Mira et al. 2007, 2010), although this increase cannot be assessed quantitatively because of many factors related to the land cover, the overpass time data acquisition, the minimum element of measure, and the equivalent soil moisture, among others.

Equation 15.3 includes the emissivity in order to provide a better estimation of the soil moisture. Traditionally, surface emissivity has been retrieved from low-resolution sensors using approaches based on vegetation indices such as the NDVI. In the simplest cases, emissivity can be obtained as a simple mean between soil and vegetation emissivities (ε_s, ε_v) using the FVC as the weighting factor:

$$E = \varepsilon_s(1 - \text{FVC}) + \varepsilon_v\text{FVC} \tag{15.5}$$

where the FVC can be retrieved from the normalized NDVI (NDVI*) according to Carlson and Ripley (1997):

$$\text{FVC} = (\text{NDVI*})^2 \tag{15.6}$$

or according to Gutman and Ignatov (1998),

$$\text{FVC} = \text{NDVI*} \tag{15.7}$$

with NDVI* calculated as

$$\text{NDVI*} = \frac{(\text{NDVI} - \text{NDVIs})}{(\text{NDVIv} - \text{NDVIs})} \tag{15.8}$$

where NDVIs and NDVIv are representative values of NDVI for soil and vegetation, respectively.

Equation 15.8 has been modified to account for the cavity effect and also to avoid the dependence on the soil emissivity, which is usually unknown. This is the case of the NDVI thresholds method (NDVI-THM) developed by Sobrino and Raissouni (2000):

$$
\begin{array}{lll}
\text{NDVI} < \text{NDVIs} & \varepsilon = a + b\rho & \\
\text{NDVIs} < \text{NDVI} < \text{NDVIv} & \varepsilon = \varepsilon_s(1 - \text{FVC}) + \varepsilon_v\text{FVC} + C & \quad(15.9) \\
\text{NDVI} > \text{NDVIv} & \varepsilon = 0.99 &
\end{array}
$$

where ρ is the reflectance in the red band and C is the cavity term, with a theoretical expression obtained from a geometrical model (Sobrino et al. 1990) but in practice assumed to be a mean value for different geometrical distributions or land covers. Adaptation of the NDVI-THM to different sensors can be found in Sobrino et al. (2008), and methods for FVC retrieval based on other vegetation indices or techniques in Jiménez-Muñoz et al. (2009).

It is important to note that emissivity methods based on NDVI approaches do not require TIR bands, because NDVI is obtained from VNIR bands (red and near-infrared). Therefore, when using Equation 15.3 for soil moisture estimations, emissivity should be retrieved from the TIR data, because the dependence on NDVI is already included in that equation. Moreover, surface emissivity is only diagnostic of surface composition and land cover (and partly of soil moisture) when it is retrieved from thermal data. Different approaches were developed to retrieve surface emissivity from multispectral TIR data (some of them reviewed in the work of Sobrino et al. 2008), being the temperature and emissivity separation (TES) algorithm (Gillespie et al. 1998), the state of the art of such a retrieval. TES algorithm requires multispectral TIR data and accurate atmospheric correction to retrieve surface emissivity and also temperature with enough accuracy: around 0.015 in terms of emissivity and 1.5 K in terms of temperature (Gillespie et al. 1998).

As an application of this method including land surface emissivity, the next section presents a case study using a different kind of measurements acquired at two different field campaigns carried out over an agricultural calibration/validation (CAL/VAL) test site.

15.2.3 Case Study 1

15.2.3.1 Test Site

The Barrax test site is widely known as one of the most important remote sensing CAL/VAL sites in Europe. In the framework of its Earth Observation Envelope Programme (EOEP), the European Space Agency (ESA) carries out a number of ground-based and airborne campaigns to support the development of geo/biophysical algorithms, calibration/validation activities, and simulation of future spaceborne EO missions. They are the Spectra Barrax Campaign (SPARC) in 2004, the Sentinel-2 and Fluorescence Experiment (SEN2FLEX) in 2005, and the Sentinel-3 Experiment (SEN3EXP). Moreover, the European Union also supported remote sensing field campaign throughout the project Exploitation of Angular Effects in Land Surface Observations from Satellites (EAGLE) in 2006. Finally, another important field campaign regarding the directional effects on remote sensing data was also financed by the Sciences Ministry of Spain with the project Earth Observation: Optical Data Calibration and Information Extraction (EODIX) in 2011. All these field campaigns are well detailed in Sobrino et al. (2008, 2009, 2011).

The Barrax test site is located in the La Mancha region, Spain. This test site is in the western part of the province of Albacete, 28 km from the capital town of Albacete (39°3′N; 2°60′W). The dominant cultivation pattern in the 10,000 ha area is approximately 65% dry land (two-thirds winter cereals, one-third fallow) and 35% irrigated crops (75% corn, 15% barley, and 10% others, including alfalfa) (Moreno et al. 2001). The climate of Barrax is of the Mediterranean type, with heavy rainfall in spring and autumn and a dry summer. It presents a high level of continentality, with sudden changes from cold months to warm months and high thermal oscillations in all seasons between maximum and minimum daily temperatures. Rainfall statistics show that the mean annual rainfall is a little more than 400 mm in most of the area (Moreno et al. 2001). The study area includes annual crops and natural surfaces such as alfalfa, wheat, oat, corn, green grass, and bare soil.

Spectrally, the soil is classified as an Inceptisol and is developed on shallowly buried flint clasts and bedrock. Spectrum of reflectance and emissivity is detailed by Sobrino et al. (2009). Figure 15.2 presents the location of Barrax.

15.2.3.2 Data Acquisition

The whole dataset considered here was acquired during the SEN2FLEX field campaign and SEN3EXP. SEN2FLEX was a campaign that combined different activities in support of initiatives related both to fluorescence experiments and to GMES Sentinel-2 (Sobrino et al. 2008, 2009). Data collected during SEN2FLEX were carried out on July 12 and 13, 2005. In the case of SEN3EXP, the data used here were acquired on June 20, 22, and 24, 2009. The data acquisitioned during SEN2FLEX and SEN3EXP are classified at *in situ* measurements (thermal and soil moisture) and AHS hyperspectral airborne images. Figure 15.3 shows an NDVI and thermal images during a SEN3EXP field campaign.

A set of thermal radiometric measurements was carried out to estimate surface emissivity and temperature, which was the main aim of these measurements. To this end, radiometric measurements were carried out in the thermal infrared region with the CIMEL CE312-2 radiometer. Its detector includes 6 bands: a broadband, 8–13 μm, and five narrower filters, 8.1–8.5, 8.5–8.9, 8.9–9.3, 10.3–11, and 11–11.7 μm. The characterization of the emissivity of the different surfaces was carried out by means of the TES algorithm (Gillespie et al. 1998) applied to ground-based measurements (Jiménez-Muñoz and Sobrino 2006; Payan and Royer 2004; Sobrino et al. 2009).

WET-Sensor and Theta Probe instruments were intercompared and several surface soil water content measurements were made. Different transects were defined and measured simultaneously to the aircraft overpass along 4 days. Accuracy of the instrument is about

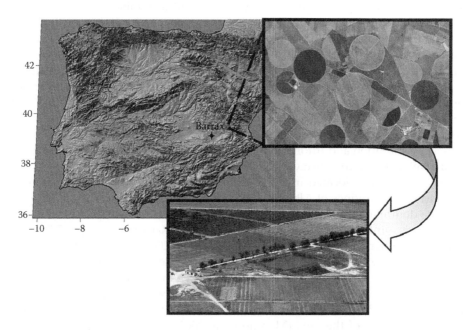

FIGURE 15.2
(**See color insert.**) Barrax Cal/Val site.

FIGURE 15.3
(See color insert.) (a) NDVI and (b) land surface temperature (LST) (in K) obtained from the AHS images during SEN3EXP field campaigning (June 22, 2009). The spatial resolution of the images is about 6 × 6 m.

± 0.01 m^3 m^{-3} from 273 to 313 K and ± 0.02 m^3 m^{-3} from 313 to 343 K. Each measurement is representative of a volume of 15 cm depth and a diameter of 10 cm. CAL/VAL measurement protocols were performed. Figure 15.4 presents the measurement point to be analyzed here.

Finally, airborne data were acquired by the Airborne Hyperspectral Scanner (AHS), which is an 80-band airborne imaging radiometer, developed and built by SensyTech Inc. (currently Argon ST, and formerly Daedalus Enterprises Inc.) and operated by the Spanish Institute for Aerospace Technology. It has 63 bands in the reflective part of the electromagnetic spectrum, 7 bands in the 3–5 μm range, and 10 bands in the 8–13 μm region. During the SEN2FLEX campaign, the AHS flights considered took place on two different dates, July 12 and 13, 2005. Data were acquired at an altitude of 983 m, resulting in a spatial resolution of 2 m. During the SEN3EXP campaign, the AHS flights took place on three different dates, June 20 (flight 1), 22 (flight 2), and 24 (flight 3), 2009, and were acquired at an altitude of 2743 m, resulting in a spatial resolution of 6 m. Table 15.1 summarizes the data used in this case study.

15.2.3.3 Methodology

The main objective of this case study is to demonstrate the usefulness of the emissivity in predicting soil moisture at high spatial resolution. The regression models were first calibrated with *in situ* soil moisture, temperature, and emissivity estimations retrieved during the SEN3EXP field campaign and also with NDVI extracted from the images themselves. Then, the equations were applied to airborne images and were compared with *in situ* soil moisture measurements.

To start with airborne inputs, surface reflectances in the visible and near-infrared (VNIR) bands of AHS were estimated to compute the NDVI. In this way, the atmospheric correction was applied through the radiative transfer code MODTRAN 4.3 (Berk et al. 1999) following the Verhoef and Bach (2003) methodology. Ts and emissivity were computed through the TES algorithm (Gillespie et al. 1998). Other technical description and thermal applications for agricultural areas using AHS imagery are described in detail by Sobrino et al. (2006, 2008).

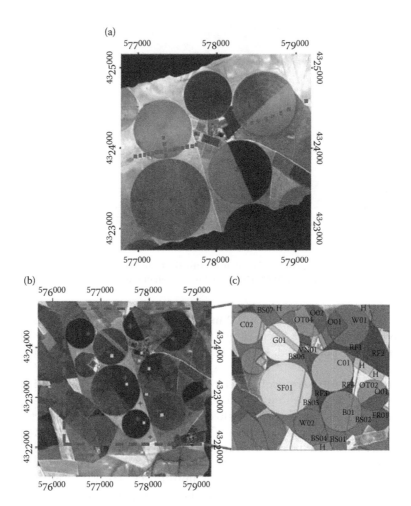

FIGURE 15.4
(See color insert.) Main study area and location (UTM coordinates) of soil moisture *in situ* measurements (a) in the SEN2FLEX campaign (image at 2 × 2 m spatial resolution) and (b) in the SEN3EXP campaign (image at 6 × 6 m spatial resolution), and (c) land use map of the SEN3EXP study area, where the fields are pointed as C (corn), G (garlic), W (wheat), OT (oat), O (onion), BS (bare soil), SF (sunflower), B (barley), VN (vine), RF (reforestation), H (harvested), and FR (fruit). Squares point to the validation points and crosses to the calibration points. (Adapted from Sobrino, J. A. et al., *Remote Sensing of Environment*, 17, 415–428, 2012.)

15.2.3.4 Soil Moisture Retrievals

Figure 15.5 represents the VI-T diagram obtained from the normalized NDVI and LST using the AHS images during SEN3EXP field campaign. Irrigated crops and bare soil area can be identified throughout the high-density points.

A simple primary analysis was performed in order to assess the influence of the emissivity. In this way, two simple equations were considered:

$$SM = a_0 + a_1 T^* + a_2 \text{NDVI}^* \tag{15.10}$$

$$SM = a_0 + a_1 T^* + a_2 \text{NDVI}^* + a_3 \varepsilon^* \tag{15.11}$$

TABLE 15.1

Summary of *In Situ* Measurements Used to Estimate Soil Moisture at Barrax Test Site

	SEN2FLEX	SEN3EXP
Dates studied	July 12 and 13, 2005	June 20, 22, and 24, 2009
Number of different crops considered in the calibration	0	11
Number of different crops considered in the validation	2	8
AHS spatial resolution	2 m	6 m
Thermal measurements	CIMEL CE-312 Transects in two crops (hot and cold points)	CIMEL CE-312 Transects in several crops
Soil moisture	Theta probe and WET-sensors Transects in two crops (hot and cold points)	Theta probe and WET-sensors Transects in several crops

These equations were calibrated from the *in situ* measurements in order to estimate their coefficient values shown in Table 15.2.

The correlation between *in situ* soil moisture measurements and the derived soil moisture is displayed in Figure 15.6. The soil moisture was estimated using Equations 15.10 and 15.11 considering the calibration data. It shows that Equation 15.10 overestimates the soil moisture in almost every case, the exception being on the grass field. However, Equation 15.11 presents better statistics regarding both the root-mean-square error (RMSE) and the correlation coefficient (*r*).

Figure 15.7 displays the correlation between the soil moisture *in situ* measurements and the soil moisture estimated applying Equations 15.1 and 15.3 to the calibration points' data. It shows that Equation 15.1 overestimates the soil moisture data for values lower than 0.1 m^3 m^{-3} (reforestation, bare soil, vineyards, and oat) and higher than 0.3 m^3 m^{-3} (grass).

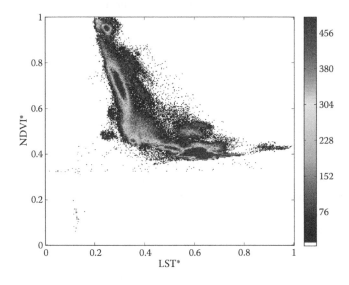

FIGURE 15.5

(See color insert.) Normalized VI-T scatterplot derived from Figure 15.4. Scale bar represents the density of points.

TABLE 15.2

Coefficients Values of Equations 15.10 and 15.11

	a_0	a_1	a_2	a_3
Equation 15.10	0.36	−0.22	−0.02	
Equation 15.11	0.36	−0.26	−0.07	−0.13

In order to make a first assessment of the equations, each graph shows the bias, standard deviation, RMSE, and correlation coefficient. It can also be deduced that considering the emissivity (Equation 15.3) decreases the RMSE and increases the correlation coefficient.

Correlation coefficient, Figure 15.8 presents the spatialized soil moisture at AHS images. In this figure, the soil moisture, NDVI, surface temperature, and emissivity images (Figure 15.8a through d) for the June 20, 2009, are also presented.

The corresponding profiles drawn in Figure 15.8a are shown in Figure 15.9, where two different transects that overpass bare soils, senescent vegetation, and different vegetated fields are plotted. In the graph of the profiles, the bare soils (transect 1) and senescent vegetation (transect 2) can be well distinguished from other surfaces owing to their low soil moisture (lower than 0.1) and NDVI, high temperature, and surface emissivity values ranging between 0.95 and 0.96. In the case of the vegetated fields overpassed by the transects, it can be observed that temperature profile is opposite to the NDVI and the soil moisture profiles. An increase in the soil moisture is correlated to the increase in the NDVI and the decrease in the temperature. Emissivity profiles show differences between bare areas and green vegetation areas. Hence the series of peaks and valleys in transect 1 (Figure 15.9a) between pixel IDs 50 and 150 corresponds to a sunflower filed cultivated in rows, so the emissivity profile in this region distinguished between the subsequent bare soil and vegetation along the plot. The increase in emissivity near pixel ID 2000 corresponds to a wheat plot with green vegetation. In the case of transect 2 (Figure 15.9b), emissivity values between pixel IDs 125 and 150 correspond to senescent vegetation over a reforestation area. Values range between 0.96 and 0.97, with *in situ* values of 0.97. In this case, emissivity values are closer to areas with green vegetation, so these two different areas are hardly distinguished in terms of emissivity. This could be a consequence of the uncertainties in the emissivity retrievals over senescent vegetation when using the TES algorithm. Note also that maximum emissivity values are below 0.98, because the two transects do not cross any fully vegetated field (see NDVI image, Figure 15.8b).

15.2.3.5 Discussion

It has been demonstrated that inclusion of emissivity improves soil moisture estimation from optical remote sensing data in comparison with previous techniques using only NDVI and Ts. However, this improvement is clearly observed when using the calibration data, although not as evident when applying equations to real data (i.e., AHS imagery). The reasons for this result can be found in the uncertainties associated to the retrieval of some parameters from image data, in particular, the surface emissivity, a key point in this study. It is well known that TES algorithm fails over fully vegetated areas, although emissivity over these kinds of surfaces is commonly known. Retrievals over senescent vegetation also can be uncertain. In fact, in this study, senescent vegetation over the reforestation area provided a low spectral contrast (b0.03) in the same way as green vegetation. This result could explain the similar values obtained in the case of senescent and green vegetation for most plots. An important step in the TES algorithm is the atmospheric correction,

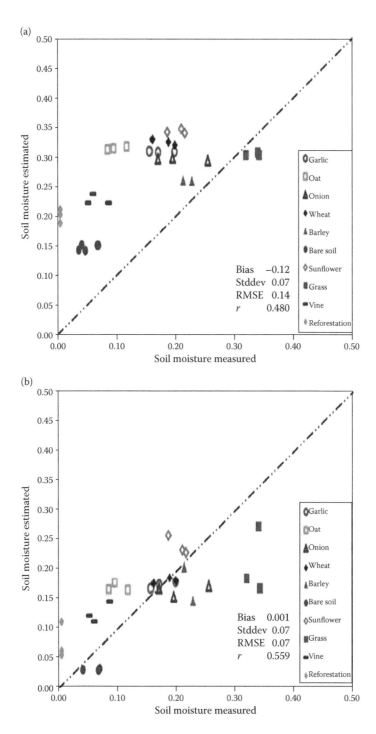

FIGURE 15.6
Soil moisture measured *in situ* versus estimated from airborne data from (a) Equation 15.10 and (b) Equation 15.11. Each graph also shows the bias, standard deviation (Stddev), root-mean-square error (RMSE), and correlation coefficient (*r*) obtained for each proposed model referred to the calibration points. (Adapted from Sobrino, J. A. et al., *Remote Sensing of Environment*, 17, 415–428, 2012.)

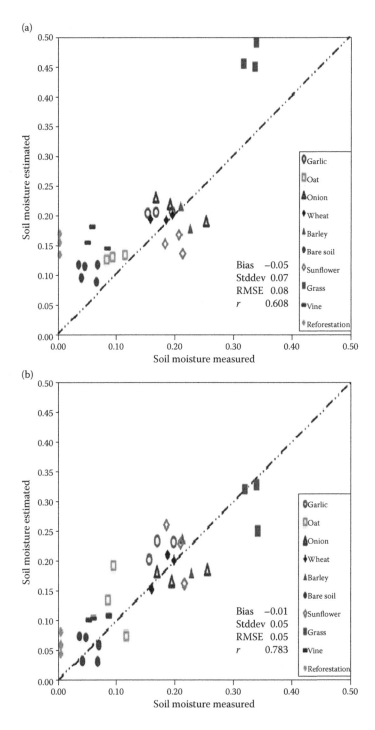

FIGURE 15.7
Soil moisture measured *in situ* versus estimated from airborne data from (a) Equation 15.1 using NDVI as a vegetation parameter and (b) Equation 15.3. Each graph also shows the bias, standard deviation (Stddev), RMSE, and correlation coefficient (*r*) obtained for each proposed model referred to the calibration points. (Adapted from Sobrino, J. A. et al., *Remote Sensing of Environment*, 17, 415–428, 2012.)

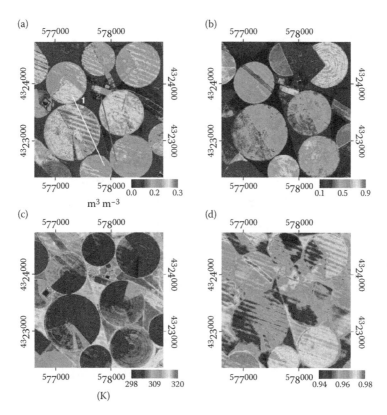

FIGURE 15.8
(See color insert.) (a) Soil moisture (m³ m⁻³), (b) NDVI, (c) land surface temperature (K), and (d) emissivity (AHS band 73, 9.15 μm) images for June 20, 2009. (Adapted from Sobrino, J. A. et al., *Remote Sensing of Environment*, 17, 415–428, 2012.)

because inputs to TES algorithm are surface radiances atmospherically corrected. Errors on atmospheric correction will affect spectral bands located in 8–9 μm in comparison to bands located in 10–12 μm, as, for example, AHS band 73 used in this chapter. Hence it has been recently demonstrated that the Water Vapor Scaling method improves traditional atmospheric correction of TIR data (Hulley and Hook 2011), so application of this technique can lead to better estimations of surface emissivities. Another important source of error is related to the noise observed in the emissivity images, which is translated to the final soil moisture outputs. This is due to electronic striping of the sensors, which, in turn, is coupled to the atmospheric correction (Gillespie et al. 2011). This effect has been analyzed in the case of AHS, and it was concluded that the removal of AHS instrumental noise is a hard task because the noise is spectrally and spatially correlated, so traditional techniques (e.g., principal components or minimum/maximum noise fraction transforms) only remove part of the noise, see for example Jiménez-Muñoz et al. 2012.

The inclusion of absolute emissivity retrieved from a single band in the soil moisture algorithm can lead to higher errors than in the case where a relative magnitude is introduced, because relative magnitudes are commonly less sensitive to errors. This could be solved, for example, by including the spectral contrast or maximum minimum difference (MMD) in the equations instead of emissivity at band 73. It has been suggested that scatterplots between NDVI and MMD allow the discrimination between bare surfaces, green vegetation, and senescent vegetation (French et al. 2000).

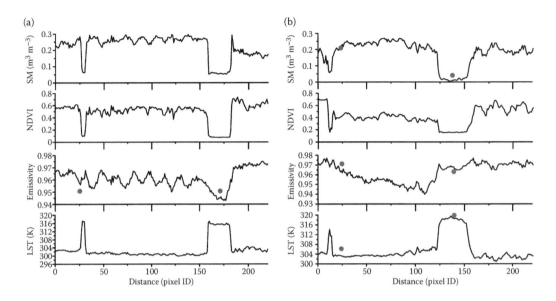

FIGURE 15.9
Profiles of the transects (a) 1 and (b) 2 displayed in Figure 15.8a. Points in the profiles refer to *in situ* measurements. (Adapted from Sobrino, J. A. et al., *Remote Sensing of Environment*, 17, 415–428, 2012.)

Figure 15.10 shows plots of soil moisture versus NDVI, surface temperature, and emissivity of AHS band 73. It shows that the relationship between soil moisture and emissivity depends on the surface considered (it is different for vegetated areas than for bare areas and also different for senescent vegetation). Figure 15.10c presents the interactions between NDVI and emissivity. It shows that the soil moisture grows up until an NDVI value of 0.5 and after that, the soil moisture tends to present a saturated value. Moreover, the soil moisture presents a relative maximum between 0.94 and 0.97, which can be attributed to the row crops or other crops under irrigation. One interesting improvement can be a previous discrimination between bare soils, green, and senescent vegetation. That could be a useful tool in order to apply different equations to different pixels (e.g., the dependence of soil moisture algorithm on NDVI could not be considered over bare soils; on the contrary, dependence of soil moisture on emissivity could be neglected over fully vegetated areas, with almost constant emissivity).

Previous work has developed similar empirical equations using the SEN2FLEX database (Fernandez et al. 2006; Paladino et al. 2007). These empirical equations are based on a lineal combination between soil moisture, Ts, and NDVI. The leaf area index (LAI) was also used as a vegetation indicator. However, the results obtained in those works could be biased since the data used to calibrate the lineal relation are limited to a few crops. This can be evidenced in the work published by Richter et al. (2009), who applied the same lineal equation to estimate soil moisture in a different agricultural site. Despite these facts, the synergic relationship between soil moisture, Ts, and VI (or other vegetation indicator) is present at merging hot and cold spots representing the extremes of the VI-T plot. The coefficients of the obtained regression, lineal or empirical, have to be tested for a wide range of land cover types in order to include a high number of land cover situations and generate a generalized relationship.

The possible interaction between soil moisture estimations at high and low spatial resolutions allows synergistic approaches to be established using the current remote sensing

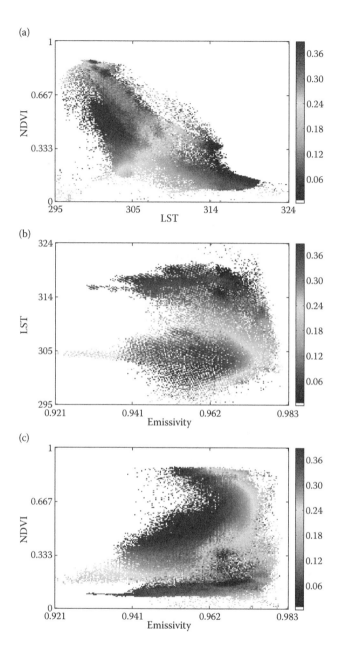

FIGURE 15.10
(**See color insert.**) Soil moisture versus (a) surface temperature and NDVI, (b) surface temperature and emissivity, and (c) NDVI and emissivity. Plots were obtained from soil moisture values derived from Equation 15.11 and surface emissivity at AHS band 73 using the image acquired June 20, 2009, during the SEN3EXP campaign. (Adapted from Sobrino, J. A. et al., *Remote Sensing of Environment*, 17, 415–428, 2012.)

techniques. The multispectral information from visible to microwave domain can generate a strong synergistic interaction between this information. For instance, several works focused the importance of obtaining soil moisture at better spatial resolution than SMOS mission retrievals (Merlin et al. 2008, 2009). In this chapter, with the proposed algorithm the soil moisture can be estimated from intermediate spatial resolution, generating more accurate results describing the soil moisture temporal and spatial variability. Additionally, the interactions between soil moisture and surface emissivity should be analyzed in order to look for a deeper understanding of the physics beneath them.

15.3   Soil Moisture and Energy Flux

The quantification of ET (latent heat flux [LE]) and of its connected variable, soil moisture is fundamental in the determination of the exchanges of energy and mass among the hydrosphere, atmosphere, and biosphere in order to understand the state of a river basin or agricultural area and their response to atmospheric forcings, thus improving applications in the field of water resources management such as definition of crop water requirements, water right concession, and real-time systems for irrigation management in a sustainable way, and also flood monitoring (Bowen 1926; Penman 1948; Brutsaert 1982; Famiglietti and Wood 1994; Sellers et al. 1996).

Generally speaking, soil moisture and ET are the key parameters in the radiative, energy, and water balance (Figure 15.11) where the central role on the soil–plant–atmosphere system exchange processes is played by the energy balance, which is the link between radiation and water budgets (Eltahir 1998). In particular, soil moisture can be estimated in different ways as a result of the interaction between precipitation and ET.

In the last 20 years the increase in availability of remote sensing data led to the development of distributed hydrological models with mass and energy balance equations (LSM) for water content and ET estimations, on both local and large scales, which use remote sensing data of land surface temperature, a connected variable to soil moisture, as input parameters, validation terms, or assimilation variables (Noilhan and Planton 1989; Shuttleworth and Wallace 1985; Famiglietti and Wood 1994; Bastiaanssen et al. 1998; Anderson et al. 1997; Corbari et al. 2011; Menenti and Choudhury 1993; Norman et al.

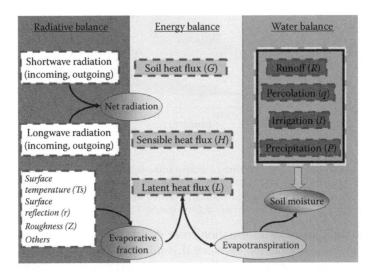

FIGURE 15.11
Soil moisture involving the radiative, energy, and water balance.

1995; Su 2002; Crow et al. 2003; Kustas et al. 2004; Minacapilli et al. 2009; Roerink et al. 2000; Sandholt et al. 2002).

The first method developed to estimate spatial maps of ET based on soil moisture estimates is from Price (1990), based on the Ts/NDVI triangular space. One of the main limitations of this method is linked to the need of independent estimates of latent heat flux on wet and dry bare soil and vegetation areas. Subsequently, Jiang and Islam (1999, 2003) proposed a modification of the triangle Ts/NDVI method computing ET with the Priestley–Taylor equation and introducing the difference between Ts and air temperature. Stisen et al. (2008) modified the Jiang and Islam (2003) approach computing the difference only on Ts measured at midday and in the early morning and adding a nonlinear relation between the Priestley–Taylor equation and NDVI in the VI-T domain. Sandholt et al. (2002) introduce a TVDI to link soil moisture and ET, even though a main limitation is the need of the whole range of soil moisture in the same satellite image. The simplified surface energy balance index (S-SEBI) evaluates the difference of Ts between the wet and dry pixels in relation to soil content and broadband albedo, which is then used to partition between latent and sensible heat fluxes (Roerink et al. 2000). Gomez et al. (2005) modified S-SEBI in order to retrieve daily latent heat fluxes based on the hypothesis that the daily evaporative fraction is similar to the instantaneous one. Sobrino et al. (2005, 2007) evaluated this method over an agricultural site in Spain with airborne and AVHRR data obtaining an accuracy of less than 1 mm day^{-1}. With respect to other VI-T methods, it has the advantage of distinguishing temperatures for wet and dry pixel with reflectance. On the one hand, the limitations are linked to the need of Ts both for bare soil and vegetation and to the hypothesis that atmospheric conditions are constant over the area limiting the use in large zones. Another possibility using the VI-T triangle method is the coupling with a SVAT model to compute soil moisture. In particular, Gillies and Carlson (1995), then modified by Gillies et al. (1997), introduced this method based on empirical correlations between soil moisture and/or vegetation fraction and land surface temperature and/or ET.

15.4 Case Study 2

A case study on ET estimate is now presented, comparing daily ET from a model based on the triangle VI-T method and a distributed hydrological energy water balance model for Barrax area during SEN2FLEX funded by the ESA (Sobrino et al. 2008; Corbari et al. 2012). Data from AHS were used, as well as ground data from a lysimeter station located in a green grass plot and an eddy covariance station in a vineyard (Sobrino et al. 2008).

In Sobrino et al. (2008), daily ET from remote sensing data is retrieved according to the methodology presented by Gomez et al. (2005), which is based on the S-SEBI model (Roerink et al. 2000) from a modified triangle VI-T method. S-SEBI directly estimates the evaporative fraction (EF) from the scatterplot between land surface temperature and albedo defining a reflectance-dependent maximum temperature for dry conditions and a minimum temperature for wet conditions. Then, this instantaneous EF is assumed to be equal to the daily values, while daily net radiation is extrapolated from instantaneous data. Instantaneous evapotranspiration (ET$_i$) is given by

$$LET_i = \Lambda_i(Rn_i - G_i) \tag{15.12}$$

where L is the latent heat of vaporization (2.45 MJ kg^{-1}), Λ is the evaporative fraction, Rn is the net radiation, and G is the soil heat flux. The subscript i refers to "instantaneous" values and the subscript d is used for "daily" values. Then, daily ET (ET$_d$) can be obtained as

$$\text{ET}_d = \frac{\Lambda_i C_{di} Rn_i}{L} \tag{15.13}$$

where C_{di} is the ratio between daily and instantaneous Rn, $C_{di} = Rn_d/Rn_i$. This was estimated from measurements of net radiation registered in a meteorological station located in the Barrax area. Instantaneous net radiation (R_{ni}) is obtained from the balance between the incoming and outgoing shortwave and longwave radiation:

$$R_{ni} = (1 - \alpha_i) \cdot R_{ci}^{\downarrow} + \varepsilon \cdot R_{gi}^{\downarrow} - \varepsilon \sigma \text{LST}_i^4 \tag{15.14}$$

where α_i is the surface albedo, $R_{ci}^{\downarrow}$ and $R_{gi}^{\downarrow}$ are, respectively, the shortwave and longwave incoming radiation (measured at meteorological stations), ε is the surface emissivity, and σ is the Stefan-Boltzmann constant (5.67 × 10^{-8} W m^{-2} K^{-4}).

The surface albedo was estimated from a weighted mean of at-surface reflectivities (ρ^{surface}) estimated with AHS VNIR atmospherically corrected bands:

$$\alpha_s = \sum_{i=1}^{20} \omega_i \rho_i^{\text{surface}} \tag{15.15}$$

where ω_i are the weight factors obtained from the solar irradiance spectrum at the top of the atmosphere. Finally, the evaporative fraction is estimated as

$$\Lambda_i = \frac{T_H - Ts}{T_H - T_{\text{LET}}} \tag{15.16}$$

where T_H and T_{LET} are two characteristic temperatures obtained from the plot of surface temperature versus surface albedo (Roerink et al. 2000).

Daily ET from AHS using the S-SEBI model is compared for 14 images and an RMSE of 1 mm day^{-1} is obtained. ET$_d$ values for near-consecutive flights are found almost the same, but ET$_d$ values along the same day are not constant, with differences ranging between 1.3 and 1.9 mm day^{-1}. This indicates that some errors are expected when a temporal integrated value of ET is obtained from instantaneous values by means of the ratio C_{di} because daily ET refers to a unique value for a certain day.

These results from S-SEBI model are compared with the outputs of a continuous distributed hydrological model, Flash-flood Event-based Spatially-distributed rainfall-runoff Transformation Energy Water Balance model (FEST-EWB), which computes all the main processes of the hydrological cycle, such as ET (Corbari et al. 2010, 2011), infiltration, surface runoff (Montaldo et al. 2007; Rabuffetti et al. 2008), flow routing, subsurface flow (Ravazzani et al. 2011), and snow dynamics (Corbari et al. 2009).

The core of the model is the system between energy and water balance equations which links latent heat flux, land surface temperature, and soil moisture. All the terms of the energy balance are functions of a representative equilibrium temperature that ensure the closing of the energy budget equation.

So ET daily cumulated values retrieved from the S-SEBI model, using as input Ts images from AHS, and computed from FEST-EWB are compared for six images (Figure 15.12).

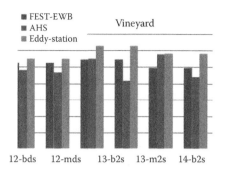 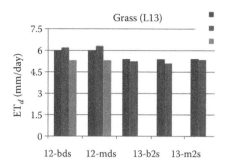

FIGURE 15.12

Comparison among S-SEBI, FEST-EWB, and ground data from eddy covariance station (eddy station) and lysimeter. (Adapted from Sobrino, J. A. et al., *Remote Sensing of Environment*, 17, 415–428, 2012; Corbari, C. et al., *International Journal of Remote Sensing*, 34(9–10), 3208–3230, 2012.)

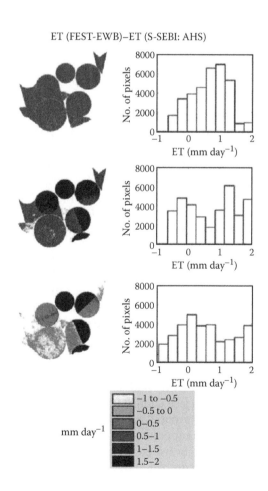

FIGURE 15.13

Maps of difference between ET from S-SEBI model with AHS and FEST-EWB at 10 m and histograms for July 12, 13, and 14. (Adapted from Sobrino, J. A. et al., *Remote Sensing of Environment*, 17, 415–428, 2012; Corbari, C. et al., *International Journal of Remote Sensing*, 34(9–10), 3208–3230, 2012.)

A good agreement between modeled and observed values is shown with similar mean and standard deviation values. FEST-EWB and S-SEBI generally tend to overestimate ET with respect to the lysimeter with a mean absolute difference of 0.69 and 0.94 mm day^{-1}, respectively. When the models are compared to eddy covariance station measurements, FEST-EWB underestimates on average -0.31 mm day^{-1} with a standard deviation of 0.16 mm day^{-1}, while S-SEBI underestimates on average -0.48 mm day^{-1} with a standard deviation of 0.31 mm day^{-1}.

The performance of the two models is also evaluated by computing the difference maps of ET for three images in order to study their spatial variability (Figure 15.13). The difference values range from -1 to 2 mm day^{-1}, with absolute mean differences equal to 0.80 mm day^{-1} for July 12, 0.95 mm day^{-1} for July 13, and 0.9 mm day^{-1} for July 14.

These obtained results are comparable with previous studies on the S-SEBI model. In fact, for example, Gomez et al. (2005) found an root square mean difference (RSMD) of 1 mm day^{-1} in daily ET estimates from airborne data in comparison to ground observations. Sobrino et al. (2005, 2007) evaluated this method over an agricultural site in Spain with airborne and AVHRR data obtaining an accuracy of less than 1 mm day^{-1}.

15.5 Summary and Conclusions

This chapter has presented a relative comprehensive review of the new synergic method in remote sensing to predict soil moisture based on VI-T physical relation. Moreover, it also described how the combination of different information sources measured by remote sensing techniques could improve the soil moisture estimations through a synergic process in contrast to using only one spectral range. Most synergic models need to strengthen their intrinsic variable relationships using a combined approach between thermal, visible, and passive microwave information. These kinds of interaction showed better results in soil moisture estimation if they are ensembled to be modeled by using a land surface modeling scheme.

Nowadays, the current use of soil moisture from passive microwave is promising for the VI-T method as a key tool for downloading or improving the soil moisture–VI-T relationship. At present, passive microwave data, which deliver the most accurate soil moisture estimation, can be successfully included in VI-T method, although the coarse spatial resolution is one of the main constraints. However, in the near future, SMAP data can complete the spatial resolution of the soil moisture product, improving the VI-T method and also the interaction between other biophysical variables such as surface emissivity, albedo, and brightness temperature in the passive microwave range or even a combination of them.

Finally, the interactions of soil moisture and VI-T applied on SVAT or land surface models allow the estimation of land surface energy fluxes. Different methodologies have been developed in literature, mainly differentiating between each other for the use of NDVI or albedo or for use of land surface temperature or of the difference between Ts during the day and night or of the difference between Ts and air temperature. These models have been independently tested against completed hydrological models and ground data showing good agreements.

Acknowledgments

We acknowledge funding from the European Union (CEOP-AEGIS, project FP7-ENV-2007-1 Proposal no. 212921) and the Ministerio de Economía y Competitividad (EODIX, project AYA2008-0595-C04-01; CEOS-Spain, project AYA2011-29334-C02-01). This work was also partially funded by Program U-INICIA VID 2012, grant U-INICIA 4/0612, University of Chile.

References

Anderson, M. C., J. M. Norman, G. R. Diak, W. P. Kustas, and J. R. Mecikalski. 1997. A two-source time-integrated model for estimating surface fluxes using thermal infrared remote sensing. *Remote Sensing of Environment*, 60, 195–216.

Bastiaanssen, W. G. M., M. Menenti, R. A. Feddes, and A. A. M. Holtslag. 1998. A remote sensing surface energy balance algorithm for land (SEBAL), 1. Formulation. *Journal of Hydrology*, 212–213, 198–212.

Berk, A., G. P. Anderson, P. K. Acharya, J. H. Chetwynd, L. S. Bernstein, E. P. Shettle, M. W. Matthew, and S. M. Adler-Golden. 1999. MODTRAN4 User's Manual (Hanscom AFB, MA: Air Force Research Laboratory, Space Vehicles Directorate).

Bowen, I. S. 1926. The ratio of heat losses by conduction and by evaporation from any water surface. *Physical Reviews*, 27, 779–787.

Brutsaert, W. 1982. *Evaporation into the Atmosphere*. D. Reidel, Norwell, MA, 299 pp.

Capehart, W. J., and T. N. Carlson. 1997. Decoupling of surface and near surface soil water content: A remote sensing perspective. *Water Resources Research*, 33(6), 1383–1395.

Carlson, T. 2007. An overview of the "Triangle Method" for estimating surface evapotranspiration and soil moisture from satellite imagery. *Sensors*, 7(8), 1612–1629.

Carlson, T., R. Gillies, and E. Perry. 1994. A method to make use of thermal infrared temperature and NDVI measurements to infer surface soil water content and fractional vegetation cover. *Remote Sensing Review*, 9, 161–173.

Carlson, T., and D. Ripley. 1997. On the relation between NDVI, fractional vegetation cover, and leaf area index. *Remote Sensing of Environment*, 62, 241–252.

Carlson, T. N., J. K. Dodd, S. G. Benjamin, and J. N. Cooper. 1981. Remote estimation of surface energy balance, moisture availability and thermal inertia. *Journal of Applied Meteorology*, 20, 67–87.

Carlson, T. N., R. R. Gillies, and T. J. Schmugge. 1995. An interpretation of methodologies for indirect measurement of soil-water content. *Agricultural and Forest Meteorology*, 77, 191–205.

Carlson, T. N., F. G. Rose, and E. M. Perry. 1984. Regional scale estimates of surface moisture availability from GOES satellite. *Agronomy Journal*, 76, 972–979.

Chauhan, N. S., S. Miller, and P. Ardanuy. 2003. Spaceborne soil moisture estimation at high resolution: A microwave optical/IR synergistic approach. *International Journal of Remote Sensing*, 24(22), 4599–4622.

Chen, J. M., B. J. Yang, and R. H. Zhang. 1989. Soil thermal emissivity as affected by its water content and surface treatment. *Soil Science*, 148(6), 433–435.

Corbari, C., G. Ravazzani, and M. Mancini. 2011. A distributed thermodynamic model for energy and mass balance computation: FEST-EWB. *Hydrological Processes*, 25, 1443–1452.

Corbari, C., G. Ravazzani, J. Martinelli, and M. Mancini. 2009. Elevation based correction of snow coverage retrieved from satellite images to improve model calibration. *Hydrology and Earth System Sciences*, 13(5), 639–649.

Corbari, C., J. A. Sobrino, M. Mancini, and V. Hidalgo. 2010. Land surface temperature representativeness in a heterogeneous area through a distributed energy-water balance model and remote sensing data. *Hydrology and Earth System Sciences*, 14, 2141–2151.

Corbari, C., J. A. Sobrino, M. Mancini, and V. Hidalgo. 2012. Mass and energy flux estimates at different spatial resolutions in a heterogeneous area through a distributed energy–water balance model and remote-sensing data. *International Journal of Remote Sensing*, 34(9–10), 3208–3230.

Crow, W. T., E. F. Wood, and M. Pan. 2003. Multiobjective calibration of land surface model evapotranspiration predictions using streamflow observations and spaceborne surface radiometric temperature retrievals. *Journal of Geophysical Research*, 108(D23), 4725.

Eltahir, E. A. 1998. A soil moisture-rainfall feedback mechanism 1. Theory and observations. *Water Resources Research*, 34(4), 765–776.

Entekhabi, D., E. Njoku, P. E. O'Neill, K. H. Kellogg, W. T. Crow, W. N. Edelstein, J. K. Entin et al. 2010. The Soil Moisture Active Passive (SMAP) mission. *Proceedings of the IEEE Transactions of the Geosciences and Remote Sensing*, 98(5), 704–716.

Famiglietti, J. S., and E. F. Wood. 1994. Multiscale modeling of spatially variable water and energy balance processes. *Water Resources Research*, 30, 3061–3078.

Fernandez, G., M. Palladino, G. D'Urso, J. Moreno, and J. C. Jimenez. 2006. Exploring soil water content dynamics from thermal observations within the SEN2FLEX experiment (CD-ROM), SEN2FLEX Workshop, WPP 271, 2006. ESA Publications Division, ESTEC, Noordwijk.

French, A. N., T. J. Schmugge, and W. P. Kustas. 2000. Discrimination of senescent vegetation using thermal emissivity contrast. *Remote Sensing of Environment*, 74, 249–254.

Friedl, M. A., and F. W. Davis. 1994. Sources of variation in radiometric surface temperature over a tall-grass prairie. *Remote Sensing of Environment*, 48, 1–17.

Gillespie, A., S. Rokugawa, T. Matsunaga, J. S. Cothern, S. Hook, and A. B. Khale. 1998. A temperature and emissivity separation algorithm for advance spaceborne thermal emission and reflection radiometer (ASTER) images. *IEEE Transactions on Geoscience and Remote Sensing*, 36, 1113–1126.

Gillespie. A. R., E. A. Abbott, L. Gilson, G. Hulley, J. C. Jiménez-Muñoz, and J. A. Sobrino. 2011. Residual errors in ASTER temperature and emissivity standard products AST08 and AST05. *Remote Sensing of Environment*, 115, 3681–3694.

Gillies, R. R., and T. N. Carlson. 1995. Thermal remote sensing of surface soil water content with partial vegetation cover for incorporation into climate models. *Journal of Applied Meteorology*, 34, 745–756.

Gillies, R. R., T. N. Carlson, J. Cui, W. P. Kustas, and K. S. Humes. 1997. A verification of the 'triangle' method for obtaining surface soil water content and energy fluxes from remote measurements of the Normalized Divergence Vegetation Index (NDVI) and surface radiant temperature. *International Journal of Remote Sensing*, 18(15), 3145–3166.

Gillies, R. R., and B. Temesgen. 2004. Coupling thermal infrared and visible satellite measurements to infer biophysical variables at the land surface. In *Thermal Remote Sensing in Land Surface Processes*. D. A. Quattrochi and J. C. Luval (Eds.), CRC Press, Boca Raton, FL, pp. 160–184.

Gomez, M., A. Olioso, J. A. Sobrino, and F. Jacob. 2005. Retrieval of evapotranspiration over the Aplilles/ReSeDA experimental site using airborne POLDER sensor and a thermal camera. *Remote Sensing of Environment*, 96, 399–408.

Gutman, G., and A. Ignatov. 1998. The derivation of the green vegetation fraction from NOAA/AVHRR data for use in numerical weather prediction models. *International Journal of Remote Sensing*, 19(8), 1533–1543.

Hulley, G. C., and S. J. Hook. 2011. Generating consistent land surface temperature and emissivity products between ASTER and MODIS data for Earth science research. *IEEE Transactions on Geoscience and Remote Sensing*, 49(9), 1304–1315.

Hulley, G. C., S. J. Hook, and A. M. Baldridge. 2010. Investigating the effects of soil moisture on thermal infrared land surface temperature and emissivity using satellite retrievals and laboratory measurements, *Remote Sensing of Environment*, 114, 1480–1493.

Idso, B., T. Schmugge, T. R. Jackson, and R. Reginto. 1975. The utility of surface temperature measurements for remote sensing of foil water studies. *Journal of Geophysical Research*, 80, 3044–3049.

Ishida, T., H. Ando, and M. Fukuhara. 1991. Estimation of complex refractive index of soil particles and its dependence on soil chemical properties. *Remote Sensing of Environment*, 38, 173–182.

Jiang, L., and S. Islam. 1999. A methodology for estimation of surface evapotranspiration over large areas using remote sensing observations. *Geophysical Research Letters*, 26, 2773–2776.

Jiang, L., and S. Islam. 2003. An intercomparison of regional latent heat flux estimation using remote sensing data. *International Journal of Remote Sensing*, 24, 2221–2236.

Jiménez-Muñoz, J. C., and J. A. Sobrino. 2006. Emissivity spectra obtained from field and laboratory measurements using the temperature and emissivity separation algorithm. *Applied Optics*, 45(27), 7104–7109.

Jiménez-Muñoz, J. C., J. A. Sobrino, and A. R. Gillespie. 2012. Surface emissivity retrieval from Airborne Hyperspectral Scanner data: Insights on atmospheric correction and noise removal. *IEEE Geoscience and Remote Sensing Letters*, 9(2), 180–184.

Jiménez-Muñoz, J. C., J. A. Sobrino, A. Plaza, L. Guanter, J. Moreno, and P. Martínez. 2009. Comparison between fractional vegetation cover retrievals from vegetation indices and spectral mixture analysis: A case study of PROBA/CHRIS data over an agricultural area. *Sensors*, 9, 768–793.

Kerr, Y. H. 2007. Soil moisture from space: Where are we? *Hydrogeology Journal*, 15, 117–120.

Kerr, Y. H., P. Waldteufel, J. P. Wigneron, J. M. Martinuzzi, J. Font, and M. Berger. 2001. Soil moisture retrieval from space: The Soil Moisture and Ocean Salinity (SMOS) mission. *IEEE Transactions on Geosciences and Remote Sensing*, 39(8), 1729–1735.

Kustas, W. P. 1990. Estimates of evapotranspiration with a one- and two-layer model of heat transfer over partial canopy cover. *Journal of Applied Meteorology*, 29, 205–223.

Kustas, W. P., F. Li, T. J. Jackson, J. H. Pruger, J. I. MacPherson, and M. Wolden. 2004. Effects of remote sensing pixel resolution on modeled energy flux variability of croplands in Iowa. *Remote Sensing of Environment*, 92(4), 535–547.

Kustas, W. P., and J. M. Norman. 1996. Use of remote sensing for evapotranspiration monitoring over and surfaces. *Hydrological Sciences Journal*, 41, 495–516.

Lambin, E. F., and D. Ehrlich. 1995. Combining vegetation indices and surface temperature for land-cover mapping at broad spatial scales. *International Journal of Remote Sensing*, 16(3), 573–579.

Lambin, E. G., and D. Ehrlich. 1996. The surface temperature-vegetation index space for land cover and land-cover change analysis. *International Journal of Remote Sensing*, 17, 463–487.

Margulis, S., J. Kim, and T. Hogue. 2005. A comparison of the triangle retrieval and variational data assimilation methods for surface turbulent flux estimation. *Journal of Hydrometeorology*, 6, 1063–1072.

McLaughlin, D. 1995. Recent developments in hydrologic data assimilation. *Reviews of Geophysics*, 33, 977–984.

Mecikalski, J. R., G. R. Diak, M. C. Anderson, and J. M. Norman. 1999. Estimating fluxes on continental scales using remotely sensed data in an atmospheric-land exchange model. *Journal of Applied Meteorology*, 38, 1352–1369.

Menenti, M., and B. J. Choudhury. 1993. Parameterization of land surface evapotranspiration using a location-dependent potential evapotranspiration and surface temperature range. In *Exchange Processes at the Land Surface for a Range of Space and Time Scales*, H. J. Bolle et al. (Eds.). *IAHS Publication*, 212, 561–568.

Merlin, O., A. Al Bitar, J. P. Walker, and Y. Kerr. 2009. A sequential model for disaggregating near-surface soil moisture observations using multi-resolution thermal sensors. *Remote Sensing of Environment*, 113(10), 2275–2284, doi:10.1016/j.rse.2009.06.012.

Merlin, O., A. Chehbouni, Y. Kerr, E. Njoku, and D. Entekhabi. 2005. A combined modeling and multispectral/multiresolution remote sensing approach for disaggregation of surface soil moisture: Application to SMOS configuration. *IEEE Transactions on Geoscience and Remote Sensing*, 43(9), 2036–2050.

Merlin, O., J. Walker, A. Chehbouni, and Y. Kerr. 2008. Towards deterministic downscaling of SMOS soil moisture using MODIS derived soil evaporative efficiency. *Remote Sensing of Environment*, 112(10), 3935–3946.

Minacapilli, M., C. Agnese, F. Blanda, C. Cammalleri, G. Ciraolo, G. D'Urso, M. Iovino et al. 2009. Estimation of actual evapotranspiration of Mediterranean perennial crops by means of remote-sensing based surface energy balance models. *Hydrology and Earth System Sciences*, 13(7), 1061–1074.

Mira, M., E. Valor, R. Boluda, V. Caselles, and C. Coll. 2007. Influence of soil water content on the thermal infrared emissivity of bare soils: Implication for land surface temperature determination. *Journal of Geophysical Research*, 112, F04003, doi:10.1029/2007JF000749.

Mira, M., E. Valor, V. Caselles, E. Rubio, C. Coll, J. M. Galve, R. Niclos, J. M. Sanchez, and R. Boluda. 2010. Soil moisture effect on thermal infrared (8–13-µm) emissivity. *Geoscience and Remote Sensing*, 48(5), 2251–2260.

Montaldo, N., G. Ravazzani, and M. Mancini, 2007. On the prediction of the Toce alpine basin floods with distributed hydrologic models. *Hydrological Processes*, 21, 608–621.

Moradkhani, H. 2008. Hydrologic remote sensing and land surface data assimilation. *Sensors*, 8, 2986–3004.

Moran, M. S., T. R. Clarke, W. P. Kustas, M. Weltz, and S. A. Amer. 1994. Evaluation of hydrologic parameters in a semiarid rangeland using remotely-sensed spectral data. *Water Resources Research*, 30, 1287–1297.

Moreno, J., A. Calera, V. Caselles, J. M. Cisneros, J. A. Martínez-Lozano, J. Meliá, F. Montero, and J. A. Sobrino. 2001. The measurement programme at Barrax. In *Proceedings of the DAISEX Final Results Workshop*, M. Wooding and R.A. Harris (Eds.), ESA Special Publication, SP-499 (Noordwijk, the Netherlands: ESA Publications, ESTEC), pp. 43–51.

Nemani, R. R., and S. W. Running. 1989. Estimation of regional surface-resistance to evapotranspiration from NDVI and thermal-IR AVHRR data. *Journal of Applied Meteorology*, 28, 276–284.

Noilhan, J., and S. Planton. 1989. A simple parameterization of land surface processes for meteorological models. *Monthly Weather Review*, 117, 536–549.

Norman, J. M., W. P. Kustas, and K. S. Humes. 1995. A two source approach for estimating soil and vegetation energy fluxes from observations of directional radiometric surface temperature. *Agricultural and Forest Meteorology*, 77, 263–293.

Olioso, A. 1992. Simulation des echanges d'energie et de masse d'un convert vegandal, dans le but derelier ia transpiration et la photosynthese anx mesures de reflectance et de temperature de surface. Doctorate Thesis. University de Montpellier II, Montpelier, France, pp. 1–260.

Olioso, A., H. Chauki, D. Courault, and J. P. Wigneron. 1999. Estimation of evapotranspiration and photosynthesis by assimilation of remote sensing data into SVAT models. *Remote Sensing of Environment*, 68, 341–356.

Owen, T. W., T. N. Carlson, and R. R. Gillies. 1998. An assessment of satellite remotely-sensed land cover parameters in quantitatively describing the climatic effect of urbanization. *International Journal of Remote Sensing*, 19, 1663–1681.

Palladino, M., G. Fernandez, G. D'Urso, and J. Moreno. 2007. Studying the relationship between superficial soil water content and observed land surface temperature with AHS data and modeling techniques within the SEN2FLEX experiment. Paper presented at European Geosciences Union 2007, Geophysical Research Abstracts 9, 08180.

Payan, A., and A. Royer. (2004). Analysis of Temperature and Emissivity Separation (TES) algorithm applicability and sensitivity. *International Journal of Remote Sensing*, 25(1), 15–37.

Penman, H. L. 1948. Natural evaporation from open water, bare soil and grass. *Proceedings of the Royal Society, A*, 193, 120–146.

Petropoulos, G., T. N. Carlson, and M. J. Wooster. 2009b. An overview of the use of the SimSphere Soil Vegetation Atmosphere Transfer (SVAT) model for the study of land-atmosphere interactions. *Sensors*, 9, 4286–4308.

Petropoulos, G., T. N. Carlson, M. J. Wooster, and S. Islam. 2009a. A review of Ts/VI remote sensing based methods for the retrieval of land surface energy fluxes and soil surface moisture. *Progress in Physical Geography*, 33(2), 224–250.

Piles, M., A. Camps, M. Vall-llossera, I. Corbella, R. Panciera, C. Rudiger, Y. H. Kerr, and J. Walker. 2011. Downscaling SMOS-derived soil moisture using MODIS visible/infrared data. *IEEE Transactions on Geosciences and Remote Sensing*, 49(9), 3156–3166.

Pinty, B., and M. Verstraete. 1992. On the design and validation of surface bidirectional reflectance and albedo model. *Remote Sensing of Environment*, 41, 155–167.

Price, J. C. 1980. The potential of remotely sensed thermal infrared data to infer surface soil moisture and evaporation. *Water Resources Research*, 16, 787–795.

Price, J. C. 1990. Using spatial context in satellite data to infer regional scale evapotranspiration. *IEEE Transactions on Geoscience and Remote Sensing*, 28, 940–948.

Rabuffetti, D., G. Ravazzani, C. Corbari, and M. Mancini. 2008. Verification of operational Quantitative Discharge Forecast (QDF) for a regional warning system—The AMPHORE case studies in the upper Po River. *Natural Hazards and Earth System Sciences*, 8, 161–173.

Ravazzani, G., D. Rametta, and M. Mancini, (2011). Macroscopic Cellular Automata for groundwater modelling: A first approach. *Environmental Modelling Software*, 26(5), 634–643.

Richter, K., M. Palladino, F. Vuolo, L. Dini, and G. D'Urso. 2009. Spatial distribution of soil water content from airborne thermal and optical remote sensing data. In *Remote Sensing for Agriculture, Ecosystems, and Hydrology*, vol. XI, Christopher M. U. Neale, and A. Maltese (Eds.), *Proc. SPIE*, 7472.

Roerink, G. J., Z. Su, and M. Menenti. 2000. SSEBI: A simple remote sensing algorithm to estimate the surface energy balance. *Physics and Chemistry of the Earth, Part B: Hydrology, Oceans and Atmosphere*, 25, 147–157.

Sandholt, I., K. Rasmussen, and J. Andersen. 2002. A simple interpretation of the surface temperature/vegetation index space for assessment of surface moisture status. *Remote Sensing of Environment*, 79, 213–224.

Shuttleworth, W. J., R. J. Gurney, A. Y. Hsu, and J. P. Ormsby. 1989. FIFE: The variation in energy partition at surface flux sites. *IAHS Publication*, 186, 67–74.

Shuttleworth, W. J., and J. S. Wallace. 1985. Evaporation from sparse crops—an energy combination theory. *Quarterly Journal of the Royal Meteorological Society*, 111, 839–855, doi:10.1256/smsqj.46909.

Sellers, P. J., D. A. Randall, G. J. Collatz, J. A. Berry, C. B. Field, D. A. Dazlich, C. Zhang, G. D. Collelo, and L. Nounoua. 1996. A revised land surface parameterisation (SiB2) for atmospheric GCMS, part 1: Model formulation. *Journal of Climate*, 9, 676–705.

Sobrino, J. A., V. Caselles, and F. Becker. 1990. Significance of the remotely sensed thermal infrared measurements obtained over a citrus orchard. *ISPRS Photogrammetric Engineering and Remote Sensing*, 44, 343–354.

Sobrino, J. A., B. Franch, C. Mattar, J. C. Jiménez-Muñoz, and C. Corbari. 2012. A method to estimate soil moisture from airborne hyperspectral scanner (AHS) and aster data: Application to SEN2FLEX and SEN3EXP campaigns. *Remote Sensing of Environment*, 17, 415–428.

Sobrino, J. A., M. Gómez, J. C. Jiménez-Muñoz, and A. Olioso. 2007. Application of a simple algorithm to estimate daily evapotranspiration from NOAA–AVHRR images for the Iberian Peninsula. *Remote Sensing of Environment*, 110, 139–148.

Sobrino, J. A., M. Gomez, J. Jimenez-Munoz, A. Olioso, and G. Chehbouni. 2005. A simple algorithm to estimate evapotranspiration from DAIS data: Application to the DAISEX campaigns. *Journal of Hydrology*, 315, 117–125.

Sobrino, J. A., J. C. Jiménez-Muñoz, G. Sòria, M. Gómez, A. Barella Ortiz, M. Romaguera, M. Zaragoza et al. 2008. Thermal remote sensing in the framework of the SEN2FLEX project: Field measurements, airborne data and applications. *International Journal of Remote Sensing*, 29(17), 4961–4991.

Sobrino, J. A., J. C. Jiménez-Muñoz, P. J. Zarco-Tejada, G. Sepulcre-Cantó, and E. de Miguel. 2006. Land surface temperature derived from airborne hyperspectral scanner thermal infrared data. *Remote Sensing of Environment*, 102, 99–115.

Sobrino, J. A., C. Mattar, J. P. Gastellu-Etchegorry, J. C. Jiménez-Muñoz, and E. Grau. 2011. Evaluation of DART 3D model in the thermal domain using satellite/airborne imagery and ground-based measurements. *International Journal of Remote Sensing*, 32(22), 7453–7477.

Sobrino, J. A., C. Mattar, P. Pardo, J. C. Jiménez-Muñoz, S. J. Hook, A. Baldridge, and F. Ibañez. 2009. Soil emissivity and reflectance spectra measurements. *Applied Optics*, 48, 3664–3670.

Sobrino, J. A., and N. Raissouni. 2000. Toward remote sensing methods for land cover dynamics monitoring: Application to Morocco. *International Journal of Remote Sensing*, 21(2), 353–366.

Stewart, J. B., W. P. Kustas, K. S. Humes, W. D. Nichols, M. S. Moran, and H. A. R. De Bruin. 1994. Sensible heat flux-radiant surface temperature relationship for semiarid areas. *Journal of Applied Meteorology*, 33, 1110–1117.

Stisen, S., I. Sandholt, A. Norgaard, R. Fensholt, and K. H. Jensen. 2008. Combining the triangle method with thermal inertia to estimate regional evapotranspiration—Applied to MSG SEVIRI data in the Senegal River basin. *Remote Sensing of Environment*, 112, 1242–1255.

Su, Z. 2002. The Surface Energy Balance System (SEBS) for estimation of turbulent heat fluxes. *Hydrology and Earth System Sciences*, 6(1), 85–99.

Van Loon, E. E., and P. A. Troch. 2001. Directives for 4D soil moisture data assimilation in hydrological modeling. In *Soil-Vegetation-Atmosphere Transfer Schemes and Large-Scale Hydrological Models*, IAHS Publication, 270, 257–267.

Watson, K. 1971. Applications of thermal modeling in the geological interpretation of IR images. In *Proceedings of the Seventh International Symposium of Remote Sensing of Environment*. ERIM. Ann Arbor, MI, pp. 2017–2041.

Wigneron, J.-P., J.-C. Calvet, T. Pellarin, A. A. Van de Griend, M. Berger, and P. Ferrazoli. 2003. Retrieving near-surface soil moisture from microwave radiometric observations: Current status and future plans. *Remote Sensing of Environment*, 85, 489–506.

16

Soil Moisture Remote Sensing and Data Assimilation

Carsten Montzka

CONTENTS

16.1 Introduction

Remote sensing for the estimation of surface soil moisture (SSM) is performed mainly with active and passive sensors working in the microwave region of the electromagnetic spectrum. Airborne microwave measurements (Hajnsek et al. 2009; Saleh et al. 2009) are often recorded for a short time period only. This makes it difficult to consider these data in numerical models in an operational way. In contrast, spaceborne sensors like ASCAT (Bartalis et al. 2007), AMSR-E (Njoku et al. 2003), SMOS (Kerr et al. 2010), or PALSAR (Takada et al. 2009) provide information on SSM from the regional to the global scale (Wagner et al. 2007) with long-term repetitive coverage. This is important for further utilization of SSM records in numerical modeling (Munoz-Sabater et al. 2012; Scipal et al. 2008). Because soil moisture was recognized as an essential climate variable (GCOS 2010), long-term time series starting in 1979 have been created for global-scale applications by merging SSM products from different active and passive microwave satellite sensors into a single dataset (Dorigo et al. 2012).

A feasible way to integrate these measurements and high-level SSM products in numerical models is by means of data assimilation (DA) (Ni-Meister 2008) (see Figure 16.1). DA, that is, correcting of model state variables such as SSM on runtime using distributed

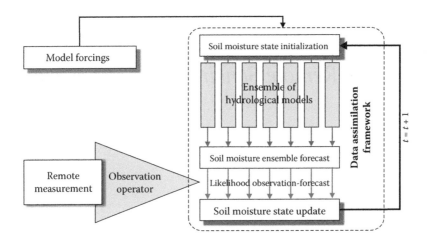

FIGURE 16.1
Framework for soil moisture DA. An ensemble of hydrological models predicts soil moisture states, which are updated by the likelihood between these predictions and observed soil moisture. The observation operator denotes the soil moisture retrieval method.

observations, might be one way to reduce predictive model uncertainty (Beven and Freer 2001; Pauwels et al. 2001; Reichle et al. 2001a). Moreover, satellite observations are recorded at specific temporal resolutions, whereas the combination of observations and model simulations in a DA framework is able to generate SSM monitoring and prediction continuous in space and time (Moradkhani and Sorooshian 2008). In order to improve soil moisture profile predictions compared to open-loop model simulations, not only the update of SSM model states but also the dual state–parameter update is possible. Typically, SSM observations derived from remote microwave systems have been used to improve the prediction of soil moisture profiles and moisture content in the soil root zone.

After an early work by Ottle and Vidalmadjar (1994), who simply integrated SSM retrievals from spaceborne optical infrared sensors into a hydrological model, the first study of remotely sensed microwave-based SSM DA was conducted by Houser et al. (1998). Then, several studies followed (Calvet and Noilhan 2000; Crosson et al. 2002; Galantowicz et al. 1999; Hoeben and Troch 2000; Li and Islam 1999; Margulis et al. 2002; Montaldo et al. 2001; Parada and Liang 2004; Pauwels et al. 2002; Reichle et al. 2001b, 2002a; Walker et al. 2002). In this chapter, recent advancements in the assimilation of remotely sensed SSM into numerical models are reviewed, both for synthetic and real-world studies. First (Section 16.2), an overview over two prominent DA techniques, namely, the ensemble Kalman filter (EnKF) and the particle filter, is given. In Section 16.3, recent works focusing on the soil moisture profile estimation by assimilation of one remotely sensed soil moisture product are discussed. In Section 16.4 it is discussed how recent studies deal with different scales in remote sensing SSM observations and numerical models. Moreover, in Section 16.5 the ability of proposed approaches to estimate model parameters for improved predictions is presented. In Section 16.6 the impact of SSM DA is reviewed for the improved estimation of further components of the hydrological cycle, for example, evapotranspiration, streamflow, and groundwater recharge. In this chapter the assimilation of single SSM datasets into hydrological models is presented only. General aspects of the assimilation of multiple SSM datasets (Barrett and Renzullo 2009; Li et al. 2010) as well as several examples about multiscale SSM DA can be found, for example, in the work of Montzka et al. (2012).

16.2 DA Theory

In this section we present a brief introduction into two prominent representatives of DA techniques, namely, the EnKF and the sequential importance resampling particle filter (SIR-PF). Both algorithms (or their derivatives) are widely used in environmental modeling. The general principle of operation will be clarified, which is important for the understanding of SSM DA applications.

DA algorithms as recursive Bayesian estimation techniques first emerged with the Kalman filter (Kalman 1960) and, later, its nonlinear version extended Kalman filter (EKF). The EnKF, a Monte Carlo implementation of the Bayesian updating, proposed by Evensen (1994) and clarified by Burgers et al. (1998), is widely used in hydrological applications. It reduces the computational demand relative to the EKF by integrating an ensemble of states from which the covariances are obtained at each update, thereby avoiding the need to linearize. The EnKF approach can be superior to the EKF, owing to the EnKF's ability to represent nonadditive model errors with fewer realizations, as well as to its flexibility in covariance modeling including horizontal error correlations (Moran et al. 2004; Reichle et al. 2002b). Similar to the Kalman filter, EnKF relies on a normal assumption of model and observation errors that may not be valid in environmental modeling (Crow and Van Loon 2006; Han and Li 2008; Teuling et al. 2006). In addition, linear updating of model states using this method reduces its applicability for highly nonlinear systems (Moradkhani et al. 2005a). These limitations can be relaxed by the use of the sequential Monte Carlo method in the form of particle filtering (PF) (Gordon et al. 1993; Moradkhani et al. 2005a). The PF differs from classical Kalman filtering methods as it can handle the propagation of non-Gaussian distributions through nonlinear models. Both PF and EnKF are Monte Carlo techniques that use samples (i.e., ensemble members or particles) to estimate the underlying probability density function (pdf) of model states and parameters. In case of limited computational power, the EnKF is able to provide a good approximation for nonlinear, non-normal land surface problems, despite its dependence on normality assumptions (Zhou et al. 2006). The general idea of sequential DA is presented in Figure 16.2.

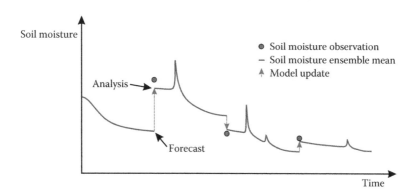

FIGURE 16.2
Simplified representation of sequential DA of soil moisture. The model is used to calculate a forecast; once an observation is available, the analysis takes place and the model is updated.

16.2.1 Ensemble Kalman Filter

The idea of ensemble-based DA methods is to calculate several model runs with different forcings, initial state conditions, parameters, or system noise. The resulting set of random state samples represents the state pdf. In a forecast step, to calculate a single realization i of the state vector X_t at time t, the model M forwards the state of the previous time X_{t-1} by

$$X_{t,i} = M[X_{t-1,i}, V_{t-1,i}] \tag{16.1}$$

where V_t is the system or process noise. The measurement of the state Y_t is related to the real state X_t by the observation equation H:

$$Y_{t,i} = H[X_{t,i}, U_{t,i}] \tag{16.2}$$

where U_t is the measurement noise. The ensemble mean of the states is calculated with

$$\overline{X_t} = \frac{1}{N} \sum_{i=1}^{N} X_{t,i} \tag{16.3}$$

N is the number of elements in the ensemble, that is, the number of realizations. In an update step, as every realization has its own expected error, the error expectation vector of the ensemble is calculated with

$$E_t = \left[X_{t,i} - \overline{X_t}, \ldots, X_{t,N} - \overline{X_t} \right] \tag{16.4}$$

The model error covariance matrix P_t is calculated as follows:

$$P_t = \frac{1}{N-1} E_t E_t^T \tag{16.5}$$

where E_t^T is the transposed matrix of E_t, that is, the reflection of E_t over its main diagonal (which runs top left to bottom right). The Kalman gain K can be calculated with

$$K_t = P_t H^T (H P_t H^T + R)^{-1} \tag{16.6}$$

R is the variance of the measurement noise. Knowing the *a priori* error covariance of the model states from Equation 16.5, the state update ensemble X_t^{up} can be obtained by

$$X_{t,i}^{up} = X_{t,i} + K_t \left[Y_{t,i} - \overline{H[X_t]} \right] \tag{16.7}$$

With other words, the EnKF represents the distribution of the system state using random samples. Instead of evolving the covariance matrix of the pdf of the state vector X with the original Kalman filter, the distribution is represented by a collection of realizations, called an ensemble. One advantage of EnKF is that advancing the pdf in time is achieved by simply advancing each member of the ensemble.

16.2.2 Particle Filter

Particle filters are the sequential or on-line analogue of Markov chain Monte Carlo batch methods. With sufficient samples, the so-called particles, they approach the Bayesian optimal estimate. The advantage is that there is no limitation to Gaussian distributed prior pdf of the model states. A particle filter can manage the propagation of non-Gaussian distribution through nonlinear models (Moradkhani et al. 2005a). Particle filters share the same forecast step with EnKF. In the analysis step, the states of the particles remain the same as during the forecast step ($X_{t,i}^{up} = X_{t,i}$), but according to their likelihood to the observation Y_t their weight (probability) $W_{t,i}$ is updated only.

The weights for the SIR-PF are calculated as follows:

$$W_{t,i} = \frac{\exp(-0.5/R)(Y_t - H[X_{t,i}])^2}{\sum\limits_{i=1}^{N}\left(\exp(-0.5/R)(Y_t - H[X_{t,i}])^2\right)} \tag{16.8}$$

The denominator in Equation 16.8 normalizes the weights. After several updates, a few particles may remain with high weights and a large fraction may remain with a weight of zero. To avoid this degeneracy of particles, a resampling step is performed. Particles with low weights have a large chance to be substituted by replicates of particles with high weights, where the probability of a selection is equal to the individual weight. After resampling, the particles are equally weighted. The posterior distribution will be presented as

$$p\left(X_t|Y_t\right) \approx \sum\limits_{i=1}^{N} W_{t,i}\delta(X_t - X_{t,i}) \tag{16.9}$$

The particle replicates may have the same values to be forwarded by the model, which is ineffective. Therefore Moradkhani et al. (2005a) recommend a minor parameter particle perturbation after each assimilation step in order to avoid sample impoverishment. More detailed discussion of the particle filter with different resampling strategies can be found, for example, in the works of Han and Li (2008), Moradkhani (2008), Moradkhani et al. (2005a), Plaza et al. (2012), Reichle (2008), and van Leeuwen (2009).

16.3 Remotely Sensed SSM DA for Soil Moisture Profile Estimation

Most spaceborne microwave sensors provide soil moisture estimates for the top soil layer (1–5 cm). However, information on the moisture condition in the root zone and subsurface layers is more critical for understanding and simulating many hydrologic processes including evapotranspiration and surface runoff (Han et al. 2012a; Vereecken et al. 2008). By the utilization of remote SSM observations in a DA framework containing a hydrological model, it is possible to enhance the predictability of soil moisture at depths greater than the penetration depth of the remote sensor (Flores et al. 2012). Therefore an estimation of the soil moisture profile is generally possible. Walker and Houser (2004) showed that for a good performance of the DA procedure the near-SSM observation error must be less than the required soil moisture forecast error. Another important prerequisite is that the moisture of the near-surface soil layer must not become decoupled from the deep soil layer (Brocca et al. 2012; Flores et al. 2012; Walker et al. 2002). In this case the soil moisture profile cannot be retrieved. The reason for this decoupling is that during extended drying periods a divergence between the drying rates at the soil surface and deeper levels occurs. Moreover, frozen soil conditions manifest in the data as low soil moisture observations, which can have a significant detrimental impact on the assimilation and need to be excluded (Draper et al. 2011).

There are basically two different options to assimilate remotely sensed SSM: (1) direct assimilation of high-level data products, that is, the state variable soil moisture or soil water indices (SWIs), and (2) direct assimilation of (georeferenced) raw or low-level sensor data (brightness temperatures or backscatter), which requires the integration of a forward model or enhanced observation operator to internally calculate soil moisture. These two approaches are discussed next.

16.3.1 High-Level Product Assimilation

The advantage of direct assimilation of high-level products is that it is not necessary to implement an additional SSM retrieval model or a complex observation operator. Here, the assimilation of SSM products (e.g., as obtained by SMOS) or SWI products (e.g., as obtained by ASCAT by the method of Wagner et al. [1999]) is discussed. However, SWI products often need to be simply converted to volumetric soil moisture in order to directly update the state vector.

Das and Mohanty (2006) used SSM retrievals from the electronically scanned thinned array radiometer (ESTAR) for assimilation into the HYDRUS-ET model by the EnKF. HYDRUS-ET was applied in a distributed, column-based way, assuming a homogeneous texture for the whole soil profile. At one site, they were able to verify the hypothesis that SSM observations carry information about the unobserved portion of the state, that is, the profile soil moisture. Moreover, the SSM measurements were particularly important during the dry-down period when the open-loop model may diverge. However, at two other sites they identified the importance of an adequate description of the soil hydraulic properties. For example, a clay layer in the lower part of a soil profile restricted the downward flux in reality and retained higher moisture than in the DA result, where this clay layer was not considered. In the mentioned paper the open-loop simulation was presented for the SSM only, but not for deeper layers. This makes it difficult to identify the impact of the DA procedure for deeper regions of the soil.

A study focusing on the impact of SSM DA on operational agricultural drought monitoring has been published by Bolten et al. (2010). The SSM was retrieved from AMSR-E X-band observations using the method by Jackson (1993). For the assimilation into a water balance model with two soil layers by the EnKF, forest regions of the SSM retrievals were masked. In order to implement a dynamic link between the surface soil and the root zone, a simplistic diffusion term was introduced allowing the model to vertically propagate anomalies of soil moisture from one layer to another. This diffusion capability results in a more realistic vertical water flow and more efficient correction of root zone soil moisture errors during the DA procedure. This study indicates that for soil moisture profile estimation an adequate hydrological model is characterized by the ability to consider cross correlations between different soil layers. However, Bolten et al. (2010) state that the intensity of the link between the layers has minor impact on the root zone soil moisture predictions and simple hydrological models can be used for soil moisture profile estimation in an SSM DA environment. Best results were obtained in grassland regions and also in croplands, where both SSM retrievals and the applied model are known to perform reasonably well. In more complex hydrological models the error correlation between the surface and deep layers has a higher impact. For example, limiting the increase in the soil moisture error correlation with the increase in the vertical distance between the respective layers improves the accuracy of the deeper soil moisture (Zhang et al. 2010).

SSM as well as SWI products have been retrieved from microwave sensors with often changing retrieval accuracy over time (see U_t in Equation 16.2). For example, with ASCAT and SMOS higher-level products this information is provided. Crow and van den Berg (2010) present the theoretical and Draper et al. (2011) show the practical impact of accurate estimation of the observing error parameters in SSM DA for enhanced predictions. Also, Ni-Meister et al. (2006) highlight the need for accurate model and observation error estimates, and Reichle and Koster (2003) investigated the importance of horizontal error correlations.

A way to further reduce the model predictive uncertainty (especially in lower soil regions by SSM DA) is to update not only the states when an observation becomes available but also several time steps before. Dunne and Entekhabi (2006) used a so-called ensemble Kalman smoother (EnKS) to assimilate SSM data obtained by ESTAR into the Noah land surface model. Here, the backward propagation of information from subsequent observations reduced the standard deviation across the ensemble, indicating increased confidence in the smoothed estimates. For SSM simulations the approach performs well during drying conditions, which is, in general, a smoothed process. In contrast, the sudden increase in SSM during precipitation is improperly smoothed out. However, deeper soil layers take an advantage from smoothed SSM assimilation, as here generally smoother soil moisture variability is present. In the deeper soil layers the time lag of recent observations considered in the update process is significant; a time lag of 1 (the most recent observation) is sufficient for SSM DA with the EnKS.

Soil moisture retrieval based on microwave measurements is often hampered by the presence of vegetation cover or topography or influenced by radio frequency interferences. Those regions and accordant pixels have to be excluded from the analysis, which leads to incomplete spatial coverage of the SSM dataset and limitations during the DA procedure (De Lannoy et al. 2009). Therefore Han et al. (2012b) considered spatial correlation characteristics of SSM in order to update also regions where no SSM data were available using a variant of the EnKF with a local analysis scheme in the community land model (CLM). The model estimations at the grid cells without observations were improved depending on observational coverage and the observational distribution over the area under investigation.

Another important issue that needs to be addressed is the alteration of the mass balance during state update. Most applications of SSM DA neglect the mass conservation laws in geophysical applications, which can be critical for some DA applications (Pan and Wood 2006; Ryu et al. 2009). In order to solve this problem, Yilmaz et al. (2012) propose the utilization of unperturbed observations in the EnKF, which leads to a suboptimal DA scheme and biased analysis error covariances. However, they found that suppressing perturbations in the EnKF consistently improves the water budget residual without significantly increasing the state errors. Similarly, Li et al. (2012) introduced a mass conservation updating scheme in the EnKF, which outperforms the conventional EnKF in estimating the deeper soil moisture profile.

16.3.2 Low-Level Product Assimilation

Brightness temperature data obtained by the already mentioned ESTAR system were used by Crow (2003) to correct for land surface model errors associated with poorly retrieved rain rates. Pathmathevan et al. (2003) assimilated ground-based brightness temperature data obtained at three different frequencies into a biophysically based model (simplified biosphere model 2: SiB2) with a variational DA approach. A radiative transfer model serves as an observation operator to calculate the simulated brightness temperature from the simulated soil moisture. They found improved agreement between the DA SSM and independent SSM observations prior to a precipitation event. However, after heavy storms, the DA SSM was found to be worse than open-loop simulations, owing to the poor quality of radio brightness data, which were affected by a water film over the vegetation and soil. The inadequate parameterization (especially soil surface roughness) of the observation operator resulted in biased brightness temperature simulations.

In the root zone, the soil moisture estimation can be improved after a certain initialization time. Failures due to the uncounted horizontal flow and model errors could be partially overcome with the DA process.

Jia et al. (2009) directly assimilate microwave brightness temperature records from descending orbit AMSR-E X-band with a dual-pass DA scheme into the soil water model of CLM. The dual-pass approach was selected in order to first optimize the parameters in the observation operator (i.e., the radiative transfer model) for the whole time period with the shuffled complex evolution method. Second, the resulting optimized soil moisture retrievals were assimilated into half-hourly resolved CLM by the EnKF. Results indicate that assimilating low-level products can significantly improve the SSM estimation and to some extent also the soil moisture estimation at deeper layers. By the optimization of the radiative transfer parameters the uncertainty in the soil moisture retrieval was able to be reduced, which introduces larger flexibility than relying on higher-level SSM products. In a similar setup, Tian et al. (2009) assimilate C- and K-band AMSR-E brightness temperatures with a dual-pass ensemble-based four-dimensional variational DA framework, showing the same propagation of the improvements in the SSM also to lower layers. It has to be considered that the adequacy of the observation operator has large impact on the success of the DA procedure. For example, Mahadevan et al. (2007) showed that for dry surfaces the observation operator needs to consider volume-scattering effects for brightness temperature estimation, which is not necessary for wet surfaces.

Recently, Flores et al. (2012) assimilated simulated SMAP backscatter data into a physics-based hydrologic model by the EnKF. They identified significant improvement in the estimation of soil moisture from the surface through soil depths of 50–75 cm, followed by significant decay in predictive skill at greater depth.

16.4 Scale Issues

In most cases the spatial resolution of remote observations does not correspond to that of the simulation models. Especially sensors designed for global applications provide data in resolutions of several tens of meters, whereas hydrological models often were developed for regional applications. The impacts of resolution of a remotely sensed SSM product on the performance of a DA procedure need to be considered (Parada and Liang 2008). In order to overcome this often neglected scale difference, several studies analyze the performance of scaling techniques prior to assimilation (Merlin et al. 2006). Another option is the utilization of the observation operator for scaling. Both options provide the possibility to integrate additional data for the scaling procedure (Montzka et al. 2012). The importance of an appropriate scaling approach is shown by Draper et al. (2009). They found that the spatial scaling issue has a greater influence than a detailed analysis of the error covariance evolution in an EKF framework.

Merlin et al. (2006) developed a downscaling algorithm to disaggregate synthetic near-SSM SMOS-type data (50-km resolution) to the scale of a distributed SVAT model with a resolution of 1 km. The disaggregation procedure used thermal and optical data remotely sensed at the resolution of the SVAT model to estimate the subpixel variability of microwave soil moisture based on a relationship between the soil skin temperature and SSM. It was found that the assimilation of the downscaled soil moisture improved the spatial distribution of SSM at the time of observation and performed better than the open-loop system or the case where coarse scale microwave soil moisture was assimilated.

Similarly, Vernieuwe et al. (2011) applied first a disaggregation step, which employs a scaling relationship between field-averaged SSM and its corresponding standard deviation. In a second step, SSM states were updated in the TOPLATS model. However, homogeneous conditions were assumed, and more work on the scaling impact of heterogeneous climate variables, topography, vegetation, and soil characteristics on the soil moisture pattern is needed. If micrometeorological, soil texture, and land cover inputs are available at a finer scale, a synthetic experiment conducted by Reichle et al. (2001a) shows that soil moisture can be satisfactorily estimated at scales finer than the resolution of the brightness images. This downscaling experiment indicates that brightness temperature images with a resolution of tens of kilometers can yield soil moisture profile estimates on a scale of a few kilometers. Further multiscale DA studies were recently reviewed by Montzka et al. (2012).

Not only the horizontal scale differences between observation and model needed attention. The reviewed publications about remotely sensed SSM DA in Section 16.3 show that mainly C- and L-band data products have been assimilated. Here, the different sensing depths between the observations and specific vertical model resolutions are often neglected. The determination of the actual sensing depth that is a function of the moisture content, temperature, and texture of the soil is still an area that needs to be investigated. Currently, only rules of thumb can be given for the sensing depth that varies between 0.1 and 0.25 of the wavelength (C, L, and P band with wavelengths of 5.7, 24, and 68 cm, respectively). In a synthetic study Zhang et al. (2005) investigated the impact of the sensing depths 2, 6, and 10 cm. They found that the observation depth does not have an obvious influence on the soil moisture profile retrieval. However, combined multifrequency microwave observations or SSM products provide information yields of the soil moisture content over different depths and may lead to more complex considerations in SSM DA. An assimilation of these datasets may enhance the description of hydrological processes in the vadose zone. The impact of such multifrequency DA approaches is not yet fully explored.

In addition to scale differences in the spatial domain, scale differences in the state space may occur with the presence of a systematic observation bias. For example, for several SSM products such a bias is reported (Lacava et al. 2012; Montzka et al. 2013). This needs to be addressed separately within the DA system, because, in general, it is better to consider the cause of a model bias directly rather than rely on an assimilation to correct it (Draper et al. 2011).

Reichle et al. (2007) found systematic differences in the climatology of AMSR-E and SSMI SSM retrievals. Assimilation of either dataset would result in systematic differences in the model prediction. Matching the cumulative distribution function (CDF) (Drusch et al. 2005) of the SSM products to the model CDF, modest enhancements of the model results compared to *in situ* observations were achieved. Similarly, Kumar et al. (2012) analyzed several approaches for bias mitigation strategies, whereas two *a priori* scaling approaches showed good results: the *a priori* scaling of observations based on standard normal deviates and based on CDF matching. These results were slightly improved by estimating the model parameters during the DA procedure without any bias mitigation (see Section 16.5). However, the CDF matching approach requires a statistical dependency between model predictions and observations. Gao et al. (2007) provide a statistical copula-based approach that offers great flexibility in fitting the marginal distributions and the dependency between datasets.

16.5 Parameter Estimation

The ability to retrieve the vertical soil moisture profile and root zone water content was found to depend strongly on the accurate knowledge of the representative depth of the SSM measurement (Walker et al. 2001), the correct identification of soil hydraulic properties (Lu et al. 2010; Montaldo and Albertson 2003), the correct representation of the physical processes such as root water uptake processes (Hoeben and Troch 2000), and the presence of decoupling phenomena (Walker et al. 2002). However, debate is still ongoing whether, for example, a correct specification of hydraulic parameters or soil water processes descriptions (e.g., Richards equation, simplified Richards equation, and capacity-based models) is more important than, for example, the assimilation frequency of SSM data. Often incorrect specification of soil water processes and parameters may lead to model bias that then needs to be corrected by introducing a bias model (de Lannoy et al. 2007). In addition, it is not clear how accurately hydraulic properties should be estimated to obtain improved retrieval of soil moisture profiles or hydrological fluxes. Das and Mohanty (2006), for example, found that in their study, Kalman filtering could not substantially improve soil moisture profile estimates in cases of incorrect parameterization of soil hydraulic properties.

The major disadvantage of most DA approaches is that they typically focus on state update only. They neglect any potential bias introduced by wrong estimates of the soil hydraulic and other relevant parameters. Recent work on PF methods (Moradkhani et al. 2005b) and methods for simultaneous optimization and DA (Vrugt et al. 2005) paved the way for the estimation of parameters within a DA approach. This may provide an elegant way to reduce the bias introduced by inaccurate parameterization within the model error of the state update equations. The potential of these methods in order to obtain improved forecasts of evapotranspiration and groundwater recharge through assimilation of soil moisture has not yet been explored in more detail. Especially the estimation of unsaturated zone hydraulic properties as a bias mitigation strategy may help to reach this goal. Earlier work of Katul et al. (1993) and Wendroth et al. (1993) used the Kalman filter technique to estimate soil hydraulic properties. Montaldo and Albertson (2003) showed that estimation of the saturated hydraulic conductivity within a multiscale DA approach based on a force-restore SVAT model may substantially reduce the model error in predicting the total root zone soil moisture content obtained from *in situ* measurements. Similarly, Montaldo et al. (2007) removed a systematic bias of a model originating from a bias in saturated hydraulic conductivity after a calibration period. Then, the Kalman filter assumption with zero mean model error was able to be recovered. The introduction of a bias term by De Lannoy et al. (2007) might also be viewed as an attempt to correct for inappropriate model specification or parameterization. Parada and Liang (2004) proposed an extension of the multiscale Kalman filtering technique (Kumar 1999) to assimilate SSM data into a hydrological based land surface model taking into account the spatial dependence and scaling properties of soil moisture but also the time dependence of observation and model errors as new data become available. Assimilating SSM fields derived from ESTAR in the framework of the Southern Great Plains experiment (SGP97), they showed that both short-term predictions of soil moisture and energy fluxes could be significantly improved. Airborne remote sensing data from the same Southern Great Plains site and additionally from a site in Oklahoma were used by Ines and Mohanty (2009) to quantify effective soil hydraulic properties. However, the radiometer SSM retrievals were corrected by the area-average *in situ* SSM in order to correct for uncertainties in the retrieval algorithm, sensor accuracy, geoprojection, etc. The reason for this was that the uncertainties in the soil moisture product can directly propagate to the derived soil hydraulic parameters at the pixel scale.

Kumar et al. (2012) showed that dual state and parameter estimation during SSM DA is able to correct for systematic bias in the observed SSM data. In this special case the bias is transferred from the SSM product to the estimated parameters. The parameters are then valid for this specific situation and should not be used in any other context.

In a synthetic study, Montzka et al. (2011) analyzed the effect of SSM DA for homogeneous clay, loam, silt, and loamy sand soils. They showed that dual state–parameter

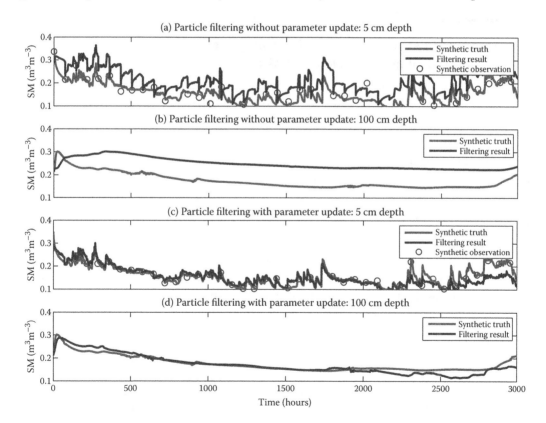

FIGURE 16.3

(See color insert.) Results of a synthetic SSM DA study by Montzka et al. (2011). (a) and (b) show synthetic true soil moisture and state update only simulations as well as synthetic SMOS observations every 3 days. Incorrect parameterization leads to modeled soil moisture drifting away after caught back to the true state. This results in biased modeled soil moisture in deeper soil regions. (c) and (d) show synthetic true and dual state-parameter updated soil moisture simulations. The bias in deeper soil regions is reduced.

update outperforms the single state update (Figure 16.3), especially for longer assimilation intervals. Lower layers, where no observations were available, benefit from SSM observations, because the corrected soil moisture content for the upper layer is propagated downward in the course of time. In loamy sand the SSM conditions have a large impact on the moisture of lower parts of the soil column. This leads to a fast correction of both states and parameters. In loam or clay soils, water is transported much more slowly, that is, a balancing of soil water gradients originating from soil moisture state update needs a longer time. Results indicate that there is a clear relation between the improvement of predictions with the help of DA and the pressure head. However, in real-world applications, the relevant soil parameters change with the soil depth so that Qin et al. (2009) were not able to simulate soil moisture in the root or deeper zone with a high accuracy. Additional results show that the uncertainty of the retrieved soil texture cannot be reduced, while the retrieved soil porosity is confined to a relatively narrow range. In order to improve the encouraging results, SSM products with higher accuracy are needed.

16.6 Improved Estimation of Further Components of the Hydrological Cycle by SSM DA

While the impact of SSM DA for the estimation of the soil moisture profile is reported to be generally positive, it still remains controversial if Earth Observation products for SSM introduce added value for further hydrological applications (Matgen et al. 2012).

16.6.1 Evapotranspiration

Improved SSM conditions may contribute to greater accuracy in simulating evaporation, because evaporation occurs mainly on the soil surface. Transpiration is related to the moisture conditions in the root zone; that is, the improvements of soil moisture profile estimation, as discussed in Section 16.3, may also improve the transpiration description (Alavi et al. 2010; Dunne and Entekhabi 2006).

One of the first attempts to use additional remotely sensed information in order to improve the retrieval of soil moisture and to improve evapotranspiration predictions was done by Pauwels et al. (2007). They used an ensemble Kalman filter within a coupled hydrology-crop growth model to assess the usefulness of measured soil moisture and LAI data in obtaining improved predictions of soil moisture and LAI at the field scale. They concluded that simulated state variables can be improved if both soil moisture and LAI are updated every 2 weeks or more. In addition, assimilation of soil moisture and LAI measurements improved evapotranspiration predictions.

Peters-Lidard et al. (2011) presented an evaluation of the impact of SSM DA in the estimation of evapotranspiration. Among other datasets, the DA integrations employ two different retrievals of the AMSR-E measurements, namely, the NASA Level 3 product (Njoku et al. 2003) and the LPRM product from the Free University of Amsterdam (Owe et al. 2008). In general, the NASA product assimilation led to a degradation of the open-loop evapotranspiration estimates, whereas the LPRM product assimilation showed modest improvements. However, particular improvements of the evapotranspiration were found during spring and summer, when increasing soil moisture stress and insolation can be expected.

Chen et al. (2011) demonstrated in synthetic twin experiments the ability of the EnKF to update simulated SSM leading to a moderate improvement of lower layer soil moisture and evapotranspiration estimates in a 341 km^2 catchment. However, these good results could not be achieved using real-world data. This might be attributed to the setup of the study. *In situ* data from a dense observation network were spatially averaged for the entire catchment and temporally aggregated to daily observations in order to simulate coarse-scale remote sensing data. Moreover, a systematic bias between observations and model predictions was eliminated by linearly rescaling their long-term mean and standard deviation to match those of the multiyear model estimation.

16.6.2 Runoff/Streamflow Prediction

Parajka et al. (2006) as well as Matgen et al. (2012) assimilated SWI data obtained by the ERS-1/-2 scatterometers and ASCAT, respectively. The latter showed that ASCAT SWI data assimilated by a PF were not able to enhance the hydrological response of a catchment in terms of a runoff prediction to rainfall events. One reason for this was that the satellite data were too coarse (25 × 25 km^2) to effectively improve model predictions for a small catchment (10.8 km^2) that only covers part of the satellite footprint. Because of the scale mismatch as well as the measurement uncertainty, ASCAT data were not able to significantly reduce the predictive model uncertainty. Similarly, Brocca et al. (2012) reported for a larger catchment (137 km^2) a small effect of ASCAT SSM DA on flood estimation. In contrast, the assimilation of the ASCAT root zone soil moisture product showed a significant impact on the runoff simulation that provides a clear improvement in the discharge modeling performance. The rescaled ASCAT SWI product valid for the soil surface is able to improve runoff predictions especially when initial soil wetness conditions were unknown (Brocca et al. 2009, 2010).

Simulated SSM has been assimilated to the soil and water assessment tool (SWAT) with the EnKF by Han et al. (2012a) with the aim to answer the question of whether SSM DA is able to improve streamflow prediction. Results show the constraint by the accuracy of precipitation and characteristics of the rainfall-runoff mechanism in the model. Assuming accurate precipitation, the magnitude of errors in soil moisture profile estimation is effected by different land use and soil types. Another application of SWAT and the EnKF is presented by Chen et al. (2011) (for more details, see also Section 16.1). While in a synthetic experiment showing moderate results in evapotranspiration estimation, deeper soil moisture, surface runoff, as well as stream flow predictions were not improved. The results of the real-world experiment were not successful. This was caused by significant underpredictions of the vertical soil water coupling in SWAT in a small catchment with annual precipitation and potential evapotranspiration of 820 and 1850 mm, respectively. In order to provide a method to solve the mentioned problems in runoff estimation after SSM DA, Crow and Ryu (2009) correct not only the soil moisture states in the Sacramento hydrological model but also the precipitation input. The positive impact of remotely sensed SSM is documented in particular for high-flow events and also for large precipitation errors.

16.6.3 Base Flow and Groundwater Recharge

The improvement of the prediction of base flow fluxes was the aim of a synthetic study conducted by Plaza et al. (2012) in a catchment in Luxembourg. The performance assimilating SSM with the PF and state update only is limited, leading to relatively good representation of the modeled soil moisture (with the presence of a bias), whereas the base flow estimation has been degraded. The dual state and parameter estimation is presented as a solution to this shortcoming by correcting the consistency between parameters and soil moisture states. This improves the performance not only in the estimation of soil moisture but also in the influence on base flow.

The use of remotely sensed soil moisture to predict groundwater recharge at larger scales has mainly focused on calibration of hydrological models through adjustment of model parameters (Jackson 2002; Scott et al. 2000). Entekhabi and Moghaddam (2007) outline the potential of spaceborne radars operating at GHz and sub-GHz frequencies to allow direct measurements of SSM in order to provide data to constrain a physically based model approach within a Kalman filtering framework. Despite the potential of remote sensing techniques in providing a high spatial and temporal coverage of SSM, these data have not yet been directly used in an operational manner to predict groundwater recharge or were generally unsuccessful (Chen et al. 2011).

16.7 Conclusions

In this review, the EnKF and the PF were briefly introduced and the state of the art in SSM DA in land surface and hydrological models was presented. The direct assimilation of both higher-level products like SSM and soil water index and low-level microwave sensor data such as brightness temperatures or backscatter is possible. By assimilating higher-level products, there is no need to apply a complex observation operator, as the observation is provided in the same quantity as the state variable of the model. However, it is recommended to consider the quality of the SSM retrieval, which is often provided with the product. The advantage of direct assimilation of low-level products is to flexibly react to local characteristics as well as the presence of a bias. Parameters of the more complex observation operator, for example, a radiative transfer model in passive microwave DA, can be adjusted during the DA procedure. This is often done in a dual-pass scheme, where the parameters are globally optimized before state assimilation. Moreover, technical characteristics of the observation, for example, the antenna pattern, can be easier implemented resulting in improved predictions.

Scaling issues in the assimilation of SSM become more and more important with the growing complexity of models and DA methods. Especially, improved predictions of deeper layer soil moisture as well as evapotranspiration, runoff, etc. require more accurate SSM information where scale differences cannot be neglected. Here, the adequate description of the parameters is important. Dual state–parameter assimilation procedures provide a valuable basis in this respect. Moreover, high accuracy levels of forcing data, in particular, precipitation, are an important prerequisite for advanced prediction of different compartments of the hydrological cycle by SSM DA.

Therefore the following issues need to be addressed by further research in order to improve the utilization of remotely sensed SSM products in DA environments:

- Synthetic studies need to be followed by real-world applications.
- Improved accuracy of SSM products or improved description of error structures in SSM products is necessary.
- Improved forcing, especially precipitation, data for hydrological models are required.
- An application of higher-resolution soil moisture products (e.g., as obtained by radar sensors) in DA studies has not been adequately addressed so far.

References

Alavi, N., A. A. Berg, J. S. Warland, G. Parkin, D. Verseghy, and P. Bartlett. 2010. Evaluating the impact of assimilating soil moisture variability data on latent heat flux estimation in a land surface model. *Canadian Water Resources Journal*, 35(2): 157–172.

Barrett, D. J., and L. J. Renzullo. 2009. On the efficacy of combining thermal and microwave satellite data as observational constraints for root-zone soil moisture estimation. *Journal of Hydrometeorology*, 10(5): 1109–1127, doi:10.1175/2009jhm1043.1.

Bartalis, Z., W. Wagner, V. Naeimi, S. Hasenauer, K. Scipal, H. Bonekamp, J. Figa, and C. Anderson. 2007. Initial soil moisture retrievals from the METOP-A Advanced Scatterometer (ASCAT). *Geophysical Research Letters*, 34(20): L20401, doi:10.1029/2007gl091088.

Beven, K., and J. Freer. 2001. Equifinality, data assimilation, and uncertainty estimation in mechanistic modelling of complex environmental systems using the GLUE methodology. *Journal of Hydrology*, 249(1–4): 11–29.

Bolten, J. D., W. T. Crow, X. W. Zhan, T. J. Jackson, and C. A. Reynolds. 2010. Evaluating the utility of remotely sensed soil moisture retrievals for operational agricultural drought monitoring. *IEEE Journal of Selected Topics in Applied Earth Observations and Remote Sensing*, 3(1): 57–66, doi:10.1109/Jstars.2009.2037163.

Brocca, L., F. Melone, T. Moramarco, and V. P. Singh. 2009. Assimilation of observed soil moisture data in storm rainfall-runoff modeling. *Journal of Hydrologic Engineers*, 14(2): 153–165, doi:10.1061/(Asce)1084-0699(2009)14:2(153).

Brocca, L., F. Melone, T. Moramarco, W. Wagner, V. Naeimi, Z. Bartalis, and S. Hasenauer. 2010. Improving runoff prediction through the assimilation of the ASCAT soil moisture product. *Hydrology and Earth System Sciences*, 14(10): 1881–1893, doi:10.5194/hess-14-1881-2010.

Brocca, L., T. Moramarco, F. Melone, W. Wagner, S. Hasenauer, and S. Hahn. 2012. Assimilation of surface- and root-zone ASCAT soil moisture products into rainfall–runoff modeling. *IEEE Transactions on Geoscience and Remote Sensing*, 50(7): 2542–2555, doi:10.1109/TGRS.2011.2177468.

Burgers, G., P. J. van Leeuwen, and G. Evensen. 1998. Analysis scheme in the ensemble Kalman filter. *Monthly Weather Review*, 126(6): 1719–1724.

Calvet, J. C., and J. Noilhan. 2000. From near-surface to root-zone soil moisture using year-round data. *Journal of Hydrometeorology*, 1(5): 393–411.

Chen, F., W. T. Crow, P. J. Starks, and D. N. Moriasi. 2011. Improving hydrologic predictions of a catchment model via assimilation of surface soil moisture. *Advances in Water Resources*, 34(4): 526–536, doi:10.1016/j.advwatres.2011.01.011.

Crosson, W. L., C. A. Laymon, R. Inguva, and M. P. Schamschula. 2002. Assimilating remote sensing data in a surface flux-soil moisture model. *Hydrological Processes*, 16(8): 1645–1662.

Crow, W. T. 2003. Correcting land surface model predictions for the impact of temporally sparse rainfall rate measurements using an ensemble Kalman filter and surface brightness temperature observations. *Journal of Hydrometeorology*, 4(5): 960–973.

Crow, W. T., and D. Ryu. 2009. A new data assimilation approach for improving runoff prediction using remotely-sensed soil moisture retrievals. *Hydrology and Earth System Sciences*, 13(1): 1–16.

Crow, W. T., and M. J. van den Berg. 2010. An improved approach for estimating observation and model error parameters in soil moisture data assimilation. *Water Resources Research*, 46, doi:10.1029/2010wr009402.

Crow, W. T., and E. Van Loon. 2006. Impact of incorrect model error assumptions on the sequential assimilation of remotely sensed surface soil moisture. *Journal of Hydrometeorology*, 7(3): 421–432.

Das, N. N., and B. P. Mohanty. 2006. Root zone soil moisture assessment using remote sensing and vadose zone modeling. *Vadose Zone Journal*, 5(1): 296–307.

De Lannoy, G. J. M., P. R. Houser, N. E. C. Verhoest, and V. R. N. Pauwels. 2009. Adaptive soil moisture profile filtering for horizontal information propagation in the independent column-based CLM2.0. *Journal of Hydrometeorology*, 10(3): 766–779, doi:10.1175/2008jhm1037.1.

De Lannoy, G. J. M., R. H. Reichle, P. R. Houser, V. R. N. Pauwels, and N. E. C. Verhoest. 2007. Correcting for forecast bias in soil moisture assimilation with the ensemble Kalman filter. *Water Resources Research*, 43(9): W09410, doi:10.1029/2006wr005449.

Dorigo, W. A., R. A. M. de Jeu, D. Chung, R. M. Parinussa, Y. Y. Liu, W. Wagner, and D. Fernandez-Prieto. 2012. Evaluating global trends (1988–2010) in harmonized multi-satellite surface soil moisture. *Geophysical Research Letters*, 39: L18405, doi:10.1029/2012GL052988.

Draper, C., J. F. Mahfouf, J. C. Calvet, E. Martin, and W. Wagner. 2011. Assimilation of ASCAT near-surface soil moisture into the SIM hydrological model over France. *Hydrology and Earth System Sciences*, 15(12): 3829–3841, doi:10.5194/hess-15-3829-2011.

Draper, C. S., J. F. Mahfouf, and J. P. Walker. 2009. An EKF assimilation of AMSR-E soil moisture into the ISBA land surface scheme. *Journal of Geophysical Research*, 114, doi:10.1029/2008jd011650.

Drusch, M., E. F. Wood, and H. Gao. 2005. Observation operators for the direct assimilation of TRMM microwave imager retrieved soil moisture. *Geophysical Research Letters*, 32(15): L15403, doi:10.1029/2005gl023623.

Dunne, S., and D. Entekhabi. 2006. Land surface state and flux estimation using the ensemble Kalman smoother during the Southern Great Plains 1997 field experiment. *Water Resources Research*, 42(1): W01407, doi:10.1029/2005wr004334.

Entekhabi, D., and M. Moghaddam. 2007. Mapping recharge from space: Roadmap to meeting the grand challenge. *Hydrogeology Journal*, 15(1): 105–116, doi:10.1007/s10040-006-0120-6.

Evensen, G. 1994. Sequential data assimilation with a nonlinear quasi-geostrophic model using Monte-Carlo methods to forecast error Statistics. *Journal of Geophysical Research*, 99(C5): 10,143–10,162.

Flores, A. N., R. L. Bras, and D. Entekhabi. 2012. Hydrologic data assimilation with a hillslope-scale-resolving model and L band radar observations: Synthetic experiments with the ensemble Kalman filter. *Water Resources Research*, 48: W08509, doi:10.1029/2011WR011500.

Galantowicz, J. F., D. Entekhabi, and E. G. Njoku. 1999. Tests of sequential data assimilation for retrieving profile soil moisture and temperature from observed L-band radiobrightness. *IEEE Transactions on Geoscience and Remote Sensing*, 37(4): 1860–1870.

Gao, H. L., E. F. Wood, M. Drusch, and M. F. McCabe. 2007. Copula-derived observation operators for assimilating TMI and AMSR-E retrieved soil moisture into land surface models. *Journal of Hydrometeorology*, 8(3): 413–429, doi:10.1175/Jhm570.1.

Global Climate Observing System (GCOS). 2010. Implementation plan for the global observing system for climate in support of the UNFCCC - 2010 Update Rep.

Gordon, N. J., D. J. Salmond, and A. F. M. Smith. 1993. Novel-approach to nonlinear non-Gaussian Bayesian state estimation. *IEE Proceedings-F Radar and Signal Processing*, 140(2): 107–113.

Hajnsek, I., T. Jagdhuber, H. Schön, and K. P. Papathanassiou. 2009. Potential of estimating soil moisture under vegetation cover by means of PolSAR. *IEEE Transactions on Geoscience and Remote Sensing*, 47(2): 442–454.

Han, E. J., V. Merwade, G. C. and Heathman. 2012a. Implementation of surface soil moisture data assimilation with watershed scale distributed hydrological model. *Journal of Hydrology*, 416: 98–117, doi:10.1016/j.jhydrol.2011.11.039.

Han, X., and X. Li. 2008. An evaluation of the nonlinear/non-Gaussian filters for the sequential data assimilation. *Remote Sensing of Environment*, 112(4): 1434–1449.

Han, X., X. Li, H. J. H. Franssen, H. Vereecken, and C. Montzka. 2012b. Spatial horizontal correlation characteristics in the land data assimilation of soil moisture. *Hydrology and Earth System Sciences*, 16(5): 1349–1363, doi:10.5194/hess-16-1349-2012.

Hoeben, R., and P. A. Troch. 2000. Assimilation of active microwave measurements for soil moisture profile retrieval under laboratory conditions. In *Proceedings of International Geoscience and Remote Sensing Symposium*, vol. I–VI, IEEE, New Brunswick, NJ, pp. 1271–1273.

Houser, P. R., W. J. Shuttleworth, J. S. Famiglietti, H. V. Gupta, K. H. Syed, and D. C. Goodrich. 1998. Integration of soil moisture remote sensing and hydrologic modeling using data assimilation. *Water Resources Research*, 34(12): 3405–3420.

Ines, A. V. M., and B. P. Mohanty. 2009. Near-surface soil moisture assimilation for quantifying effective soil hydraulic properties using genetic algorithms: 2. Using airborne remote sensing during SGP97 and SMEX02. *Water Resources Research*, 45.

Jackson, T. J. 1993. Measuring surface soil-moisture using passive microwave remote-sensing. 3. *Hydrological Processes*, 7(2): 139–152.

Jackson, T. J. 2002. Remote sensing of soil moisture: Implications for groundwater recharge. *Hydrogeology Journal*, 10(1): 40–51, doi:10.1007/s10040-001-0168-2.

Jia, B. H., Z. H. Xie, X. J. Tian, and C. X. Shi. 2009. A soil moisture assimilation scheme based on the ensemble Kalman filter using microwave brightness temperature. *Science China Earth Sciences, Series D*, 52(11): 1835–1848, doi:10.1007/s11430-009-0122-z.

Kalman, R. E. 1960. A new approach to linear filtering and prediction problems. *Journal of Basic Engineering*, 82(1): 35–45.

Katul, G. G., O. Wendroth, M. B. Parlange, C. E. Puente, M. V. Folegatti, and D. R. Nielsen. 1993. Estimation of in situ hydraulic conductivity function from nonlinear filtering theory. *Water Resources Research*, 29(4): 1063–1070.

Kerr, Y. H., P. Waldteufel, J. P. Wigneron, S. Delwart, F. Cabot, J. Boutin, M. J. Escorihuela et al. 2010. The SMOS mission: New tool for monitoring key elements of the global water cycle. *Proceedings of the IEEE*, 98(5): 666–687.

Kumar, P. 1999. A multiple scale state-space model for characterizing subgrid scale variability of near-surface soil moisture. *IEEE Transactions on Geoscience and Remote Sensing*, 37(1): 182–197.

Kumar, S. V., R. H. Reichle, K. W. Harrison, C. D. Peters-Lidard, S. Yatheendradas, and J. A. Santanello. 2012. A comparison of methods for a priori bias correction in soil moisture data assimilation. *Water Resources Research*, 48: W03515, doi:10.1029/2010wr010261.

Lacava, T., P. Matgen, L. Brocca, M. Bittelli, N. Pergola, T. Moramarco, and V. Tramutoli. 2012. A first assessment of the SMOS soil moisture product with in situ and modeled data in Italy and Luxembourg. *IEEE Transactions on Geoscience and Remote Sensing*, 50(2): 1612–1622.

Li, B., D. Toll, X. Zhan, and B. Cosgrove. 2012. Improving estimated soil moisture fields through assimilation of AMSR-E soil moisture retrievals with an ensemble Kalman filter and a mass conservation constraint. *Hydrology and Earth System Sciences*, 16(1): 105–119, doi:10.5194/hess-16-105-2012.

Li, F. Q., W. T. Crow, and W. P. Kustas. 2010. Towards the estimation root-zone soil moisture via the simultaneous assimilation of thermal and microwave soil moisture retrievals. *Advances in Water Resources*, 33(2): 201–214.

Li, J., and S. Islam. 1999. On the estimation of soil moisture profile and surface fluxes partitioning from sequential assimilation of surface layer soil moisture. *Journal of Hydrology*, 220(1–2): 86–103.

Lu, H. S., X. L. Li, Z. B. Yu, R. Horton, Y. H. Zhu, Z. C. Hao, and L. Xiang. 2010. Using a H-infinity filter assimilation procedure to estimate root zone soil water content. *Hydrological Processes*, 24(25): 3648–3660, doi:10.1002/Hyp.7778.

Mahadevan, P., T. Koike, H. Fujii, K. Tamagawa, X. Li, and I. Kaihotsu. 2007. Modification and application of the satellite-based land data assimilation scheme for very dry soil regions using AMSR-E images: Model validation for Mongolia—A CEOP data platform. *Journal of Meteorological Society of Japan*, 85A: 243–260.

Margulis, S. A., D. McLaughlin, D. Entekhabi, and S. Dunne. 2002. Land data assimilation and estimation of soil moisture using measurements from the Southern Great Plains 1997 Field Experiment. *Water Resources Research*, 38(12).

Matgen, P., F. Fenicia, S. Heitz, D. Plaza, R. de Keyser, V. R. N. Pauwels, W. Wagner, and H. Savenije. 2012. Can ASCAT-derived soil wetness indices reduce predictive uncertainty in well-gauged areas? A comparison with in situ observed soil moisture in an assimilation application. *Advances in Water Resources*, 44: 49–65, doi:10.1016/j.advwatres.2012.03.022.

Merlin, O., A. Chehbouni, Y. H. Kerr, and D. C. Goodrich. 2006. A downscaling method for distributing surface soil moisture within a microwave pixel: Application to the Monsoon '90 data. *Remote Sensing of Environment*, 101(3): 379–389.

Montaldo, N., and J. D. Albertson. 2003. Temporal dynamics of soil moisture variability: 2. Implications for land surface models. *Water Resources Research*, 39(10): 1275.

Montaldo, N., J. D. Albertson, and M. Mancini. 2007. Dynamic calibration with an ensemble Kalman filter based data assimilation approach for root-zone moisture predictions. *Journal of Hydrometeorology*, 8(4): 910–921, doi:10.1175/Jhm582.1.

Montaldo, N., J. D. Albertson, M. Mancini, and G. Kiely. 2001. Robust simulation of root zone soil moisture with assimilation of surface soil moisture data. *Water Resources Research*, 37(12): 2889–2900.

Montzka, C., H. R. Bogena, L. Weihermüller, F. Jonard, C. Bouzinac, J. Kainulainen, J. E. Balling et al. 2013. Brightness temperature and soil moisture validation at different scales during the SMOS Validation Campaign in the Rur and Erft catchments, Germany. *IEEE Transactions on Geoscience and Remote Sensing*, 51(3): 1728–1743, doi:10.1109/TGRS.2012.2206031.

Montzka, C., H. Moradkhani, L. Weihermüller, H.-J. Hendricks Franssen, M. Canty, and H. Vereecken. 2011. Hydraulic parameter estimation by remotely-sensed top soil moisture observations with the particle filter. *Journal of Hydrology*, 399(3–4): 410–421, doi:10.1016/j.jhydrol.2011.01.020.

Montzka, C., V. R. N. Pauwels, X. Han, H.-J. Hendricks Franssen, and H. Vereecken. 2012. Multiscale and multivariate data assimilation in terrestrial systems: A review. *Sensors*, 12: 16,291–16,333, doi:10.3390/s121216291.

Moradkhani, H. 2008. Hydrologic remote sensing and land surface data assimilation. *Sensors*, 8(5): 2986–3004.

Moradkhani, H., K. L. Hsu, H. Gupta, and S. Sorooshian. 2005a. Uncertainty assessment of hydrologic model states and parameters: Sequential data assimilation using the particle filter. *Water Resources Research*, 41(5): W05012, doi:10.1029/2004WR003604.

Moradkhani, H., and S. Sorooshian. 2008. General review of rainfall-runoff modeling: Model calibration, data assimilation, and uncertainty analysis. In *Hydrological Modelling and the Water Cycle—Coupling the Atmospheric and Hydrological Models*, S. Sorooshian et al. (Eds.). Springer, Berlin, pp. 1–24.

Moradkhani, H., S. Sorooshian, H. V. Gupta, and P. R. Houser. 2005b. Dual state-parameter estimation of hydrological models using ensemble Kalman filter. *Advances in Water Resources*, 28(2): 135–147.

Moran, M. S., C. D. Peters-Lidard, J. M. Watts, and S. McElroy. 2004. Estimating soil moisture at the watershed scale with satellite-based radar and land surface models. *Canadian Journal of Remote Sensing*, 30(5): 805–826.

Munoz-Sabater, J., A. Fouilloux, and P. de Rosnay. 2012. Technical implementation of SMOS data in the ECMWF Integrated Forecasting System. *IEEE Geoscience and Remote Sensing Letters*, 9(2): 252–256, doi:10.1109/Lgrs.2011.2164777.

Ni-Meister, W. 2008. Recent advances on soil moisture data assimilation. *Physical Geography*, 29(1): 19–37.

Ni-Meister, W., P. R. Houser, and J. P. Walker. 2006. Soil moisture initialization for climate prediction: Assimilation of scanning multifrequency microwave radiometer soil moisture data into a land surface model. *Journal of Geophysical Research*, 111(D20): D20102, doi:10.1029/2006jd007190.

Njoku, E. G., T. J. Jackson, V. Lakshmi, T. K. Chan, and S. V. Nghiem. 2003. Soil moisture retrieval from AMSR-E. *IEEE Transactions on Geoscience and Remote Sensing*, 41(2): 215–229, doi:10.1109/Tgrs.2002.808243.

Ottle, C., and D. Vidalmadjar. 1994. Assimilation of soil-moisture inferred from infrared remote-sensing in a hydrological model over the Hapex-Mobilhy Region. *Journal of Hydrology*, 158(3–4): 241–264.

Owe, M., R. de Jeu, and T. Holmes. 2008. Multisensor historical climatology of satellite-derived global land surface moisture. *Journal of Geophysical Research*, 113(F1), doi:10.1029/2007jf000769.

Pan, M., and E. F. Wood. 2006. Data assimilation for estimating the terrestrial water budget using a constrained ensemble Kalman filter. *Journal of Hydrometeorology*, 7(3): 534–547.

Parada, L. M., and X. Liang. 2004. Optimal multiscale Kalman filter for assimilation of near-surface soil moisture into land surface models. *Journal of Geophysical Research*, 109(D24).

Parada, L. M., and X. Liang. 2008. Impacts of spatial resolutions and data quality on soil moisture data assimilation. *Journal of Geophysical Research-Atmospheres*, 113(D10): D24109.

Parajka, J., V. Naeimi, G. Bloschl, W. Wagner, R. Merz, and K. Scipal. 2006. Assimilating scatterometer soil moisture data into conceptual hydrologic models at the regional scale. *Hydrology and Earth System Sciences*, 10(3): 353–368.

Pathmathevan, M., T. Koike, X. Li, and H. Fujii. 2003. A simplified land data assimilation scheme and its application to soil moisture experiments in 2002 (SMEX02). *Water Resources Research*, 39(12): 1341, doi:10.1029/2003wr002124.

Pauwels, V. R. N., R. Hoeben, N. E. C. Verhoest, and F. P. De Troch. 2001. The importance of the spatial patterns of remotely sensed soil moisture in the improvement of discharge predictions for small-scale basins through data assimilation. *Journal of Hydrology*, 251(1–2): 88–102.

Pauwels, V. R. N., R. Hoeben, N. E. C. Verhoest, F. P. De Troch, and P. A. Troch. 2002. Improvement of TOPLATS-based discharge predictions through assimilation of ERS-based remotely sensed soil moisture values. *Hydrological Processes*, 16(5): 995–1013.

Pauwels, V. R. N., N. E. C. Verhoest, G. J. M. De Lannoy, V. Guissard, C. Lucau, and P. Defourny. 2007. Optimization of a coupled hydrology-crop growth model through the assimilation of observed soil moisture and leaf area index values using an ensemble Kalman filter. *Water Resources Research*, 43(4): W04421, doi:10.1029/2006wr004942.

Peters-Lidard, C. D., S. V. Kumar, D. M. Mocko, and Y. D. Tian. 2011. Estimating evapotranspiration with land data assimilation systems. *Hydrological Processes*, 25(26): 3979–3992, doi:10.1002/Hyp.8387.

Plaza, D. A., R. De Keyser, G. J. M. De Lannoy, L. Giustarini, P. Matgen, and V. R. N. Pauwels. 2012. The importance of parameter resampling for soil moisture data assimilation into hydrologic models using the particle filter. *Hydrology and Earth System Sciences*, 16(2): 375–390, doi:10.5194/hess-16-375-2012.

Qin, J., S. L. Liang, K. Yang, I. Kaihotsu, R. G. Liu, and T. Koike. 2009. Simultaneous estimation of both soil moisture and model parameters using particle filtering method through the assimilation of microwave signal. *Journal of Geophysical Research*, 114: D15103.

Reichle, R. H. 2008. Data assimilation methods in the Earth sciences. *Advances in Water Resources*, 31(11): 1411–1418.

Reichle, R. H., D. Entekhabi, and D. B. McLaughlin. 2001a. Downscaling of radio brightness measurements for soil moisture estimation: A four-dimensional variational data assimilation approach. *Water Resources Research*, 37(9): 2353–2364.

Reichle, R. H., and R. D. Koster. 2003. Assessing the impact of horizontal error correlations in background fields on soil moisture estimation. *Journal of Hydrometeorology*, 4(6): 1229–1242.

Reichle, R. H., R. D. Koster, P. Liu, S. P. P. Mahanama, E. G. Njoku, and M. Owe. 2007. Comparison and assimilation of global soil moisture retrievals from the Advanced Microwave Scanning Radiometer for the Earth Observing System (AMSR-E) and the Scanning Multichannel Microwave Radiometer (SMMR). *Journal of Geophysical Research*, 112(D9): D09108.

Reichle, R. H., D. B. McLaughlin, and D. Entekhabi. 2001b. Variational data assimilation of microwave radiobrightness observations for land surface hydrology applications. *IEEE Transactions on Geoscience and Remote Sensing*, 39(8): 1708–1718.

Reichle, R. H., D. B. McLaughlin, and D. Entekhabi. 2002a. Hydrologic data assimilation with the ensemble Kalman filter. *Monthly Weather Review*, 130(1): 103–114.

Reichle, R. H., J. P. Walker, R. D. Koster, and P. R. Houser. 2002b. Extended versus ensemble Kalman filtering for land data assimilation. *Journal of Hydrometeorology*, 3(6): 728–740.

Ryu, D., W. T. Crow, X. W. Zhan, and T. J. Jackson. 2009. Correcting unintended perturbation biases in hydrologic data assimilation. *Journal of Hydrometeorology*, 10(3): 734–750, doi:10.1175/2008jhm1038.1.

Saleh, K., Y. H. Kerr, P. Richaume, M. J. Escorihuela, R. Panciera, S. Delwart, G. Boulet et al. 2009. Soil moisture retrievals at L-band using a two-step inversion approach (COSMOS/NAFE'05 experiment). *Remote Sensing of Environment*, 113(6): 1304–1312.

Scipal, K., M. Drusch, and W. Wagner. 2008. Assimilation of a ERS scatterometer derived soil moisture index in the ECMWF numerical weal-her prediction system. *Advances in Water Resources*, 31(8): 1101–1112, doi:10.1016/j.advwatres.2008.04.013.

Scott, R. L., W. J. Shuttleworth, T. O. Keefer, and A. W. Warrick. 2000. Modeling multiyear observations of soil moisture recharge in the semiarid American Southwest. *Water Resources Research*, 36(8): 2233–2247.

Takada, M., Y. Mishima, and S. Natsume. 2009. Estimation of surface soil properties in peatland using ALOS/PALSAR. *Landscape and Ecological Engineering*, 5(1): 45–58.

Teuling, A. J., R. Uijlenhoet, F. Hupet, E. E. van Loon, and P. A. Troch. 2006. Estimating spatial mean root-zone soil moisture from point-scale observations. *Hydrology and Earth System Sciences*, 10(5): 755–767.

Tian, X. J., Z. H. Xie, A. Dai, C. X. Shi, B. H. Jia, F. Chen, and K. Yang. 2009. A dual-pass variational data assimilation framework for estimating soil moisture profiles from AMSR-E microwave brightness temperature. *Journal of Geophysical Research*, 114: D16102, doi:10.1029/2008jd011600.

van Leeuwen, P. J. 2009. Particle filtering in geophysical systems. *Monthly Weather Review*, 137(12): 4089–4114.

Vereecken, H., J. A. Huisman, H. Bogena, J. Vanderborght, J. A. Vrugt, and J. W. Hopmans. 2008. On the value of soil moisture measurements in vadose zone hydrology: A review. *Water Resources Research*, 44: W00D06.

Vernieuwe, H., B. De Baets, J. Minet, V. R. N. Pauwels, S. Lambot, M. Vanclooster, and N. E. C. Verhoest. 2011. Integrating coarse-scale uncertain soil moisture data into a fine-scale hydrological modelling scenario. *Hydrology and Earth System Sciences*, 15(10): 3101–3114, doi:10.5194/hess-15-3101-2011.

Vrugt, J. A., C. G. H. Diks, H. V. Gupta, W. Bouten, and J. M. Verstraten. 2005. Improved treatment of uncertainty in hydrologic modeling: Combining the strengths of global optimization and data assimilation. *Water Resources Research*, 41(1): W01017.

Wagner, W., G. Bloschl, P. Pampaloni, J. C. Calvet, B. Bizzarri, J. P. Wigneron, and Y. Kerr. 2007. Operational readiness of microwave remote sensing of soil moisture for hydrologic applications. *Nordic Hydrology*, 38(1): 1–20.

Wagner, W., G. Lemoine, and H. Rott. 1999. A method for estimating soil moisture from ERS scatterometer and soil data. *Remote Sensing of Environment*, 70(2): 191–207, doi:10.1016/S0034-4257(99)00036-X.

Walker, J. P., and P. R. Houser. 2004. Requirements of a global near-surface soil moisture satellite mission: Accuracy, repeat time, and spatial resolution. *Advances in Water Resources*, 27(8): 785–801.

Walker, J. P., G. R. Willgoose, and J. D. Kalma. 2001. The Nerrigundah dataset: Soil moisture patterns, soil characteristics, and hydrological flux measurements. *Water Resources Research*, 37(11): 2653–2658.

Walker, J. P., G. R. Willgoose, and J. D. Kalma. 2002. Three-dimensional soil moisture profile retrieval by assimilation of near-surface measurements: Simplified Kalman filter covariance forecasting and field application. *Water Resources Research*, 38(12): 1301.

Wendroth, O., G. G. Katul, M. B. Parlange, C. E. Puente, and D. R. Nielsen. 1993. A nonlinear filtering approach for determining hydraulic conductivity functions in-field soils. *Soil Science*, 156(5): 293–301.

Yilmaz, M. T., T. DelSole, and P. R. Houser. 2012. Reducing water imbalance in land data assimilation: Ensemble filtering without perturbed observations. *Journal of Hydrometeorology*, 13(1): 413–420, doi:10.1175/Jhm-D-11-010.1

Zhang, S. W., H. R. Li, W. D. Zhang, C. J. Qiu, and X. Li. 2005. Estimating the soil moisture profile by assimilating near-surface observations with the ensemble Kalman filter (EnKF). *Advances in Atmospheric Sciences*, 22(6): 936–945.

Zhang, S. W., X. B. Zeng, W. D. Zhang, and M. Barlage. 2010. Revising the ensemble-based Kalman filter covariance for the retrieval of deep-layer soil moisture. *Journal of Hydrometeorology*, 11(1): 219–227, doi:10.1175/2009jhm1146.1.

Zhou, Y. H., D. McLaughlin, and D. Entekhabi. 2006. Assessing the performance of the ensemble Kalman filter for land surface data assimilation. *Monthly Weather Review*, 134(8): 2128–2142.

17

Scaling and Filtering Approaches for the Use of Satellite Soil Moisture Observations

Luca Brocca, Florisa Melone, Tommaso Moramarco,
Wolfgang Wagner, and Clement Albergel

CONTENTS

17.1 Introduction

Soil moisture observations from satellite sensors are becoming widely available at the global scale with a good accuracy and a spatial–temporal resolution suitable for many applications (Wagner et al. 2007; Seneviratne et al. 2010). However, the climatology of satellite-derived and modeled/*in situ* soil moisture observations can be very different because of uncertainties affecting both data sources (Koster et al. 2009; Entekhabi et al. 2010). Moreover, the difference between the spatial extent of *in situ* (~0.1 m²) and coarse-resolution satellite (~600 km²) observations is very large. Because of the high soil moisture spatial variability (Famiglietti et al. 2008; Brocca et al. 2010b), point measurements are expected to be not representative of the actual absolute value (i.e., in $m^3\ m^{-3}$) of average soil moisture at the satellite pixel scale. Anyhow, the well-known scaling properties of soil moisture (Vachaud et al. 1985; Brocca et al. 2012b) have showed that *in situ* measurements can capture the large-scale temporal dynamics. Another matter is that satellite sensors are able to monitor only a very thin soil layer (less than 5 cm), providing soil moisture information that is difficult to be assimilated into hydrological and meteorological models (Brocca et al. 2010a, 2012a; Dharssi et al. 2011).

For all these reasons, satellite data usually are required to be scaled and/or filtered before their actual use within hydrological or meteorological models (e.g., Scipal et al. 2008). Previous studies have adopted different approaches for scaling/filtering satellite data for the comparison with *in situ* observations (Brocca et al. 2011; Albergel et al. 2010, 2012) as well as for their use in modeling approaches (Dharssi et al. 2011; Draper et al. 2011; Brocca et al. 2012a). Likely, the most widely used scaling technique is the cumulative distribution function (CDF) matching approach (e.g., Drusch et al. 2005) that allows matching the complete CDF of satellite and *in situ* observations by applying a nonlinear operator. However, the stability of the CDF matching operator parameters over time could not be assured and only a few studies addressed this issue (Matgen et al. 2012). Several studies employed linear techniques (Brocca et al. 2010a; Jackson et al. 2010; Dharssi et al. 2011) that are expected to be more robust as they require a lower number of parameters.

As regards filtering approaches, an important contribution was given by Wagner et al. (1999), who suggested the use of an exponential filter to turn the time series of surface measurements into a signal that is able to capture the dynamics of the lower soil layer. In recent years, this approach has been tested with both simulated and measured data, providing good results, and has been extensively used to improve the description of the root zone soil moisture in rainfall/runoff applications (e.g., Manfreda et al. 2011; Brocca et al. 2011, 2012; Matgen et al. 2012). Other filtering methods include more physically based approaches (Manfreda et al. 2012) or the simple use of the moving average (Draper et al. 2009). Notwithstanding, for most of works in the scientific literature that have employed scaling and filtering techniques, neither their intercomparison nor a detailed validation has been still carried out.

The main purpose of this chapter is, first, to review the different approaches that have been developed in the scientific literature for scaling/filtering satellite data. Successively, their evaluation and intercomparison are carried out by using satellite and *in situ* soil moisture data. Specifically, satellite observations are obtained through the Advanced Scatterometer (ASCAT) and the Advanced Microwave Scanning Radiometer (AMSR) sensors by using the TUWIEN change detection (Wagner et al. 1999) and the Land Parameter Retrieval Model (LPRM; Owe et al. 2008) algorithms, respectively. In principle, the scaling and filtering techniques can be also applied to high-resolution satellite time series, thus reducing the scale differences with *in situ* observations. Anyhow, we employed coarse-scale satellite imagery, as, because of its high time resolution, it is better suited to the temporal analysis performed in this chapter. Long-term (2007–2010) and high-quality *in situ* data for three locations across the world, characterized by different climatic and physiographic conditions, are employed as case studies.

17.2 Scaling and Filtering Techniques Methods: Overview

This section presents the scaling and filtering techniques commonly used in the comparison between satellite and *in situ* soil moisture time series. We note that the same approaches are also used when satellite data have to be employed within modeling approaches (Dharssi et al. 2011; Brocca et al. 2012a; de Rosnay et al. 2013), for instance, through data assimilation techniques.

17.2.1 Scaling Techniques

Systematic differences between remote sensing–derived and site-specific data of soil moisture prevent an absolute agreement between the two time series. Consequently, comparison of the remotely sensed and site-specific time series is often aided by normalizing the remotely sensed data to match the distribution of ground data. In this study, four different techniques are reviewed: (1) linear regression correction, (2) CDF matching approach, (3) linear rescaling, and (4) Min/Max correction.

17.2.1.1 Linear Regression

The linear regression method is based on the application of a regression equation between satellite, SAT, and *in situ* observed, OBS, time series:

$$SAT_{REG} = a\,SAT + b \tag{17.1}$$

where SAT_{REG} is the regression corrected satellite time series and a and b are parameters obtained through the least squares method.

17.2.1.2 CDF Matching

The CDF matching approach (e.g., Reichle and Koster 2004; Drusch et al. 2005) can be considered as an enhanced nonlinear technique for removing systematic differences between two datasets. Through this method, the satellite time series is rescaled in such a way that its CDF matches the CDF of *in situ* measurements. There are several ways to apply this method. Among them, one is based on, first, the computation of the differences in soil moisture values between the corresponding elements of each ranked dataset. Then, these differences are plotted against the satellite data and a fitting function f is used to calculate the CDF-corrected soil moisture datasets, SAT_{CDF} (Brocca et al. 2011):

$$SAT_{CDF} = SAT + f(SAT) \tag{17.2}$$

In this study, a fifth-order polynomial function is used for f as it was found suitable to fit the relationship between the ranked differences and the satellite data.

17.2.1.3 Linear Rescaling

In the linear rescaling approach (Draper et al. 2009; Brocca et al. 2010a), satellite time series are forced to have the same mean μ and standard deviation σ of *in situ* time series:

$$SAT_{RES} = \frac{\left[SAT - \mu(SAT)\right]}{\sigma(SAT)}\sigma(OBS) + \mu(OBS) \tag{17.3}$$

where SAT_{RES} is the linearly rescaled satellite time series.

17.2.1.4 Min/Max Correction

For the application of the Min/Max correction method, the maximum and minimum values of *in situ* data are chosen to define the upper and lower values of the satellite time series through the following equation:

$$SAT_{MinMax} = \frac{\left[SAT - min(SAT)\right]}{\left[max(SAT) - min(SAT)\right]}\left[max(OBS) - min(OBS)\right] + min(OBS) \quad (17.4)$$

where SAT_{MinMax} is the scaled satellite time series with the Min/Max approach and min( ) and max( ) are the minimum and maximum operators, respectively. Therefore satellite and *in situ* observations have the same maximum and minimum values. We note that the quality of *in situ* observations is tested before the application of the Min/Max correction method. Specifically, the outliers are removed by considering the upper $CI^+_{95\%}$ and lower $CI^-_{95\%}$, 95% limits of the confidence interval. Therefore, assuming normal distribution of data, the limits are defined as follows (Albergel et al. 2010):

$$CI^+_{95\%} = \mu(OBS) + 1.96\sigma(OBS) \quad (17.5)$$

$$CI^-_{95\%} = \mu(OBS) - 1.96\sigma(OBS) \quad (17.6)$$

17.2.2 Filtering Techniques

The satellite soil moisture products, usually based on C- and X-band observations (for ASCAT and AMSR sensors), are representative of a layer depth of less than 5 cm (Schmugge 1983; Escorihuela et al. 2010). However, *in situ* observations are usually available for deeper soil layers (>5 cm) and modeled data are representative for the root zone (e.g., >1 m for hydrological models). Therefore significant differences in the sensing depth between satellite and *in situ* sensors are present. The effect of these discrepancies could be minimized by applying filtering techniques allowing us to infer root zone soil moisture data from surface measurements. Three different approaches are here analyzed: (1) exponential filter, (2) nonlinear exponential filter, and (3) moving average.

17.2.2.1 Exponential Filter

The semiempirical approach, also known as the exponential filter, proposed by Wagner et al. (1999) has been widely adopted to obtain a root zone soil moisture product [soil water index (SWI)] as a function of the surface soil moisture (SSM) product directly obtained through satellite sensors. This method depends on a single parameter T, named characteristic time length and representing the timescale of soil moisture variation. The recursive formulation of the method relies on (Albergel et al. 2008)

$$SWI_n = SWI_{n-1} + K_n [SSM (t_n) - SWI_{n-1}] \quad (17.7)$$

with both SWI and the gain K_n at time t_n given by

$$K_n = \frac{K_{n-1}}{K_{n-1} + e^{-\left(\frac{t_n - t_{n-1}}{T}\right)}} \tag{17.8}$$

where the gain K_n ranges between 0 and 1. For the initialization of the filter, K_0 and SWI_0 are set to 1 and $SSM(t_0)$, respectively.

17.2.2.2 Nonlinear Exponential Filter

To take account of the nonlinearity of the percolation process (Brocca et al. 2008) the T parameter may be assumed varying with $SWI(t)$ as

$$T = T_s [SWI (t)]^{-n} \tag{17.9}$$

where T_s represents the minimum value of T corresponding to saturated soil conditions [$SWI(t) = 1$] and n is a parameter allowing to increase T by decreasing $SWI(t)$. Through Equation 17.9, T assumes lower values for higher wetness conditions as it can be expected for gravity-driven flow because drainage rate increases with soil moisture. It has to be noticed that for $n = 0$, T becomes a constant as in the original approach (Equation 17.8). The exponential filter incorporating Equation 17.9 is named nonlinear exponential filter.

17.2.2.3 Moving Average

Alternatively, the most common technique for filtering time series is based on the computation of the moving average for a fixed number of days (Draper et al. 2009). This approach is employed in this chapter for testing if simple methods can be also used for filtering satellite data.

17.3 Case Studies Datasets

17.3.1 *In situ* Soil Moisture Observations

In the past decade, huge efforts were made to make available *in situ* soil moisture observations in contrasting biomes and climate conditions. The International Soil Moisture Network (ISMN; Dorigo et al. 2011), a new data hosting center where globally available ground-based soil moisture measurements are collected, harmonized, and made available to users shows clear evidence of how much relevant the availability of such data is for the scientific community.

Three sites over three different continents and characterized by contrasting climatic condition are selected: OZNET-Y03 in Australia (hereinafter named OZNET for sake of simplicity; Smith et al. 2012), AMMA-Benin in Africa (AMMA; Pellarin et al. 2009), and SMOSMANIA-Creon d'Armagnac (SMOSM; Albergel et al. 2008). These sites are chosen considering the quality of the dataset, the availability of long-term data (>3 years), and the possibility to have both surface (5 cm) and root zone measurements: 30 cm for SMOSM, 60 cm for OZNET, and 100 cm for AMMA. Figure 17.1 shows the soil moisture time series for the three investigated sites considering the common period 2007–2010. The seasonal

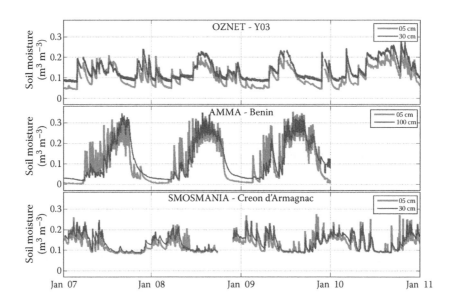

FIGURE 17.1
(See color insert.) Time series of soil moisture *in situ* observations for the three sites used in this study.

pattern is evident for all sites, mainly for AMMA in Africa because of the arid climate coupled with the effects of monsoons. It has to be stated that all datasets have been downloaded from the ISMN Web site at http://www.ismn.geo.tuwien.ac.at/.

17.3.2 Satellite Soil Moisture Observations

Two satellite soil moisture products are used as case studies to test the different scaling/filtering approaches derived by two different sensors: ASCAT and AMSR.

The ASCAT sensor is a real-aperture radar measuring backscatter at C band (5.255 GHz) in VV polarization. The equipment, mounted on the Metop satellite, scans the globe in a push-broom mode by six side-looking antennae and has been operating since October 2006. The swath width is approximately 550 km, and a global coverage is achieved in ~1.5 days. The basic equipment sampling distance is 12.5 km. Soil moisture is retrieved from the ASCAT backscatter measurements using a time series–based change detection approach previously developed for the ERS-1/2 scatterometer by Wagner et al. (1999) and then applied to ASCAT by Bartalis et al. (2007). The ASCAT-TUWIEN (hereinafter referred as ASCAT) SSM product is expressed in relative terms (percent of saturation between 0 and 100%) and is representative of the first few centimeters of the soil (less than 5 cm).

The AMSR sensor, on board NASA's Aqua satellite, is a passive microwave radiometer operating at 6.9 GHz (C band) and five higher frequencies (including 36.5-GHz Ka band) since May 2002. The temporal resolution is nearly daily (as ASCAT), and the AMSR products are usually mapped on a 0.25° grid (EASE grid projection, ~25 km). Among the different algorithms developed for retrieving soil moisture from AMSR data, the Land Parameter Retrieval Model (LPRM; Owe et al. 2008) is adopted here. LPRM considers the dual polarized channel (either 6.9 or 10.6 GHz) for the retrieval of both SSM and vegetation optical depth. The AMSR-LPRM (hereinafter AMSR) SSM product is expressed in volumetric terms ($m^3 m^{-3}$), and it is also representative of the first few centimeters of the soil (less than 5 cm). In accordance with Brocca et al. (2011), only ascending AMSR passes are used here.

17.4 Results and Discussions

The approaches described in Section 17.2 are applied separately considering surface (5 cm depth) and root zone (greater than 30 cm) *in situ* observations for the scaling and filtering techniques, respectively. Indeed, on the one hand, the scaling techniques allow only matching the variability of *in situ* and satellite data, and the shallowest *in situ* measurements are the best candidates for this purpose. On the other hand, filtering methods are addressed to estimate root zone soil moisture data, and hence observations for deeper layers have to be considered. For the calibration of the filtering methods performed in the following, the maximization of the correlation between satellite and *in situ* time series is used as an objective function.

17.4.1 Scaling Techniques

The purpose of this section is twofold: (1) to evaluate and contrast the reliability of the different scaling techniques when their parameters are calibrated (calibration period) and (2) to assess the robustness of the methods by applying them to a testing dataset (validation period). For that, the observation period is split in two parts, with the same length, the first for calibration and the second for validation. The length of these periods is 2 years for OZNET and SMOSM and 1.5 years for AMMA. A further calibration of the parameters for the second period (validation period) is also done to understand the loss in the performance due to the noncalibration of the parameters. By considering also a period for the validation of the techniques, we partly addressed the issue of the absence of *in situ* observations (or their availability for different locations). Table 17.1 summarizes the performance of all methods in both the calibration and validation periods in terms of correlation coefficient *R* and root-mean-square difference (RMSD). By considering the calibration period, ASCAT outperforms AMSR for AMMA and SMOSM sites and the opposite for OZNET. Among the scaling methods (see average values at the end of Table 17.1), the CDF matching (nonlinear method) slightly outperforms the other linear approaches in terms of *R* (mainly for AMSR) but provides slightly worse results in terms of RMSD. Among the linear methods, the linear regression performs better than the others (in terms of RMSD), even though the linear rescaling provides very similar results; the Min/Max correction shows the lower performance. Unexpectedly, in the validation period the CDF matching approach provides the best results both in terms of *R* and RMSD, even though it requires a larger number of parameters (6 in this study versus 2 for the linear methods) to be calibrated. Overall, in the validation period, both satellite sensors provide reasonable results with average *R* and RMSD values in the range 0.690–0.923 and 0.020–0.082 $m^3 m^{-3}$, respectively.

Figure 17.2 shows an example for OZNET site of the applications of the scaling approaches in both calibration and validation. Visually, the differences between the linear regression and the CDF matching methods are not significant, while the MIN/MAX shows some inconsistencies, mainly in the validation period. All the methods, anyhow, are able to follow reasonably the dynamic of *in situ* observation thus being able to remove the systematic differences between satellite and *in situ* time series. It is worth noting that this result has been achieved also thanks to the good accuracy of both satellite products.

Moreover, by recalibrating the parameters on the validation period (see "recalibration in the validation period" in Table 17.1), the improvement in the performance is quite low with, on average, an increase in *R* of only 3% for the CDF matching approach and a decrease in RMSD of 13%, leaving out the Min/Max correction that provides higher RMSD if the parameter is recalibrated. Therefore the different scaling methods can be considered

TABLE 17.1

Comparison between the Performance of the Different Scaling Methods for the Three Investigated Sites and Two Satellite Sensors ASCAT and AMSR

Site	Sample Size	Linear Regression		CDF Matching		Linear Rescaling	Min/Max Correction
		R	RMSD	R	RMSD	RMSD	RMSD
ASCAT—Calibration Period							
OZNET	659	0.655	0.033	0.671	0.034	0.035	0.042
AMMA	373	0.898	0.037	0.902	0.037	0.038	0.052
SMOSM	677	0.679	0.021	0.663	0.023	0.023	0.039
ASCAT—Validation Period							
OZNET	662	0.789	0.032	0.795	0.032	0.031	0.035
AMMA	409	0.919	0.046	0.923	0.041	0.064	0.046
SMOSM	742	0.794	0.024	0.800	0.023	0.023	0.028
ASCAT—Recalibration in the Validation Period							
OZNET	662	0.789	0.030	0.798	0.031	0.032	0.034
AMMA	409	0.919	0.040	0.923	0.039	0.040	0.065
SMOSM	742	0.794	0.020	0.805	0.020	0.021	0.035
AMSR—Calibration Period							
OZNET	529	0.726	0.028	0.742	0.029	0.030	0.030
AMMA	328	0.671	0.061	0.717	0.062	0.067	0.069
SMOSM	317	0.584	0.024	0.577	0.027	0.027	0.029
AMSR—Validation Period							
OZNET	518	0.715	0.040	0.753	0.041	0.034	0.048
AMMA	307	0.749	0.070	0.765	0.065	0.082	0.071
SMOSM	576	0.709	0.026	0.690	0.027	0.026	0.025
AMSR—Recalibration in the Validation Period							
OZNET	518	0.715	0.034	0.835	0.028	0.036	0.044
AMMA	307	0.749	0.065	0.803	0.061	0.070	0.070
SMOSM	576	0.709	0.023	0.706	0.025	0.025	0.030
Average Values—Calibration Period							
ASCAT	570	0.744	0.030	0.745	0.031	0.032	0.044
AMSR	391	0.660	0.038	0.679	0.039	0.041	0.043
Average Values—Validation Period							
ASCAT	604	0.834	0.034	0.839	0.032	0.039	0.036
AMSR	467	0.724	0.045	0.736	0.044	0.047	0.048
Average Values—Recalibration in the Validation Period							
ASCAT	604	0.834	0.030	0.842	0.030	0.031	0.045
AMSR	467	0.724	0.041	0.781	0.038	0.044	0.048

Note: R, correlation coefficient; RMSD, root-mean-square difference in $m^3\ m^{-3}$. The R-values for the linear rescaling and Min/Max correction methods are the same as the linear regression approach (linear operators).

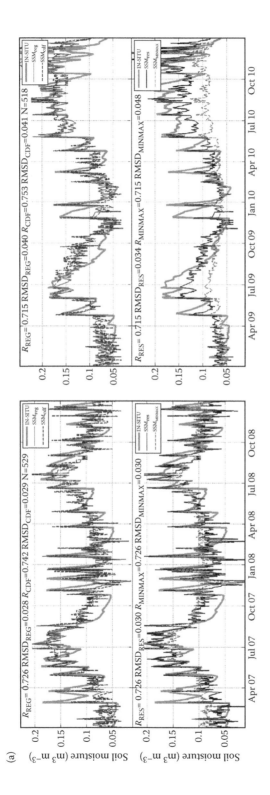

FIGURE 17.2

(See color insert.) Comparison for the OZNET-Y03 site in Australia between *in situ* observations, IN-SITU, and scaled satellite data, SSM, with the different scaling methods and two satellite soil moisture products: (a) AMSR and (b) ASCAT. The calibration period is shown in the first column, while the validation is shown in the second one (REG, linear regression; CDF, CDF matching; RES, linear rescaling; MINMAX: Min/Max correction; R, correlation coefficient; RMSD, root-mean-square difference in m³ m⁻³).

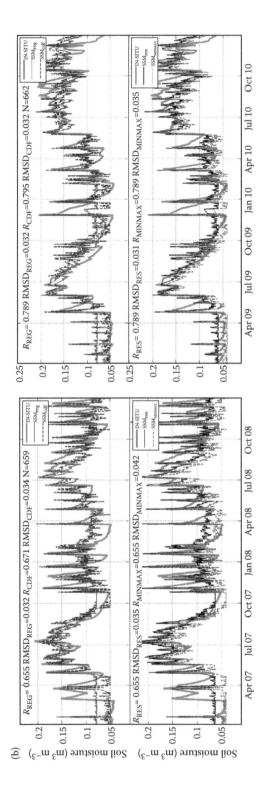

FIGURE 17.2 (Continued)

Comparison for the OZNET-Y03 site in Australia between *in situ* observations, IN-SITU, and scaled satellite data, SSM, with the different scaling methods and two satellite soil moisture products: (a) AMSR and (b) ASCAT. The calibration period is shown in the first column, while the validation is shown in the second one (REG, linear regression; CDF, CDF matching; RES, linear rescaling; MINMAX: Min/Max correction; R, correlation coefficient; RMSD, root-mean-square difference in $m^3 \, m^{-3}$).

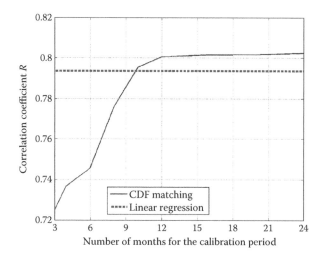

FIGURE 17.3

Correlation coefficient in the validation period varying the number of months used for the calibration period considering ASCAT satellite data and SMOSM *in situ* observations.

robust because their parameter values do not vary too much between the calibration and validation periods. Analyzing all the obtained results shown in Table 17.1, it can be surmised that, comparatively, they are quite similar for all sites and satellite sensors thus being not too much influenced by climatic conditions and satellite sensor accuracy.

To test the influence of the length of the calibration period on the results, Figure 17.3 shows the variability of R in the validation period, with the number of months used for the calibration of the CDF matching approach considering the SMOSM site as an example (similar results are also obtained for the other two sites). The R value for the linear methods is also shown for comparison. On the one hand, it is evident that at least 1 year of data is needed to obtain robust results with the CDF matching. If a lower number of months is available, the linear approaches provide better findings.

17.4.2 Filtering Techniques

By considering *in situ* observations for the root zone, the three filtering techniques are contrasted considering the whole period of data. A split sampling test is also done, but the performance scores in the two splitted periods (calibration and validation) are basically the same, and they are not shown in this section for the sake of simplicity. After filtering, the linear regression scaling approach, Equation 17.1, is used to match the variability of satellite and *in situ* observations. The performance of the three methods and their parameter values for all sites and satellite products are reported in Table 17.2. As before, the performances of the two satellite products are very similar with average R and RMSD values in the range 0.834–0.857 and 0.027–0.031 $m^3 m^{-3}$, respectively. Among the different techniques, the two exponential filters provide, on average, comparable findings, while the moving average performs the worst. However, the performances vary in accordance with the investigated site. Indeed, for AMMA site the R value increases from 0.911 to 0.935 when the nonlinear exponential filter is adopted. This result is expected because for the AMMA site, the soil completely dries between December and April and then a fast increase is observed with the arrival of the monsoon season (see Figure 17.4). However, the rapid increase in SSM is not

TABLE 17.2

Comparison between the Performance of the Different Filtering Methods for the Three Investigated Sites and Two Satellite Sensors ASCAT and AMSR

	Exponential Filter			Nonlinear Exponential Filter				Moving Average		
Site	R	RMSD	T	R	RMSD	T_s	n	R	RMSD	N
ASCAT										
OZNET	0.82	0.026	10.2	0.822	0.025	6.29	0.39	0.815	0.026	15
AMMA	0.911	0.039	7.9	0.935	0.034	2.30	1.56	0.911	0.039	13
SMOSM	0.813	0.022	8.2	0.813	0.022	7.54	0.10	0.775	0.023	13
AMSR										
OZNET	0.881	0.022	7.0	0.883	0.022	0.10	0.13	0.879	0.022	11
AMMA	0.875	0.045	10.8	0.875	0.045	0.83	0.39	0.859	0.048	11
SMOSM	0.804	0.022	30.1	0.803	0.022	17.30	0.10	0.783	0.023	30
Average Values										
ASCAT	0.848	0.029	8.8	0.857	0.027	5.38	0.68	0.834	0.029	13.7
AMSR	0.853	0.030	16.0	0.854	0.030	6.08	0.21	0.840	0.031	17.3

Note: R, correlation coefficient, RMSD, root-mean-square difference in $m^3\ m^{-3}$. The parameter values of the three methods are also shown: T, T_s, and N are expressed in days.

quickly reflected in the root zone time series (soil layer of 100 cm) owing to the low infiltration rate in very dry conditions. Partly, the nonlinear exponential filter is able to reproduce this behavior having higher T values during dry conditions and hence simulating a lower infiltration rate. On the contrary, the classical exponential filter fails in the transition period between dry and wet conditions owing to a too fast increase in soil moisture (Figure 17.4). A slight increase in the performance is also observed at the OZNET site while the same

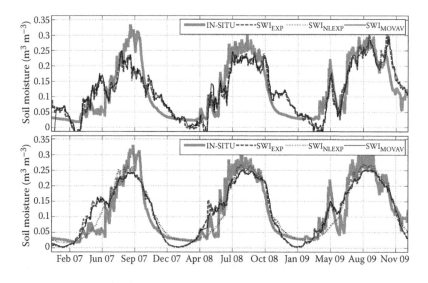

FIGURE 17.4

(See color insert.) Comparison for the AMMA-Benin site in Africa between *in situ* observations, IN-SITU, and filtered satellite data, SWI, with the different filtering algorithms and two satellite soil moisture products: (top) AMSR and (bottom) ASCAT. EXP, exponential filter; NLEXP, nonlinear exponential filter; MOVAV, moving average.

performance is obtained with the two exponential filters at SMOSM. We also underline that for the SMOSM site the value of the *T* parameter for AMSR soil moisture product is so high because of the lower agreement of AMSR data with this dataset. Therefore the filter tends to smooth as much as possible the satellite time series to reproduce, at least, the seasonal soil moisture pattern but completely missing the small timescale soil moisture variability.

17.5 Conclusions

The most common scaling and filtering approaches used for matching satellite observations with *in situ* or modeled data have been here described and analyzed in terms of performance. Specifically, four scaling (linear regression, CDF matching, linear rescaling, Min/Max correction) and three filtering (exponential filter, nonlinear exponential filter, moving average) methods are investigated and compared with each other. On the basis of the results obtained in this study, the following conclusions can be drawn:

- Among all the reviewed scaling techniques, the CDF matching approach performs slightly better than the linear methods. This is also true in the validation period notwithstanding the higher number of parameters involved in the CDF operator (Table 17.1 and Figure 17.2).
- The Min/Max correction method is found to be less robust and accurate with respect to the other linear approaches. The linear regression and rescaling methods show very similar findings (Table 17.1).
- A period of at least 1 year is required to correctly calibrate the parameters of the CDF matching method (Figure 17.3).
- The classical and the nonlinear exponential filters provide very similar results in reproducing the root zone soil moisture dynamic from surface measurements, whereas the moving average method gives lower performance (Table 17.2).
- For the AMMA soil moisture dataset, characterized by an abrupt transition between dry and wet conditions, the nonlinear exponential filter is found to reproduce better than the *in situ* observations, allowing the simulation of the variability in the infiltration rate occurring at the beginning of the monsoon season (Figure 17.4).

The obtained results can be considered a first attempt to evaluate the accuracy and robustness of the different scaling/filtering techniques. However, a large number of *in situ* observations have to be analyzed to obtain more general results that can be applied at a global scale. Moreover, the analysis should also consider modeled data for different soil layers in order to optimize the use of satellite data within modeling approaches and hence for improving operational applications.

References

Albergel, C., Calvet, J.-C., de Rosnay, P., Balsamo, G., Wagner, W., Hasenauer, S., Naemi, V., Martin, E., Bazile, E., Bouyssel, F., and Mahfouf, J.-F., 2010: Cross-evaluation of modelled and remotely sensed surface soil moisture with in situ data in southwestern France. *Hydrology and Earth System Sciences*, 14, 2177–2191.

Albergel, C., de Rosnay, P., Gruhier, C., Muñoz-Sabater, J., Hasenauer, S., Isaksen, L., Kerr, Y., and Wagner, W., 2012: Evaluation of remotely sensed and modelled soil moisture products using global ground-based in situ observations. *Remote Sensing of Environonment*, 118, 215–226.

Albergel, C., Rüdiger, C., Pellarin, T., Calvet, J.-C., Fritz, N., Froissard, F., Suquia, D., Petitpa, A., Piguet, B., and Martin, E., 2008: From near-surface to root-zone soil moisture using an exponential filter: an assessment of the method based on in situ observations and model simulations. *Hydrology and Earth System Sciences*, 12, 1323–1337.

Bartalis, Z., Wagner, W., Naeimi, V., Hasenauer, S., Scipal, K., Bonekamp, H., Figa, J., and Anderson, C., 2007: Initial soil moisture retrievals from the METOP-A Advanced Scatterometer (ASCAT). *Geophysical Research Letters*, 34, L20401.

Brocca, L., Hasenauer, S., Lacava, T., Melone, F., Moramarco, T., Wagner, W., Dorigo, W. et al., 2011: Soil moisture estimation through ASCAT and AMSR-E sensors: An intercomparison and validation study across Europe. *Remote Sensing of Environment*, 115(12), 3390–3408.

Brocca, L., Melone, F., and Moramarco, T., 2008: On the estimation of antecedent wetness conditions in rainfall-runoff modelling. *Hydrological Processes*, 22(5), 629–642.

Brocca, L., Melone, F., Moramarco, T., and Morbidelli, R., 2010b: Spatial-temporal variability of soil moisture and its estimation across scales. *Water Resources Research*, 46, W02516.

Brocca, L., Melone, F., Moramarco, T., Wagner, W., Naeimi, V., Bartalis, Z., and Hasenauer, S., 2010a: Improving runoff prediction through the assimilation of the ASCAT soil moisture product. *Hydrology and Earth System Sciences*, 14, 1881–1893.

Brocca, L., Moramarco, T., Melone, F., Wagner, W., Hasenauer, S., and Hahn, S., 2012a: Assimilation of surface and root-zone ASCAT soil moisture products into rainfall-runoff modelling. *IEEE Transactions on Geoscience and Remote Sensing*, 50(7), 2542–2555.

Brocca, L., Tullo, T., Melone, F., Moramarco, T., and Morbidelli, R., 2012b: Catchment scale soil moisture spatial-temporal variability. *Journal of Hydrology*, 422–423, 71–83.

de Rosnay, P., Drusch, M., Vasiljevic, D., Balsamo, G., Albergel, C., and Isaksen, L., 2013: A simplified extended Kalman filter for the global operational soil moisture analysis at ECMWF. *Quarterly Journal of the Royal Meteorological Society*, 139(674), 1199–1213, doi:10.1002/qj.2023.

Dharssi, I., Bovis, K. J., Macpherson, B., and Jones, C. P., 2011: Operational assimilation of ASCAT surface soil wetness at the Met Office. *Hydrology and Earth System Sciences*, 15, 2729–2746.

Draper, C., Mahfouf, J.-F., Calvet, J.-C., Martin, E., and Wagner, W., 2011: Assimilation of ASCAT near-surface soil moisture into the SIM hydrological model over France. *Hydrology and Earth System Sciences*, 15, 3829–3841.

Draper, C., Walker, J. P., Steinle, P., De Jeu, R. A. M., and Holmes, T. R. H., 2009: An evaluation of AMSR-E derived soil moisture over Australia. *Remote Sensing of Environonment*, 113(4), 703–710.

Dorigo, W. A., Wagner, W., Hohensinn, R., Hahn, S., Paulik, C., Drusch, M., Mecklenburg, S., van Oevelen, P., Robock, A., and Jackson, T., 2011: The International Soil Moisture Network: A data hosting facility for global in situ soil moisture measurements. *Hydrology and Earth System Sciences*, 15, 1675–1698.

Drusch, M., Wood, E. F., and Gao, H., 2005: Observation operators for the direct assimilation of TRMM microwave imager retrieves soil moisture. *Geophysical Research Letters*, 32, L15403.

Entekhabi, D., Reichle R. H., Koster, R. D., and Crow, W. T., 2010: Performance metrics for soil moisture retrieval and application requirements. *Journal of Hydrometeorology*, 11, 832–840.

Escorihuela, M. J., Chanzy, A., Wigneron, J. P., and Kerr, Y. H., 2010: Effective soil moisture sampling depth of L-band radiometry: A case study. *Remote Sensing of Environonment*, 114, 995–1001.

Famiglietti, J. S., Ryu, D., Berg, A. A., Rodell, M., and Jackson, T. J., 2008: Field observations of soil moisture variability across scales. *Water Resources Research*, 44, W01423.

Jackson, T. J., Cosh, M. H., Bindlish, R., Starks, P. J., Bosch, D. D., Seyfried, M., Goodrich, D. C., Moran, M. S., and Du, J., 2010: Validation of Advanced Microwave Scanning Radiometer soil moisture products. *IEEE Transactions on Geoscience and Remote Sensing*, 48(12), 4256–4272.

Koster, R. D., Guo, Z., Yang, R., Dirmeyer, P. A., Mitchell, K., and Puma, M. J., 2009: On the nature of soil moisture in land surface models. *Journal of Climate*, 22, 4322–4335.

Manfreda, S., Brocca, L., Moramarco, T., Melone, F., Sheffield, J., and Fiorentino, M., 2012: Predicting root zone soil moisture using surface data. *Geophysical Research Abstracts*, 14, EGU2012-8055.

Manfreda, S., Lacava, T., Onorati, B., Pergola, N., Di Leo, M., Margiotta, M. R., and Tramutoli, V., 2011: On the use of AMSU-based products for the description of soil water content at basin scale. *Hydrology and Earth System Sciences*, 15, 2839–2852.

Matgen, P., Heitz, S., Hasenauer, S., Hissler, C., Brocca, L., Hoffmann, L., Wagner, W., and Savenije, H. H. G., 2012: On the potential of METOP ASCAT-derived soil wetness indices as a new aperture for hydrological monitoring and prediction: A field evaluation over Luxembourg. *Hydrological Processes*, 26, 2346–2359.

Owe, M., De Jeu, R. A. M., and Holmes, T. R. H., 2008: Multi-sensor historical climatology of satellite-derived global land surface moisture. *Journal of Geophysical Research*, 113, F01002.

Pellarin, T., Tran, T., Cohard, J.-M., Galle, S., Laurent, J.-P., de Rosnay, P., and Vischel, T., 2009: Soil moisture mapping over West Africa with a 30-min temporal resolution using AMSR-E observations and a satellite-based rainfall product. *Hydrology and Earth System Sciences*, 13, 1887–1896.

Reichle, R. H., and Koster, R. D., 2004: Bias reduction in short records of satellite soil moisture. *Geophysical Research Letters*, 31, L19501.

Schmugge, T. J., 1983: Remote sensing of soil moisture: Recent advances. *IEEE Transactions on Geoscience and Remote Sensing*, GE21, 145–146.

Scipal, K., Drusch, M., and Wagner, W., 2008: Assimilation of a ERS scatterometer derived soil moisture index in the ECMWF numerical weather prediction system. *Advances in Water Resources*, 31, 1101–1112.

Seneviratne, S. I., Corti, T., Davin, E. L., Hirschi, M., Jaeger, E. B., Lehner, I., Orlowsky, B., and Teuling, A. J., 2010: Investigating soil moisture-climate interactions in a changing climate: A review. *Earth-Science Reviews*, 99(3–4), 125–161.

Smith, A. B., Walker, J. P., Western, A. W., Young, R. I., Ellett, K. M., Pipunic, R. C., Grayson, R. B., Siriwardena, L., Chiew, F. H. S., and Richter, H., 2012: The Murrumbidgee soil moisture monitoring network dataset. *Water Resources Research*, 48, W07701.

Vachaud, G., Passerat de Silans, A., Balabanis, P., and Vauclin, M., 1985: Temporal stability of spatially measured soil water probability density function. *Soil Science Society of American Journal*, 49, 822–828.

Wagner, W., Lemoine, G., and Rott, H., 1999: A method for estimating soil moisture from ERS scatterometer and soil data. *Remote Sensing of Environonment*, 70, 191–207.

Wagner, W., Naeimi, V., Scipal, K., de Jeu, R., and Martinez-Fernandez J., 2007: Soil moisture from operational meteorological satellite. *Hydrogeology Journal*, 15(1), 121–113.

18

Selection of Performance Metrics for Global Soil Moisture Products: The Case of ASCAT Product

Clement Albergel, Luca Brocca, Wolfgang Wagner, Patricia de Rosnay, and Jean-Christophe Calvet

CONTENTS

18.1 Introduction

The importance of soil moisture (SM) in the global climate system has recently been outlined by the Global Climate Observing System (GCOS; http://www.wmo.int/pages/prog/gcos/) by endorsing SM as an essential climate variable. It is a crucial variable for numerical weather and climate prediction and plays a key role in hydrological processes. A good representation of SM conditions can therefore help to improve forecasting of precipitation, droughts and floods, climate projections, and predictions. For many applications, global- or continental-scale SM maps are needed.

In recent years, active and passive microwave remote sensing offered the unique opportunity to obtain SM estimates at a global scale with adequate spatial–temporal resolution and accuracy. Indeed, microwave remote sensing is able to provide quantitative information about the water content of a shallow near surface layer (Schmugge 1983), particularly in the low-frequency microwave region from 1 to 10 GHz. Apart from a few days of L-band radiometric observations on Skylab between June 1973 and January 1974, current or past spaceborne microwave radiometers have been operating at frequencies above 5 GHz. Among them are the Scanning Multichannel Microwave Radiometer (SMMR), which operated on Nimbus 7 between 1978 and 1987 (6.6 GHz and above), followed by the Special Sensor Microwave Imager (SSM/I starting in 1987 at 19 GHz and above) and the TRMM Microwave Imager (TMI) on the Tropical Rainfall Measuring Mission (TRMM) satellite (1997, at 10.7 GHz and above), the Advanced Microwave Scanning Radiometer for the Earth Observing System (AMSR-E on the Aqua satellite from 6.9 to 89.0 GHz), Windsat (from 6.8 to 37 GHz), and the scatterometer on board the European Remote Sensing Satellite

(ERS-1&2, 5.3 GHz). More recently, the Soil Moisture and Ocean Salinity (SMOS) mission, the first dedicated SM mission, was launched in November 2009 (Kerr et al. 2010). It consists of a spaceborne L-band (~1.42 GHz, 21 cm) interferometric radiometer able to provide global surface SM estimates at a spatial resolution of about 40 km, with a sampling time of 2–3 days. Another sensor, the Advanced Scatterometer ASCAT onboard METOP-A (launched in 2006, followed by METOP-B in September 2012), produces SM estimates with a spatial resolution of 50 and 25 km (resampled to 25- and 12.5-km grids in the swath geometry). ASCAT is a C-band radar operating at 5.255 GHz (Bartalis et al. 2007), and it is the first sensing satellite to provide SM product in operation and in near-real time (NRT). There is also the short-coming Soil Moisture Active/Passive (SMAP) mission programmed by NASA scheduled for launch in 2014.

One important aspect of the environmental variables retrieval from space process is the evaluation of their performance. A usual step for evaluating such variables is to determine whether their behavior matches the observations. Hence *in situ* measurements of SM are a highly valuable source of information for evaluating SM estimates from remote sensing. Choice of performance metrics to be used is also of crucial interest; it is governed by the nature of the variable itself and is influenced by the characteristics of the science application and its sensitivity to the considered variables (Stanski et al. 1989). No single metric or statistic can capture all the attributes of environmental variables. Some are robust in respect to some attributes while insensitive with respect to other (Entekhabi et al. 2010). This chapter aims to provide first an overview of the different approaches employed in validating satellite-derived estimates of SM content focusing in particular in the description of the statistics most often used to quantify pattern similarity and differences between SM retrieval from space and ground measurements.

After a description of the metrics, a case study making use of *in situ* ground measurements from the Soil Moisture Observing System-Meteorological Automatic Network Integrated Application (SMOSMANIA) network of Météo-France and from the Red de Medición de la Humedad del Suelo (REMEDHUS) network in Spain to evaluate ASCAT SM retrieval from space is presented, and the results are discussed.

18.2 Validation of Remotely Sensed SM Products: Overview of Approaches

This section presents the most commonly used approaches for the validation of SM operational products. First, it focuses on the main metrics to evaluate SM retrieval accuracy: the correlation coefficient (R), the root-mean-square difference (RMSD), the bias (Bias), and the unbiased root-mean-square difference (ubRMSD). Also, the normalized standard deviation (SDV) and the centered unbiased RMSD (E) between satellite and *in situ* observations are presented. It proposes then a brief description of other approaches used in the recent literature to evaluate SM products.

The statistic most often used to quantify pattern similarity is the correlation coefficient R, expressed by

$$R = \frac{\frac{1}{N}\sum_{n=1}^{N}\left(\text{satellite}_n - \overline{\text{satellite}}\right)\left(in\ situ_n - \overline{in\ situ}\right)}{\sigma_{\text{satellite}}\sigma_{in\ situ}} \tag{18.1}$$

where $\sigma_{satellite}$ and $\sigma_{in\ situ}$ are the standard deviation of satellite and *in situ* SM, respectively. N is the length of the record at a given station; satellite and *in situ* represent the averaged satellite and *in situ* measurement values, respectively. R ranges between -1 and $+1$, reaching its maximum value of $+1$ when for all n values $\left(satellite_n - \overline{satellite}\right) = \alpha\left(in\ situ_n - \overline{in\ situ}\right)$, α being a positive constant. However, it does not mean that both the satellite and *in situ* dataset are identical (unless $\alpha = 1$) but that the two fields have the same centered pattern. The correlation coefficient does not account for additive difference or difference in proportionality; that is, it does not allow determining whether satellite and *in situ* measurements have the same amplitude of variation and/or the same average values. The p value (Schervish 1996), a measure of the correlation significance, should be also calculated. It indicates the significance of the test; if it is small (e.g., below 0.05), it means that the correlation is not a coincidence. Only cases with p values below 0.05 (within the 95% confidence interval error) are usually considered (e.g., Albergel et al. 2012).

The statistic most often used to quantify differences in two fields is the RMSD expressed as follows:

$$RMSD = \sqrt{\frac{1}{N}\sum_{n=1}^{N}\left(in\ situ_n - satellite_n\right)^2} \tag{18.2}$$

It is a quadratic scoring rule that gives the average magnitude of errors, weighted according to the square of the error. The rationale for using the RMSD instead of root-mean-square error is to emphasize that, as well as for the satellite retrieval, *in situ* data may contain errors (instrumental and representativeness), so they cannot be considered as "true" SM values. RMSD will penalize deviation of the estimate with respect to the *in situ* SM. Biases in the mean and/or differences in the amplitude of fluctuations of satellite retrieval will lead to high RMSD values.

The bias indicating the average direction of the deviation from observed values is also useful. It is expressed by

$$Bias = \frac{1}{N}\sum_{n=1}^{N}\left(in\ situ_n - satellite_n\right) \tag{18.3}$$

With this convention, a positive bias indicates that satellite underestimates *in situ* SM and a negative bias indicates that satellite exceeds them.

As the RMSD might be severely compromised if there are biases in either the mean or the amplitude of fluctuation of the satellite retrieval, one may use the unbiased RMSD (ubRMSD), expressed by

$$ubRMSD = \sqrt{\frac{1}{N}\sum_{n=1}^{N}\left\{\left[\left(satellite_n - \overline{satellite_n}\right) - \left(in\ situ_n - \overline{in\ situ_n}\right)\right]^2\right\}} \tag{18.4}$$

Equations 18.2 through 18.4 are related through

$$ubRMSD^2 = RMSD^2 - Bias^2 \tag{18.5}$$

Additionally, the SDV and the centered unbiased RMSD (*E*) between satellite and *in situ* observations present valuable information and should also be carried out. SDV is the ratio between satellite and *in situ* standard deviations; it gives the relative amplitude, while *E* quantifies errors in the pattern variations. It does not include any information on biases because the means of the fields are subtracted before computing second-order errors. SDV and *E* are expressed by Equations 18.6 and 18.7, respectively:

$$\text{SDV} = \sigma_{\text{Ascat}}/\sigma_{in\ situ} \tag{18.6}$$

$$E^2 = (\text{RMSD}^2 - \text{Bias}^2)/\sigma_{in\ situ}^2 \tag{18.7}$$

The main reason for computing the two last scores is that *R*, *E*, and SDV can be displayed on a single two-dimensional diagram (Taylor diagram), and this helps with the interpretation of the results. *R*, SDV, and *E* are complementary but not independent as they are related by Equation 18.8 (Taylor 2001)

$$E^2 = \text{SDV}^2 + 1 - 2\text{SDV} \cdot R \tag{18.8}$$

Taylor diagrams are used to represent these three statistics using a single two-dimensional plot. In a Taylor diagram the SDV is displayed as a radial distance and *R* as an angle in the polar plot. *In situ* data are represented by a point located on the *x* axis at *R* = 1 and SDV = 1; the distance to this point represents *E*.

Usually, SM time series shows a strong seasonal pattern that could artificially increase the agreement between satellite and *in situ* observations in terms of *R*. Therefore, to avoid seasonal effects, the analysis of anomaly time series should be also carried out. In particular, monthly anomaly time series can be computed considering the difference from the mean for a sliding window of five weeks, and the difference is scaled to the standard deviation. For each satellite or *in situ* SM estimate at day *i*, a period *F* is defined, with *F* = [*i* − 17 days, *i* + 17 days] (corresponding to a 5-week window). If at least five measurements are available in this period, the average SM value and the standard deviation are calculated. The anomaly (Ano), which is dimensionless, is then computed for both *in situ* and satellite data, and it is given by

$$\text{Ano}(i) = \frac{\text{SM}(i) - \overline{\text{SM}(F)}}{\text{stdev}(\text{SM}(F))} \tag{18.9}$$

Therefore, for a comprehensive analysis of satellite product performance, *R* should be computed for both the absolute and anomaly time series.

Other metrics, not considered in this study, such as the Willmott's index of agreement (Willmott et al. 1985) and its refined version (*dr*, Equation 18.10; Willmott et al. 2011) can also provide a good estimation of SM product performances:

$$dr = \begin{cases} 1 - \dfrac{\displaystyle\sum_{n=1}^{N}\left|\text{satellite}_n - in\,situ_n\right|}{c\displaystyle\sum_{n=1}^{N}\left|in\,situ_n - \overline{in\,situ}\right|}, \text{ when} \\[4pt] \displaystyle\sum_{n=1}^{N}\left|\text{satellite}_n - in\,situ_n\right| \leq c\displaystyle\sum_{n=1}^{N}\left|in\,situ_n - \overline{in\,situ}\right| \\[8pt] \dfrac{c\displaystyle\sum_{n=1}^{N}\left|in\,situ_n - \overline{in\,situ}\right|}{\displaystyle\sum_{n=1}^{N}\left|\text{satellite}_n - in\,situ_n\right|} - 1, \text{ when} \\[4pt] \displaystyle\sum_{n=1}^{N}\left|\text{satellite}_n - in\,situ_n\right| > c\displaystyle\sum_{n=1}^{N}\left|in\,situ_n - \overline{in\,situ}\right| \end{cases} \qquad (18.10)$$

with $c = 2$. Equation 18.10 indicates the sum of the magnitudes of the differences between satellites estimate (of SM) and observed deviations about the observed mean relative to the sum of the magnitudes of a perfect estimate (satellite$_n$ = $in\,situ_n$, for all n) and observed deviation about the observed mean. The range of dr is [–1,1]; (1) a value of 0.5 indicates that the sum of the error magnitudes is one half of the sum of the perfect estimate deviation and observed deviation magnitudes; (2) a value of 0 indicates that the sum of the magnitudes of the errors and the sum of the perfect estimate deviation and observed deviation magnitudes are equivalent; and (3) a value of –0.5 indicates that the sum of the error magnitudes is twice the sum of the perfect estimate deviation and observed deviation magnitudes. A value of one indicates a perfect match between the satellite estimate and the observation.

Using *in situ* measurements, the quality of remotely sensed surface SM can be accurately assessed for the locations of the stations. By their nature, such assessment does not provide spatially complete error fields that are important for understanding the variable product quality across different environment. That is why, besides approach of satellite data compared to local *in situ* SM measurements, Land Surface Model (LSM) estimates can also be used (Albergel et al. 2010), even though the interpretation of the results is hampered by the accuracy of the reference dataset (model itself and its inputs). The large-scale nature of LSM estimates is more representative of the scales of remotely sensed products; also, they are available at a global scale. They can be used for direct comparison and as inputs of methods such as error propagation that can provide a more global view of the uncertainty of retrieved SM. Such techniques assess the uncertainty of SM estimates resulting from errors in the input variables (Dorigo et al. 2010). For instance, Scipal et al. (2008) introduced the triple collocation method to assess the accuracy of satellite-based SM products. It is based on the assumption that a SM estimate is related to the hypothetical but unknown real SM content through an additive and multiplicative bias component and a random error term (Stoffelen 1998); the error structure of three independent datasets can be resolved if the errors are uncorrelated. It allows a simultaneous estimation of the error structure and the cross calibration of a set of at least three linearly related datasets with uncorrelated errors (Stoffelen 1998). Scipal et al. (2008) applied the triple collocation method to a combination of TRMM-TMI radiometer data, ERS scatterometer data, and modeled ERA-Interim reanalysis (ECMWF latest reanalysis; Dee et al. 2011) SM.

They obtained realistic error estimates and were able to successfully distinguish spatial error trends of retrieved SM. Triple collocation is a promising method to estimate the error structures of global SM datasets. Zwieback et al. (2012) showed that the robustness of the triple collocation is closely related to the strict application of its basic assumption.

The above-mentioned approaches consider the satellite retrieval as the final product to be evaluated. For an all-inclusive validation study, product developers might also consider a sensitivity analysis of the algorithm retrieval. It can be defined as the process of determining the effect of changing the value of one or more inputs of the algorithm and observing the effect that this has on the retrieved SM. This technique has been adopted, for instance, for the SMOS calibration–validation and retrieval plan (European Space Agency 2010). A good overview of SA can be found in the work of Saltelli (2002). There are mainly two different ways to proceed in performing a sensitivity analysis: (1) the so-called local sensitivity analysis methods, which examine the sensitivity of retrieval outputs to inputs by varying the inputs around their expected "nominal" values and observing the effect on the retrieval, and (2) the global sensitivity analysis (GSA), where sensitivity is measured by generating samples across the range of interest of parameter values. Such methods apportion the variability in output to the variability of the input parameters when they vary over their whole uncertainty domain that is generally described using their assigned probability densities.

The general procedure for obtaining sensitivity measures using GSA methods follows the steps described below (e.g., Saltelli et al. 2000; Schwieger 2004): (1) define the probability distributions for each of the retrieval input parameters, (2) generate a training sample from the previously defined input probability distributions, (3) evaluate the output of interest (target quantity, e.g., SM in this case) using the generated sample, (4) analyze the output variance and quantitatively estimate the contribution of each of the inputs to the algorithm output, and (5) complete the sensitivity analysis of the output variance in relation to the input variation. The main approach used to perform step 5 is based on a direct decomposition of the output variance into factorial terms, called importance measures. The next sections focus on demonstrating how the most common performance metrics reviewed above are used for validating SM estimates from ASCAT operational product against *in situ* observations.

18.3 *In Situ* SM Observations

While in the 1990s, records of *in situ* SM measurements were available for only a few regions and were often for a very short period, huge efforts were made in the past decade to make available such observations in contrasting biomes and climate conditions. The International Soil Moisture Network (ISMN; http://ismn.geo.tuwien.ac.at/; Dorigo et al. 2011), a new data hosting center where globally available ground-based SM measurements are collected, harmonized, and made available to users, shows clear evidence of how relevant the availability of such data is for the scientific community. This section describes two networks used in this study, the SMOSMANIA network in southwestern France (Calvet et al. 2007; Albergel et al. 2008) and the REMEDHUS network in Spain (Sánchez et al. 2012); they are illustrated in Figure 18.1.

SMOSMANIA is a long-term data acquisition effort of profile SM under way in southwestern France at 12 automated weather stations. It was developed in order to validate remote sensing and model SM estimates. The SMOSMANIA network is based on the existing automatic weather station network of Météo-France. It is the first time that automatic

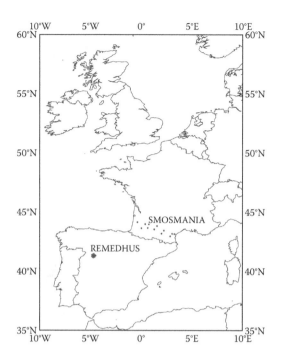

FIGURE 18.1
Map illustrating stations from the SMOSMANIA network located in southwestern France forming a 400-km transect between the Atlantic Ocean and the Mediterranean Sea and the one of REMEDHUS network in Spain (Duero basin).

measurements of SM have been integrated in an operational meteorological network. The stations were chosen to form a Mediterranean–Atlantic transect following the marked climatic gradient between the two coastlines. The locations of the chosen stations are in relatively flat areas (mountainous areas are avoided as much as possible), and the altitude of the highest station is 538 m above sea level. The three most eastward stations are representative of a Mediterranean climate. The observations from this well-monitored network were extensively used for the validation of modeled and satellite-derived SM, including ASCAT and SMOS (Albergel et al. 2009, 2010, 2012; Brocca et al. 2011; Parrens et al. 2012). The SMOSMANIA network showed to have a great potential to monitor the quality of satellite-derived SM products. Four SM probes were horizontally installed per station at four different depths: 5, 10, 20, and 30 cm. The ThetaProbe ML2X of Delta-T Device was chosen because it has successfully been used during previous long-term campaigns of Météo-France and because it can easily be interfaced with the automatic stations. Figure 18.1 illustrates the location of the 12 stations of the SMOSMANIA networks. Figure 18.2 presents *in situ* SM measurements from the second most westward station (Urgons) of this network at the four considered depths over 2007–2011. From Figure 18.2 it is possible to see the variability of SM decreasing with depth.

Twenty stations from the REMEDHUS network in Spain (Figure 18.1) are available through the ISMN Web site. This network is located in a central sector of the Duero basin; the climate is semiarid continental Mediterranean, and the land use is predominantly agricultural with some patchy forest. This area is mainly flat, ranging from 700 to 900 m above sea level. Each stations has been equipped with capacitance probes (HydraProbes, Stevens) installed horizontally at a depth of 5 cm. Analysis of soil sample was carried out

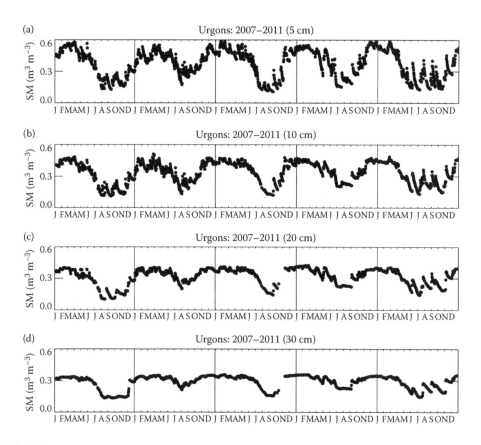

FIGURE 18.2

In situ SM measurement time series from the second most westward station (Urgons) of the SMOSMANIA network in southwestern France at the four considered depths between 2007 and 2011: (a) 5 cm, (b) 10 cm, (c) 20 cm, and (d) 30 cm.

to verify the capacitance probes and to assess soil properties at each station (Sánchez et al. 2012). This network provides continuous measurements of SM on an hourly basis. It has already been used for the evaluation of remotely sensed SM products such as from AMSR-E (Wagner et al. 2007; Brocca et al. 2011), ERS (Ceballos et al. 2005), ASCAT (Brocca et al. 2011), and SMOS (Albergel et al. 2012; Sánchez et al. 2012).

18.4 ASCAT SM Retrieval

Like its predecessors ERS-1&2 scatterometers, ASCAT is a real-aperture radar instrument measuring radar backscatter with very good radiometric accuracy and stability (Bartalis et al. 2007). It operates in the C band (5.255 GHz) and observes the surface of the Earth with a spatial resolution of circa 50 or 25 km. Measurements occur on both sides of the subsatellite track; thus two 550-km-wide swaths of data are produced. Because ASCAT operates continuously, more than twice of the ERS scatterometer coverage is provided. The SM retrieval algorithm is based on change detection of the backscatter signal. On both

TABLE 18.1

Summary of Recent Evaluation Studies of ASCAT Remotely Sensed Surface SM against *in Situ* Data

Studies	Location of *in Situ* Measurements	Period Considered	R	R Anomaly	RMSD (m³ m⁻³)	Bias
Albergel et al. 2009	Southwestern France	April–September 2007	0.58	0.53	0.255/0.06	-0.005
Albergel et al. 2010	Southwestern France	2007–2008	0.59	NA	0.237/0.057	-0.079
Brocca et al. 2010	Italy	April 2007 to May 2008	0.67 (0.92[c])	0.58 (0.70[c])	0.264/NA	NA
Brocca et al. 2011[a]	Spain, France, Luxembourg	2007	0.71 (0.81[c])	0.52 (0.64[c])	0.143/0.039	NA
Draper et al. 2011[b]	Southwestern France	May 2007 to April 2010	0.70	0.62	NA	NA
Matgen et al. 2012	Luxembourg	2007–2008	0.50 (0.86[c])	NA	0.25/0.062	0.04
Parrens et al. 2012	South of France	2010	0.64	0.55	0.86/NA	-0.03
Albergel et al. 2012	Europe (Spain, France, Italy and Germany), USA, Australia (southeast), Western Africa (Benin, Niger, Mali)	2010	0.53	0.71 (winter, spring) 0.46 (summer) 0.50 (autumn)	0.255/0.08	-0.068

Note: Recent evaluation studies used a different ASCAT dataset (see Section 18.4). Readers are referred to the considered studies for more information. NA stands for not applicable.

[a] Systematic differences between ASCAT and *in situ* measurements were removed using a cumulative distribution approach (nonlinear technic) prior to the comparison.

[b] Systematic differences between ASCAT and modeled SM were removed using a cumulative distribution approach (nonlinear technic) prior to the comparison.

[c] Filtered and bias corrected ASCAT data.

sides of METOP-A, ASCAT produces a triplet of backscattering coefficients (σ^0) from the three different antenna beams. A σ^0 measurement is the result of averaging several radar echoes. Backscatter is registered at various incidence angles, and it is possible to determine the yearly cycle of the backscatter–incidence angle relationship. This is an essential prerequisite for correcting seasonal vegetation effects. The estimated incidence-angle dependency function is then used for normalization of backscattering coefficients to a reference angle chosen as 40°. The latter is scaled using the lowest and the highest values ever measured, representing the driest and the wettest conditions. It represents the degree of saturation of the topmost soil layer and is given in percent ranging from 0 (dry) to 100% (wet). This measure is complemented by an error estimate, derived by error propagation of the backscatter noise (ranging from 0 to 100%, covering instrument noise, speckle, and azimuthal effects).

The release of ASCAT data has been the basis for numerous validation studies that verified the quality of the SM retrieval by comparing them to *in situ* data. Results from eight studies (Albergel et al. 2009, 2010, 2012; Brocca et al. 2010, 2011; Draper et al. 2011; Parrens et al. 2012; Matgen et al. 2012) are summarized in Table 18.1.

A first ASCAT SM dataset was evaluated over southwestern France (Albergel et al. 2009) with encouraging results. This dataset used change detection parameters derived from the analysis of multi-annual backscatter time series using ERS data over a 15-year long period. Brocca et al. (2010, 2011) then used a new ASCAT dataset developed by deriving the change detection model parameter from the analysis of ASCAT time series of the 2007–2008 period (Hahn and Wagner 2011). This new dataset better compares with *in situ* measurements and is able to provide reliable SM estimates across different test sites in Europe, with reduced noise with respect to the initial ASCAT SM product. Other consistent findings of evaluating ASCAT-derived SM over Europe can be found in the works of Draper et al. (2011), Parrens et al. (2012), and Matgen et al. (2012) who also used the new dataset. In the work of Albergel et al. (2012) the operational ASCAT product available in NRT via EUMETSAT and disseminated to the Numerical Weather Prediction community was used; it is a similar dataset as the one described by Albergel et al. (2009). The following section makes use of the reprocessed ASCAT dataset.

18.5 Results and Discussion

ASCAT surface SM estimates represent a relative measure of the SM in the first centimeters of soil, and it is given in percent (index between 0 and 100). *In situ* data from both the SMOSMANIA and REMEDHUS network are expressed as the volumetric fraction of water in a given soil depth (m^3 water per m^{-3} soil). Hence, to enable a fair product comparison, the SM datasets are scaled between [0,1] using their own maximum and minimum values over the whole period of time (2007–2011). For each station location available, correlation coefficient (applied to both scaled, *R*, and anomaly, *R_ANO*, time series), RMSD, bias, unbiased RMSD, SDV, and the centered unbiased RMSD (*E*) are computed between *in situ* observations and ASCAT SM retrievals. Averaged statistical scores are presented in Table 18.2.

Figure 18.3 illustrates the two SM products used in this case study for one station from (Figure 18.3a) the SMOSMANIA network (Creon d'Armagnac) and (Figure 18.3b) the REMEDHUS network (Paredinas) over the 2007–2011 period. From Figure 18.3 one may

TABLE 18.2

Comparison between ASCAT Products and the Scaled *in Situ* SM (5 cm) Measured at 12 Ground
Stations from the SMOSMANIA Network in Southwestern France and 20 Ground Stations from
the REMEDHUS Network in Spain for 2007–2011

	2007–2011						
	R	*R_ANO*	RMSD (m³ m⁻³)	Bias	ubRMSD	SDV	*E*
SMOSMANIA 12 stations	0.66	0.53	0.226/0.070	0.006	0.191	1.03	0.72
REMEDHUS 17 stations	0.62	0.42	0.249/0.072	−0.056	0.230	1.60	2.07

Note: Correlation coefficient (scaled, *R*, and anomaly, *R_ANO*, time series), root-mean-square difference
(RMSD), bias (*in situ* minus ASCAT), unbiased root-mean-square difference (ubRMSD), normalized
standard deviation (SDV), and the centered unbiased RMSD (*E*) are presented. Scores are presented for
significant correlations with *p* values < 0.05.

appreciate the ability of ASCAT to represent the SM annual variability; the annual cycle
is well represented. From time to time, the SM retrievals display a significant bias; nevertheless, peaks and troughs are well represented. Figure 18.4 presents anomaly time series
derived from ASCAT satellite measurements and from *in situ* observations for the station
Creon d'Armagnac. Most peaks and troughs are well represented; ASCAT also accurately
captures the SM short-term variability.

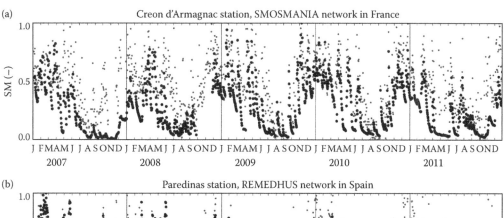

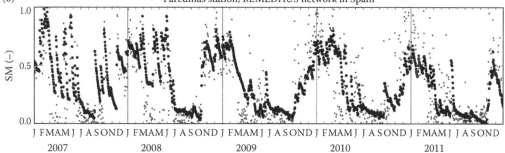

FIGURE 18.3

(See color insert.) Temporal evolution of ASCAT scaled surface SM estimates (red dots) compared to (a) the *in
situ* observations at 5 cm from the Creon d'Armagnac station of the SMOSMANIA network in southwestern
France and (b) the *in situ* observations at 5 cm from the Paredinas station of the REMEDHUS network in Spain
over 2007–2011.

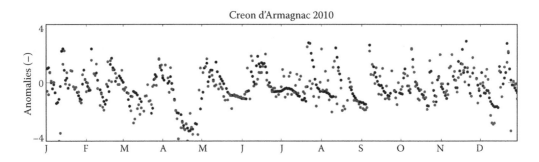

FIGURE 18.4
(See color insert.) Temporal evolution of surface SM anomalies at the Creon d'Armagnac station of the SMOSMANIA network in southwestern France: ASCAT (red dots) and *in situ* observations at 5 cm (black dots) for the year 2010.

For the station of the SMOSMANIA network, over 2007–2011, all p values are below 0.05, indicating that all correlations are significant. Correlation values range from 0.34 to 0.80 with an average value of 0.66 and a standard deviation of 0.12. One station (Mouthoumet) located in a rather mountainous (538 m above sea level) and forested area has a poor correlation value of 0.34. Its particular location can explain the poor results; not considering this station leads to an average R value of 0.69. Biases range from −0.199 to 0.165 with an average of 0.006. No systematic dry or wet bias is observed with five stations (over 12) with negative values (i.e., ASCAT values higher than *in situ* measurements). The RMSD ranges from 0.177 to 0.292, with an average value of 0.226. The RMSD represents the relative error of the SM dynamical range. With an observed average dynamic range of 0.33 $m^3\,m^{-3}$ for the SMOSMANIA network at a depth of 5 cm and an average RMSD value of 0.226, an estimate of the average error of the SM retrieval over 2007–2011 is about 0.070 $m^3\,m^{-3}$ in volumetric terms. On average, ubRMSD is 0.191, SDV is 1.03, and E is 0.72. SDV is very close to 1, as it is a ratio between ASCAT and *in situ* standard deviation (Equation 18.5); it means that their variabilities are similar. Results presented above give an overview of the quality of ASCAT to represent the annual scale of SM. To address the ability of ASCAT to capture the short-term SM variability, anomaly time series was derived and correlations were computed for the anomaly time series also. Correlations of normalized time series (0.66 on average) are larger than those for the monthly anomaly time series (0.53 on average). The good level of correlation of the volumetric time series is explained by seasonal variations, which are suppressed in monthly anomalies.

For the REMEDHUS network, 17 stations (over 18) have p values below 0.05. Correlation values are ranging from 0.38 to 0.75, with an average value of 0.62 and a standard deviation of 0.09. Biases range from −0.194 to 0.087 with an average of −0.056. Fourteen stations (over 17) present negatives biases. The RMSD ranges from 0.196 to 0.312, with an average value of 0.249. An estimate of the average error of the SM retrieval over 2007–2011 is about 0.072 $m^3\,m^{-3}$ in volumetric terms. On average, ubRMSD is 0.230, SDV is 1.60 (i.e., the variability of ASCAT is higher than the one of *in situ* measurements), and E is 2.07.

Figure 18.5 shows one Taylor diagram, illustrating the statistics from the comparison between ASCAT SM products and *in situ* data for the stations from (1) the SMOSMANIA network (red dots) and (2) the REMEDHUS network (green dots). The 2007–2011 period is considered. These results indicate the good range of correlations of ASCAT with most values being above 0.50 and 0.90. The red dashed line represents a SDV value of one, as SDV is the ration between ASCAT and *in situ* standard deviation (Equation 18.5); a symbol

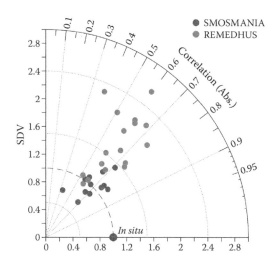

FIGURE 18.5
(See color insert.) Taylor diagram illustrating the statistics of the comparison between ASCAT and *in situ* surface SM for the stations of the SMOSMANIA network in southwestern France (red dots) and the REMEDHUS network in Spain (green dots) for 2007–2011. Only stations with significant correlations (*p* values < 0.05) are considered. Each symbol indicates the correlation value (angle), the normalized SDV (radial distance to the origin point), and the normalized centered root-mean-square error *E* (distance to the red point marked "*In situ*").

(dot in this case) located below this line indicates that ASCAT presents less variability than the *in situ* measurements (SDV < 1). There is no particular tendency for the SMOSMANIA network as symbols representing the stations (red dots) are on both sides of this line. However, the green dots representing stations from REMEDHUS network are most often than not above this line; as expressed above by the averaged scores, the variability of the ASCAT SM product is higher than the variability of the *in situ* data for this network.

Results presented here are in line with recent published evaluations of ASCAT against *in situ* observations (e.g., Parrens et al. 2012). *In situ* SM can be used to monitor the quality of satellite estimates (Albergel et al. 2010). They can be considered highly accurate, but they are representative of a limited integration volume; spatial variability of SM is very high and can vary from centimeters to meters. Precipitation, evapotranspiration, soil texture, topography, vegetation, and land use could either enhance or reduce the spatial variability of SM, depending on how it is distributed and combined with other factors (Famiglietti et al. 2008). Several studies have confirmed, however, that SM measured at a specific location is correlated with the mean SM content derived from coarser resolution products. SM variations in space and time can be related to small- and large-scale components (Entin et al. 2000). The large-scale component is related to the atmospheric forcing (precipitation and evaporation processes) and the small-scale component is mainly due to soil properties, land cover attributes, and local topography. The temporal stability concept proposed by Vachaud et al. (1985) indicates that SM patterns tend to persist in time and therefore that SM observed at a single point is often highly correlated with the mean SM content over an area. Additionally, to the representativeness issue, the sampling depth might also induce discrepancies. ASCAT provides a relative measure of the SM content of the soil sensed by C-band microwaves (0.5–2 cm), while *in situ* data at a depth of 5 cm were used in this case study. The upper layer of the soil is more subjected to fast drying/rewetting; hence SM variations are more pronounced. During a rainfall event, this can lead to a temporal shift between the time when the upper layer SM increases and the time required by water to percolate to 5 cm. This effect tends to decrease the statistical scores, particularly R of the anomaly time series. Also, a decoupling of the 0.5–2 cm layer with the 5-cm observations may develop at day time. One may note, however, that an estimation of SM in the deeper layer of soil can be obtained from satellite measurements using filtering techniques (e.g., the exponential filter; Wagner et al. 1999; Albergel et al. 2008; Brocca et al. 2011) and data assimilation (Entekhabi et al. 1994; Walker et al. 2001).

18.6 Conclusions

The validation of remotely sensed surface SM is challenging as there is no agreed on standard or best practices yet on how such validation should be done amongst the scientific community. Despite the known issues dealing with the comparison between satellite data and *in situ* observations (e.g., representativeness of point scale measurements and sampling depth), the latter are very useful for the monitoring of satellite-based SM products, as shown in the recent literature. Even if local *in situ* observations of SM do not measure the same quantity as coarse-resolution remotely sensed products, significant correlations can be obtained between the two measures. A complementary approach for the evaluation is to consider LSM estimates; their large-scale nature is more representative of the one of remotely sensed products. They suffer, however, from uncertainties related to the model itself and its input data. Like *in situ* measurements, they can be used for a direct comparison with remotely sensed data and also in more advanced statistical methods such as the triple collocation, which were recently introduced in the field of satellite-derived SM data.

In this chapter the most common metrics found in the literature to evaluate SM retrieval accuracy were presented: the correlation coefficient (R), the RMSD, the bias, and the unbiased RMSD. They were used along with the SDV and the centered unbiased RMSD (E) to evaluate the ability of ASCAT remotely sensed surface SM to match *in situ* observations from two networks in southwestern France (SMOSMANIA) and Spain (REMEDHUS) over a 5-year period (2007–2011). Correlations were computed for both absolute and anomaly time series. It is underlined that they should always be computed together and significant cases only (p value < 0.05 for the 95% confidence interval) should be considered. While R calculated on absolute values indicates the ability of ASCAT product to represent the annual SM cycle, R calculated on anomaly time series (R_ANO) gives an indication about its ability to represent SM short-term variability (e.g., to detect precipitation event). Specifically, the latter metric, R_ANO, can be more significant if satellite SM observations are assimilated into hydrological or meteorological modeling. In this case, a bias correction is usually performed before the assimilation, and hence the capability of satellite data to correctly represent the deviation from the mean is more useful (e.g., Koster et al. 2009; Brocca et al. 2012). Therefore the selection of the performance metric to be used is also dependent on the specific application for which the data need to be employed (Entekhabi et al. 2010). RMSD provides useful information; however, for evaluating quantities with different units, one may prefer the unbiased RMSD, ubRMSD, or the centered unbiased RMSD (E), which does not consider the possible bias and takes into account also the variability of *in situ* observations. R, E, and SDV have the advantage of being easily displayed in a single diagram (the so-called Taylor diagram) from which the capability of the evaluated SM product to match the behavior of the observations is easy to interpret.

Acknowledgment

Authors thank the EUMETSAT Satellite Application Facility on Support to Operational Hydrology and Water Management (HSAF) for their funding support.

References

Albergel, C., J.-C. Calvet, P. de Rosnay, G. Balsamo, W. Wagner, S. Hasenauer, V. Naemi, E. Martin, E. Bazile F. Bouyssel, and J.-F. Mahfouf, 2010: Cross-evaluation of modelled and remotely sensed surface soil moisture with in situ data in southwestern France, *Hydrol. Earth Syst. Sci.*, 14, 2177–2191, doi:10.5194/hess-14-2177-2010.

Albergel C., P. de Rosnay, C. Gruhier, J. Muñoz-Sabater, S. Hasenauer, L. Isaksen, Y. Kerr, and W. Wagner, 2012: Evaluation of remotely sensed and modelled soil moisture products using global ground-based in situ observations, *Remote Sens. Environ.*, 118, 215–226, doi:10.1016/j.rse.2011.11.017.

Albergel, C., C. Rüdiger, D. Carrer, J.-C. Calvet, N. Fritz, V. Naeimi, Z. Bartalis, and S. Hasenauer, 2009: An evaluation of ASCAT surface soil moisture products with in situ observations in southwestern France, *Hydrol. Earth Syst. Sci.*, 13, 115–124, doi:10.5194/hess-13-115-2009.

Albergel, C., C. Rüdiger, T. Pellarin, J.-C. Calvet, N. Fritz, F. Froissard, D. Suquia, A. Petitpa, B. Piguet, and E. Martin, 2008: From near-surface to root-zone soil moisture using an exponential filter: An assessment of the method based on in situ observations and model simulations, *Hydrol. Earth Syst. Sci.*, 12, 1323–1337, doi:10.5194/hess-12-1323-2008.

Bartalis, Z., W. Wagner, V. Naeimi, S. Hasenauer, K. Scipal, H. Bonekamp, J. Figa, and C. Anderson, 2007: Initial soil moisture retrievals from the METOP-A advanced Scatterometer (ASCAT), *Geophys. Res. Lett.*, 34, L20401, doi:10.1029/2007GL031088.

Brocca, L., S. Hasenauer, T. Lacava, F. Melone, T. Moramarco, W. Wagner, W. Dorigo et al., 2011: Soil moisture estimation through ASCAT and AMSR-E sensors: An intercomparison and validation study across Europe, *Remote Sens. Environ.*, 115(12), 3390–3408.

Brocca, L., F. Melone, T. Moramarco, W. Wagner, and S. Hasenauer, 2010: ASCAT soil wetness index validation through in situ and modelled soil moisture data in central Italy, *Remote Sens. Environ.*, 114(11), 2745–2755, doi:10.1016/j.rse.2010.06.009.

Brocca, L., T. Moramarco, F. Melone, W. Wagner, S. Hasenauer, and S. Hahn, 2012: Assimilation of surface and root-zone ASCAT soil moisture products into rainfall-runoff modelling, *IEEE Trans. Geosci. Remote Sens.*, 50(7), 2542–2555.

Calvet, J.-C., N. Fritz, F. Froissard, D. Suquia, A. Petitpa, and B. Piguet, 2007: In situ soil moisture observations for the CAL/VAL of SMOS: The SMOSMANIA network. In *International Geoscience and Remote Sensing Symposium*, IEEE, New Brunswick, NJ, doi:10.1109/IGARSS.2007.4423019.

Ceballos, A., K. Scipal, W. Wagner, and J. Martinez-Fernandez, 2005: Validation of ERS scatterometer-derived soil moisture data in the central part of the Duero-Basin, Spain. *Hydrol. Process.*, 19(8), 1549–1566.

Dee, D. P., S. M. Uppala, A. J. Simmons, P. Berrisford, P. Poli, S. Obayashi, U. Andrae et al., 2011: The ERA-Interim reanalysis: Configuration and performance of the data assimilation system, *Q. J. R. Meteorol. Soc.* 137, 553–597, doi:10.1002/qj.828.

Dorigo, W., K. Scipal, R. M. Parinussa, Y. Y. Liu, W. Wagner, R. A. M. de Jeu, and V. Naeimi, 2010: Error characterisation of global active and passive microwave soil, *Hydrol. Earth Syst. Sci.*, 14, 2605–2616, doi:10.5194/hess-14-2605-2010.

Dorigo, W. A., W. Wagner, R. Hohensinn, S. Hahn, C. Paulik, M. Drusch, S. Mecklenburg, P. van Oevelen, A. Robock, and T. Jackson, 2011: The International Soil Moisture Network: A data hosting facility for global in situ soil moisture measurements, *Hydrol. Earth Syst. Sci.*, 15, 1675–1698, doi:10.5194/hess-15-1675-2011.

Draper, C., J.-F. Mahfouf, J.-C. Calvet, E. Martin, and W. Wagner, 2011: Assimilation of ASCAT near-surface soil moisture into the French SIM hydrological model, *Hydrol. Earth Syst.*, 15, 3829–3841, doi:10.5194/hess-15-3829-2011.

Entin, J. K., A. N. Robock, K. Y. Vinnikov, S. E. Hollinger, S. Liu, and A. Namkhai, 2000: Temporal and spatial scales of observed soil moisture variations in the extratropics, *J. Geophys. Res.*, 105(D9), 11,865–11,877.

Entekhabi, D., H. Nakamura, and E. G. Njoku, 1994: Solving the inverse problem for soil moisture and temperature profiles by sequential assimilation of multifrequency remotely sensed observations, *IEEE Trans. Geosci. Remote Sens.*, 32, 438–448.

Entekhabi D., R. H. Reichle, R. D. Koster, and W. T. Crow, 2010: Performance metrics for soil moisture retrieval and application requirements, *J. Hydrometeorol.*, 11, 832–840, doi: 10.1175/2010JHM1223.1.

European Space Agency, 2010: SMOS calibration validation & retrieval plan, Ref. SO-PL-ESA-SY-3898. Available at http://earth.esa.int/pub/ESA_DOC/R9-SMOS-VRT-plan.pdf (Accessed March 12, 2012).

Famiglietti, J. S., D. Ryu, A. A. Berg, M. Rodell, and T. J. Jackson, 2008: Field observations of soil moisture variability across scales, *Water Resour. Res.*, 44, W01423, doi:10.1029/2006WR005804.

Hahn, S., and W. Wagner, 2011: Characterisation of calibration-related errors on the initial METOP ASCAT soil moisture product, *Geophys. Res. Abs.*, 13, EGU2011-3178.

Kerr, Y. H., P. Waldteufel, J.-P. Wigneron, S. Delwart, F. Cabot, J. Boutin, M.-J. Escorihuela et al., 2010: The SMOS mission: New tool for monitoring key elements of the global water cycle, *Proc. IEEE*, 98(5), 666–687.

Koster, R. D., Z. Guo, R. Yang, P. A. Dirmeyer, K. Mitchell, and M. J. Puma, 2009: On the nature of soil moisture in land surface models, *J. Climate*, 22, 4322–4335.

Matgen, P., S. Heitz, S. Hasenauer, C. Hissler, L. Brocca, L. Hoffmann, W. Wagner, and H. H. G. Savenije, 2012: On the potential of MetOp ASCAT-derived soil wetness indices as a new aperture for hydrological monitoring and prediction: A field evaluation over Luxembourg, *Hydrol. Process.*, 26(15), 2346–2359.

Parrens, M., E. Zakharova, S. Lafont, J.-C. Calvet, Y. Kerr, W. Wagner, and J.-P. Wigneronm, 2012: Comparing soil moisture retrievals from SMOS and ASCAT over France, *Hydrol. Earth Syst. Sci.*, 16, 423–440.

Rüdiger, C., J.-C. Calvet, C. Gruhier, T. Holmes, R. De Jeu, and W. Wagner, 2009: An intercomparison of ERS-Scat and AMSR-E soil moisture observations with model simulations over France, *J. Hydrometeorol.*, 10(2), 431–447, doi:10.1175/2008JHM997.1.

Sánchez, N., J. Martínez-Fernández, A. Scaini, and C. Pérez-Gutiérrez, 2012: Validation of the SMOS L2 soil moisture data in REMEDHUS network (Spain), *IEEE Trans. Geosci. Remote Sens.*, 50, 1602–1611.

Saltelli, A., 2002: Sensitivity analysis for importance assessment, *Risk Analysis*, 22(3), 579–590, doi:10.1111/0272-4332.00040.

Saltelli, A., K. Chan, and E. M. Scott, 2000: *Sensitivity Analysis*. John Wiley, New York.

Schervish, M. J., 1996: P values: What they are and what they are not, *Am. Stat.*, 50(3), 203–206, doi:10.2307/2684655.

Schwieger, V., 2004. Variance-based sensitivity analysis for model evaluation in Engineering Surveys. INGEO 2004 and FIG Regional Central and Eastern European Conference on Engineering Surveying Bratislava, Slovakia, November 11–13, 2004.

Schmugge, T. J., 1983: Remote sensing of soil moisture: Recent advances, *IEEE Trans. Geosci. Remote Sens.*, GE21, 145–146.

Scipal, K., T. Holmes, R. de Jeu, V. Naeimi, and W. Wagner, 2008: A possible solution for the problem of estimating the error structure of global soil moisture datasets, *Geophys. Res. Lett.*, 35, L24403, doi:10.1029/2008gl035599.

Stanski, H. R., L. J. Wilson, and W. R. Burrows, 1989: Survey of common verifications methods in meteorology. *WMO World Weather Watch Tech. Rep. 8*, WMO/TD 358, 114 pp.

Stoffelen, A. 1998: Toward the true near-surface wind speed: Error modeling and calibration using triple collocation, *J. Geophys. Res.*, 103, 7755–7766, doi:10.1029/97jc03180.

Taylor, K. E., 2001: Summarizing multiple aspects of model performance in a single diagram, *J. Geophys. Res.*, 106, 7183–7192.

Vachaud, G., A. Passerat de Silans, P. Balabanis, and M. Vauclin, 1985: Temporal stability of spatially measured soil water probability density function, *Soil Sci. Soc. Am. J.*, 49, 822–828.

Wagner, W., G. Lemoine, and H. Rott, 1999: A method for estimating soil moisture from ERS scatterometer and soil data, *Remote Sens. Environ.*, 70, 191–207.

Wagner, W., V. Naeimi, K. Scipal, R. de Jeu, and J. Martinez-Fernandez, 2007: Soil moisture from operational meteorological satellite, *Hydrogeol. J.*, 15(1), 121–113.

Walker, J. P., G. R. Willgoose, and J. D. Kalma, 2001: One-dimensional soil moisture profile retrieval by assimilation of near-surface measurements: A simplified soil moisture model and field application, *J. Hydrometeorol.*, 2, 356–373.

Willmott, C. J., S. G. Ackleson, R. E. Davis, J. J. Fedema, K. M. Klink, D. R. Legates, J. O'Donnell, and C. M. Rowe, 1985: Statistics for the evaluation and comparison of model, *J. Geophys. Res.*, 90(C5), 8995–9005.

Willmott, C. J., S. M. Robeson, and K. Matsuura, 2011: A refined index of model performance, *Int. J. Climatol.*, 32, 2088–2094, doi:10.1002/joc.2419.

Zwieback, S., K. Scipal, W. Dorigo, and W. Wagner, 2012. Structural and statistical properties of the collocation technique for error characterization, *Nonlinear Processes Geophys.*, 19, 69–80.

Section IV

Challenges and Future Outlook

19

Global-Scale Estimation of Land Surface Heat Fluxes from Space: Current Status, Opportunities, and Future Directions

Matthew McCabe, William Kustas, Martha Anderson,
Cezar Kongoli, Ali Ershadi, and Christopher R. Hain

CONTENTS

19.1 Introduction

While considerable progress has been made in the development of global flux products from space (see Chapter 10), there remain a number of issues that either limit the application of these data to their fullest extent or provide an inherent constraint on the accuracy achievable. This is particularly true when using remote sensing–based information but is also pertinent to those approaches driven by reanalysis or produced via other model-based meteorological data streams or outputs, such as land surface, regional, or global climate model (GCM) simulations. These issues fall largely (but not exclusively) into two categories: (1) model physics and structural limitations (including the provision of information to drive the models) and (2) model assessment and interpretation of simulations.

In terms of describing the model physical processes, most techniques used in global flux estimation are based on the Penman–Monteith (see Chapter 10) or a related form, due in large part to the relative simplicity of the approach, but also the availability of the needed forcing data: a key requirement in broad and general application of any model approach. However, the necessary compromise between simplicity and complexity that is required for global application means that some processes will either not be described or be described poorly. For instance, estimating evaporation over wetlands and marshes, forests, arid lands, and snow-covered surfaces presents unique challenges that are unlikely to be represented within a single modeling scheme. Yet, when considered globally, these (and other) land surface types comprise a significant component of the Earth's terrestrial surface. Developing multimodel or ensemble type modeling approaches that consider the inherent variability and complexity of the land surface and meteorological conditions may provide one means of addressing some of these deterministic model-based limitations.

Even though we have current capacity to produce long time series of global flux retrievals (see Chapter 10), considerable research effort is required to assess and evaluate the fidelity of these simulations. Unfortunately, there is no strong foundation from which to undertake such product evaluation, particularly as relates to large-scale and long time series products. The traditional approach to model evaluation has its heritage in small-scale, relatively short time period simulation, where a single stream flow record or flux tower response is sufficient to assess a model's reproduction. Once the scale of retrieval moves beyond the tower-based pixel level, the capability to evaluate model output becomes much more complicated. Issues not just of spatial scale but also of temporal resolution come into play, making a reasoned assessment of model simulations a major undertaking. The fact that there are no accepted set of tools with which to undertake such assessments further complicates the matter. For example, in thoroughly evaluating global, multiscale, multimodel flux simulations, consideration of (1) the statistical distribution of the data (do the responses reflect available observations?); (2) the spatial structure of the fields (do patterns reflect reality or are large-scale structures present?); or (3) the responses with other linked hydrological processes (i.e., are they "hydrologically consistent"?) should all form some element of the assessment process—but this is rarely the case.

Clearly, there remain considerable challenges and uncertainties in the development of global flux retrievals—at least those in which some confidence of their accuracy can be placed. However, this is not to say that there are not also great opportunities that come from analyzing the global products generated by these models. These opportunities are not just in advancing scientific techniques or in the development of new modeling tools and approaches for product development and assessment but also in the provision of fundamental knowledge and characterization of key Earth system processes, particularly in describing their variability, change, and trend. Such information has the potential to reshape our understanding of hydrological and climatological sciences by offering a truly global representation of these processes.

The following contribution seeks to describe some of the key issues requiring consideration in the development and use of global surface flux retrievals. It also provides some comment on the opportunities and needed areas of research to advance our current capability. It is by no means an exhaustive examination but represents an alternative perspective on some of the issues requiring attention, beyond those often discussed in the literature related to this topic. For a far more comprehensive examination of challenges and issues related to the estimation of evaporation, the reader is referred to recent reviews by Kalma et al. (2008) and Wang and Dickinson (2012).

19.2 The Global Energy and Water Exchanges Landflux Initiative

The GEWEX Data and Assessments Panel (GDAP) initiated the LandFlux Project in 2007 with an aim to develop a multidecadal global reference terrestrial surface turbulent heat flux dataset primarily for evaporation but also providing coverage of the sensible heat flux at the land surface. The principal purposes of the LandFlux dataset is to fulfill the GDAP goal of guiding the "… production and evaluation of long term, globally complete atmospheric and consistent water and energy budget datasets." LandFlux represents an effort to develop one of the last remaining elements of an observationally based suite of products that seek to describe the water and energy cycles. Once accomplished, the complete dataset will allow the capacity for independent assessment of global water and energy cycles. Further details of these efforts can be found at www.gewex.org/GDAP.html.

One of the other major motivations behind this effort was the need to fill a gap in the capacity to evaluate a range of global flux estimates, particularly those being delivered as part of GCM, reanalysis, and land surface modeling (LSM) activities. As part of this effort, the LandFlux-EVAL project was initiated to assist in evaluating and comparing recently developed global datasets and to provide a benchmark for the upcoming LandFlux product and other applications (see http://www.iac.ethz.ch/url/research/LandFlux-EVAL) (Jiménez et al. 2011; Mueller et al. 2011). The work of Jiménez et al. (2011) compares many of the approaches that have been described in the preceding sections (i.e., Fisher et al. 2008; Wang and Liang 2008; Jiménez et al. 2009; Jung et al. 2010; Sheffield et al. 2010). More recent (and ongoing) LandFlux comparisons have included the algorithms by Mu et al. (2007), Zhang et al. (2010), and Miralles et al. (2011b).

Results of these early comparisons demonstrate that large differences still exist between products (see Section 10.3.1), despite the fact that most of the methodologies have been successfully validated using eddy covariance measurements from flux tower networks. The absolute performance of these different algorithms is yet to be explored, and to a certain extent, the creation of future, high-quality blended evaporation datasets relies on the investigation of this performance. One of the major limitations has been the lack of a common set of input data with which to run the models and the existence of a comprehensive set of meteorological measurements with which to compare estimates against (see Section 10.4.3). Toward addressing some of these limitations, the initial LandFlux efforts have recently been supported by the launch of an ESA-funded project referred to as the Water Cycle Multi-mission Observation Strategy—EvapoTranspiration (WACMOS-ET). As one of its main objectives, WACMOS-ET is preparing a reference input dataset that can be used by project partners and other interested collaborators to facilitate the running and validation of existing algorithms under an internally consistent forcing framework, an activity that will undoubtedly assist the advancement of accurate global flux estimation.

Regardless of the challenges, with community participation the overall LandFlux effort has provided considerable insights into the range and variability of flux retrievals globally and seasonally, increased the understanding of the limits and sensitivities of common flux estimation approaches, and helped guide the production of a Version 0 product, which will allow for enhanced assessment of globally distributed terrestrial heat fluxes (see Sections 10.3.1 and 10.3.2).

19.3 Issues and Opportunities in Global Flux Retrieval

The following is a short review of some of the issues and opportunities that have arisen as a result of efforts to provide improved representation of terrestrial surface heat fluxes at global scales. While certainly not an exhaustive or comprehensive list, it is aimed more at identifying those key gaps or issues that have been identified as part of recent community efforts in evaluating flux approaches. Likewise, some opportunities, both in terms of using these new data and also in research needs to improve on these, are briefly discussed.

19.3.1 Meteorological and Surface Temperature Data Needs

From a global modeling perspective, one of the key challenges in developing retrievals is the provision of adequate (and accurate) forcing datasets with which to drive model simulations. For those models relying on thermal infrared land surface temperature retrievals, the challenges are even more acute. Many previous studies have examined the role of surface temperature on heat flux estimation, with issues such as spatial and temporal resolution (McCabe and Wood 2006; Li et al. 2008) and land surface heterogeneity and atmospheric (i.e., clouds and water vapor) influences (Raupach and Finnigan 1995; Kustas et al. 2004; Van Niel et al. 2012) confounding retrievals. No detailed discussion on these is presented here, as the topic has been extensively covered in the existing literature. It is worth noting, however, that the global flux community has (rightly or wrongly) largely shifted toward Penman–Monteith or Priestley–Taylor type schemes as a means of sidestepping the issue, because these do not rely on the provision of a surface temperature. For those approaches that do require surface temperature, the problems of accurate and repeated land surface temperature remain. The Global Landsurface Evaporation—the Amsterdam Methodology (GLEAM) (Miralles et al. 2011a) approach has incorporated microwave retrievals as a means of reducing the issues of cloud contamination, but the accuracy and representativeness of large-scale temperature responses are still present. Alternatively, data assimilation approaches coupled to water balance models greatly reduce the need for frequent land surface temperature observations and could be a tool for interpolating between cloudy conditions (Crow et al. 2008).

In terms of evaluation and intercomparison of flux retrievals, an additional challenge is the development of a common forcing dataset. Disentangling the issues of within-model variability from those related to input uncertainty is an ongoing concern. Indeed, this principal has been a guiding directive in the development of the GDAP integrated water and energy cycle product. The GDAP integrated product will be deliberately derived from a common forcing data source and include a consistent temporal period and spatial resolution across all of the products. It is expected that this will then allow an independent, observationally based assessment of global water and energy cycle closure: this is something that has not been achieved to date and that represents a grand challenge to the hydrological community.

The increasing availability of gridded meteorological sets (Sheffield et al. 2006), output from numerical weather prediction schemes (Evans and McCabe 2010), and data from reanalysis runs has aided the proliferation of derived water and energy cycle products that use these as a basis to drive process models. While adding value to studies of the hydrological cycle, they also raise the issue of forcing inconsistencies and model sensitivities. For models of evaporation, one of the most important drivers is the radiation inputs, and a number of authors have illustrated the sensitivity of simulations to the source of radiative forcing (Jiménez et al. 2011). Likewise, for those schemes that rely on surface–atmosphere temperature gradients, the influence of uncertainties in the land surface temperature can be severe. Estimation of the surface temperature, whether remotely or *in situ*, is complicated by issues of spatial and temporal resolution, land surface heterogeneity, and atmospheric and land surface effects, all of which combine to make its retrieval challenging.

Furthermore, influencing both radiative forcing and the linked surface temperature response is the confounding impact of cloud cover, which inevitably increased the degree of variation in radiative flux reaching the surface and has an ill-defined impact on surface fluxes (Catherinot et al. 2011; Jiménez et al. 2012). Understandably, most evaporation models and approaches tend to ignore this response, even though it has the potential to affect the accuracy of retrievals significantly.

Further investigation on the impact of forcing type and the influence of uncertainties in forcing variables is required. Indeed, this may provide an alternative technique to understanding model retrieval accuracy: that is, if input uncertainty can be reasonably defined, the application of statistical approaches should then be able to determine subsequent model output uncertainty. Some efforts to address this question in relation to flux retrieval are currently underway (see Kavetski et al. 2006a; Ershadi et al. 2013a).

19.3.2 Issues of Temporal and Spatial Resolution

Identifying the spatial and temporal resolution for global product development is as much a matter of developer choice and experience as it is guided by the intended application of the product itself. The general preference, if not necessarily the need, is for increasingly higher spatial and temporal resolution. Ultimately, the timescale and space scale will be dictated by the availability of the forcing data, although techniques for disaggregating products in both of these domains exist and are routinely applied. In terms of the spatial resolution, the options are inevitably driven by the experience (or foresight) of the product developer(s): global and climate related application will generally provide course spatial (tens to hundreds of kilometers) and temporal (monthly to annually) specifications, while regional hydrological and water resource or agricultural driven products will seek higher spatial (tens to thousands of meters) and temporal (subhourly to daily) resolution.

Ignoring the ever-diminishing constraints with respect to computational power, storage, and bandwidth (for product delivery), the relevance and importance of process scale remain underappreciated. That is, determining what is the appropriate scale for the process under consideration is generally a secondary consideration. As a result, the questions of scale as relates to evaporative fluxes revolve around land surface heterogeneity issues, the expansion in time of retrievals (Crago and Brutsaert 1996; Van Niel et al. 2012), or the interpolation in space (and time) of forcing data and model output (Van Niel et al. 2011; Ershadi et al. 2013b). Answering these questions is complicated by the lack of an appropriate scaling theory in the hydrological sciences, so the default response is higher retrieval resolution, regardless of whether it is needed, warranted, or even meaningful.

What is required, particularly in terms of evaluation of multimodel responses, is an understanding of the spatial and temporal variability of the forcing data, model output, and also models themselves. This is particularly important in determining precisely where (and when) model schemes are appropriate to employ. Understanding how different models of evaporation scale in space (in terms of varying the model resolution) and time (aggregating either the input data or model output to different temporal scales) requires further investigation (Ershadi et al. 2013b). Assessing models on their consistency (or more likely differences) in reproducing probability density functions across scales may be one approach to examine this, but there are sure to be numerous others measures that can provide insight into model performance, consistency, and reliability across scales.

The assumption (at least in application) of scale invariance in the modeling approaches (see Section 10.2.2) is not generally supported by the underlying theoretical considerations of the techniques. One of the advantages of pursuing global application of these different evaporation schemes is that they can be readily assessed more broadly, albeit beyond the site-specific application that developers often intended. The findings of such analyses and intercomparison studies will provide deeper insight into the general application of these approaches and also help to define what are the appropriate space scale and timescale at which to develop and apply these products.

19.3.3 Mapping of Sublimation and Evaporation over Snow-Dominated Landscapes

The sublimation of snow (conversion of snow particles from ice to vapor) and the evaporation of melted snow are important components of the hydrology of many snow-dominated regions. Snow sublimation is a key component of the Arctic water balance, typically representing between 10% and 50% of total winter precipitation but capable of representing nearly 100% (Liston and Sturm 2004). It is also a significant component of the water budget in many seasonally snow-covered coniferous forests (Molotch et al. 2007), exceeding 30% of total snowfall (Montesi et al. 2004), as well as an important aspect of Alpine water balances, where it can account for nearly 20% of total annual snowfall (MacDonald et al. 2010). Snow sublimation occurs from static surfaces (snow on the ground and snow intercepted by forest canopy) as well as via blowing snow. It is a complex process that depends on various environmental factors such as air temperature, humidity, and wind speed variations associated with changing weather patterns and space-dependent variations related to local surface roughness, vegetation, aspect, and proximity to open-water bodies.

Despite the progress made over the past three to four decades in the science and the modeling of evapotranspiration by remote sensing (see Chapters 3 and 10), substantial gaps remain in measuring, understanding, and modeling evaporative losses associated with frozen precipitation (Shuttleworth 2007). The state of knowledge on sublimation from snow-vegetation surfaces is particularly poor. Even now, canopy interception processes are poorly understood, and the evolution of intercepted snow on canopies is inadequately described. Questions remain as to how sublimation rates vary with forest density, canopy type, and atmospheric conditions (Liston and Sturm 2004). There is also very little understanding and modeling of evaporation from snow under vegetation canopies, along with evaporation of melting snow under patchy conditions (Shuttleworth 2007). Over snow-covered landscapes, a few measurements of latent heat fluxes exist for model validation. Measurements over snow-vegetation surfaces present even greater challenges (Molotch et al. 2007; Marks et al. 2008; Flerchinger et al. 2012) and thus validation datasets are limited.

As a consequence, studies on the modeling of snow evaporation/sublimation using remote sensing–based methods are very limited. In a recent contribution, Vinukollu et al. (2011b) applied three process-based models that use remote sensing data to estimate global evapotranspiration: a thermal-based single source energy budget model (SEBS; see Chapter 10), a Penman-Monteith based approach, and a Priestley-Taylor based approach (see Chapter 10 for further details of these models). In this instance, latent heat flux over snow was calculated for all the models based on a modified Penman equation, but sublimation fluxes (for subfreezing air temperatures) were not considered. Liu et al. (2003) incorporated sublimation from forest canopy-intercepted snow and snow on the ground in a remote sensing–based process model, referred to as the Boreal Ecosystem Productivity Simulator, to estimate evapotranspiration over all the Canadian landmass. However, evapotranspiration validation results over the snow components were not reported. Bowling et al. (2004) incorporated a blowing snow model in the variable infiltration capacity (VIC) macroscale hydrology model. The VIC model, however, does not use remote sensing inputs and is difficult to validate with routine ground measurements.

Given the challenges described above and the importance of snow in the global water balance, it is imperative that remote sensing–based process models for estimating latent heat fluxes over snow be developed and tested. An example of a more physically based remote sensing model for evaluating snow evaporation–sublimation was recently reported by Kongoli et al. (2013) and Kustas et al. (2012). Model parameterizations were adopted from the physically based two-source energy balance (TSEB) modeling scheme (Norman et al. 1995), which estimates fluxes from the vegetation and surface substrate separately using remotely sensed measurements of land surface temperature. TSEB has been applied routinely over snow-free surfaces over the continental United States using geostationary weather satellite data and a time-integrated boundary layer growth modeling system for monitoring evapotranspiration (Anderson et al. 2007a) and drought (Anderson et al. 2007b, 2010). With the addition of a snow module, the application of TSEB modified for snow (TSEBS) can be extended to snow-covered areas, which are currently masked or removed from these routine analyses.

A detailed assessment of the TSEBS model at the micrometeorological scale was made recently by comparing the turbulent heat fluxes computed by TSEBS with half-hourly field measurements from eddy covariance systems at two sites in a rangeland ecosystem in southwestern Idaho: one site dominated by aspen vegetation and the other by sagebrush. Comparisons between modeled and measured turbulent heat showed good agreement at both sites, demonstrating the robustness of TSEBS parameterizations for estimating turbulent fluxes over snow-dominated landscapes. The model behavior was also consistent with previous studies that indicate the occurrence of upward sensible heat fluxes during daytime owing to solar heating of vegetation limbs and branches, which often exceeds the downward sensible heat flux driving the snowmelt (e.g., Bewley et al. 2010). However, model simulations over aspen trees showed that the upward sensible heat flux could be reversed for a lower canopy fraction due to the dominance of downward sensible heat flux over snow. This indicates that reliable vegetation and snow cover fraction inputs to the model are needed for estimating fluxes over snow-covered landscapes.

Modifications of existing models or the development of new remote sensing–based processed models for estimating snow evaporation/sublimation are urgently needed in order to improve regional and global evapotranspiration estimates and for the capability of monitoring the effects of global climate change on regional hydrology and evapotranspiration. For example, Strack et al. (2003) show that changing the land-cover specification in the Regional Atmospheric Modeling System from one that was masked by the snow (crop stubble) to one that had protruding vegetation (shrubs) produced sensible heat flux increases as large as 80 W m^{-2} during the day. The increased sensible heat flux produced a 60°C increase in afternoon air temperatures and a 200–300 m increase in the afternoon boundary layer height. Such changes at the regional scale in local climate conditions are likely to accelerate changes in vegetation composition and induce feedback affecting evapotranspiration rates and components of the hydrologic cycle, causing irreversible changes to the environment, elements of which are already appearing in tundra ecosystems (McCartney et al. 2006).

19.3.4 Assessing the Capabilities of GCMs

One of the purposes of simulating the Earth system with GCMs is to assess the impact of increasing CO_2 on future climate evolution. Uncertainties in such climate simulations are strongly linked to the representation of land processes through LSMs (Seneviratne et al. 2010). Comprehensive studies on the influence of uncertainties in LSMs on predicted global temperatures into the future are still missing, but evaporation has already been identified as one of the critical variables (Crossley et al. 2000; Boé and Terray 2008) and the influences of changes in evaporation on surface temperature have been shown to be large, especially in transitional climate regions and in regions shifting from humid to transitional climate conditions (Seneviratne et al. 2006, 2012; Wei et al. 2011). Because of the recognition of the importance of evapotranspiration for the climate system as well as for climate models, its representation in LSMs and reanalyses has been and still is the subject of continual improvements.

Acknowledging the important role that evaporation plays within the climate system highlights the need for independent datasets with which to assess the many GCM simulations that are available to monitor and predict Earth system behavior. In a recent effort at examining this issue, Mueller et al. (2011) compared a range of observationally forced model sets (process models, LSMs, and reanalysis) with Intergovernmental Panel on Climate Change (IPCC) Fourth Assessment Report (AR4) GCM simulations. The authors found that the range and variety of models tended to agree broadly at the global scale but exhibited large regional differences. Indeed, in their global averages, the IPCC AR4 simulations generally showed a smaller spread in comparison to the other models.

Determining the fidelity of global (and regional) climate models is perhaps one of the most important applications of these independent datasets, particularly in terms of evaluating the expected range and variability of climate model response relative to the process-driven simulations. In this example in particular, the role and influence of spatial and temporal scale will play an integral role, highlighting again the linked nature of exploiting these potential opportunities by overcoming current limitations.

19.3.5 Evaluation and Assessment of Surface Heat Flux Products

One of the recurring elements of this chapter is the role of product assessment and evaluation of global flux retrievals. Efforts by Vinukollu et al. (2011a,b) have shown that so-called operational analysis models reproduce the latitudinal profile of the climatologically inferred evaporation but with a tendency to underestimate this quantity in midlatitudes. These types of analyses highlight that climatologically inferred evaporation (see Section 10.3.3), as an estimate for large basin and global evaporation, can provide a useful surrogate or complementary source of data for actual *in situ* evaluation of remote sensing estimates.

Another technique to assess the fidelity of remotely sensed flux retrievals is via the concept of "hydrological consistency," as detailed by McCabe et al. (2008). The idea is conceptually similar to the water budget analysis in that it seeks to exploit the expected behavior of the hydrological cycle. The approach is suited to the temporal analysis of trends in retrievals, as opposed to examining quantitative changes. Essentially, the technique requires independent measures of linked hydrological components (i.e., soil moisture, rainfall, and storage changes) to assess whether changes in observed elements are consistent with the variable being examined; that is, at the pixel scale, a rainfall response should be identified by a subsequent soil moisture change, which, in turn, should illicit an increased evaporation signal, relative to a drier period. Given the paucity of *in situ* observation data and the inherent uncertainties in quantitative retrievals, a range of evaluation approaches are preferred to any single assessment.

Ongoing evaporation intercomparison exercises (such as those described in Section 10.3) are useful in revealing general trends and agreements between and within models. However, in evaluation of the multimodel evaporation products, it is important to explicitly identify the errors and uncertainties that are inherent in input forcing, model parameters, model structure, and observed response. There are a number of existing methods that can aid in this task. For example, the "triple collocation" (Stoffelen 1998; Janssen et al. 2007; Scipal et al. 2008) approach has previously been used for an assessment of the errors in soil moisture datasets. Miralles et al. (2011a) implemented the triple collocation approach to assess the error structure of GLEAM, using a reanalysis evaporation estimate (MERRA) and a PM-based approach (Sheffield et al. 2010). However, the application of triple collocation for error assessment of evaporation datasets is complicated due to the requirement of three independent datasets for the analyses, a requirement that is often challenging for surface fluxes.

A potentially useful approach for intercomparison and joint evaluation of evaporation products is the Bayesian family of methods (Gelman et al. 2004). Methods such as the Bayesian Total Error Analysis (Kavetski et al. 2003, 2006b) have been widely used in the quantification of the input, model, and response errors in conceptual hydrological models (Renard et al. 2008, 2010). However, their use in examining evaporation models is limited, due mainly to the complexity of the process calculations. They do, however, represent a powerful technique that can explicitly account for the different sources of error that propagate through model simulations (Ershadi et al. 2013b).

Perhaps the major obstacle toward achieving robust and accurate globally distributed evaporation estimates is the lack of long-term and high-quality ground-based validation data. The issue is even more pronounced for evaporation estimation, as many of the key inputs (i.e., land surface temperature and radiation components) also lack effective globally distributed *in situ* validation sites. The need for validation sites that adequately reflect a range of climate zones, surface conditions, and atmospheric states severely restricts the capacity to which techniques can be robustly assessed.

19.3.6 Prediction from Global to Landscape and Field Scales

In our efforts to generate and evaluate global evapotranspiration products, there is a significant scale that is often overlooked but which can have tremendous impact on model estimates of evaporation and surface energy balance, namely, the scale of variability in land use and land cover (Anderson et al. 2004, 2012). Model intercomparisons at field-scale resolutions (e.g., <100 m) allow identification of specific land cover conditions (e.g., fallow fields, sparse shrubs, riparian areas, and forested patches) associated with significant discrepancies between different model formulations (Timmermans et al. 2007; Choi et al. 2009; Tang et al. 2011). In addition, at this resolution most ground-based observational footprints (i.e., measurements from eddy covariance systems) are well resolved, and a scale appropriate comparison with observed fluxes can be conducted. Direct comparisons between observations and model estimates at coarser scales (see Figure 19.1) are often confounded by subpixel heterogeneity in surface conditions and can really only provide general guidance as to actual model performance (Anderson et al. 2007a).

In contrast to the scale invariance assumed in most flux model applications, only a few modeling systems provide a methodology that explicitly bridges the gap between estimates at the land use/field scale and landscape to regional scale flux distributions (Norman et al. 2003; Anderson et al. 2011). Such approaches have focused mainly on thermal infrared (TIR) based techniques (Kalma et al. 2008). The land surface temperature retrieved using TIR remote sensing provides proxy information about surface moisture status at scales unattainable through other remote sensing wavebands (e.g., microwave) or modeling approaches. As described in Chapter 8, using the existing suite of TIR imaging satellites,

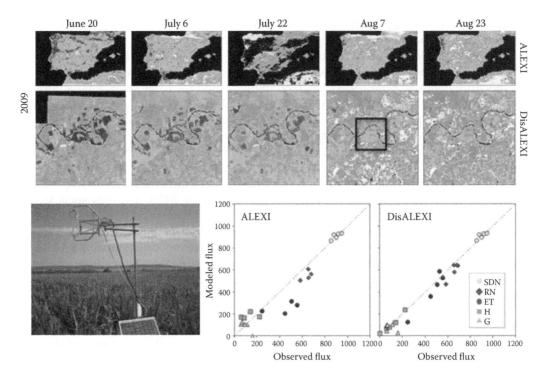

FIGURE 19.1

(See color insert.) Example of ALEXI and DisALEXI retrievals over Spain, illustrating the importance of land cover classification and changing land use on flux estimation. The ALEXI model (10-km pixels, indicated by the black box) underpredicts the flux values when compared to *in situ* eddy covariance data as a result of including un-irrigated land in the pixel footprint, which is resolved by DisALEXI (30-m pixels). Eddy covariance observations are located within the irrigated field and adjacent to a river.

evaporation maps can be routinely generated at subfield scales with sensors like Landsat or ASTER, at kilometer scales with MODIS or AVHRR, and at larger scales (>3 km) from geostationary platforms.

Such multiscale mapping algorithms can serve as an effective benchmarking tool in future global model intercomparison studies (see Section 10.3.2). Application over flux tower sites can facilitate upscaling of surface flux observations to the scale of the global modeling grids, effectively taking into account subpixel variations in the surface flux distribution. Over regions where the coarse-scale global models show significant discrepancy (e.g., including arid and cold climate zones and tropical forests; see Section 10.3.1), high-resolution diagnostic modeling may help to better understand local conditions that are causing divergent estimates and how these conditions can be better addressed within individual modeling frameworks.

Analyses of temporal variability from diagnostic TIR-based assessments at relatively fine scales allow identification of unknown sources of resiliency to moisture deficits that may not be well captured in coarser remote sensing and prognostic models. For example, relatively low variability in TIR-based retrievals of evaporative flux is observed in the Mississippi River Basin in comparison with surroundings, presumably because of access to shallow ground water and intensive irrigation activity (Anderson et al. 2011). Such dynamics may not be accurately reflected in assessments based on modeled water balance.

A pilot project generating global TIR-based surface flux maps using the Geoland2 LST and insolation data is currently underway for comparison with other global flux datasets generated under the GEWEX LandFlux initiative (Anderson et al. 2011). This implementation will use the Atmosphere–Land Exchange Inverse (ALEXI) modeling system (Anderson et al. 2007a) applied to geostationary datasets at ~20-km spatial resolution and an associated flux disaggregation algorithm (DisALEXI; Norman et al. 2003) that uses higher-resolution TIR imagery from polar orbiting platforms to downscale to the field or flux tower footprint scale. This will provide an additional TIR remote sensing model for the global LandFlux intercomparison suite, as well as a means for upscaling observations to the LandFlux model grid cell. ALEXI comparisons may also help to identify areas globally where evapotranspiration becomes partially decoupled from precipitation (e.g., due to groundwater contributions) or where land surface influences (such as irrigation practices) exert a strong influence on the local and regional scale flux (see Figure 19.1).

19.4 Concluding Comments

Regardless of the myriad issues in their production, the significant advances in global flux retrieval over the last few years have provided an expanded capacity to understand water cycle behavior, hydrological trends, and potential impacts of change in a new and exciting manner. The estimation of global fluxes for climatological studies is already well established, with a range of datasets available for retrospective analysis and comparison against LSM, reanalysis, and GCM output. The issues requiring attention so that accurate, near real-time retrievals can develop toward operational applications are certainly not insurmountable. However, it will require a concerted effort involving models and observations representing fluxes at multiple temporal and spatial resolutions to make significant advances in developing reliable global flux products. Although in their infancy and while not without problems, the development of these products, especially when driven by strong community need and guided by consistent protocols and assessment strategies, will provide a powerful capacity to expand our knowledge on the Earth and its hydrological regimes into the future.

Intercomparison studies between different modeling approaches over a range of ecosystem conditions, such as those currently implemented in the framework of the LandFLux GEWEX project, present an opportunity to quantify the relative strengths and limitations between models, as well as providing insight into their predictive capacity. To this end, the availability of ground observational networks, such as those from FLUXNET (see Chapter 1), provides the needed long-term measurements of relevant land surface parameters required to challenge model simulations and demand ongoing community support. This is important, as such data assist not only in validating satellite-derived methods and operational products but also in better understanding the relevant physical processes occurring across different observation scales (local, regional, continental) that are captured by remote sensing instruments.

Exploration of synergistic approaches for multisensor retrieval of hydrological variables represents a much needed focus of research. Data complementarity and interchangeability can assist in developing approaches that better address unresolved aspects of flux estimation, such as spatial and temporal scaling issues and also the impacts of land surface heterogeneity. With the planned launch of a number of new systems, including the SENTINELS

2/3 and Landsat 8, a capacity gap in high-resolution thermal infrared observations will be filled. The launch of these new systems may provide the impetus and opportunity for the development of synergistic multisensor/multiresolution techniques to advance the field of flux observation from space.

References

Anderson, M.C., Allen, R.G., Morse, A., and Kustas, W.P. (2012). Use of Landsat thermal imagery in monitoring evapotranspiration and managing water resources. *Remote Sensing of Environment, 122,* 50–65.

Anderson, M.C., Hain, C., Wardlow, B., Pimstein, A., Mecikalski, J.R., and Kustas, W.P. (2010). Evaluation of drought indices based on thermal remote sensing of evapotranspiration over the continental United States. *Journal of Climate, 24,* 2025–2044.

Anderson, M.C., Kustas, W.P., and Norman, J.M. (2007a). Upscaling flux observations from local to continental scales using thermal remote sensing. *Agronomy Journal, 99,* 240–254.

Anderson, M.C., Kustas, W.P., Norman, J.M., Hain, C.R., Mecikalski, J.R., Schultz, L., Gonzalez-Dugo et al. (2011). Mapping daily evapotranspiration at field to continental scales using geostationary and polar orbiting satellite imagery. *Hydrology and Earth System Sciences, 15,* 223–239.

Anderson, M.C., Norman, J.M., Mecikalski, J.R., Otkin, J.A., and Kustas, W.P. (2007b). A climatological study of evapotranspiration and moisture stress across the continental United States based on thermal remote sensing: 1. Model formulation. *Journal of Geophysical Research, 112,* D10117.

Anderson, M.C., Norman, J.M., Mecikalski, J.R., Torn, R.D., Kustas, W.P., and Basara, J.B. (2004). A multiscale remote sensing model for disaggregating regional fluxes to micrometeorological scales. *Journal of Hydrometeorology, 5,* 343–363.

Bewley, D., Essery, R., Pomeroy, J., and Ménard, C. (2010). Measurements and modelling of snowmelt and turbulent heat fluxes over shrub tundra. *Hydrology and Earth System Sciences, 14,* 1331–1340.

Boé, J., and Terray, L. (2008). Uncertainties in summer evapotranspiration changes over Europe and implications for regional climate change. *Geophysical Research Letters, 35,* L05702.

Bowling, L.C., Pomeroy, J.W., and Lettenmaier, D.P. (2004). Parameterization of blowing-snow sublimation in a macroscale hydrology model. *Journal of Hydrometeorology, 5,* 745–762.

Catherinot, J., Prigent, C., Maurer, R., Papa, F., Jiménez, C., Aires, F., and Rossow, W. (2011). Evaluation of "all weather" microwave-derived land surface temperatures with in situ CEOP measurements. *Journal of Geophysical Research, 116,* D23105.

Choi, M., Kustas, W.P., Anderson, M.C., Allen, R.G., Li, F., and Kjaersgaard, J.H. (2009). An intercomparison of three remote sensing-based surface energy balance algorithms over a corn and soybean production region (Iowa, U.S.) during SMACEX. *Agricultural and Forest Meteorology, 149,* 2082–2097.

Crago, R.D., and Brutsaert, W. (1996). Daytime evaporation and the self-preservation of the evaporative fraction. *Journal of Hydrology, 178,* 241–255.

Crossley, J., Polcher, J., Cox, P., Gedney, N., and Planton, S. (2000). Uncertainties linked to landsurface processes in climate change simulations. *Climate Dynamics, 16,* 949–961.

Crow, W., Kustas, W., and Prueger, J. (2008). Monitoring root-zone soil moisture through the assimilation of a thermal remote sensing-based soil moisture proxy into a water balance model. *Remote Sensing of Environment, 112,* 1268–1281.

Ershadi, A., McCabe, M.F., Evans, J.P., Mariethoz, G., and Kavetski, D. (2013a). A Bayesian analysis of sensible heat flux estimation: Quantifying uncertainty in meteorological forcing to improve model prediction. *Water Resour. Res., 49,* doi:10.1002/wrcr.20231.

Ershadi, A., McCabe, M.F., Evans, J.P., and Walker, J.P. (2013b). Effects of spatial aggregation on the multi-scale estimation of evapotranspiration. *Remote Sensing of Environment, 131,* 51–62.

Evans, J.P., and McCabe, M.F. (2010). Regional climate simulation over Australia's Murray-Darling basin: A multitemporal assessment. *J. Geophys. Res., 115*, D14114.

Fisher, J.B., Tu, K.P., and Baldocchi, D.D. (2008). Global estimates of the land-atmosphere water flux based on monthly AVHRR and ISLSCP-II data, validated at 16 FLUXNET sites. *Remote Sensing of Environment, 112*, 901–919.

Flerchinger, G.N., Reba, M.L., and Marks, D. (2012). Measurement of surface energy fluxes from two rangeland sites and comparison with a multilayer canopy model. *Journal of Hydrometeorology, 13*, 1038–1051.

Gelman, A., Carlin, J.B., Stern, H.S., and Rubin, D.B. (2004). *Bayesian Data Analysis,* 2nd ed. Chapman and Hall, London.

Janssen, P.A.E.M., Abdalla, S., Hersbach, H., and Bidlot, J.-R. (2007). Error estimation of buoy, satellite, and model wave height data. *Journal of Atmospheric and Oceanic Technology, 24*, 1665–1677.

Jiménez, C., Prigent, C., and Aires, F. (2009). Toward an estimation of global land surface heat fluxes from multisatellite observations. *J. Geophys. Res, 114*, D06305.

Jiménez, C., Prigent, C., Catherinot, J., Rossow, W., Liang, P., and Moncet, J.L. (2012). A comparison of ISCCP land surface temperature with other satellite and in situ observations. *Journal of Geophysical Research, 117*, D08111.

Jiménez, C., Prigent, C., Mueller, B., Seneviratne, S.I., McCabe, M.F., Wood, E.F., Rossow, W.B. et al. (2011). Global intercomparison of 12 land surface heat flux estimates. *J. Geophys. Res., 116*, D02102.

Jung, M., Reichstein, M., Ciais, P., Seneviratne, S.I., Sheffield, J., Goulden, M.L., Bonan, G., Cescatti, A., Chen, J., and de Jeu, R. (2010). Recent decline in the global land evapotranspiration trend due to limited moisture supply. *Nature, 467*, 951–954.

Kalma, J., McVicar, T., and McCabe, M. (2008). Estimating land surface evaporation: A review of methods using remotely sensed surface temperature data. *Surveys in Geophysics, 29*, 421–469.

Kavetski, D., Franks, S.W., and Kuczera, G. (2003). Confronting input uncertainty in environmental modelling, in calibration of watershed models. *Water Science and Application, 6*, 49–68.

Kavetski, D., Kuczera, G., and Franks, S.W. (2006a). Bayesian analysis of input uncertainty in hydrological modeling: 1. Theory. *Water Resources Research, 42*, W03407.

Kavetski, D., Kuczera, G., and Franks, S.W. (2006b). Bayesian analysis of input uncertainty in hydrological modeling: 2. Application. *Water Resources Research, 42*, W03408.

Kongoli, C., Kustas, W.P., Anderson, M., Alfieri, J., Flerchinger, G., and Marks, D. (2013). Evaluation of a two source snow-vegetation energy balance model for estimating surface energy fluxes in a rangeland ecosystem. *Journal of Hydrometeorology*, in review.

Kustas, W.P., Kongoli, C., Anderson, M., Alfieri, J., Flerchinger, G., and Marks, D. (2012). The utility of a thermal-based two-source energy balance model for estimating surface energy fluxes over a snow-dominated landscape. Paper presented at AGU Fall Meeting. San Francisco.

Kustas, W.P., Li, F., Jackson, T.J., Prueger, J.H., MacPherson, J.I., and Wolde, M. (2004). Effects of remote sensing pixel resolution on modeled energy flux variability of croplands in Iowa. *Remote Sensing of Environment, 92*, 535–547.

Li, F., Kustas, W.P., Anderson, M.C., Prueger, J.H., and Scott, R.L. (2008). Effect of remote sensing spatial resolution on interpreting tower-based flux observations. *Remote Sensing of Environment, 112*, 337–349.

Liston, G.E., and Sturm, M. (2004). The role of winter sublimation in the Arctic moisture budget. *Nordic Hydrology, 35*, 325–334.

Liu, J., Chen, J.M., and Cihlar, J. (2003). Mapping evapotranspiration based on remote sensing: An application to Canada's landmass. *Water Resources Research, 39*, 1189.

MacDonald, M.K., Pomeroy, J.W., and Pietroniro, A. (2010). On the importance of sublimation to an alpine snow mass balance in the Canadian Rocky Mountains. *Hydrology and Earth System Sciences, 14*, 1401–1415.

Marks, D., Winstral, A., Flerchinger, G., Reba, M., Pomeroy, J., Link, T., and Elder, K. (2008). Comparing simulated and measured sensible and latent heat fluxes over snow under a pine canopy to improve an energy balance snowmelt model. *Journal of Hydrometeorology, 9*, 1506–1522.

McCabe, M.F., and Wood, E.F. (2006). Scale influences on the remote estimation of evapotranspiration using multiple satellite sensors. *Remote Sensing of Environment, 105*, 271–285.

McCabe, M.F., Wood, E.F., Wójcik, R., Pan, M., Sheffield, J., Gao, H., and Su, H. (2008). Hydrological consistency using multi-sensor remote sensing data for water and energy cycle studies. *Remote Sensing of Environment, 112*, 430–444.

McCartney, S.E., Carey, S.K., and Pomeroy, J.W. (2006). Intra-basin variability of snowmelt water balance calculations in a subarctic catchment. *Hydrological Processes, 20*, 1001–1016.

Miralles, D.G., De Jeu, R.A.M., Gash, J.H., Holmes, T.R.H., and Dolman, A.J. (2011a). Magnitude and variability of land evaporation and its components at the global scale. *Hydrology and Earth System Sciences, 15*, 967–981.

Miralles, D.G., Holmes, T.R.H., De Jeu, R.A.M., Gash, J.H., Meesters, A.G.C.A., and Dolman, A.J. (2011b). Global land-surface evaporation estimated from satellite-based observations. *Hydrology and Earth System Sciences, 15*, 453–469.

Molotch, N.P., Blanken, P.D., Williams, M.W., Turnipseed, A.A., Monson, R.K., and Margulis, S.A. (2007). Estimating sublimation of intercepted and sub-canopy snow using eddy covariance systems. *Hydrological Processes, 21*, 1567–1575.

Montesi, J., Elder, K., Schmidt, R.A., and Davis, R.E. (2004). Sublimation of intercepted snow within a subalpine forest canopy at two elevations. *Journal of Hydrometeorology, 5*, 763–773.

Mu, Q., Heinsch, F.A., Zhao, M., and Running, S.W. (2007). Development of a global evapotranspiration algorithm based on MODIS and global meteorology data. *Remote Sensing of Environment, 111*, 519–536.

Mueller, B., Seneviratne, S.I., Jimenez, C., Corti, T., Hirschi, M., Balsamo, G., Ciais, P. et al. (2011). Evaluation of global observations-based evapotranspiration datasets and IPCC AR4 simulations. *Geophysical Research Letters, 38*, L06402.

Norman, J.M., Anderson, M.C., Kustas, W.P., French, A.N., Mecikalski, J., Torn, R., Diak, G.R., Schmugge, T.J., and Tanner, B.C.W. (2003). Remote sensing of surface energy fluxes at 10^1-m pixel resolutions. *Water Resour. Res., 39i*, 1221.

Norman, J.M., Kustas, W.P., and Humes, K.S. (1995). Source approach for estimating soil and vegetation energy fluxes in observations of directional radiometric surface temperature. *Agricultural and Forest Meteorology, 77*, 263–293.

Raupach, M.R., and Finnigan, J.J. (1995). Scale issues in boundary-layer meteorology: Surface energy balances in heterogeneous terrain. *Hydrological Processes, 9*, 589–612.

Renard, B., Kavetski, D., Kuczera, G., Thyer, M., and Franks, S.W. (2010). Understanding predictive uncertainty in hydrologic modeling: The challenge of identifying input and structural errors. *Water Resources Research, 46*, W05521.

Renard, B., Kuczera, G., Kavetski, D., Thyer, M., and Franks, S. (2008). Bayesian total error analysis for hydrologic models: Quantifying uncertainties arising from input, output and structural errors. In *International Conference on Water Resources and Environment Research*, 4th ed., pp. 608–619. Engineers Australia, Modbury, Adelaide, South Australia.

Scipal, K., Holmes, T., de Jeu, R., Naeimi, V., and Wagner, W. (2008). A possible solution for the problem of estimating the error structure of global soil moisture datasets. *Geophys. Res. Lett., 35*, L24403.

Seneviratne, S.I., Corti, T., Davin, E.L., Hirschi, M., Jaeger, E.B., Lehner, I., Orlowsky, B., and Teuling, A.J. (2010). Investigating soil moisture–Climate interactions in a changing climate: A review. *Earth-Science Reviews, 99*, 125–161.

Seneviratne, S.I., Lüthi, D., Litschi, M., and Schär, C. (2006). Land–atmosphere coupling and climate change in Europe. *Nature, 443*, 205–209.

Seneviratne, S.I., N., N., D., E., Goodess, C.M., Kanae, S., Kossin, J., Luo, Y., Marengo, J., McInnes, K., Rahimi, M., and Reichstein, M. (2012). Changes in climate extremes and their impacts on the natural physical environment. In *Managing the Risks of Extreme Events and Disasters to Advance Climate Change Adaptation*, C.B. Field et al. (Eds.). Intergovernmental Panel on Climate Change, Geneva.

Sheffield, J., Goteti, G., and Wood, E.F. (2006). Development of a 50-year high-resolution global data-set of meteorological forcings for land surface modeling. *Journal of Climate, 19*, 3088–3111.

Sheffield, J., Wood, E.F., and Munoz-Arriola, F. (2010). Long-term regional estimates of evapotranspiration for Mexico based on downscaled ISCCP data. *Journal of Hydrometeorology, 11*, 253–275.

Shuttleworth, W.J. (2007). Putting the "vap" into evaporation. *Hydrology and Earth System Sciences, 11*, 210–244.

Stoffelen, A. (1998). Toward the true near-surface wind speed: Error modeling and calibration using triple collocation. *Journal of Geophysical Research, 103*, 7755–7766.

Strack, J.E., Pielke Sr., R.A., and Adegoke, J. (2003). Sensitivity of model-generated daytime surface heat fluxes over snow to land-cover changes. *Journal of Hydrometeorology, 4*, 24–42.

Tang, Q., Vivoni, E.R., Muñoz-Arriola, F., and Lettenmaier, D.P. (2011). Predictability of evapotranspiration patterns using remotely sensed vegetation dynamics during the North American Monsoon. *Journal of Hydrometeorology, 13*, 103–121.

Timmermans, W.J., Kustas, W.P., Anderson, M.C., and French, A.N. (2007). An intercomparison of the Surface Energy Balance Algorithm for Land (SEBAL) and the Two-Source Energy Balance (TSEB) modeling schemes. *Remote Sensing of Environment, 108*, 369–384.

Van Niel, T.G., McVicar, T.R., Roderick, M.L., van Dijk, A.I.J.M., Beringer, J., Hutley, L.B., and van Gorsel, E. (2012). Upscaling latent heat flux for thermal remote sensing studies: Comparison of alternative approaches and correction of bias. *Journal of Hydrology, 468–469*, 35–46.

Van Niel, T.G., McVicar, T.R., Roderick, M.L., van Dijk, A.I.J.M., Renzullo, L.J., and van Gorsel, E. (2011). Correcting for systematic error in satellite-derived latent heat flux due to assumptions in temporal scaling: Assessment from flux tower observations. *Journal of Hydrology, 409*, 140–148.

Vinukollu, R.K., Sheffield, J., Wood, E.F., Bosilovich, M.G., and Mocko, D. (2011a). Multimodel analysis of energy and water fluxes: Intercomparisons between operational analyses, a land surface model, and remote sensing. *Journal of Hydrometeorology, 13*, 3–26.

Vinukollu, R.K., Wood, E.F., Ferguson, C.R., and Fisher, J.B. (2011b). Global estimates of evapotranspiration for climate studies using multi-sensor remote sensing data: Evaluation of three process-based approaches. *Remote Sensing of Environment, 115*, 801–823.

Wang, K., and Dickinson, R.E. (2012). A review of global terrestrial evapotranspiration: Observation, modeling, climatology, and climatic variability. *Rev. Geophys., 50*, RG2005.

Wang, K., and Liang, S. (2008). An improved method for estimating global evapotranspiration based on satellite determination of surface net radiation, vegetation index, temperature, and soil moisture. In *Geoscience and Remote Sensing Symposium*, pp. III-875–III-878. IEEE, New Brunswick, NJ.

Wei, Y., Langford, J., Willett, I.R., Barlow, S., and Lyle, C. (2011). Is irrigated agriculture in the Murray Darling Basin well prepared to deal with reductions in water availability? *Global Environmental Change, 21*, 906–916.

Zhang, Y., Leuning, R., Hutley, L.B., Beringer, J., McHugh, I., and Walker, J.P. (2010). Using long-term water balances to parameterize surface conductances and calculate evaporation at 0.05 spatial resolution. *Water Resources Research, 46*, W05512.

20

Operations, Challenges, and Prospects of Satellite-Based Surface Soil Moisture Data Services

Wolfgang Wagner, Sebastian Hahn, Julia Figa, Clement Albergel, Patricia de Rosnay, Luca Brocca, Richard de Jeu, Stefan Hasenauer, and Wouter Dorigo

CONTENTS

20.1 Introduction

Building upon the progress made in algorithmic research and improvements in sensor technologies, operational soil moisture products have increasingly become available over the past decade (Wagner et al. 2007b). The first global satellite-based soil moisture dataset that was freely shared with the user community was derived from backscatter measurements collected by the C-band scatterometer (ESCAT) on board the ERS-1 and ERS-2 satellites operated by the European Space Agency (ESA). It was first released in 2002 (Scipal et al. 2002) and after that it has been updated irregularly (Reimer et al. 2012). The ESCAT soil moisture dataset has been produced and distributed by the microwave remote sensing team of the Vienna University of Technology (TU Wien), and given that funding relied principally on research programs, it is a classical research product, that is, software development, dataset updating, and user support have all been done on a best effort basis, not yet having benefited from the more integrated and documented approach followed in operations. Despite the lack of formal operational user service support, over 350 scientific users worldwide have so far requested and received the ESCAT data. This has been the basis for numerous validation studies that verified the quality of the ESCAT soil moisture retrievals by comparing them to *in situ* and model data (Pellarin et al. 2006) and experiments testing the usability of these data in diverse application areas such as runoff forecasting (Brocca et al. 2009), numerical weather prediction (Zhao et al. 2006), yield modeling (de Wit and van Diepen 2007), greenhouse gas accounting (Verstraeten et al. 2010), climate studies (Künzer et al. 2009), and ground water modeling (Sutanudjaja 2012).

The free data policy adopted for the ESCAT soil moisture data was crucial for making the potential of active microwave scatterometry for soil moisture retrieval better known to a larger science community (Wagner et al. 2007a). This positive experience is likely to have spurred some other university teams to start distributing their satellite soil moisture dataset in a free and open manner as well. One of these was the microwave remote sensing team of the Vrije Universiteit Amsterdam (VUA), who started distributing soil moisture data derived from passive microwave radiometers in 2007. They use the Land Parameter Retrieval Model (LPRM) developed in cooperation with NASA to retrieve soil moisture from multifrequency microwave radiometers such as Advanced Microwave Scanning Radiometer–Earth Observing System (AMSR-E) (Owe et al. 2008) or Windsat (Parinussa et al. 2012). Like the ESCAT product of TU Wien, the LPRM soil moisture data of VUA is a classical research product that has found widespread use in the science community (Bolten and Crow 2012).

The increasing acceptance of these and other science products was the basis for making the first steps toward operational soil moisture data services. The first of these services was implemented for the AMSR-E instrument by NASA, starting the distribution of the data through the National Snow and Ice Data Center in 2003. Since then this dataset has been regularly processed with a polarization ratio algorithm as described by Njoku et al. (2003) with a short latency in the order of days. The online documentation of the dataset has been kept up to date, and open, straightforward data access has been provided. The quality and consistency of this data service ensured a wide popularity of the product. Yet, except over some test sites in the United States (Jackson et al. 2010), validation results for this product have, in general, not been as good as for the LPRM retrievals of VUA or the ESCAT data of TU Wien (Brocca et al. 2011; Rüdiger et al. 2009). NASA reacted to this situation by funding further research and development work to improve the polarization ratio algorithm and by implementing the LPRM algorithm in the Goddard Earth Sciences Data and Information Services Center (GES DISC). The LPRM soil moisture products from GES DISC included soil moisture retrievals and their associated errors (Parinussa et al. 2011b) from the Tropical Rainfall Measuring Mission (TRMM) observations and AMSR-E. Unfortunately, the implementation was realized only a few months before the failure of AMSR-E in October 2011. Therefore this operational service has not yet reached a broader user community. For completeness, it should also be noted that the Japan Aerospace Exploration Agency (JAXA) implemented several different retrieval algorithms for AMSR-E. Yet, this service never gained much popularity probably because too little information about the product characteristics and data access conditions was made available.

The next important step in the development toward operational soil moisture data services was undertaken by the European Organisation for the Exploitation of Meteorological Satellites (EUMETSAT). On the basis of the positive experiences made with ESCAT, EUMETSAT decided to implement a fully operational near-real-time (NRT) soil moisture processing chain for the successor instrument of ESCAT, the Advanced Scatterometer

(ASCAT) (Bartalis et al. 2007). ASCAT is flown on a series of three Meteorological Operational (METOP) satellites, which are planned to span the period from October 2006 to beyond 2020. The ASCAT soil moisture service was developed by EUMETSAT in cooperation with TU Wien, and operational NRT dissemination of ASCAT soil moisture data started in December 2008 (Wagner et al. 2010). METOP-B was launched in September 2012, while METOP-A is still in orbit and functional. ASCAT soil moisture data services will be available from both instruments, thereby significantly improving the spatiotemporal coverage (Wagner et al. 2013).

Finally, in November 2009, not long after the first operational ASCAT soil moisture data became available, ESA launched the Soil Moisture and Ocean Salinity (SMOS) satellite. SMOS is an experimental satellite that was designed for the purpose of soil moisture measurements over land (Kerr et al. 2010), and consequently, the release of the SMOS soil moisture data had been much awaited for by the soil moisture science community. The first SMOS soil moisture data were released in 2010, and since then many interesting studies comparing SMOS, ASCAT, and/or AMSR-E soil moisture data have been carried out (Albergel et al. 2012; Brocca et al. 2011; Wanders et al. 2012).

With SMOS, ASCAT, Windsat, and AMSR-2 (the successor of AMSR-E) in space and the launch of the next dedicated soil moisture satellite called Soil Moisture Active/Passive (SMAP) on the horizon (Entekhabi et al. 2010), operational soil moisture data services have become a reality. As discussed in Chapter 4, the measurement concepts and retrieval algorithms differ quite widely for these different satellite missions; yet, the basic challenges faced by the providers and users of the corresponding soil moisture data services are very similar. In this chapter we discuss these challenges based on the example of the 25-km ASCAT surface soil moisture products distributed by EUMETSAT and TU Wien. Even though many aspects are ASCAT-specific, the discussion is kept on a rather general level in order to ensure that it is also of relevance for the other satellite soil moisture monitoring services. This chapter will conclude with a discussion of the prospects of operational soil moisture services.

20.2 ASCAT Soil Moisture Service Operations

Operational aspects of satellite soil moisture data services are hardly discussed in the scientific literature. Nonetheless, it is important to describe the implementation of these services as this has an impact on the characteristics and quality of the data. In this section we therefore discuss the implementation of the ASCAT surface soil moisture data services of EUMETSAT and TU Wien, which have been part of EUMETSAT's Satellite Application Facility in Support to Operational Hydrology and Water Management (H-SAF) since 2012. Further value-added ASCAT soil moisture products such as the ASCAT Soil Water Index (SWI), assimilated profile soil moisture products, and a disaggregated 1-km product obtained by downscaling the ASCAT data using static scaling coefficients from synthetic aperture radar (SAR) data are also available (Wagner et al. 2013), but these are not discussed here.

20.2.1 Near-Real-Time Swath Products of EUMETSAT

The ASCAT surface soil moisture data services provide a suite of data products that all share the same physical basis (i.e., from a physical point of view all products should be identical) but come in different flavors (spatiotemporal sampling, data latency, data format, and consistency) in order to meet the requirements of different user communities. The basic EUMETSAT products were designed to meet the needs of the Numerical Weather Prediction (NWP) user community. For that user group, one of the most important requirements was to receive the data in NRT with a maximum delay of 130 min after data acquisition (Wagner et al. 2010) and in a format commonly used by NWP centers, following World Meteorological Organization (WMO) standards. In NWP data assimilation, each individual measurement can be considered as an independent observation and it is furthermore favored to use data as close as possible to the original measurement, while because of the size of currently used NWP grids, spatial resolution is less of a concern. Consequently, the basic EUMETSAT ASCAT surface soil moisture data product is orbit swath-based, has a spatial resolution of approximately 50 km, and is sampled on a regular 25-km grid along and across orbit swaths. It should be noted that the resolution and sampling characteristics are directly inherited from the parent Level 1B Normalized Radar Cross Section (NRCS) ASCAT product so that no additional resampling errors are introduced. In order to achieve the latency requirements, products are sliced up in batches of data corresponding to 3 min of ASCAT sensed data and formatted in the Binary Universal Form for the Representation of meteorological data, which is a binary data format maintained by WMO. An example for such a 3-min data slice is shown in Figure 20.1. After processing the data within EUMETSAT's Central Application Facility with software called Water Retrieval Package–Near-Real-Time (WARP-NRT) the data are distributed over the WMO Global Telecommunications System and via EUMETCast, a multicast dissemination system based on commercial telecommunication geostationary satellites and operated by EUMETSAT. In parallel to the 25-km sampled product a 12.5-km product with an approximate resolution of 25–30 km and a somewhat higher noise level is also produced operationally.

Both products are additionally archived and available in the EUMETSAT Data Center, where the original 3-min data patches are collated to form a complete orbit to reduce the

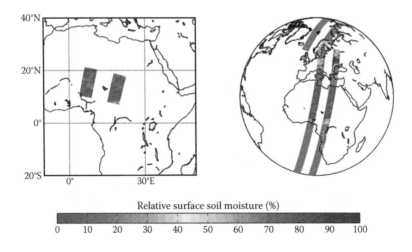

FIGURE 20.1
(See color insert.) (Left) Example of an NRT ASCAT surface soil moisture data product, representing a 3-min data take. This data slice was acquired on April 1, 2012, 0845:01 UTC over Africa. (Right) Example of a full orbit ASCAT surface soil moisture data product as available from EUMETSAT's data archive is shown. The orbit was acquired between April 1, 2012, 0833:00 UTC and 1014:59 UTC.

number of small data files. EUMETSAT thus offers four distinct data products: 3-min NRT data file and full-orbit off-line data files at both 25- and 12.5-km sampling, respectively. The archived data are very useful to carry out off-line analysis and investigations, but because of occasional updates in the operational NRT products, they may be inhomogeneous. In support to climate research, EUMETSAT plans to carry out retrospective reprocessing campaigns for all the ASCAT missions in order to provide a consistently processed data record and up-to-date instrument calibration. This effort is already being initiated for METOP-A, where 6 years of scatterometer measurements are currently available. Therefore harmonized ASCAT swath-based soil moisture products will become available in the future.

20.2.2 Off-Line Time Series Product of TU Wien

In addition to the four data products of EUMETSAT, TU Wien produces and distributes a 25-km ASCAT surface moisture data product stored as time series for a fixed global sinusoidal grid with a spacing of approximately 12.5 km. The data are available in a user-defined binary format. The motivation for this product is that many users are interested in having ASCAT surface soil moisture time series over relatively small study regions, sometimes even only over one particular location. In such a situation, the EUMETSAT products with their irregular spatiotemporal sampling (the exact orbit swath may vary from one orbit cycle to the next, causing a quasi-random shift of the latitude/longitude coordinates of the swath file) are impractical, requiring from the users major processing efforts to subset and resample the data to their areas of interest. An example of a time series product is shown in Figure 20.2.

The TU Wien time series product is not available in NRT, but it is generally of better quality than the EUMETSAT NRT swath products. This is because it is always based on the latest version of the retrieval algorithms due to the development and implementation cycle as adopted by TU Wien and EUMETSAT. As illustrated in Figure 20.3, the software development cycle starts with researchers—typically students pursuing their master or PhD degrees—investigating the ASCAT data and exploring new algorithms over selected test sites in order to improve one or more processing steps in the complete ASCAT retrieval scheme. Once an algorithm is found to perform well, it is tested globally to investigate its scientific quality and to check its computational performance. It may happen that even

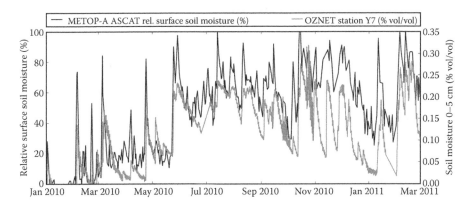

FIGURE 20.2
Example of an ASCAT surface soil moisture time series as produced and distributed by TU Wien. The ASCAT time series is compared with *in situ* measurements taken over the station Y7 from the Australian *in situ* network OZNET. For a description of the OZNET network, see Smith et al. (2012).

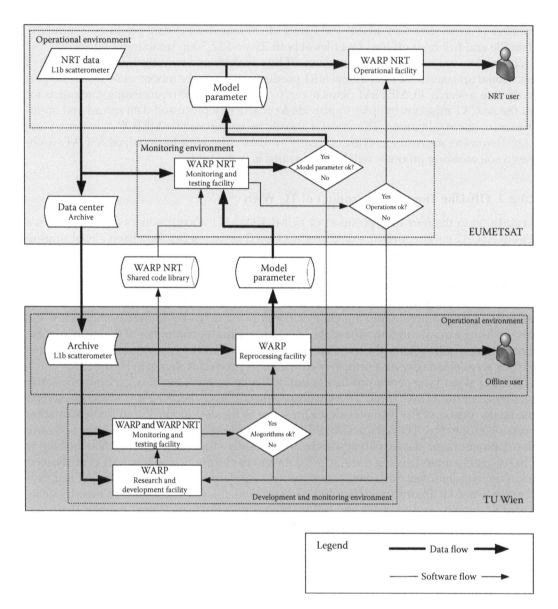

FIGURE 20.3
Development and operational system engineering model for the METOP ASCAT soil moisture products distributed in NRT by EUMETSAT and off-line by TU Wien within the framework of the EUMETSAT Satellite Application Facility on support to operational hydrology and water management (H-SAF).

though an algorithm performs well from a scientific point of view, it is abandoned because processing of the historic time series would take too long (several weeks or months). Such computationally demanding algorithms may be reconsidered in the future when more powerful computers or distributed processing systems become available.

Once an algorithm is found to improve the quality of the soil moisture retrievals at reasonable processing costs, it is committed to the latest version of a software package called Water Retrieval Package (WARP). From then on it is being used in the reprocessing of the

ASCAT backscatter time series in order to derive consistent surface soil moisture time series and a set of optimally estimated model parameters. These model parameters are not only important for WARP itself, but also they are used by EUMETSAT as one of the inputs to the NRT processing software WARP-NRT (Figure 20.3). The use of these empirically derived model parameters ensures that the NRT retrieval is robust and computationally very fast. However, one caveat of this approach is that changes in the calibration of the ASCAT backscatter data may—if unaccounted for—lead to artifacts in the NRT products. This was the case at the start of the ASCAT NRT service of EUMETSAT (Hahn et al. 2012) but has significantly improved since then by putting procedures in place to ensure that changes in the backscatter calibration are accounted for in the soil moisture processor through the use of NRCS back-calibration tables (Section 20.4.1).

20.3 Maturity of ASCAT Soil Moisture Products

Given that the radiometric and measurement geometry characteristics of ESCAT and ASCAT are very similar, the development of an ASCAT soil moisture service could draw directly from the experiences with the ESCAT soil moisture product (Bartalis et al. 2007). The initial implementation of the ASCAT soil moisture data services was therefore reasonably fast, with EUMETSAT declaring the NRT ASCAT soil moisture service operational in December 2008, only about 2 years after the launch of METOP-A. Since then, various aspects of this service have been improved step by step. Nonetheless, the transition from a science to a fully operational Earth Observation product is more demanding than one might expect. Consequently, even though the ASCAT soil moisture products have now been available for several years, they still cannot be considered to be fully mature from a service point of view. In this section we illustrate this point by using a maturity model first proposed by Bates and Barkstrom (2006) and later refined by Bates and Privette (2012), herein after referred to as Bates maturity model.

20.3.1 Bates Maturity Model

In Earth Observation (EO) science the concept of the maturity of data services has not yet been widely adopted. With the emergence of science-driven EO data services it becomes, however, increasingly important to introduce maturity indices that characterize the process of moving from a basic research dataset to a sustained and routinely generated product. The objectives of such maturity indices are to produce an easily understood way of characterizing the maturity of EO data services and to help identifying areas needing improvement. The first objective is mostly of importance for prospective new users of EO data services; the second for the service providers themselves. While many providers (such as EUMETSAT) develop and operate EO data services according to their own well-defined service requirements, there is so far only one science-driven maturity model that has gained more widespread acceptance in the international EO community, namely, the Bates maturity model.

The Bates maturity model has its origin in discussions between NASA and the National Oceanic and Atmospheric Administration (NOAA) on how to transition data services dealing with the generation of climate data records (CDRs) from NASA to NOAA for long-term sustained operations. As a result of this heritage, the Bates maturity model in some respects may not be fully adequate for characterizing European data services designed for nonclimate applications. Nevertheless, the latest version of this maturity index (V4.0 from December 20, 2011) as used here has already benefited from several international assessments such as ones organized by the World Climate Research Programme or the Global Climate Observing System. These assessments have broadened the scope and usability of the Bates maturity model significantly, and we therefore apply it here also to the NRT ASCAT soil moisture data services, despite some remaining caveats for our specific purpose.

The latest version of the Bates maturity model considers six different thematic areas, namely, *Software Readiness* (stability of code), *Metadata* (amount and compliance with international standards), *Documentation* (description of the processing steps and algorithms), *Product Validation* (data quality in space and time), *Public Access* (availability of data and code), and *Utility* (uses by broader community). Each of these thematic areas is further split up into three to five subthemes that specify the best practices in these different domains. The maturity of each thematic areas and each subtheme is rated with a number from one (basic) to six (mature). Advice is given that EO data products of the overall maturity levels 1 and 2 are used only for research purposes but not for decision making. Maturity levels 3 and 4 indicate initial operational capabilities, implying that the data products may tentatively be used in decision making. Full operational capabilities, allowing the use of the data products in decision making, are reached in maturity levels 5 and 6 (Bates and Privette 2012).

20.3.2 Maturity Analysis

The results of our maturity analysis are presented in Table 20.1 for the NRT swath products of EUMETSAT produced with the WARP-NRT software and in Table 20.2 for the off-line time series product of TU Wien produced with WARP. Our approach in this assessment was to select the appropriate ratings for each of the subthemes, then calculate the mean subtheme rating for each of the six thematic areas, and finally round the mean rating to the next lower or next higher number depending on how well the description of the maturity levels fitted to the actual situation. To make Tables 20.1 and 20.2 readable, they contain not only the rating for each thematic area and its subthemes but also the corresponding verbal description. For example, for the case of the EUMETSAT product shown in Table 20.1, *Portability*, which is a subtheme of *Software Readiness*, receives the rather high maturity level 5 as the software WARP-NRT is: "Operational: can be systematically and routinely run by 3rd party." Tables 20.1 and 20.2 display the six thematic areas at the top row and their ratings at the last two bottom rows, the rest of the rows display the subcategories. Staying with the example *Software Readiness* of WARP-NRT, the average rating for all subthemes is 3.6, but as "Moderate code changes [*are still*] expected," we selected the comparatively low overall maturity level of 3.

The outcome of this assessment of the maturity of the two ASCAT products groups provided some interesting insights. For EUMETSAT's NRT products it turned out that, in terms of the categories *Public Access* and *Utility*, a maturity level of 5 has been reached; that is, in these two aspects the products can indeed be regarded fully operational. However, in the other categories it fares less well. While for the categories *Software Readiness*, *Documentation*, and *Product Validation* the scores are 3 and 4 (2 times), respectively, it only reaches the low maturity level 2 for *Metadata*. This demonstrates that still substantial efforts will be required to improve the overall system engineering (software, database management, and documentation) of the NRT ASCAT soil moisture service of

TABLE 20.1

Bates Maturity Matrix for ASCAT Surface Soil Moisture Orbit Product Produced by EUMETSAT Using the Software WARP-NRT 3.1.0

Category	Software Readiness	Metadata	Documentation	Product Validation	Public Access	Utility
Subcategory	Documentation	File level	C-ATBD	Independent validation	Archive	Data usage if TCDR
Description	Overview and process descriptions complete; program headers and README complete	Complete pixel or grid-point level metadata; metadata sufficent to reproduce the data independent of external assistance	Updated and public	At least 5 comparisons to models, in situ data, or other independent products as available and appropriate to particular CDR; differences in results understood	Data, documentation, and source code archived and available to the public	Regular requests for data
Rating	3	5	5	5	3[a]	6
Subcategory	Portability	Collection Level	OAD	Uncertainty (for TCDRs)	Updates to Record	Societal Sector Decision Support Systems
Description	Operational: can be systematically and routinely run by 3rd party	Limited	Not evaluated	Biases and errors identified and documented	Done systematically and operationally as dictated by availability of new input data	Potential benefits published
Rating	5	2	3	3	6	4
Subcategory	Numerical Reproducibility	Standards	Process Flow Chart	Quality Flag	Version	Citations in peer-reviewed literature
Description	3rd party output within machine rounding errors	Not evaluated	Public	Masks applied as appropriate (e.g., land masks, and cloud masks); algorithm failures identified	Under NCDC version control	Citations of product occurring
Rating	5	1	3	4	4[b]	5
Subcategory	Meets coding standards		Peer Reviewed Docs Describing Algorithm and Product	Operational monitoring		Feedback to CDRP
Description	Passes review against minimum CDRP coding standards1		Paper on product published	Operational monitoring in place		Comments received
Rating	3		5	5		5
Subcategory	Security					
Description	PI knows of no security problems					
Rating	2					
Description	Moderate code changes expected	Research grade	Public C-ATBD; Draft Operational Algorithm Description (OAD); Peer-reviewed publication on algorithm; paper on product submitted	Uncertainty estimated over widely distributed times/location by multiple investigators; Differences understood	Record is archived and publicly available with associated uncertainty estimate; Known issues public. Periodically updated	May be used in applications by other investigators; assessments demonstrating positive value
Overall Rating	3	2	4	4	5	5

[a] Source code is not available.

[b] Under control of EUMETSAT (instead of NCDC).

TABLE 20.2

Bates Maturity Matrix for ASCAT Surface Soil Moisture Time Series Product Produced by TU Wien Using the Software WARP 5.5

	Software Readiness	Metadata	Documentation	Product Validation	Public Access	Utility
Subcategory	Documentation	File level	C-ATBD	Independent validation	Archive	Data usage if TCDR
Description	Process descriptions updated; history of changes added	Limited	Public	At least 5 comparisons to models, *in situ* data, or other independent products as available and appropriate to particular CDR; differences in results understood	Data available through PI	Regular requests for data
Rating	4	2	3	5	2	6
Subcategory	Portability	Collection Level	OAD	Uncertainty (for TCDRs)	Updates to Record	Societal Sector Decision Support Systems
Description	Not evaluated	Limited	Not evaluated	Biases and errors identified and documented	As done by PI	Potential benefits published
Rating	1	2	1	3	2	4
Subcategory	Numerical Reproducibility	Standards	Process Flow Chart	Quality flag	Version	Citations in peer-reviewed literature
Description	Not evaluated	Not evaluated	Created and reviewed	Masks applied as appropriate (e.g., land masks, and cloud masks); algorithm failures identified	Versioning by PI	Citations of product occurring
Rating	1	1	2	4	2	5

			Peer-Reviewed Docs Describing Algorithm and Product			
Subcategory	**Meets Coding Standards**		**Peer-Reviewed Docs Describing Algorithm and Product**	Operational monitoring		Feedback to CDRP
Description	Not evaluated		Paper on product published	Incomplete		Mostly positive comments received
Rating	1		5	2		6
Subcategory	**Security**					
Description	Not evaluated					
Rating	1					
Description	**Significant code changes expected**	Research grade	Public C-ATBD; Peer-reviewed publication on algorithm	Uncertainty estimated for some locations	Limited data availability to develop familiarity	May be used in applications by other investigators; assessments demonstrating positive value
Overall Rating	2	2	3	3	2	5

EUMETSAT, while from a scientific point of view the product can be considered to be relatively mature. This, however, does not mean that it can be recommended to scale back scientific research in favor of system engineering—far from that! The reason is that with increasing scientific maturity of an algorithm it becomes, in general, increasingly difficult and hence time consuming to further improving it (Figure 20.4). This is an expression of the fact that algorithms, even if unchanged in their core characteristics, tend to become more complex over time in order to capture the physical world more realistically. This is also the case for the ASCAT soil moisture retrieval algorithm, which has become increasingly complex over the years by adding, among other improvements, algorithms for correcting azimuthal effects (Bartalis et al. 2007), improved error propagation models (Naeimi 2009), and advanced freeze/thawing detection methods (Naeimi et al. 2012; Zwieback et al. 2012a). The development of the ASCAT algorithms can thus be seen to follow the so-called Pareto principle, also known as the 80-20 rule, which states that roughly 80% of the effects come from 20% of the causes.

The observations made with regard to the maturity of the EUMETSAT products are even more pronounced in the case of the off-line time series product of TU Wien. For this latter product, only in the category *Utility* is a maturity level of 5 reached. All other categories have low maturity levels of 2 or 3, indicating that this product is still very much a research product with significant deficits particularly with respect to all system engineering components, most importantly *Software Readiness*, *Metadata*, and *Public Access*. This is a reflection not only of the research orientation of the service provider (i.e., TU Wien) but also of the type of funding that has so far been available to develop this service. However, it should also be noted that the WARP software used by TU Wien is much more extensive than the WARP-NRT software used by EUMETSAT. While the first provides a research platform to investigate new algorithms and processes, the latter was purposefully designed to be slim and computationally efficient to meet the timeliness requirements of the NWP community. Therefore the efforts required for improving the maturity level of the categories *Software Readiness*, *Metadata*, and *Documentation* are much higher for WARP than they are for WARP-NRT. Fortunately, in 2012 the TU Wien time series product became one of the products of H-SAF, providing a funding source for supporting system engineering activities.

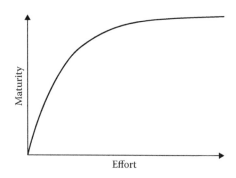

FIGURE 20.4
Schematic illustration of the relationship between the scientific maturity of an algorithm and the efforts needed to improve it.

20.4 Challenges Faced by Providers and Users of Soil Moisture Services

The previous section illustrated the many and complex tasks behind operational EO data services based on the example of the ASCAT soil moisture services provided by EUMETSAT in cooperation with TU Wien. An analysis of the maturity of the ASCAT services showed that at this point in time their weakest elements are related to a backlog of general system engineering tasks (software, metadata, data supply, documentation, etc.), while the science behind the products is already quite mature as evidenced by the large number of validation and application studies published in the peer review literature (Wagner et al. 2013). However, there are of course many outstanding challenges requiring further scientific research and development activities. In the following, we discuss some of these challenges, namely, those related to the instrument calibration, product validation, algorithmic improvements, masking and quality flagging, data assimilation, and application development.

20.4.1 Instrument Calibration

The calibration and validation of a satellite instrument and its derived data are some of the major tasks that have to be performed before satellite data are ready for further use. Stability of the data quality needs to be sustained in time through monitoring the instrument calibration and if necessary correcting any possible drifts. In the case of ASCAT, several measures are in place to ensure a stable instrument calibration, probably making it the best calibrated satellite sensor currently used for soil moisture retrieval. On one hand, the instrument design relies on an internal calibration system, which allows any internal instrument changes that might affect the transmitted power to be accounted for in near real time. On the other hand, an external calibration system allows the provision of an absolute calibration reference, with respect to which any further changes in the measurement system, such as variations of the antenna gain due to instrument aging, can be detected and corrected for. This absolute calibration is carried out by measuring campaigns involving the use of ground transponders, which are operated to transmit a signal that is received by the ASCAT instrument during selected satellite overpasses. The position and cross section of this signal being precisely known, comparison with that received by ASCAT allows to precisely determining the gain and the pointing of each of its antennas (Anderson et al. 2012; Figa-Saldaña et al. 2002). External calibration campaigns are carried out periodically every 2 to 3 years and each time provide an independent snapshot of the instrument absolute calibration with an uncertainty within 0.1 dB in backscatter (Wilson et al. 2010). An update of the absolute calibration is therefore not a continuous correction but introduced as a step, after careful validation of its occurrence over natural targets with known backscatter statistics, such as the rainforest or the ocean.

The retrieval of geophysical parameters from the instrument data are in many cases tuned to, or empirically derived from, a given instrument calibration. This is also the case of the ASCAT surface soil moisture data generated by WARP-NRT, which at the time of writing (December 2012) relies on a set of empirically derived model parameters using ASCAT data with an instrument calibration corresponding to the years 2007 and 2008. Introducing a new instrument calibration into the processing is therefore not straightforward and a good understanding of the expected impact of a calibration update on the geophysical parameter, in this case soil moisture, is very important. For ASCAT, NRCS changes in any of the antenna beams of the order of ±0.2 dB result in changes in surface

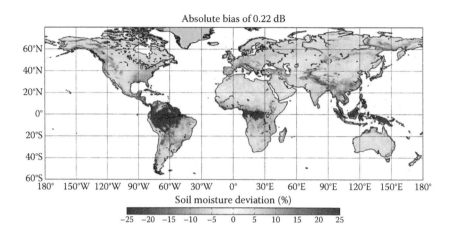

FIGURE 20.5
(See color insert.) Simulated soil moisture errors assuming that the absolute calibration bias is constantly 0.22 dB all over the world. (Adapted from Hahn, S. et al., *IEEE Transactions on Geoscience and Remote Sensing, 50*, 2556–2565, 2012.)

soil moisture index of the order of approximately 6%–8% for most parts of the globe (see Figure 20.5), as has been shown in previous studies (Hahn et al. 2012). Areas with higher impact (up to 20%) are observed, but they correspond to tropical forested areas where the backscatter response stems from the canopy instead. As a conclusion, the effect of a change in calibration of one of the beams on the order of ±0.1 dB is difficult to observe as a soil moisture change. On the contrary, a permanent and consistent shift in the calibration of all beams beyond ±0.1 dB would, however, be problematic for climate change applications.

As a general guideline, introducing in the processing chain observed instrument absolute calibration changes in the NRCS over ±0.2 dB for any given beam or of ±0.1 dB consistently for all beams should trigger an update of the empirically derived model parameters. However, this requires reprocessing of the soil moisture time series, which cannot be done regularly, nor in near real time. What has been done in such a case, during the introduction by EUMETSAT of the 2010 NRCS absolute calibration in the operational Level 1 backscatter processing, is to adopt at the same time a so-called back-calibration in the WARP-NRT soil moisture processor. This back-calibration consists of compensating the calibration changes back to the previous calibration state in order to keep the output soil moisture series consistent and unaffected. The full benefit of the NRCS calibration update can only be taken on board through reprocessing of the soil moisture time series, at which time the NRT products can be brought up to date with the state-of-the-art instrument calibration. The frequency of such reprocessing efforts depends on the rate of calibration changes. For the ASCAT mission the analysis of the external calibration campaigns suggests that this frequency is 2 to 3 years.

20.4.2 Product Validation

The validation of the ASCAT soil moisture data is, like for any other remotely sensed soil moisture product, challenging, as there is no reference dataset that can be considered to represent the truth. When the satellite data are compared to *in situ* soil moisture measurements, the main problems are the significant scale gap (area representative 25–50 km data vs. point-like *in situ* data) and differences in soil layers (thin remotely sensed topsoil layer vs. *in situ* probes installed at depths of 5 cm or more), and when compared to modeled soil moisture data, uncertainties stem from the land surface model itself and its input data, for example, precipitation (Crow et al. 2012). Therefore scientists have put forward many different validation strategies involving, for example, the implementation of densely instrumented validation sites (Jackson et al. 2010), the organization of extensive field and airborne campaigns (Delwart et al. 2008), the simultaneous comparison to *in situ* and modeled soil moisture data (Matgen et al. 2012), and advanced statistical methods such as the triple collocation method (Zwieback et al. 2012b) or the *R* metric (Crow et al. 2010). There are important lessons to be learned from all of these approaches, and clearly each approach lets us better understand the strengths and weaknesses of the satellite soil moisture data. This understanding is essential for improving the retrieval algorithms, the associated error estimates, and the quality flags as discussed in the next sections.

In many cases the different validation approaches complement and/or reinforce each other (Brocca et al. 2011; Parinussa et al. 2011a). Nonetheless, it also happens that the validation results obtained with two or more different validation approaches may disagree. There is also often a disagreement on how the obtained results are to be interpreted, which is, in particular, the case when the evidence that points to an error in the satellite retrievals is believed to be weak or in contradiction to previous results. In such situations the providers of satellite-based soil moisture data services tend to await further validation results, before engaging in algorithmic research and development to solve the identified problem.

Another problem often encountered is that results from two or more independent validation studies cannot be directly compared to each other. This is a consequence of the lack of agreed standards and best practices in the validation of satellite soil moisture data. Consequently, there are not currently any internationally agreed values for the retrieval error of ASCAT, SMOS, AMSR-E, or any of the other suited satellite sensors. This is particularly a problem for users who thus lack an important guidance for their decisions on which soil moisture data to use for which application. It is also a problem for space agencies that require target numbers for designing the next generation of EO sensors. This lack of best practices in the validation of satellite soil moisture dataset is therefore currently an important topic for several international organizations such as the Global Climate Observing System, the Global Energy and Water Cycle Experiment, the Committee on Earth Observation Satellites, and the intergovernmental Group on Earth Observation. Thanks to their efforts first international initiatives such as the International Soil Moisture Network could already be launched successfully (Dorigo et al. 2011). Nonetheless, the path toward universally agreed best practices is probably still quite long, requiring the active participation of data providers and users alike.

20.4.3 Algorithmic Improvements

Once sufficient empirical and/or theoretical evidence has been brought together to conclude that an algorithm, or part thereof, is suboptimal or wrong under particular environmental circumstances, data providers can start investigating new algorithms to solve this problem. Such scientific investigations can have many different outcomes. One outcome may be that the reasons behind the algorithmic problem become better understood but that, unfortunately, no solution to solve it is yet known. Another equally unsatisfying outcome may be that an algorithmic solution may exist but that it does not fit in the overall retrieval strategy applied for a given satellite soil moisture data product. In both cases, the only alternative may be to provide error fields and quality flags to mask observations affected by this issue. Finally, the scientific investigation may of course also be successful, providing an algorithm prototype that is found to work satisfactorily over the selected study sites. One may think that from here on the transfer to the operational processing framework should be reasonably fast, but unfortunately this is, in general, not the case.

As described in Section 20.2 and illustrated in Figure 20.4 for the case of the ASCAT soil moisture data services, substantial challenges are faced regarding the operational implementation and development of scientific algorithms. Independent and flexible algorithm implementations such as scientific prototypes are fundamental to the validation of the first integrated operational prototype. In the case of ASCAT soil moisture WARP-NRT, coded in C++ for performance reasons, an independent implementation in IDL was used to verify and validate its implementation. The complete process of developing the first WARP-NRT operational prototype initially off-line, validating it with respect to the scientific prototype, and integrating it into the operational environment took at least a year, which is not unusual from other experiences at EUMETSAT with similar developments.

The integration of further algorithm improvements in an already mature operational implementation can also be time consuming. Once new algorithmic improvements have been successfully implemented and validated in the scientific development environment, a smooth migration to the operational system is not always a straightforward task either. For example, differences in the software environment, operating system, programming language, or code library used in the algorithm development with respect to those used in operations may lead to problems such as incompatibility of interfaces, inconsistency of scientific results, or difficulties in system configuration and/or operability. At that time, the establishment of a common development and testing framework is therefore recommended in order to allow a controlled migration and validation of new improvements. In this context, a common shared code library constitutes one of the basic elements of such a framework, together with a version control tool, in order to keep a common track of changes. This setup allows running the same piece of code in different environments while at the same time producing results that can be directly compared, thereby reducing the costs and complexity of the overall software development, integration, validation, and implementation process. All necessary validation and testing can already be done in the off-line environment before it is committed into the operational one and the integration time is shorter, reducing the risk of having to troubleshoot recoding errors.

Once the new algorithmic updates have been integrated successfully into the operational environment, the next challenge is to prepare the users for the upcoming modifications in the product. They might use the NRT product in an operational fashion and switching to a new version is not possible instantaneously without having a strong impact on downstream data services. In this case, it is recommended to set up, when possible, a dissemination stream of the new product, in parallel to the currently operational one. This parallel dissemination phase may last several weeks, allowing users to adapt their processing environment to the modifications coming with the new product.

20.4.4 Masking and Quality Flags

The accuracy of satellite soil moisture data is affected by many environmental conditions, most notably the density of the vegetation, the presence of urban areas and open water surfaces, and the occurrence of snow and frost. To make the measurements useful to users, it is thus important to attach to each soil moisture value additional data fields describing its reliability, accuracy, and proficiency. Without any ancillary data the quality of the product remains doubtful causing exclusion for specific applications. For example, Numerical Weather Prediction (NWP) centers desire a suitable set of quality flags, which allows automatic selection of data of sufficient quality for assimilation. Hence quality screened data are preferred at the cost of some data loss. In case of the ASCAT swath-based soil moisture product operationally produced by EUMETSAT several quality indicators are included. The necessity of these indicators comes from the fact that the retrieval is performed for every NRCS measurement acquired over land without exception, although in certain situations, for example, when snow or open water dominates the satellite footprint, a successful retrieval of soil moisture is hardly possible or even impossible. Thus an attentive masking of unreliable data is very important.

The indicators can be grouped into quality flags and advisory flags with respect to their functionality. The quality flags are directly derived from the ASCAT data and describe the intrinsic quality of the soil moisture retrieval (e.g., noise, processing, and correction flags). On the other side, the advisory flags support the user in judging the validity of the soil moisture product in those situations that could not have been identified during the retrieval because of certain limitations in the geophysical model (e.g., snow, frozen soil, topography, inundation, or wetland). Therefore these indicators originate from external datasets and complement the quality flags. However, the identification of reliable external datasets is a challenging task and depends primarily on temporal and spatial data availability. Furthermore, also property rights and operational readiness need to be clarified for those external datasets. In the simplest form, quality flags can be defined as probabilities, which are based on analysis of historical data. This elementary but practical approach overcomes some of the aforementioned issues for the NRT products. However, those probabilistic flags should not be used in favor of actual reanalysis data if, for example, users are interested in analyzing only historic time series. In other words, whenever possible the best reference dataset available shall be used in order to mask unreliable measurements, keeping in mind the limitations of the reference data as well.

20.4.5 Data Assimilation

Satellite-based surface soil moisture data are of high interest for data assimilation applications. Sensors such as ASCAT, AMSR-E, and SMOS achieve global coverage in less than 3 days, which ensures that the variability of the surface soil moisture content in time and space is captured. When EUMETSAT's ASCAT surface soil moisture product was declared operational in 2008, it was the first ever operational satellite soil moisture product. Despite the increasing availability of satellite-based soil moisture products from a range of sensors (such as SMOS and AMSR-E), ASCAT remains the only operational product of soil moisture from space for data assimilation. Its short latency makes it relevant for NRT applications, including NWP and hydrological forecast. Several operational meteorological centers, such as the Met Office, Météo-France, and ECMWF, already use ASCAT surface soil moisture data for monitoring and/or data assimilation either in operational mode (de Rosnay et al. 2012b; Dharssi et al. 2011) or for research applications (Mahfouf 2010).

Bias correction is a crucial component of data assimilation systems that ensures data assimilation corrections for random errors in the models. Most advanced data assimilation systems, which were developed by the NWP community to analyze atmospheric variables (e.g., 4D-Var approaches), rely on adaptive bias correction approaches as described in the review paper by Dee (2006). In contrast, global land data assimilation systems are decoupled from the atmospheric data assimilation systems (de Rosnay et al. 2012b), and they use simpler stand-alone bias correction approaches (e.g., Draper et al. 2011). Therefore discontinuities in the land surface model or in the observational characteristics require updating bias correction parameters used by the data assimilation systems. Hence, for

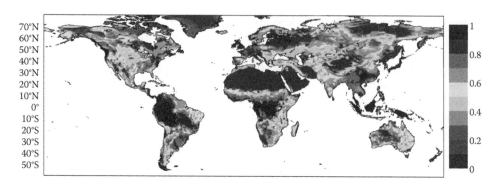

FIGURE 20.6
(See color insert.) EUMETSAT H-SAF ASCAT root zone soil moisture in m³ m⁻³ on November 25, 2012. ASCAT root zone soil moisture profile retrieval relies on ASCAT surface soil moisture data assimilation in the ECMWF land surface data assimilation system specifically adapted for the H-SAF retrieval of root zone soil moisture. (From EC MWF VT: Sunday November 25, 2012 00UTC Surface: H14 H-SAF CDOP. © 2014 EUMETSAT.)

NRT applications, excellent communication and synchronization between data providers and users are critical to ensure that satellite product upgrades are well prepared before implementation and dissemination. In the future, land surface data assimilation systems will develop further in order to account for adaptive bias correction approaches similar to those used in atmospheric 4D-Var. This will make it possible to account for upgrades in surface-related satellite product more easily.

Although different data assimilation approaches are used in the community, as described by de Rosnay et al. (2012a), mainly based on Extended Kalman Filter (de Rosnay et al. 2012b; Draper et al. 2011), Ensemble Kalman Filter (Draper et al. 2012), or simple nudging (Dharssi et al. 2011; Scipal et al. 2008), they all rely on similar bias correction and quality control approaches. Appropriate use of ASCAT quality flags enables to select the best quality observations in preprocessing, which can then be used in an optimal way in the data assimilation schemes. For all these approaches, the land surface model used in the data assimilation scheme describes the physical processes that control land–atmosphere interactions, including vertical transfer of soil moisture between the surface and root zone reservoirs. Beside NWP and hydrological forecast applications, data assimilation of ASCAT surface soil moisture constitutes a physically based root zone soil moisture retrieval approach in which the observed surface soil moisture is propagated downward by a land surface model infiltration processes. In the context of the EUMETSAT H-SAF project ASCAT root zone soil moisture profile (Figure 20.6) is produced in NRT, based on ASCAT data assimilation in the ECMWF land surface data assimilation system (Albergel et al. 2012; de Rosnay et al. 2012b).

20.4.6 Application Development

Users interested in using satellite soil moisture data for developing specific applications such as, for example, flood prediction or agricultural management, face many scientific and technical challenges. Therefore the path from, first, considering using satellite soil moisture data to successfully applying them in specific applications normally takes years. The first challenge for users lies probably in the difficulty of changing their perception about the exploitable information content of coarse-resolution (25–50 km) satellite data such as provided by ASCAT, SMOS, and AMSR-E. Indeed, for many small- to medium-scale applications (<400 km²), the spatial resolution of these soil moisture products is usually considered not sufficient, and hence many scientists simply do not take these products into account. However, this position is not justified for all small- to medium-scale applications. For instance, Brocca et al. (2012a) demonstrated that the ASCAT soil moisture product can be a valuable additional information for improving runoff prediction in small to medium catchments (<200 km²) (see Figure 20.7). Similar conclusions have been obtained by Brocca et al. (2012b) for the prediction of the displacement rate of a well-monitored landslide in central Italy. As a matter of fact, the high temporal resolution, which is going to be further increased with the recent launch of ASCAT-B, is the main added value of the ASCAT soil moisture product from the application viewpoint. Another challenge faced by users is to select the most suitable satellite product for their application. This is not straightforward because several different data products are, in general, available from just one instrument (as was discussed for ASCAT in Section 20.2); with more satellites the diversity of products in terms of their physical content and format specifications increases substantially.

Moreover, not only the spatial–temporal coverage but also the resolution and the long-term availability of the different satellite soil moisture products are important information for the users, as well as understanding how to assimilate the data into their models. In that context, guidelines about the modeling structure (one or more soil layers, equations coupling surface and root zone layers, etc.) that should be used for optimizing the use of satellite soil moisture data can be very useful. One important aspect is that many applications require the knowledge of soil moisture for the root zone, that is, for a layer depth of 1–1.5 m, while satellite data are only available for a surface layer of less than 5 cm. In these cases, the SWI product (Albergel et al. 2008; Wagner et al. 1999) or assimilated profile soil moisture data might be the better choice for many users. Nonetheless, they should be fully aware on how these value-added products are obtained, for example, in the case of SWI how to select the value of the involved parameter T (characteristic time length).

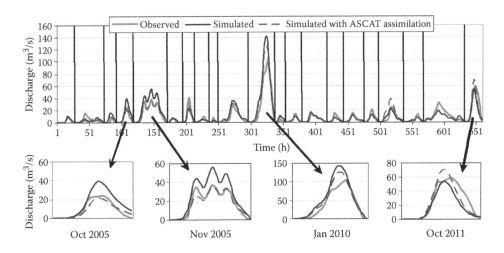

FIGURE 20.7
(See color insert.) Comparison between observed and simulated discharge by assimilating the ASCAT-derived Soil Water Index for a sequence of 21 flood events that occurred in the Niccone basin (central Italy); the simulated discharge without assimilation is also shown as reference. The enlargement for four selected flood events is also shown in the lower panels where it highlighted the higher benefit due to the assimilation for the October and November 2005 events.

A further challenge that limits the use of satellite data by users working in environmental problems is also related to the format and the data volume. Indeed, many users are not used to work with large datasets (tens of gigabytes to terabytes), which is why it is important for satellite data providers to find the right choice between the number of meta data fields and size of the soil moisture data files. With respect to the ease of use, the availability of files in more widely known formats, for example, GEOTIFF, NETCFD, or ASCII files, may increase the number of users potentially able to use the data themselves.

20.5 Conclusions and Prospects

In this chapter the operations and challenges of satellite-based soil moisture services have been discussed based on the example of the ASCAT soil moisture data services operated by EUMETSAT in cooperation with TU Wien. Thanks to the heritage from the scatterometer flown on board of ERS-1 and ER2-2, these services are long-standing and relatively mature when compared to other satellite soil moisture data services such as the ones based on SMOS or AMSR-E. Nonetheless, as an assessment of the Bates maturity index suggests, even the ASCAT services do not yet meet all the criteria of fully operational EO data services as defined by Bates and Privette (2012). Among other aspects, moderate to significant code changes are still expected, as validation efforts and scientific investigations of new algorithms continue to show the need to further improve the ASCAT soil moisture data. Additionally, as much of the initial development of these services relied on research funding, there are still a few general systems engineering tasks pending, for example, related to the handling of metadata or the completeness of the documentation. Fortunately, the services became part of the EUMETSAT Satellite Application Facility (SAF) on Support to Operational Hydrology and Water Management (H-SAF) in 2012, which provides a stable framework for improving the operational aspects of these services step by step in the coming years.

The experiences from the ASCAT soil moisture data services illustrate that the efforts required to move from a pure science-based dataset to an operational EO data service are much more extensive than what most scientists would expect. Scientists should be mindful of the complexity of the retrieval approach they choose, as with an increasing complexity of the algorithms the costs of the operational implementation and running of these services increase quickly. The same is true if data from several input data sources are needed. High algorithmic complexity and input data requirements may also imply that, once the EO service is operational, algorithm developers lose their scientific agility as any change in the algorithms may generate overproportionally high costs in the implementation. This helps understanding why the algorithms used for generating the different satellite soil moisture products do, in general, not undergo any significant changes any longer once the services are operational. In fact, if the required changes are too significant, then it may be better to abandon the original algorithms altogether, replacing them with new algorithms, although that normally requires substantial effort in software implementation.

Fortunately, the basic soil moisture retrieval algorithm of the ASCAT soil moisture NRT services is a mathematically simple change detection algorithm that only requires the ASCAT data as input (Wagner et al. 1999). External data sources are currently only used for data masking and quality control. As a result, it has been possible to implement these services relatively quickly, with the NRT ASCAT soil moisture data service being declared operational only about 2 years after the launch of METOP-A. With an increasing understanding of the strength and weaknesses of the product, it will now also be possible to improve selected parts of the algorithm and processing chain step by step. Planned improvements relate to the introduction of snow and frost detection algorithms, an adaption of the algorithm over deserts and barren land, and enhanced vegetation correction methods. These algorithmic updates will require the use of external data sources such as NWP forecast fields to support the detection of freeze/thaw events or the use of satellite derived leaf area index data to improve the modeling of vegetation effects. Such improvements will have a strong impact on the complexity of the ASCAT processing environment, which is why any complex algorithmic upgrade will only be considered if positive effects on the quality of the data are clearly demonstrated.

Most of the challenges faced by the ASCAT soil moisture data service equally apply to the other satellite soil moisture services (SMOS, AMSR-E, etc.). Fortunately, there is a lot to be learned from each, which will ensure that all of these data services will improve step by step. The most obvious benefit from a cooperation of the different satellite teams lies in the joint validation of the satellite products, eventually leading to more standardized validation protocols. Furthermore, the insights gained from joint validation work will help to better characterize the instrument calibration, the scientific algorithms, and the service specifications. It can thus be expected that the overall quality of the soil moisture services (retrieval accuracy, error characterization, accessibility, ease of use, etc.) based on ASCAT, SMOS, AMSR-E, and other microwave instruments will improve significantly in the coming years.

A direct consequence of these service improvements is that the soil moisture data retrieved from the different satellite instruments will become more similar, a trend that could already be noted in the last few years (de Jeu et al. 2008). This will not only facilitate the use of these data but also open the possibility to create a range of merged soil moisture products designed for specific applications. The good correspondence between some soil moisture datasets has in fact already made it possible to merge these to create long-term soil moisture data series suitable for climate change studies (Dorigo et al. 2012; Liu et al. 2012; Wagner et al. 2012). Furthermore, one can expect to see more and more soil moisture datasets derived from multi-instrument observations that aim to exploit the complementary information content of active and passive remote sensing data acquired in the visible, infrared, and microwave domain of the electromagnetic spectrum (Kolassa et al. 2012; Piles et al. 2011). All this taken together shows that the prospects of satellite-based soil moisture data services are extremely good and that, consequently, the number of scientific and operational users of these services can be expected to grow rapidly in the next years.

Acknowledgments

The ASCAT soil moisture data services are part of EUMETSAT's Satellite Application Facility in Support to Operational Hydrology and Water Management (H-SAF). The authors further acknowledge the financial support by the Austrian Space Application Programme (ASAP) through the GMSM project and by the European Space Agency (ESA) through the ESCAT project.

References

Albergel, C., de Rosnay, P., Gruhier, C., Muñoz-Sabater, J., Hasenauer, S., Isaksen, L., Kerr, Y., and Wagner, W. (2012). Evaluation of remotely sensed and modelled soil moisture products using global ground-based in situ observations. *Remote Sensing of Environment, 118*, 215–226.

Albergel, C., Rüdiger, C., Pellarin, T., Calvet, J.-C., Fritz, N., Froissard, F., Suquia, D., Petitpa, A., Piguet, B., and Martin, E. (2008). From near-surface to root-zone soil moisture using an exponential filter: An assessment of the method based on in-situ observations and model simulations. *Hydrology and Earth System Sciences, 12*, 1323–1337.

Anderson, C., Figa, J., Bonekamp, H., Wilson, J.J.W., Verspeek, J., Stoffelen, A., and Portabella, M. (2012). Validation of backscatter measurements from the advanced scatterometer on MetOp-A. *Journal of Atmospheric and Oceanic Technology, 29*, 77–88.

Bartalis, Z., Scipal, K., and Wagner, W. (2006). Azimuthal anisotropy of scatterometer measurements over land. *IEEE Transactions on Geoscience and Remote Sensing, 44*, 2083–2092.

Bartalis, Z., Wagner, W., Naeimi, V., Hasenauer, S., Scipal, K., Bonekamp, H., Figa, J., and Anderson, C. (2007). Initial soil moisture retrievals from the METOP-A Advanced Scatterometer (ASCAT). *Geophysical Research Letters, 34*, L20401.

Bates, J.J., and Barkstrom, B.R. (2006). A maturity model for satellite-derived climate data records. In *14th Conference on Satellite Meteorology and Oceanography*, (pp. 2.11: 11–14. Amercian Meteorological Society, Boston.

Bates, J.J., and Privette, J.L. (2012). A maturity model for assessing the completeness of climate data records. *Eos, 93*, 441.

Bolten, J.D., and Crow, W.T. (2012). Improved prediction of quasi-global vegetation conditions using remotely-sensed surface soil moisture. *Geophysical Research Letters, 39*, L19406.

Brocca, L., Hasenauer, S., Lacava, T., Melone, F., Moramarco, T., Wagner, W., Dorigo, W. et al. (2011). Soil moisture estimation through ASCAT and AMSR-E sensors: An intercomparison and validation study across Europe. *Remote Sensing of Environment, 115*, 3390–3408.

Brocca, L., Melone, F., Moramarco, T., and Morbidelli, R. (2009). Antecedent wetness conditions based on ERS scatterometer data. *Journal of Hydrology, 364*, 73–87.

Brocca, L., Moramarco, T., Melone, F., Wagner, W., Hasenauer, S., and Hahn, S. (2012a). Assimilation of surface- and root-zone ASCAT soil moisture products into rainfall-runoff modeling. *IEEE Transactions on Geoscience and Remote Sensing, 50*, 2542–2555.

Brocca, L., Ponziani, F., Moramarco, T., Melone, F., Berni, N., and Wagner, W. (2012b). Improving landslide forecasting using ASCAT-derived soil moisture data: A case study of the Torgiovannetto Landslide in central Italy. *Remote Sensing, 4*, 1232–1244.

Crow, W.T., Berg, A.A., Cosh, M.H., Loew, A., Mohanty, B.P., Panciera, R., De Rosnay, P., Ryu, D., and Walker, J.P. (2012). Upscaling sparse ground-based soil moisture observations for the validation of coarse-resolution satellite soil moisture products. *Reviews of Geophysics, 50*, RG2002.

Crow, W.T., Miralles, D.G., and Cosh, M.H. (2010). A quasi-global evaluation system for satellite-based surface soil moisture retrievals. *IEEE Transactions on Geoscience and Remote Sensing, 48*, 2516–2527.

De Jeu, R., Wagner, W., Holmes, T., Dolman, H., van de Giesen, N.C., and Friesen, J. (2008). Global soil moisture patterns observed by space borne microwave radiometers and scatterometers. *Surveys in Geophysics, 29*, 399–420.

De Rosnay, P., Balsamo, G., Albergel, C., Muñoz-Sabater, J., and Isaksen, L. (2012a). Initialisation of land surface variables for Numerical Weather Prediction. *Surveys in Geophysics*, doi:10.1007/s10712-012-9207-x.

De Rosnay, P., Drusch, M., Vasiljevic, D., Balsamo, G., Albergel, C., and Isaksen, L. (2012b). A simplified Extended Kalman Filter for the global operational soil moisture analysis at ECMWF. *Quarterly Journal of the Royal Meteorological Society*, doi:10.1002/qj.2023.

De Wit, A., and van Diepen, C. (2007). Crop model data assimilation with the Ensemble Kalman filter for improving regional crop yield forecasts. *Agricultural and Forest Meteorology, 146*, 38–56.

Dee, D.P. (2006). Bias and data assimilation. *Quarterly Journal of the Royal Meteorological Society, 131,* 3323–3343.

Delwart, S., Bouzinac, C., Wursteisen, P., Berger, M., Drinkwater, M., Martin-Neira, M., and Kerr, Y.H. (2008). SMOS validation and the COSMOS campaigns. *IEEE Transaction on Geoscience and Remote Sensing, 46,* 695–704.

Dharssi, I., Bovis, K.J., Macpherson, B., and Jones, C.P. (2011). Operational assimilation of ASCAT surface soil wetness at the Met Office. *Hydrology and Earth System Sciences, 15,* 2729–2746.

Dorigo, W., de Jeu, R., Chung, D., Parinussa, R., Liu, Y., Wagner, W., and Fernández-Prieto, D. (2012). Evaluating global trends (1988–2010) in harmonized multi-satellite surface soil moisture. *Geophysical Research Letters, 39,* L18405.

Dorigo, W.A., Wagner, W., Hohensinn, R., Hahn, S., Paulik, C., Xaver, A., Gruber, A., Drusch, M., Mecklenburg, S., van Oevelen, P., Robock, A., and Jackson, T. (2011). The International Soil Moisture Network: A data hosting facility for global in situ soil moisture measurements. *Hydrology and Earth System Sciences, 15,* 1675–1698.

Draper, C., Mahfouf, J.F., Calvet, J.C., Martin, E., and Wagner, W. (2011). Assimilation of ASCAT near-surface soil moisture into the SIM hydrological model over France. *Hydrology and Earth System Sciences, 15,* 3829–3841.

Draper, C.S., Reichle, R.H., De Lannoy, G.J.M., and Liu, Q. (2012). Assimilation of passive and active microwave soil moisture retrievals. *Geophysical Research Letters, 39,* L04401.

Entekhabi, D., Njoku, E.G., O'Neill, P.E., Kellog, K.H., Crow, W.T., Edelstein, W.N., Entin, J.K. et al. (2010). The Soil Moisture Active Passive (SMAP) mission. *Proceedings of the IEEE, 98,* 704–716.

Figa-Saldaña, J., Wilson, J.J.W., Attema, E., Gelsthorpe, R., Drinkwater, M.R., and Stoffelen, A. (2002). The advanced scatterometer (ASCAT) on the meteorological operational (MetOp) platform: A follow on for European wind scatterometers. *Canadian Journal of Remote Sensing, 28,* 404–412.

Hahn, S., Melzer, T., and Wagner, W. (2012). Error assessment of the initial near real-time METOP ASCAT surface soil moisture product. *IEEE Transactions on Geoscience and Remote Sensing, 50,* 2556–2565.

Jackson, T.J., Cosh, M.H., Bindlish, R., Starks, P.J., Bosch, D.D., Seyfried, M., Goodrich, D.C., Moran, M.S., and Du, J.Y. (2010). Validation of Advanced Microwave Scanning Radiometer soil moisture products. *IEEE Transactions on Geoscience and Remote Sensing, 48,* 4256–4272.

Kerr, Y., Waldteufel, P., Wigneron, J.-P., Delwart, S., Cabot, F., Boutin, J., Escorihuela, M.-J. et al. (2010). The SMOS mission: New tool for monitoring key elements of the global water cycle. *Proceedings of the IEEE, 98,* 666–687.

Kolassa, J., Aires, F., Polcher, J., Prigent, C., Jiménez, C., and Pereira, J.M. (2012). Soil moisture retrieval from multi-instrument observations: Information content analysis and retrieval methodology. *Journal of Geophysical Research,* doi:10.1029/2012JD018150.

Künzer, C., Zhao, D., Scipal, K., Sabel, D., Naeimi, V., Bartalis, Z., Hasenauer, S., Mehl, H., Dech, S., and Wagner, W. (2009). El Niño southern oscillation influences represented in ERS scatterometer-derived soil moisture data. *Applied Geography, 29,* 463–477.

Liu, Y.Y., Dorigo, W.A., Parinussa, R.M., de Jeu, R.A.M., Wagner, W., McCabe, M.F., Evans, J.P., and van Dijk, A.I.J.M. (2012). Trend-preserving blending of passive and active microwave soil moisture retrievals. *Remote Sensing of Environment, 123,* 280–297.

Mahfouf, J.-F. (2010). Assimilation of satellite-derived soil moisture from ASCAT in a limited-area NWP model. *Quarterly Journal of the Royal Meteorological Society, 136,* 784–798.

Matgen, P., Heitz, S., Hasenauer, S., Hissler, C., Brocca, L., Hoffmann, L., Wagner, W., and Savenije, H.H.G. (2012). On the potential of MetOp ASCAT-derived soil wetness indices as a new aperture for hydrological monitoring and prediction: A field evaluation over Luxembourg. *Hydrological Processes, 26,* 2346–2359.

Naeimi, V. (2009). Model improvements and error characterization for global ERS and METOP scatterometer soil moisture data. PhD Thesis. Institute of Photogrammetry and Remote Sensing, Vienna University of Technology, Vienna, Italy.

Naeimi, V., Paulik, C., Bartsch, A., Wagner, W., Kidd, R., Park, S.E., Elger, K., and Boike, J. (2012). ASCAT Surface State Flag (SSF): Extracting information on surface freeze/thaw conditions from backscatter data using an empirical threshold-analysis algorithm. *IEEE Transactions on Geoscience and Remote Sensing, 50*, 2566–2582.

Njoku, E.G., Jackson, T.J., Lakshmi, V., Chan, T.K., and Nghiem, S.V. (2003). Soil moisture retrieval from AMSR-E. *IEEE Transactions on Geoscience and Remote Sensing, 41*, 215–229.

Owe, M., de Jeu, R., and Holmes, T. (2008). Multisensor historical climatology of satellite-derived global land surface moisture. *Journal of Geophysical Research, 113*, F01002.

Parinussa, R.M., Holmes, T.R.H., and De Jeu, R.A.M. (2012). Soil moisture retrievals from the WindSat spaceborne polarimetric microwave radiometer. *IEEE Transactions on Geoscience and Remote Sensing, 50*, 2683–2694.

Parinussa, R.M., Holmes, T.R.H., Yilmaz, M.T., and Crow, W.T. (2011a). The impact of land surface temperature on soil moisture anomaly detection from passive microwave observations. *Hydrology and Earth System Sciences, 15*, 3135–3151.

Parinussa, R.M., Meesters, A., Liu, Y.Y., Dorigo, W., Wagner, W., and de Jeu, R.A.M. (2011b). Error estimates for near-real-time satellite soil moisture as derived from the Land Parameter Retrieval model. *IEEE Geoscience and Remote Sensing Letters, 8*, 779–783.

Pellarin, T., Calvet, J.-C., and Wagner, W. (2006). Evaluation of ERS scatterometer soil moisture products over a half-degree region in southwestern France. *Geophysical Research Letters, 33*, L17401.

Piles, M., Camps, A., Vall-Llossera, M., Corbella, I., Panciera, R., Rudiger, C., Kerr, Y.H., and Walker, J. (2011). Downscaling SMOS-derived soil moisture using MODIS visible/infrared data. *IEEE Transactions on Geoscience and Remote Sensing, 49*, 3156–3166.

Reimer, C., Melzer, T., Kidd, R., and Wagner, W. (2012). Validation of the enhanced resolution ERS-2 scatterometer soil moisture product. In *Geoscience and Remote Sensing Symposium 2012 (IGARSS'2012)*, IEEE, New Brunswick, NJ.

Rüdiger, C., Holmes, T., Calvet, J.-C., de Jeu, R., and Wagner, W. (2009). An intercomparison of ERS-Scat and AMSR-E soil moisture observations with model simulations over France. *Journal of Hydrometeorology, 10*, 431–447.

Scipal, K., Drusch, M., and Wagner, W. (2008). Assimilation of a ERS scatterometer derived soil moisture index in the ECMWF numerical weather prediction system. *Advances in Water Resources, 31*, 1101–1112.

Scipal, K., Wagner, W., Trommler, M., and Naumann, K. (2002). The global soil moisture archive 1992–2000 from ERS scatterometer data: First results. In *Geoscience and Remote Sensing Symposium 2002 (IGARSS'2002)*, pp. 1399–1401. IEEE, New Brunswick, NJ.

Smith, A.B., Walker, J.P., Western, A.W., Young, R.I., Ellett, K.M., Pipunic, R.C., Grayson, R.B., Siriwardena, L., Chiew, F.H.S., and Richter, H. (2012). The Murrumbidgee Soil Moisture Monitoring Network dataset. *Water Resources Research, 48*, W07701.

Sutanudjaja, E.H. (2012). The use of soil moisture remote sensing products for large-scale groundwater modeling and assessment. PhD Thesis. Department of Physical Geography, Utrecht University, Utrecht, Netherlands.

Verstraeten, W.W., Veroustraete, F., Wagner, W., van Roey, T., Heyns, W., Verbeiren, S., and Feyen, J. (2010). Remotely sensed soil moisture integration in an ecosystem carbon flux model. The spatial implication. *Climatic Change, 103*, 117–136.

Wagner, W., Bartalis, Z., Naeimi, V., Park, S.-E., Figa-Saldana, J., and Bonekamp, H. (2010). Status of the METOP ASCAT soil moisture product. In *IEEE Geoscience and Remote Sensing Symposium (IGARSS'2010)*, pp. 276–279. IEEE, New Brunswick, NJ.

Wagner, W., Blöschl, G., Pampaloni, P., Calvet, J.-C., Bizzarri, B., Wigneron, J.-P., and Kerr, Y. (2007a). Operational readiness of microwave remote sensing of soil moisture for hydrologic applications. *Nordic Hydrology, 38*, 1–20.

Wagner, W., Dorigo, W., de Jeu, R., Fernandez, D., Benveniste, J., Haas, E., and Ertl, M. (2012). Fusion of active and passive microwave observations to create an Essential Climate Variable data record on soil moisture. In *XXII ISPRS Congress*, pp. 315–321. ISPRS, Melbourne, Australia.

Wagner, W., Hahn, S., Kidd, R., Melzer, T., Bartalis, Z., Hasenauer, S., Figa, J. et al. (2013). The ASCAT soil moisture product: Specifications, validation results, and emerging applications. *Meteorologische Zeitschrift, 22*, 5–33.

Wagner, W., Lemoine, G., and Rott, H. (1999). A method for estimating soil moisture from ERS scatterometer and soil data. *Remote Sensing of Environment, 70*, 191–207.

Wagner, W., Naeimi, V., Scipal, K., de Jeu, R., and Martinez-Fernandez, J. (2007b). Soil moisture from operational meteorological satellites. *Hydrogeology Journal, 15*, 121–131.

Wanders, N., Karssenberg, D., Bierkens, M., Parinussa, R., de Jeu, R., van Dam, J., and de Jong, S. (2012). Observation uncertainty of satellite soil moisture products determined with physically-based modeling. *Remote Sensing of Environment, 127*, 341–356.

Wilson, J.J.W., Anderson, C., Baker, M.A., Bonekamp, H., Figa Saldaña, J., Dyer, R.G., Lerch, J.A. et al. (2010). Radiometric calibration of the Advanced Wind Scatterometer Radar ASCAT carried onboard the METOP-A satellite. *IEEE Transaction on Geoscience and Remote Sensing, 48*, 3236–3255.

Zhao, D., Su, B., and Zhao, M. (2006). Soil moisture retrieval from satellite images and its application to heavy rainfall simulation in eastern China. *Advances in Atmospheric Sciences, 23*, 299–316.

Zwieback, S., Bartsch, A., Melzer, T., and Wagner, W. (2012a). Probabilistic fusion of Ku- and C-band scatterometer data for determining the freeze/thaw state. *IEEE Transactions on Geoscience and Remote Sensing, 50*, 2583–2594.

Zwieback, S., Scipal, K., Dorigo, W., and Wagner, W. (2012b). Structural and statistical properties of the collocation technique for error characterization. *Nonlinear Processes in Geophysics, 19*, 69–80.

Index

Page numbers followed by f and t indicate figures and tables, respectively.